The Satellite Experimenter's Handbook

Martin Davidoff, PhD, K2UBC

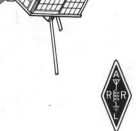

Published by
The American Radio Relay League
225 Main Street, Newington, CT 06111

Table of Contents

Foreword

OSCAR I, Amateur Radio's first satellite, was launched into orbit in December 1961. A small battery-powered box, OSCAR I continually transmitted the Morse code identifier HI to amateur receivers on earth until its battery ran down. A tremendous achievement for Amateur Radio in the early days of the Space Age, the successful mission was to be but the first of many.

The resourcefulness, ingenuity and skill of the Amateur Radio satellite community in the years since have made a fascinating story. From the California garage and basement workshops of the '60s to the cooperative international projects of the '80s, amateurs have pursued the dream of reliable, predictable, long-distance and long-duration radio communication on VHF and higher frequencies. Each successive OSCAR has been one more step toward the realization of that dream. The many successful launches, including AMSAT-OSCAR 13, Fuji-OSCAR 20 and a number of Microsats, have further expanded the communications capabilities of Amateur Radio. Yesterday's dreams are today's reality.

You are a part of that reality! From setting up a modest ground station and communicating through the "birds," to gaining an understanding of some of the more advanced concepts of satellite orbits and tracking, this expanded and updated second edition of *The Satellite Experimenter's Handbook* provides the information you'll need. In addition to amateur satellites, it also covers weather and TV satellites, two areas which continue to gather new followers. Whether you're a beginner, an old hand at satellite work or a student of space science, this book is your launch vehicle into the fascinating journey of Amateur Radio in space.

David Sumner, K1ZZ
ARRL Executive Vice President
August 1990

Dedication

Dedicated to the unsung heroes of the amateur satellite program, those who have made the most selfless contributions, the family members of the volunteer workers, and especially to one such person whose good-natured tolerance over the past two decades has made this book possible, Linda.

Acknowledgments

In 1979 AMSAT (the Radio Amateur Satellite Corporation) and the ARRL (the American Radio Relay League) agreed to jointly sponsor the production of a comprehensive handbook on radio amateur satellites. The primary goals for the book included: documenting Amateur Radio space achievements, serving as a beginners' manual and presenting information for the more advanced amateur. AMSAT would provide technical support to the author, and the ARRL would contribute expertise in editing, production and distribution. You're reading the results of this cooperative effort.

As the writer, I wish to acknowledge the overwhelming support I've received from both AMSAT and the ARRL. In reality, it's individuals, not groups, that provide assistance and I am indebted to a great many people who helped me pull it all together. While it's impossible to mention everyone personally, I would like to publicly thank three people without whose personal encouragement I would never have undertaken this project: Jan King and Perry Klein of AMSAT, and Linda, my spouse since the days of OSCAR III.

Over the years, while working on this book and on other projects for the amateur satellite program, I've felt extremely grateful for having the opportunity to help turn the amateur satellite dream into the amateur satellite reality. Few scientists ever get the chance to take part in a project as exciting and satisfying, or to work with individuals as dedicated, persevering and bright. To all, my sincerest thanks.

Martin Davidoff, K2UBC
December 1989

About the Author

Martin R. Davidoff, K2UBC, has been a licensed radio amateur since 1956, the year before the launch of Sputnik I. An Amateur Extra licensee, he is a life member of both AMSAT and ARRL. After earning a PhD in physics (experimental, solid state) at Syracuse University in 1971, Davidoff was employed by the Illinois Institute of Technology Research Institute where he was involved with satellite communications systems. He is now a Professor of Mathematics and Engineering at Catonsville Community College in Catonsville, Maryland, where he has taught since 1972. From 1975-1978 he directed a science education project for the National Science Foundation focusing on the use of satellites by college-level science and engineering educators. As part of that project he authored *Using Satellites in the Classroom: A Guide for Science Educators*, published in 1978. He also currently serves as Senior AMSAT Science Education Advisor.

To the Reader

Challenging or frustrating, fascinating or confusing—no matter how they're described, space satellites certainly have added a new dimension to Amateur Radio. The *Satellite Experimenter's Handbook* was written to provide information on spacecraft built by, and for, radio amateurs. In addition, it discusses weather, TV-broadcast and other satellites of interest to amateurs. Before beginning, I'd like to comment on the objectives of the book, the organization of the material, the underlying philosophy, and the major changes that you'll encounter in the second edition.

The *Satellite Experimenter's Handbook* is designed to provide amateurs having a background in radio communications and electronics with an introduction to those aspects of satellite systems that are likely to be of most interest to them. Since there is a great deal of material to cover, decisions must be constantly made on what to include and what to leave out. I've tried to focus on satellite topics, on material that's not readily available elsewhere, and on material that will not become quickly out of date. Most satellite communicators find that good access to information adds to their operating pleasure. Information sources include packet bulletin boards, weekly nets (HF and satellite), periodicals published by AMSAT and the ARRL, the old standby ARRL manuals, and the proceedings of the yearly AMSAT technical symposiums. This text is meant to serve as a detailed introduction and reference manual to be used in conjunction with the up-to-date specialized sources just mentioned.

The *Satellite Experimenter's Handbook* is divided into three main sections. Part I (Chapters 1-4) covers the history of Amateur Radio activities (and radio amateurs) in space. Part II (Chapters 5-10) provides the information needed by those starting out in satellite communications. Taken together, Parts I and II form a beginner's manual. Part III (Chapters 11-16) presents material on special technical topics for serious experimenters, including those who are contemplating building spacecraft or spacecraft subsystems.

If you're anxious to start communicating via satellite you may be tempted to skip the history and jump right into Chapter 5. Try to avoid this temptation, for the "story" contains important basic technical information. By the time you finish Chapter 4, you'll find that you have painlessly acquired a solid technical background in satellite system fundamentals and have become comfortable with satellite jargon.

Although the beginner needs little more than a basic knowledge of communications systems to start communicating via satellite, the advanced experimenter encounters a much bigger hurdle. When you look into spacecraft design, you encounter a wide range of disciplines: physics (basic physics, geophysics, astrophysics), mechanical engineering (structures, materials, heat flow) and electrical engineering (communications systems, propagation, control systems, digital electronics), and so on.

In the past, the serious experimenter found it difficult to acquire knowledge about satellite systems because much of the information was buried in the advanced scientific texts and journals of the various fields involved. One had to first find the information and then try to digest it. This book attempts to reduce both of these barriers by providing the reader with key references and a wide-ranging introduction to satellite systems. If I have been successful, you should have the background needed to understand the cited references by the time you turn to them.

Although I've tried to boil many complex topics down to simple terms, there are times where too much simplification can cloud fundamental ideas. Whenever such a conflict arose, clarity of fundamentals was always given precedence. As a result, you may encounter some relatively advanced mathematics in certain sections, but this material can be skipped over if you're mainly interested in the "how" and not the "why."

The references cited throughout this book were selected carefully for their clarity and significance. Although your local library may not stock all the items listed, you'll find that most are willing to arrange interlibrary loans and/or obtain photocopies of articles for you at a small charge. You may also wish to check whether local colleges and universities will grant you library access. Many institutions now offer such privileges to members of the local community, either free or for a modest fee. If an institution near you has an engineering department, be sure to investigate this possibility.

The problem of choosing units can be a sticky one. I've chosen to emphasize the English system (for example, miles instead of kilometers) in the early sections of the book, the "beginners manual," when the values quoted are meant to convey a rough feeling for size. In the later chapters, when detailed computations are illustrated, the MKS system is used because it's easier to work with. This leaves me open to charges of inconsistency, but I believe it best serves the needs of readers.

The second edition of the *Satellite Experimenter's Handbook* differs from the first in several significant ways. It has, of course, been extensively updated to reflect the fact that several new spacecraft have been launched while others have ceased operating since the last edition was published. Sample calculations and illustrations throughout the text have been changed to include, or focus on, current spacecraft. Details have been added on the Phase 4 and Microsat programs. The material on tracking has been expanded to include considerably more information on classical orbital elements and computer methods. Several small computer programs in BASIC illustrating fundamental tracking algorithms have been added. A key design objective of these programs was to present the underlying algorithms as clearly as possible. The discussion of the nature of circular polarization and its practical implications has been significantly expanded. A new method for comparing the effectiveness of antennas used at terrestrial stations in conjunction with low-altitude satellites is presented. And for those tracking AMSAT-OSCAR 13 with an Oscarlocator, we've provided ground track overlay tracing masters that will take you past 1995.

If you spot any errors or would like to make suggestions for a new edition, I can be reached at the following address:

Martin Davidoff
Department of Mathematics and Computer
  Science
Catonsville Community College
Catonsville, MD 21228
USA

I'm sorry that I can't offer to respond to detailed requests for information, but I will try to answer brief requests if an SASE is enclosed. Hope to hear you on the birds.

73,
Martin Davidoff, K2UBC

Chapter 1

Enter the Space Age

With this launch, OSCAR I brought Amateur Radio into the space age. It was the world's first nongovernmental satellite—and the last built with $26 worth of surplus parts!

The Space Age, marked from the launch of Sputnik I on October 4, 1957, is now over 30 years old. More than half the earth's current residents were born after it began. During this time a considerable amount of money has been spent on space exploration and the development of related technologies. The investment has clearly begun to pay off—earth satellites are the prime example. For more than a decade most international telecommunications have been handled by satellite.[1-3] Daily data provided by spacecraft have revolutionized our ability to forecast weather, predict crop yields and monitor the environment. Satellites also contribute significantly to terrestrial navigation, scientific exploration, TV broadcasting and natural-resource management. And, their unique ability to monitor compliance with arms limitation treaties has contributed to international political stability.[4] Parking spots in certain desirable regions of the sky have already become scarce.[5] Perhaps in the 21st century financial support for the United Nations will come from parking meters—leases on communications rights to orbital slots.

Satellites have already modified the way we think about ourselves, the earth, and the civilizations that inhabit it. Pictures of the earth, taken from space, have probably had a greater impact on both our awareness of this planet's limited resources and our need to pull together if we want our spaceship to continue sustaining life than all the words ever written on the subject.[6] Isaac Asimov sees satellites as causing profound, positive changes in personal communications between earth's inhabitants—changes comparable in magnitude to the creation of speech, writing and printing. One result he expects is that "The earth for the first time will be knit together on a personal and not a governmental level."[7]

RADIO LINKS

The radio signals linking satellites and ground stations (stations on, or near, the surface of the earth) are central to most satellite systems. These radio links can provide information about the spacecraft's operation and environment and form the basis of satellite communications systems. Therefore, it's important that we be aware of some basic properties of radio waves right from the beginning of our work.

The common expression "line-of-sight" is probably most familiar in the context of light waves. In essence, it means that a person (A) who is looking at an object (B) can see the object only if nothing is in the way (that is, if the straight line path joining A and B is unobstructed). Radio signals generally adhere to the line-of-sight principle.

Taking this analogy between light and radio waves to its logical end suggests that it's impossible for two ground stations more than a few hundred miles apart to communicate using radio (or light) waves since the earth blocks the direct line joining them. Such a conclusion, however, is false for two important reasons. First, the line-of-sight principle is based on the assumption that events are taking place in a vacuum; realistically, it must be modified for events in the earth's atmosphere.[8] Second, we can use our ingenuity to get around (no pun intended) the principle. Just as we use mirrors to see around corners, we can use anything that reflects radio waves to bend them around obstacles. Most non-satellite, long-distance radio communication between two terrestrial stations does, in fact, involve the reflection of radio waves off the radio mirror known as the ionosphere (Fig 1-1).

Radio reflectors are important both to those interested in long-distance communications and to scientists probing the structure of the earth's near environment. What acts as a radio mirror? Various candidates have been investigated in great detail; a partial list is given in Table 1-1. Much of the important experimental work with the re-

Table 1-1

Some Passive Reflectors of Radio Waves

Reflector (description)	Height above earth (miles)	Max communications distance (single-hop) (miles)	Freq of interest (MHz)
1) Layers of the earth's ionosphere			
a) F_2 layer (ultraviolet radiation from sun causes ionization of air molecules)	200-300	2500	<100
b) E layer (ultraviolet radiation from sun)	35-70	1400	<250
c) E layer (ionized trails left by meteors as they burn up)	35-70	1400	<450
2) Aurora (ionized particles emitted by sun trapped by earth's magnetic field near north and south poles)	50-60	1200	<500
3) Moon (has been used successfully by amateurs at frequencies indicated)	240,000	12,000	50-10,500
4) Large balloon-like artificial satellites with metallic coating (see discussion of Echo I and Echo II in Chapter 2)	100-200	2000	>30

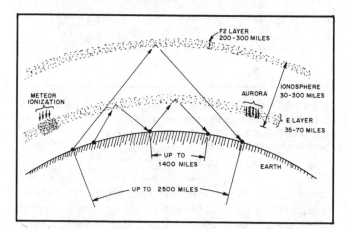

Fig 1-1—When various layers of the ionosphere are activated by solar radiation, solar particles, meteors or other means, they can reflect radio waves of certain wavelengths. Communication paths often involve more than one hop.

flectors listed in Table 1-1 was conducted by radio amateurs (more details and references will be found in Chapters 2 and 3). To evaluate the communications effectiveness of each reflector it must be compared to alternative modes of communicating: other reflectors, telephone cables, mail service or terrestrial microwave links. Each of the reflectors mentioned exhibits one or more serious drawbacks including erratic, unpredictable behavior, the need for extremely high transmitter power and large, complex antennas for both transmission and reception, and excessive system (all transmitters plus receivers) cost.

The Satellite Relay

In the mid 1950s, before the first artificial satellite was in orbit, Arthur C. Clarke published an article detailing how a satellite relay station would enable terrestrial stations to communicate over large distances.[9] In an independent analysis, John Pierce, a physicist at Bell Telephone Labora-

tories, came to a similar conclusion: Active satellite relays could have a significant positive impact on long-distance communication.[10] A relay system using satellites would be most reliable if the radio frequencies employed were not affected by the ionosphere.

A relay station can take many forms. Consider the linear transponder. A linear transponder is an electronic device that receives a small slice of the radio frequency spectrum, greatly amplifies the strength of signals within the entire slice, and then retransmits the signals in another portion of the spectrum. This is accomplished so quickly (within microseconds) that, insofar as users are concerned, it's instantaneous. The Amateur Radio transponders used aboard satellites can handle a large number of signals of various types at the same time, with the power of each received signal being multiplied by roughly 10^{13} times (130 dB) before being transmitted back to earth. Other types of relay stations are designed to receive and store messages (usually in digital form). These messages can then be played back at a later time as the spacecraft passes over a different region of the earth. Such devices are often called digital store-and-forward repeaters.

This brief overview of the basic properties of radio waves and the active-relay concept provides the background needed to take a more in-depth look at satellite radio links.

Satellite Radio Links

The radio signals between satellites and ground stations are often categorized as downlinks (signals from a spacecraft to a ground station—Fig 1-2a), uplinks (signals originating at the ground and directed to a satellite—Fig 1-2b), and broadcast or communication links (which involve both uplinks and downlinks—Fig 1-2c). The simplest downlink, a continuous tone, can be useful to ground stations tracking the satellite and to experimenters studying radio propagation or investigating the ionosphere. A more complex downlink beacon can be used to convey

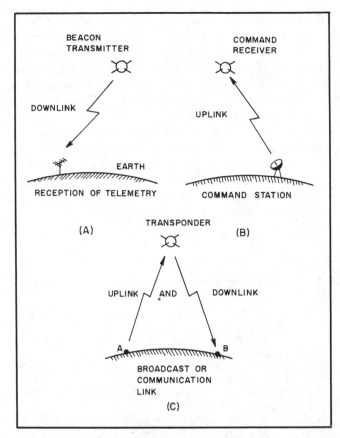

Fig 1-2—Satellite radio links.

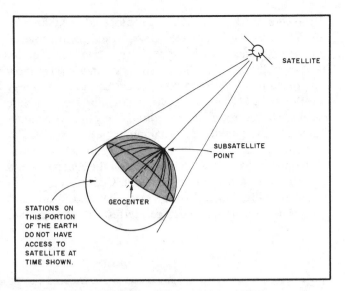

Fig 1-3—Only those terrestrial stations with unobstructed line-of-sight paths to a satellite are in range.

telemetry information (measurements made by sensors aboard the spacecraft) to interested ground stations. A number of different techniques are used for modulating telemetry beacons (superimposing the telemetry data on the radio signal). We'll look at some of these techniques later.

Uplinks can be used to control the operation of a satellite. For example, if a particular satellite's design permits, we could reprogram an on-board computer from earth to adjust the spacecraft's attitude (orientation in space), or to turn a beacon off temporarily to conserve energy. Ground stations equipped to control a spacecraft are called command stations.

Uplinks and downlinks are used together in many applications. For example, ground station A (Fig 1-2c) may send a message to ground station B (over a non-line-of-sight path) by way of a satellite relay. The number of ground stations equipped to use a particular satellite relay may range from only a few to tens of thousands or more. For example, consider the 4-GHz direct satellite-to-home TV systems currently operating in the US. Station A (Fig 1-2c) is in this case a central TV transmitter, and station B represents the millions of homes capable of receiving the TV signals. This example illustrates a broadcast link. If both stations in Fig 1-2c are equipped to transmit and receive, we have a communications link. A multiple-access relay satellite of the type built by radio amateurs can be used for two-way communications by a large number of ground stations simultaneously.

In order for two ground stations to communicate

through a single satellite relay, both must be in range of the spacecraft (that is, the line-of-sight path between each station and the satellite must be unobstructed). In addition, the satellite's antenna pattern must be broad enough to include both ground stations. Fig 1-3 shows that all stations within range of a satellite at a given instant lie inside the circle formed by an imaginary cone whose curved surface just grazes the earth. The axis of the cone goes through the satellite, the subsatellite point (the point on the surface of the earth directly below the satellite) and the geocenter

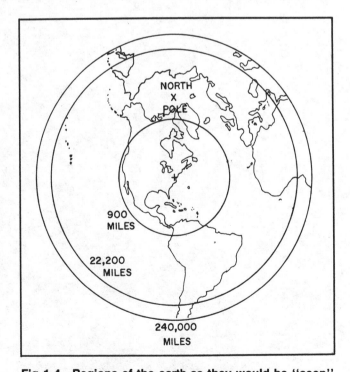

Fig 1-4—Regions of the earth as they would be "seen" by a satellite located directly over Catonsville, Maryland, USA (39° N, 76° W). The three distances chosen correspond to the moon (240,000 miles), AMSAT-OSCAR 13 at its highest point (22,200 miles) and AMSAT-OSCAR 7 (900 miles).

(the center of the earth). A satellite located 240,000 miles over Catonsville, Maryland, USA, would be in range of everything enclosed by the outer circle of Fig 1-4 —essentially half the planet. The two inner circles show the portions of the earth that would be visible at lower altitudes: 22,000 miles and 900 miles. For most satellite orbits, the position of the subsatellite point constantly moves as the satellite travels along its path in space. For certain special orbits, however, the subsatellite point remains fixed at a specific longitude on the equator. We'll discuss these geostationary orbits in Chapter 11.

From Figs 1-3 and 1-4 it's clear that maximum communications distance (the largest terrestrial distance over which it's possible to relay a radio message by a linear transponder) increases as satellite height increases, but it can never exceed half the circumference of the earth. (Graphs and mathematical expressions relating satellite height to maximum communications distance are presented in Chapter 12.) In contrast, a digital store-and-forward system can support communication between two stations that can't see the satellite at the same time. Of course, the communication doesn't take place in real-time (immediately).

The basic concepts just presented underlie the development of communications systems using artificial earth satellites. From the beginning, Amateur Radio has played a significant role.

Notes

[1] B. I. Edelson, "Global Satellite Communications," *Scientific American*, Vol 236, no. 2, Feb 1977, pp 58-69, 73.

[2] B. I. Edelson and L. Pollack, "Satellite Communications," *Science*, Vol 195, no. 4283, 18 Mar 1977, pp 1125-1133.

[3] H. L. Van Trees, E. V. Hoversten and T. P. McGarty, "Communications Satellites: Looking to the 1980s," *IEEE Spectrum*, Dec 1977, pp 43-51.

[4] T. R. McDonough, *Space: The Next Twenty-Five Years* (NY: Wiley, 1987), p 64ff. "A Proposal for an International Arms Verification and Study Center" by Jerome Wiesner (Science Advisor to President J. F. Kennedy), MIT, 1986.

[5] W. L. Morgan, "Satellite Utilization of the Geosynchronous Orbit," *COMSAT Technical Review*, Vol 6, no. 1, 1976, pp 195-205.

[6] A. C. Clarke, "Beyond Babel," *UNESCO Courier*, Mar 1970, pp 32, 34-37.

[7] I. Asimov, "The Fourth Revolution," *Saturday Review*, Oct 24, 1970, pp 17-20.

[8] H. S. Brier and W. I. Orr, *VHF Handbook* (Wilton, CT: Radio Publications, 1974), Chapter 3.

[9] A. C. Clarke, "Extra-Terrestrial Relays," *Wireless World,* Vol 51, no. 303, Oct 1945, Chapter 3.

[10] J. R. Pierce, "Orbital Radio Relays," *Jet Propulsion*, Vol 25, no. 4, Apr 1955, p 153.

Chapter 2

The Early Days

THE DAWN OF THE SPACE AGE

In early October the warmth of the morning sun is very welcome in Tyuratam, a village in the USSR 150 miles northeast of the Sea of Aral. But this particular morning, as the countdown for Sputnik I nears its climax, the sun is hardly noticed. Day turns to night as the inevitable technical hitches are dealt with until finally, late in the evening (GMT) on 4 October 1957: Zazhiganiye (blastoff). The Space Age begins.

Shortly after midnight (GMT), a BBC radio operator at a monitoring station near London notes the appearance of a strange "beep-beep-beep..." Something about the unfamiliar signal attracts his attention—though the average strength is gradually increasing, rapid fading is superimposed. The signal's frequency is drifting slowly downward, and direction-finding equipment shows the azimuth of the source to be changing rapidly. Only one conclusion is possible —the signal is coming from an artificial space satellite.

These events may not sound very spectacular today, but they generated almost unimaginable excitement back in 1957. Within minutes the news flashed round the world. In Washington, DC, it was still early in the evening. Following a week of meetings focusing on the IGY (International Geophysical Year), the Soviet embassy was holding a reception for many of the senior scientists involved. Lloyd Berkner, president of the US IGY co-ordinating group, was paged and informed of the BBC observation and given a report of the launch just released by TASS (the Soviet news service) that identified the spacecraft as Sputnik I. When he returned to the cocktail party, Berkner announced the event to the scientists present and his Russian hosts.[1-3] It certainly must have been a lively party.

Newspaper accounts report that the world responded with "surprise and elation." In retrospect, the elation is understandable, but the surprise element seems a little misplaced. The June 1957 issue of *Radio* (Радио), a widely distributed Soviet journal on practical electronics, stated that a sputnik (the Russian word for satellite) would soon be launched. The column provided information on the projected launch date (late September), the transmitter frequencies (20.005 and 40.010 MHz) and the type of modulation. The Russians again announced their plans at

international scientific meetings in Barcelona and Washington later that summer. The sense of surprise certainly didn't arise from Soviet secrecy.

One of the transmitters on Sputnik I operated just above the 20-MHz frequency used by the United States and other countries for a worldwide network of high-power radio transmitters sending standard time signals. Therefore, the hundreds of thousands of radio amateurs and shortwave listeners owning radio receivers capable of picking up the time broadcasts were able to listen for the spacecraft. Signals from the satellite were generally so loud that they could be tuned in on even the simplest of these receiving sets. The Soviets' choice of frequencies was obviously no accident. Their interest in amateur reports was clear: "Since radio amateur observations will be of a mass character they can secure extremely important data on the satellite's flight and the state of the ionosphere."[4]

While the 20-MHz signal of Sputnik I showed the world that very simple ground stations could be used to monitor satellite signals, this frequency was not suitable for reliably sending large amounts of information on a satellite's performance, or its environment, back to earth. Future satellites would use higher frequencies and more sophisticated techniques for forwarding telemetry data to improve reliability. As a result, it would become increasingly difficult, even for the amateur scientist with extensive Amateur Radio experience, to monitor government satellite programs directly.

Determining the actual launch time of Sputnik I has proved difficult. When a careful search through the *New York Times* and various scientific periodicals over the weeks and months following launch failed to turn up the desired information, we had to resort to a combination of "reverse engineering" and detective work. The details are covered in the last section of Chapter 14.

The US Enters Space

Barely four months on the heels of Sputnik I, on 31 January 1958, the United States launched Explorer I, its first successful satellite. Explorer I contained a scientific instrument package to measure radiation levels in space. The payload was designed by a group under the direction of Dr James Van Allen of the University of Iowa. When the spacecraft began operating, most of the instruments were reading completely off scale. The immediate reaction of the scientists was that some sort of catastrophic failure had occurred in the satellite electronics. After a painstaking but unsuccessful search to pinpoint where the equipment had failed, their gloom gradually turned to elation. The scientists were forced to conclude that the instruments were actually operating properly; radiation of such unexpected levels had been encountered that the instruments had been driven into saturation. With soaring spirits, the team of scientists began the task of mapping the belts of radiation that would later bear Van Allen's name.[5]

Radio Amateurs and Space

Over the years, radio amateurs have taken an active and important part in space-related investigations. Many of the pioneers in radio astronomy were also hams. During the late '30s and early '40s, the exploratory studies and comprehensive radio sky maps prepared by Dr Grote Reber (W9GFZ)—using a home-built 32-foot diameter parabolic antenna in his Wheaton, Illinois, backyard—stimulated the birth of a new branch of basic science, later called radio astronomy. As this is being written, a half century later, his research continues. In 1985, while radio amateurs around the world were talking with astronaut Tony England (WØORE), orbiting the earth on the Spacelab-2 mission, Dr Reber was making measurements of signals from space at 1.7 MHz. Such signals normally cannot penetrate the ionosphere, but a 16-second burn of the orbital maneuvering rockets aboard Spacelab-2 opened a temporary window, providing a unique opportunity to view cosmic medium frequency signals for nearly four hours.[6]

On 27 January 1953 Ross Bateman (W4AO) and William L. Smith (W3GKP) beamed radio signals at the moon and succeeded in hearing echoes.[7] And in the late '50s, as mentioned, thousands of radio amateurs monitored signals from early Soviet and American satellites. What would radio amateurs do next?

In April 1959 Don Stoner (W6TNS), a widely known and well-respected electronics experimenter, writing in *CQ*, suggested that amateurs undertake the construction of a relay satellite.[8] Stoner was looking far beyond placing a simple beacon in orbit; he was proposing that hams build a spacecraft containing a transponder capable of supporting two-way communications. Many said Stoner was fantasizing. After all, construction of the first government-supported satellite to use the proposed techniques (Telstar I, launched in July 1962) hadn't even begun, FM repeaters were virtually unknown in the Amateur Radio community, and most experimenters had little or no experience with the newfangled devices called transistors. But these fantasies were the disciplined dreams of an intelligent and far-sighted thinker. Although Stoner's comments were couched in humorous terms, he was serious. His note provided the spark that would lead, not many years later, to radio amateurs placing an operational, active relay satellite in orbit—but that's getting ahead of our story.

In 1960, imaginations fired by the Stoner article, a group of radio amateurs in Sunnyvale, California, organized the OSCAR Association (*O*rbiting *S*atellite *C*arrying *A*mateur *R*adio). The aims of this pioneering club included both building amateur satellites and obtaining launches. It wasn't clear which goal would be more difficult. To arrange a launch, the US government would have to be convinced that amateur satellites could serve a useful function in one or more of the following areas: scientific exploration, technical development, disaster communications, and scientific or technical education. One important factor was to help both radio amateurs and the scientific community. Most large satellites are mated to rockets having excess lift capacity; it's simpler and cheaper to ballast a rocket with dead weight than to reduce the thrust. As a result, it was possible to add secondary payloads to many missions at very little cost. Over the years, many scientific and amateur satellites have hitchhiked into space piggybacking on primary payload missions.

OSCAR I

After two years of effort by members of the OSCAR Association, the first radio amateur satellite—OSCAR I—was ready and scheduled for launch. Weighing in at 10 pounds, the spacecraft contained a 140-mW beacon at 145 MHz transmitting a simple, repetitive message at a speed controlled by a sensor responding to the internal satellite temperature. Fig 2-1 compares the OSCAR I beacon to the 10-mW beacons flown on Explorer I (the first US satellite) and an early US Vanguard mission. Although OSCAR I did not contain a transponder, it was a significant first step toward that goal. The events and emotions surrounding the beginning of OSCAR I's 22-day sojourn in space were beautifully captured in a classic *QST* article by Bill Orr (W6SAI).[9] We let Bill tell the story.

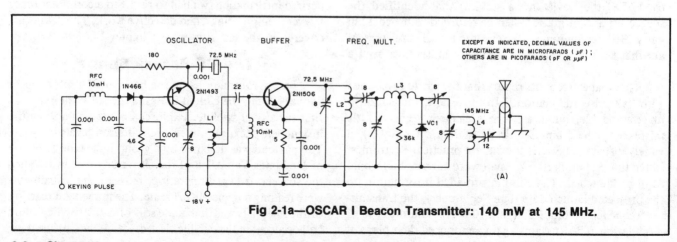

Fig 2-1a—OSCAR I Beacon Transmitter: 140 mW at 145 MHz.

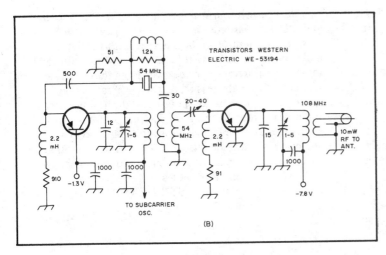

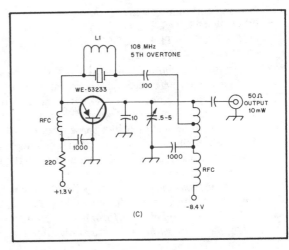

Fig 2-1b—Explorer I Beacon Transmitter: 10 mW at 108 MHz.

Fig 2-1c—Vanguard Beacon Transmitter: 10 mW at 108 MHz.

The spirit of adventure lies buried in every man's soul. Strike the spark and ignite the soul and the impossible is accomplished. So it was on December 12, 1901 on a chill, Newfoundland morning. The first self-proclaimed radio amateur, Guglielmo Marconi, bent intently over his crude receiving instruments and heard the letter "S" transmitted across the stormy Atlantic Ocean, from a station in Cornwall.

The spirit of adventure again made its mark sixty years later on December 12, 1961. The locale this time was an experimental aerospace base on the border of the Pacific Ocean. A group of radio amateurs saw launched into orbit the first Amateur Radio space satellite. Born in a burst of flame, the 10-pound, home-made beacon satellite transmitted to the world that the spirit of adventure and quest that drove Marconi down the road of history was still goading the radio amateur in his eternal search after the mysteries of nature. This is the story of a small portion of that quest.

Sixty Years of Radio Amateur Communication

Marconi to the OSCAR Satellite
By William I. Orr, W6SAI

February, 1959: The radio amateur gazed thoughtfully for a moment at the white paper in his typewriter. Suddenly his fingers sprang into action and the keys flashed the fateful words, "Currently being tested is a solar powered six- to two-meter transistor repeater which could be ballooned over the Southwest. Can anyone come up with a spare rocket for orbiting purposes? . . . 73, Don, W6TNS."[†] He slapped the page from the typewriter, setting in motion a chain of events that conclusively proved that truth is indeed stranger than fiction.

The local time is 0200 on a cold, starless 1961 December morning. The location is Vandenberg Air Force Base, California. It is a cheerless, predawn moment. Inside the reinforced block house, the combined USAF and contractor crews are busy at work. The blockhouse walls are lined with TV monitoring screens. Along one side is the launch control console. Communications, radar and propellant monitors are on; talkers and other intercommunications people are at their stations. The key personnel are locked in unison by a single communications net. All wear headsets and microphones

[†]"Semiconductors," *CQ*, Apr 1959, p 84.

. . .T MINUS TEN AND STILL COUNTING. . .
Tension builds up as moment of launch nears. (Left to right): Capt. Turner (USAF); Bill, W6SAI; Ray, W6MLZ; Dos, WØTSN; and Chuck, K6LFH. Chuck talks to OSCAR Control Center, WA6GFY, to make sure that traffic net to South Pole is ready for acquisition of OSCAR as it passes on initial revolution. (Photo: USAF)

so that they can use their hands freely. A complex network permits several simultaneous conversations. The outpouring of this network culminates in a teletype transmission to the Program Director located 170 miles away in Inglewood, California. The RTTY channel springs to life and begins to clatter: ... FM 6565TH TEST WING VAFB CALIF TO SSD LOSA CALIF THIS IS A CONTINUOUS MESSAGE ... R MINUS 500 AND COUNTING ...

In the cold night illuminated by a thousand lamps, the Agena-Thor aerospace vehicle sits on the reinforced launching pad. Known as Discoverer XXXVI, this intricate, calm, sophisticated spire of brute power awaits the command to hurl itself into space. From it will eject man-made satellites, orbiting the earth hundreds of miles above. One of these will be of great interest to the radio amateur. It is OSCAR.

Of the thousands of readers of Don Stoner's article, none was struck more forcibly than Fred Hicks, W6EJU, of Campbell, California. An old-timer in the communications game, Fred was now employed by a large missile contractor in the San Francisco Bay area. Fred had been present in the blockhouse at Vandenberg for the first six Discoverer launches. To Fred goes all credit for grasping the true nature of Don's message, and interpreting it in terms of the full spirit of Amateur Radio.

Fred dropped the magazine on his desk, pushed aside a cup of coffee and reached for the telephone. He dialed a number and listened to the automatic stepping switches go through their complicated dance in the earpiece of the instrument. "Hello, Chuck? ... Hey, buddy, did you read Don Stoner's article this month? ... Well, he said in effect that the radio hams could build a satellite if they could only find somebody to launch it for them ..." The voice on the phone crackled. "Right! That's what I was thinking. Why don't you drop Don a line and get this thing organized? ... If old K6LFH and W6EJU and their buddies can't do the job, why, nobody can!" Fred chuckled to himself as he hung up the phone. Chuck was right. Why not build a ham satellite? The idea wasn't so crazy after all. A lot could be learned from such a device. The satellite would ... it would ... well ... Fred suddenly realized that such a simple beguiling idea could not be defined and would entail a lot of work and planning to even begin to be coherent. Obviously it was a fine project for a club, or group of hams. One ham couldn't handle this "brainbuster." As HPM, The Old Man, might have said, "It

Vandenberg control and tracking station pinpoints the Discoverer as it races in orbit around the earth at 18,000 miles per hour. OSCAR satellite follows its own orbit at approximately the same speed as the parent satellite. Orbital data is plotted on boards at the rear of the room from the acquisition and control consoles in the foreground. (Photo: Lockheed)

was an idea without a handle to grab it." ... Truly, W6EJU was blessed with the spark of adventure.

The count down begins at R minus 500 minutes and is divided into more than 20 tasks. More than 1500 separate instructions must be given from the launch console before the vehicle is ready for the great voyage into space. Guidance checks, polarity and phasing checks, vehicle erection, re-check of destruct systems, orbital electronics and control checks, propellant tank checks, telemetry operational checks, and satellite operational checks must go on in infinite, precise detail. The voice of the teletype chatters endlessly ...
... R MINUS 350 AND STILL COUNTING ...

13 October, 1959

"Dear Don:

I remember you wrote an article for CQ some time ago that described a small transistorized two-meter station, and appealed for 'anyone with a space vehicle, please?' ... Though I do not hold out any too much hope for this, I will do my best to interest certain parties ... please send me the exact weight of the installation and space it occupies ... Actually, the 'Discoverer' is ideally suited to such a ham project. I will sound out the local hams ... look for me on 14,285 kc ... 73 and I certainly hope we can pull this off! ... Fred, W6EJU."

The die was cast. The spark of adventure had found fuel and was burning brightly. The fateful letter was on the way, was in the mail. It would start a thousand minds dreaming and planning, and the concept would eventually involve high-level decisions in the US government. Now, at this moment in time it was a gossamer, a fancy that might be lightly discarded as a mere exercise of the imagination. (After all, why not? Would not a home-made satellite be yet another convincing proof that Amateur Radio was indeed in the public interest, convenience, and necessity? At the very least it would be a self-educational program, introducing the great body of amateurs to space communications. Of course.)

Bob Herrin, K4RFP/6 (Launch Operations Manager), was listening on the countdown net in the communications and control laboratory, at the launch site. He joshed a few words with other technicians and engineers, intent upon their tasks. The package had been carefully placed into its egg-crate-shaped compartment in the Agena second stage of the immense vehicle a few days earlier. Soon the package would fall into line in the check-off procedure that was now running at a rapid pace. Would the antenna erect itself? Would the squib fire the spring that would place the 10-pound satellite into a free orbit of its own? Would the compact, transistorized beacon spring into life, as it had done thousands of times in the shacks of the builders? Or would OSCAR I merely become a footnote in the history pages of Amateur Radio? R MINUS 180 AND HOLDING FOR FIFTEEN MINUTES chattered the teletype.

Bob looked up and his heart jumped. Even though he was an old hand at the launching game, the sound of the "hold" announcement never failed to affect him. "I hope it's only a technical hold," he wished to himself as he continued with his duties. He noticed that the black sky was breaking in the East. Daylight was near.

It was always easier in daylight, for some reason ...
R MINUS 180 AND RESUMING COUNT ...

15 October, 1959

"Dear Fred:

To say I was elated to receive your letter would be the understatement of the year. However, before I allow myself to get too excited, I am going to submit a proposal to you and see what happens ... As you say, I hope we can pull (or is it push) this thing off. Best regards, Don."

The radio amateurs seated around the conference table grinned as Fred, W6EJU, Chairman, read the message. The first meeting of the OSCAR Committee was about to be called to order. There were: Chuck Towns, K6LFH; Bernie Barrick, W6OON; Stan Benson, K6CBK; and Nick Marshall, W6OLO. These amateurs are the trailblazers into space in the year 1959!

In Los Angeles, Don Stoner had many conversations with Ray Meyers, W6MLZ, and Henry Richter, W6VZT. Gradually a concept of a suitable radio satellite package was being pounded out. The phone bill between W6TNS and W6EJU began to grow to alarming proportions, supplemented by sideband schedules on 7 Mc. Don suggested that the rapidly growing group of hams be called the OSCAR Association: Orbital Satellite Carrying Amateur Radio! A natural name. So was OSCAR born in spirit.

At 7 AM Bill, W6SAI, rolled over in bed in the BOQ at Vandenberg Air Force Base, California. He reached across and shook Chuck, K6LFH, awake. "0700 local time," he said as Chuck turned his face to the wall and tried to go back to sleep. "We meet the press at 0800, and go to the pad at 1000. Today we'll either be heroes or tramps!" Chuck sat up in bed and looked at his watch. "The count down started at about two AM," he said. "They must be down to about R minus 180 by now." ... R MINUS 180 AND COUNTING ... CLEAR AREA TO LOAD FUEL ... CHECK LOG TO DETERMINE FINAL ULLAGE REQUIREMENTS ...

The tension in the blockhouse was quietly growing. A charged atmosphere punctuated by short commands and remarks served only to emphasize the quick passage of time. The sun would rise in a few moments and the air was growing warmer. A cool, mild breeze was coming in from the Pacific and the sky, which was not yet red, was a flat steel color. An Air Police helicopter hovered briefly by the launching site then slanted away on some mysterious mission, its huge rotor chopping the air. The Discoverer stood waiting, a white tall spire, gleaming dully in the giant light of dawn, yet bathed on all sides by spotlights. Soon it would burst into space.

21 October, 1960

"Federal Communications Commission:

We thank you for your comments regarding our proposed OSCAR program and will attempt herein to clarify our objectives ... The former OSCAR Committee has been reorganized as the Project OSCAR Association ... the Board of Directors have approved the project plans ... the proposed satellite will be transmitting in the 2-meter amateur band, and will be electronically keyed ... it will have a restricted life of perhaps

20 days ... Fred H. Hicks, W6EJU, for Project OSCAR."

26 September, 1960

"Dear Mr. Hicks:

This will acknowledge receipt of your letter regarding Project OSCAR ... It appears that, with the exception of the requirement for positive control of the transmitter by the station licensee, you may be able to meet the other rule requirements in question ... you realize that this project must receive the sanction of the other government agencies before final approval could be granted ... Ben F. Waple, Acting Secretary, FCC."

By now the OSCAR Association had grown to the point where items of hardware could be built and tested for the proposed satellite. Project volunteers had been assigned jobs and an OSCAR mailing list was created. Because of the press of business, W6EJU turned the chairmanship of the OSCAR program over to Mirabeau ("Chuck") Towns, K6LFH, to implement and carry on the ultimate dream of having an Amateur Radio station in orbit about the earth. For it was only a dream ...

"Really, Mr. Towns. I admit the idea has some merit to it, but I do not see what earthly good it would do to have a bunch of amateurs engage in such an effort. After all, the government has spent millions of dollars in establishing exotic tracking stations ... really, now, let's be serious for a moment ..."

Bill, W6SAI, looked dully at the plate of congealed eggs and the cup of cold coffee. "To heck with breakfast," he said to Chuck. "I'm too excited to eat." The other amateurs were equally elated: Don Stoner, W6TNS, who had been invited to the launch to see his dream come true; Goodwin L. Dosland, WØTSN, President of ARRL; and Ray Meyers, W6MLZ, Director of the Southwestern Division, ARRL. Absent because of illness was Harry Engwicht, W6HC, Director of the Pacific Division, ARRL. Two hundred miles to the north Fred, W6EJU, now acting as Operations Director, and the complete OSCAR Tracking network were standing by, waiting to flash word of OSCAR orbit to waiting radio amateurs. "Let's get the show on the road," said "Dos," reaching for his overcoat. "It's almost ten

Ready to go! OSCAR completes its qualification tests with flying colors! At final check-out are (left to right): Gail Gangwish; Nick Marshall, W6OLO; Don Stoner, W6TNS; Chuck Towns, K6LFH; and Fred Hicks, W6EJU.

minutes to eight and we have to attend the pre-launch press meeting.''

The radioteletype chattered its endless song ...

... R MINUS 150 AND COUNTING ... CLEAR AREA TO LOAD OXIDIZER ... CHECK ULLAGE REQUIREMENTS BEFORE ZEROING FLOW METER ...

10 November, 1960

"John Huntoon, ARRL

As I have mentioned to you, a proposal has been made to place an amateur satellite in orbit, using a future space vehicle as a 'piggy-back' carrier ... a need exists for strong, amateur leadership from a group that represents a majority of the amateurs, rather than a small, local club. I believe that the only organization that can truly represent the amateur in this matter is ARRL. Without ARRL sponsorship, the amateur satellite program will wither and die ... 73, Bill, W6SAI.''

In the meantime, OSCAR had enlisted additional support. George Jacobs, W3ASK, Propagation and Space Communications Editor of CQ, had volunteered to be the Washington, DC contact man for Project OSCAR. George spent many hours discussing the project with sympathetic officials of the FCC and the State Department. He tried to discover what conditions must be met by such a unique undertaking in order to receive approval from key government officials, some of whom had only a hazy concept of the ideals and dreams of the radio amateur. George worked in close collaboration with John Huntoon, General Manager of ARRL. Finally, in the early spring of 1961, after a trip to HQ by K6LFH and W6SAI for a conference with League officials, the ARRL adopted Project OSCAR, granting its endorsement to the project and providing important, vital backing in the name of the amateurs of the United States.

The launch site was atop a scrubby sand dune in a far corner of Vandenberg AFB. A jolting Air Force bus crossed innumerable sand dunes and washes, carrying the amateurs and reporters who would soon observe the launch. Dry bush dotted the rough landscape. Suddenly, the Discoverer atop the launch pad was visible on the horizon. It stood majestically alone, surrounded by lesser objects that emphasized its size. It was a clear white, with the motto "United States" emblazoned on it. A single plume of evaporating liquid oxygen curled lazily from one side. There was no movement about the vehicle, and the area seemed deserted and asleep. The bus, loaded with newspaper, radio and TV reporters and the group of radio amateurs, ground to a halt atop a small plateau about five hundred yards from the launch site. The riders dismounted and slowly walked to a clear spot from which the Discoverer rocket was in clear view. At one corner of the plateau stood a small gasoline generator, a communications truck, a table with a battery of telephones, and a portable loud speaker plugged into the base communications system. ... R MINUS 80 AND STILL COUNTING ...

The Air Force Thor booster, standing on the launching pad, had completed the touchy fueling operation in which thousands of pounds of RP-1 (a souped-up version of aircraft jet fuel) and LOX (liquid oxygen)

had been pumped into it. On top of the booster, the 25-foot-long Agena brought the total height of the satellite-vehicle combination to 81 feet. The sun was climbing higher in the sky and the wind had died down now, and the site was clear and warm. ... R MINUS 50 AND STILL COUNTING ... TANK PRESSURES CHECKED ... DESTRUCT SQUIBS ARMED ... RECORDERS ARE ON ...

"Why do you employ an 'R' count instead of a 'T' count?" asked W6SAI of Captain Barbato (USAF), the Public Information Officer.

"The R-count is in minutes and is used up to about minus 10 minutes. At that time we switch to the T-count, which is run in minutes or seconds," explained the Captain. The communications truck gave notice from the Missile Flight Safety Officer that the range was clear, and that it was clear to launch.

31 July, 1961

"Secretary of State, US State Department:

The American Radio Relay League, the national nonprofit membership association of Amateur Radio operators, requests the cooperation of the Department of State concerning space communication and experimentation by radio amateurs. A group of skilled radio amateurs on the West coast, which is incorporating as the Project OSCAR Association, has designed and constructed communications equipment suitable for launch into orbit. The Association is nonprofit and is entirely noncommercial and nonmilitary. It is affiliated with and has the full support of the American Radio Relay League ... an informal session was held in Washington recently, with the following results:

a) Air Force representatives stated that Project OSCAR has been approved by HQ AFSC for incorporation in the Discoverer series of launchings, subject to coordination with other interested government agencies ... it is our hope that the information contained herein will be sufficient to enable the Department of State now to undertake the procedure outlined and agreed to at the meeting—that is, to solicit the formal concurrence of the several agencies concerned in this matter so that the project may go forward ... (signed) John Huntoon, General Manager, ARRL.''

Simultaneously, the Project OSCAR Communications link was being organized under the direction of Tom Lott, VE2AGF/W6. It was desired to have early acquisition of the OSCAR satellite by a responsible party, so various amateurs were contacted at the South Pole bases by Captain David Veazey, W4ABY, USN, Assistant for Communications, Special Projects Office. Dave promised to arrange a suitable amateur tracking station to be set up on the Antarctic continent by the KC4 hams to flash back word of OSCAR, once it achieved orbit.

The crowd at the Discoverer site had grown to a small army. General Francis H. Griswold, K3RBA, Director of the National War College, Washington, DC, had arrived. In addition, a group of scientists from California Institute of Technology had heard of the launch, and had interrupted their important work to watch the world's first homemade Amateur Radio satellite hurled into orbit. ... T MINUS 30 AND COUNT-

Directors of the Project OSCAR Association. Left to right: Fred Hicks, W6EJU; Bill Orr, W6SAI; Harley Gabrielson, W6HEK; Tom Lott, VE2AGF/W6; Chuck Towns, Jr, K6LFH (Chairman); B. Barrick, W6OON; Dick Esneault, W4IJC/W6; Harry Workman, K6JTC; and Nick Marshall, W6OLO. Not present at the time the photo was taken were Stan Benson, K6CBK; Jerre Crosier, W6IGE; Harry Engwicht, W6HC; and M. K. Caston, WA6MSO.

ING . . . REPORTING WILL BE BY EYEBALL AND F.M. RADAR AFTER LIFT-OFF . . . TERMINAL COUNT WILL START AT T MINUS 11 MINUTES . . . GUIDANCE LOCK ON COMPLETE . . . BTL READY AND STANDING BY FOR LAUNCH . . . RANGE GREEN . . . T MINUS 20 AND COUNTING . . .

The sky had clouded over and a slight overcast settled down above the poised bird. "Do you require a clear sky for launch?" asked Ray, W6MLZ. "No," replied the Public Information Officer. "This overcast won't affect the launch."

Now the news service wires were open, and Chuck, K6LFH, placed a long-distance call to the OSCAR control center, WA6GFY. Was everything ready in Sunnyvale? . . . Good . . . Good . . . South Pole link through W4ABY and KC4USB is open . . . W6EJU at the other end of the land line queried as to the exact time of launch . . . "Sorry, Fred, can't announce the time until after lift-off . . ." Fred laughed, "I can tell from the sound of your voice it will be within a *very* few minutes," he said. As if to verify his words, the communications speaker over Chuck's shoulder blared into the telephone, "T minus 16 and counting!!!"

15 September, 1961

"John Huntoon, ARRL,

Reference is made to your letter of July 31, 1961, requesting the cooperation of the Department of State concerning space communication and experimentation by radio amateurs, specifically with respect to 'Project OSCAR.'

"In reply I am pleased to inform you, after consultation on this subject with other interested agencies of the Government, that the Department perceives no objection to the carrying out of Project OSCAR . . . For the Secretary of State: Edwin M. Martin, Assistant Secretary."

T MINUS 14 AND COUNTING . . . ONE MINUTE UNTIL START OF TERMINAL COUNT . . . TERMINAL COUNT WILL START ON MARK . . . MARK . . . PHASE ONE PROCEEDING NORMAL . . . PHASE ONE COMPLETE . . . PHASE TWO PROCEEDING NORMAL . . .

Don, W6TNS, plugged his tape recorder into the ac outlet of the portable generator. Bill, W6SAI, climbed atop a sand dune immediately behind the plateau. The Air Force men looked to their recording cameras and the babble of voices on the press telephones rose in pitch. The Air Police helicopter scooted overhead, looping about the press area, and inquisitively shot behind a sand dune. The pulsating beat of its rotor could be heard above the noise of the preparations.

. . . The teletype pounded on in a relentless beat . . . PHASE FOUR PROCEEDING NORMAL . . . ORBITAL STAGE TLM AND BEACON BEING VERIFIED . . . FUELING COMPLETE . . . MAIN SAFETY RECEIVERS INTERNAL . . . PHASE FOUR COMPLETE . . . PHASE FIVE PROCEEDING NORMAL . . .

Suddenly "Dos," WØTSN, laughed out loud.

"What's so funny, Dos?" asked Don. "The incongruity of the situation just struck me," said Dos. "Here I am, a radio ham and an attorney, on a launching pad in California! It's 14 below zero in Minnesota and a judge and jury are in recess until I return! Who would imagine I'd be here today watching OSCAR fly?"

Who indeed? There were many doubters and some who had damned the project with faint praise. Many times the future of the OSCAR Project looked black, as some insurmountable road block loomed ahead. The support of interested amateurs was great comfort in such moments:

PAØVF: It is with much interest that amateurs in the Netherlands were reading of Project OSCAR . . . we thank you for your kind information . . .

GM3NQB: . . . those with whom I have talked are tremendously interested . . .

VU2NR: . . . I would be quite happy to make any kind of observations required in regard to OSCAR . . . good luck!

LU9HAT: . . . please send me information . . . I am a member of the local amateur satellite observers' group . . .

ZS3G: Send us full details, as we intend building equipment for OSCAR . . .

Indeed, there were those who believed in OSCAR. Actually, many more than was known at the time. These amateurs knew the spirit of adventure, too.

. . . PHASE FIVE PROCEEDING NORMAL . . . ORBITAL STAGE ON INTERNAL POWER . . . BOOSTER AND BTL ON INTERNAL POWER . . . ENGINE SLEW COMPLETE . . .

The missile stood silent, awaiting the final seconds before the powerful motor would burst forth. The culmination of months of work of thousands of people was rapidly approaching a climax. The atmosphere was tense on the plateau. People spoke to each other now in half-whispers, as the newsmen unfolded the story into telephones. "Put me on the air now . . . launch will be in about ten seconds."

High-speed cameras near the launch site were now whirling and the telescopic cameras at the plateau were aimed at the bird. The master tape in the Communications Center was recording every action and sound. The air was literally charged with electricity. Oblivious to the tension, Discoverer XXXVI resembled a giant finger, pointed serenely at the heavens. Within its giant frame, the tiny OSCAR package waited ... the teletype went mad with speed ... ... LAUNCHER CLEAR TO FIRE ... CLEAR TO LAUNCH ... RANGE CLEAR TO LAUNCH ... ON MARK WILL BE T MINUS TWO SECONDS ... MARK ...

November 3, 1961

"John Huntoon, ARRL:

I am pleased to advise that the Air Force will undertake to place in orbit an OSCAR package in conjunction with a military space vehicle launching. Our Space Systems Division has been instructed to accomplish the OSCAR package launching at the earliest feasible date on a non-interference basis to the performance or mission of the launch carrier vehicle ... Please be assured of the complete cooperation by the Air Force toward successful accomplishment of this amateur experiment ... (Signed) Joseph V. Charyk, Under Secretary of the Air Force."

... LIFT-OFF ...

A brilliant flash of red-orange flame burst from the Discoverer. An awesome outpouring of sound marks the birth of space flight. The roar splits into frightful stridencies that beat upon the men as ocean waves attack the land with hurricane force. The red-orange ball of fire grows with astounding speed as the solemn silver shape rises on a plume of flame. Slowly, but with astounding acceleration, the flame grows, with the Discoverer at its head. The shouts of the observers are lost in the forest of noise. Now Discoverer is free of the land: It glories in its upward flight ... faster and faster ... the track of flame marks its progress into the heavens ... the program control starts to tilt the vehicle in the proper direction out over the Pacific Ocean ... the teletype could once again be heard tapping out history ... ... GOING UP ... LOOKS GOOD ... STILL CLIMBING ... ON COURSE ... ON AZIMUTH ... ON COURSE ...

And so, on December 12, 1961, at 2042 GMT, Discoverer XXXVI was launched into orbit, carrying into separate orbit OSCAR I, guided in its flight into history by the thoughts and prayers of thousands of radio amateurs who stand on the threshold of tomorrow.

OSCAR I was an overwhelming success. More than 570 amateurs in 28 countries forwarded observations to the Project OSCAR data reduction center. The observations provided important information on radio propagation through the ionosphere, the spacecraft's orbit and satellite thermal design. The OSCAR I mission clearly demonstrated that amateurs are capable of (1) designing and constructing reliable spacecraft, (2) tracking satellites and (3) collecting and processing related scientific and engineering information. Because of its low altitude, OSCAR I only remained in orbit for 22 days before burning up as it re-entered the earth's atmosphere.

OSCAR I was the first auxiliary package to eject from

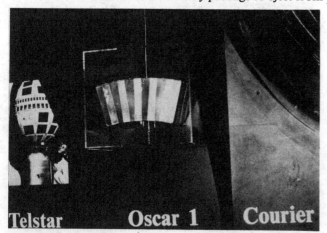

The OSCAR I that was launched in 1961 has long since burned on re-entry through the atmosphere. This version, the actual backup to the first OSCAR, now resides in the Hall of Satellites at the Smithsonian's National Air and Space Museum in Washington, DC.

the parent spacecraft and go on its own way. The stress-analyzed, mechanically and thermally balanced ejection mechanism was powered by a $1.15 spring from Sears.[10] When scientific groups wanted to put up their own free-flying spacecraft the Air Force suggested they study the OSCAR I design. For additional information on this spacecraft, see notes 11 and 12, and Appendix A (which summarizes the amateur satellite program).

OSCAR II

OSCAR II was successfully launched on 2 June 1962, barely six months after OSCAR I. They were very similar, both structurally and electrically. Despite severe time pressures, however, results from the OSCAR I flight led to a number of improvements in OSCAR II. These included (1) changing the surface thermal coatings to achieve a cooler internal spacecraft environment, (2) modifying the sensing system so the satellite temperature could be measured accurately as the batteries decayed, and (3) lowering the transmitter power output to 100 mW to extend the life of the onboard battery. Fig 2-2 shows the thermal history of OSCARs I and II. The rapid rise in temperature of OSCAR II in its final orbits was most probably caused by aerodynamic heating (friction from air molecules) as the spacecraft re-entered the atmosphere. The final telemetry reports from orbit 295, 18 days after launch, indicated an internal spacecraft temperature of 54 °C; the outer shell was probably over 100 °C by this time.[13]

OSCAR*

Along with OSCARs I and II, OSCAR* was designed, built and tested by Chuck Smallhouse (WA6MGZ) and Orv

Dalton (K6VEY). Dimensionally, it was interchangeable with the earlier OSCARs, but it contained a 250-mW beacon with phase-coherent keying. Because of the success of its predecessors, OSCAR* was never launched, as workers decided to focus their efforts on the first relay satellite—OSCAR III.[14]

WHAT PRICE SUCCESS?

People often ask, "What did OSCAR I cost?" It's impossible to give a simple answer to this question. The most expensive commodity involved—the technical expertise of the radio amateurs who designed and built the spacecraft—was donated. Almost all the parts used were donated. Use of testing and machine shop facilities: donated. The main out-of-pocket expenses—long-distance phone calls, gasoline for local travel, technical books, and so on, were absorbed by the volunteers. As satellites became more complex (for example, OSCAR III), this situation had to change. In April 1962 the OSCAR Association formally incorporated as Project OSCAR, Inc, and began soliciting memberships nationally (and distributing a newsletter for interested experimenters) to help finance future satellites.

SPACE COMMUNICATION I

During the early days of the Space Age, before the first active relay satellites were launched (see Table 2-1), the US government was prudently investigating other space communication techniques. Two projects—ECHO and West Ford—were of special interest to amateurs. Project ECHO placed in orbit large (90- to 125-foot diameter) "balloons" with aluminized Mylar surfaces capable of reflecting radio signals. The West Ford project was an attempt to create an artificial reflecting band around the earth by injecting hundreds of millions of needle-like copper dipoles into orbit. Radio amateurs were quick to realize that Projects ECHO and West Ford were, by their nature, free access (that is, anyone, anywhere could use the reflecting surface without requesting the permission of the US government). As always, amateurs were willing to try bouncing signals off almost anything—a large balloon, the moon, newly discovered scientific phenomena such as ionized trails left by satellites, or an Amateur Radio spacecraft.[15,16]

Project ECHO. ECHO A-10, the first in the series, never attained orbit. ECHO 1, which followed, though used successfully for communication by high power, non-amateur experimenters, did not enable amateur communications. Interest continued because ECHO 2 looked more promising as a reflector; it was to be larger and lower than its predecessor.[17] Radio path loss calculations at 144 MHz suggested that communications might be possible by amateurs running the legal power limit (1 kW) and using large antennas. The launch of ECHO 2, originally planned for 1962, didn't occur until 1964. In the interim, both radio amateurs and the government tested their first active relay spacecraft. The overwhelming success of active relays led to the demise of Project ECHO, but not before return signals were obtained from ECHO 2 at 144 MHz by Bill

Conkel (W6DNG) and Claude Maer (WØIC).[18] Rapid fading and weak signals, however, prevented two-way communications. Meanwhile, radio amateurs were refocusing their interest on other passive reflecting surfaces—the West Ford needles and the moon.

Project West Ford. Because a mechanical malfunction occurred in the dipole ejection mechanism, the first West Ford mission (October 1961) was a failure. A second test in 1963 successfully demonstrated that a belt of needles could support communication between very high power (far above amateur levels) ground stations. The needles decayed from orbit quickly, as expected. The program was discontinued because several scientific organizations seriously warned against the possible undesirable side effects of Project West Ford on future active satellite relays, the manned space program, radio astronomy and even the weather. Also, by this time, the advantages of active satellite relays had been demonstrated sufficiently.

Moonbounce. Radio amateurs have successfully communicated by using the moon, a natural satellite of earth, as a passive reflector on all amateur bands between 50 MHz and 10 GHz (see Table 2-2). Although moonbounce communication, often called EME (earth-moon-earth), has always taken high power, large antennas and super receivers, it continues to have a special attraction to radio amateurs. Today, most EME activity is concentrated on 144 MHz and 432 MHz. Signals are weak at best, but system performance seems to improve continually and two-way SSB contacts on 432 MHz are not uncommon. On 2 meters, W5UN, using his 48-Yagi array known as the MBA (Mighty Big Antenna), contacted over 1000 different stations in 70 countries via the moon between 1985 and 1988, several running as little as 150 W to a single long-boom Yagi.[19] On 70 cm, K2UYH reports having had QSOs with more than 440 individual stations. Both of these big guns, and many others, have earned WAS and WAC via the moon. Cumulative small improvements in station performance related to preamp performance, antenna design and digital signal processing may one day change EME from a marginal mode to a highly reliable one.[20]

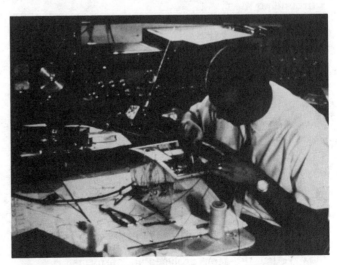

Lance Ginner, K6GSJ, prepares the first OSCAR in his garage workshop in California.

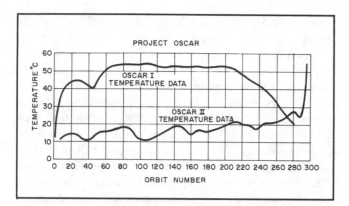

Fig 2-2—A comparison of the OSCAR I and OSCAR II temperature curves as derived from the telemetered data logged by nearly 1000 tracking stations. Modifications to the thermal design of OSCAR II based on the flight of the first satellite provided a relatively constant temperature at a more desirable lower level until the spacecraft began to rapidly re-enter.

Data processing in days of yore, W6HEK, W6MKE and K6BHN coded the OSCAR II telemetry data onto punched cards for processing on the IBM computer surrounding them.

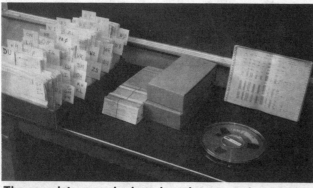

The raw data, punched cards, printouts and permanent magnetic-tape storage were state of the art in the early '60s. Today, the same analyses are performed on personal computers and programmable calculators by individual satellite users throughout the world.

Satellite Scatter. Another, albeit not very well known, space communications medium was being investigated by amateurs. In 1958, Dr John Kraus (W8JK), director of the Ohio State University Radio Observatory, noted that certain terrestrial HF beacon signals increased in strength and changed in other characteristic ways as low-altitude satellites passed nearby. He attributed the enhancements to reflection off a trail of short-lived ionized particles caused by the passing spacecraft.[21] Capitalizing on this effect, amateurs were able to locate (or confirm the positions of) several silent (non-transmitting) US and Soviet satellites by monitoring signals from WWV.[22]

Two electrical engineering students, Perry Klein (W3PK) and Ray Soifer (W2RS), read Kraus's work and decided to see if the effect—high-frequency satellite scatter—would support communication. Calculations showed that 21 MHz was the optimal amateur frequency for tests. Their positive results received national publicity in the news media, but signals using amateur power levels proved only marginal for practical communications purposes.[23]

1963 Space Conference. The OSCAR program nearly came to a screeching halt when the ITU (International Telecommunication Union) convened the 1963 Geneva Space Conference. A motion submitted by the United

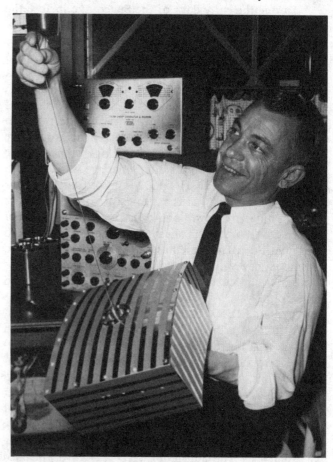

Chuck Towns, K6LFH, in his own garage workshop with OSCAR II. Though the Amateur Radio Satellite Program had its roots in basements and garages, the strictest professional standards were always maintained for the final spacecraft to pass rigorous testing by the various launch agencies.

Kingdom proposed the addition of a footnote in the 144-146 MHz worldwide amateur allocation authorizing amateur space satellites. The USSR delegation emphatically rejected the proposal and stated its opposition to any amateur space activity. Luckily, representatives of the International Amateur Radio Union (IARU) and Project OSCAR were present at the meeting, and they were able to mobilize support of the delegations which were sympathetic to amateur satellite activities, notably those of the US, Canada and the UK. A *QST* article describing the conference specifically mentions the importance of a report by Bill Orr, W6SAI, representing Project OSCAR, in helping to neutralize the Soviet position. The incident had a happy ending—the proposed footnote authorizing amateur satellite operation in the 144-146 MHz band was adopted. And, amateurs had learned a critical lesson— the radio amateur satellite program could be suddenly terminated if they didn't become actively involved in international regulatory activities.[24]

OSCAR III

Even as OSCAR II lifted off the launch pad, work was underway on OSCAR III, a far more complex satellite with the communications capabilities Don Stoner had dared speculate about years earlier. OSCAR III carried a

50-kHz-wide, 1-W transponder that received radio signals near 146 MHz and retransmitted them, greatly amplified, back to earth near 144 MHz. The transponder was designed so that it would enable radio amateurs with modest equip-

OSCAR III in full-dress. Note that the two 2-meter dipole antennas are constructed of flexible steel carpenter's rule material. The dark "checkerboard" areas are the solar-cell panels that are used as a battery backup and the springs shown on the top face were used to separate the spacecraft from the launch vehicle.

Table 2-1

Early Communications Satellites (Comsats)

All carried active, real-time transponders except for SCORE, the ECHO "balloons," Courier 18 and the West Ford needles.

Satellite	Launch date	Perigee/apogee (miles)	Comments
SCORE	18 December 1958	115/914	Often referred to as first comsat. However, it carried only a taped message for playback. It could *not* be used for relaying signals.
ECHO A-10	13 May 1960	—	Passive comsat (mylar balloon) failed to orbit (NASA)
ECHO 1	12 August 1960	941/1052	First successful passive comsat
Courier 1B	4 October 1960	586/767	First successful active comsat employed store-and-forward message system (non real-time)
Midas 4	21 October 1961	2058/2324	West Ford dipoles, failed to disperse
Telstar 1	10 July 1962	593/3503	First active real-time comsat (AT&T)
Relay 1	13 December 1962	819/4612	(RCA)
Syncom 1	14 February 1963	21,195/22,953	Electronics failure (NASA)
Telstar 2	7 May 1963	604/6713	
—	9 May 1963	2249/2290	West Ford dipoles, successful
Syncom 2	26 July 1963	22,062/22,750	First successful comsat in stationary orbit
Relay 2	21 Jan 1964	1298/4606	
ECHO 2	25 Jan 1964	642/816	Last passive comsat; first joint program with USSR
Syncom 3	19 August 1964	22,164/22,312	
LES 1	11 February 1965	1726/1744	
OSCAR III	9 March 1965	565/585	First radio amateur active real-time comsat
Early Bird (INTELSAT I)	6 April 1965	21,748/22,733	First commercial comsat
Molniya 1A	23 April 1965	309/24,470	First Soviet comsat
LES 2	6 May 1965	1757/9384	
Molniya 18	14 October 1965	311/24,855	
OSCAR IV	21 December 1965	101/20,847	First radio amateur high altitude comsat; partial launch failure

ment (normally only effective over distances under 200 miles) to communicate over paths ranging up to 3000 miles. In addition to the transponder, OSCAR III contained two beacon transmitters. One provided a continuous carrier for tracking and propagation studies, the other telemetered three critical spacecraft parameters: temperature and terminal voltage of the main battery and temperature of the transponder's final amplifier.[25-28]

Because of their low initial orbits, OSCARs I and II

Radio Amateur Club of TRW members K6MWR, Dave Moore, W6ZPX and W6RTG make the final adjustments to the OSCAR IV satellite package, here mounted in its launch cradle.

remained in space only a short time before re-entering the atmosphere and burning up. A simple battery was therefore an adequate power supply to support these spacecraft for the expected mission duration. OSCAR III, however, was being placed in a higher orbit, where it would remain considerably longer. Since weight constraints severely limited the spacecraft's battery complement, consideration was given to using solar cells on this mission. Their cost and availability, and the additional complexity required of the spacecraft, precluded this approach. Nonetheless, a small bank of solar cells was used to back up the battery that powered the beacons. OSCAR III was the first amateur spacecraft to employ solar power, though only to a limited extent. To give some perspective to this achievement, it should be noted that solar cells were then recent technology, having been invented only in 1954.[29]

Following the successful launch of OSCAR III (9 March 1965), the transponder operated for 18 days, during which time about 1000 amateurs in 22 countries were heard through it. A number of long-distance communications were reported, including Massachusetts to Germany, New Jersey to Spain, and New York to Alaska. The transponder clearly demonstrated that the concepts of free-access and multiple-access satellites would work. *Free access* means that anyone with the proper government license may uplink through the spacecraft without charge and without prior notification. *Multiple access* means that a large number of ground stations can use the spacecraft simultaneously if they cooperate in choosing frequencies and limiting power levels. The telemetry beacon, working off its own

Table 2-2

Radio Amateur EME Milestones (two-way QSOs except as noted)

Date	Band	Stations	Initial QST Reference
Jul 21, 1960	1.2 GHz	W6HB, W1BU	Sep 1960, cover, pp 10-11 62-66, 158
Apr 12, 1964	144 MHz	W6DNG, OH1NL	Jun 1964, p 95
May 20, 1964	430 MHz	W1BU, KP4BPZ	Jul 1964, p 105. Arecibo 1000-ft hemispherical dish in use at KP4BPZ
Apr 15, 1967	430 MHz	W2IMU, HB9RG	Jun 1967, pp 92-93
Mar 16, 1970	220 MHz	WB6NMT, W7CNK	May 1970, p 83
Oct 19, 1970	2304 MHz	W3GKP, W4HHK	Dec 1970, pp 92-93
Jul 30, 1972	50 MHz	WA5HNK & W5SXD, K5WVX & W5WAX	Sep 1972, p 91
Apr 5, 1987	3456 MHz	KD5RO & WA5TNY, W7CNK & KA5JPD	Jun 1987, p 61
Apr 24, 1987	5760 MHz	W7CNK & KA5JPD, WA5TNY & KD5RO	May 1988, p 67
Jan 22, 1988	902 MHz	K5JL, WA5ETV	Apr 1988, p 74
Aug 27, 1988	10 GHz	WA5VJB & KF5N WA7CJO & KY7B	Nov 1988, p 64

One-Way Reception of Echoes

Date	Band	Stations	Initial QST Reference
Jan 10, 1946	112 MHz	First report of EME echoes (non amateur)	T. Clark, ref 20
Jan 27, 1953	144 MHz	W3GKP & W4AO	Mar 53, pp 11-12, 116; Apr 53, p 56
Early 1963	28 MHz	W6UGL receives echoes using 1200-ft antenna at Stanford University	Sep 1963, p 20

battery and the solar cells, functioned for several months.[30]

Success was to bring new challenges. It was clear that unless radio amateurs just wanted to replay yesterday's triumphs, their future satellites would need major changes. First and foremost, operating lifetimes would have to increase by 10 or 100 times to justify the major effort and expense needed to build the sophisticated spacecraft designs being considered.

OSCAR IV

While OSCAR III was being completed, amateurs were presented with a launch opportunity aboard a Titan III-C rocket headed for a circular orbit 21,000 miles above the earth. At this height almost half the planet would be within range of the spacecraft at any time. OSCARs I, II and III had been designed to operate at lower altitudes (under 700 miles), where a transponder's output power, spacecraft antennas and attitude stabilization are much less critical. Building a spacecraft for a higher orbit, even with plenty of time and resources, is a formidable challenge. But time wasn't available: The projected launch date was roughly a year away and the Project OSCAR crew was deeply involved in readying OSCAR III for its flight. It appeared that this once-in-a-lifetime offer might have to be passed up. Several members of the TRW (Thompson-Ramo-Woolridge) Radio Club of Redondo Beach, California, recognized the uniqueness of the opportunity and decided to undertake the project even though the constraints seemed overwhelming. (More than two decades have passed since the launch of OSCAR IV, and a similar launch opportunity has never again materialized.)

To meet the time schedule, the spacecraft would have to be kept as simple as possible—just a transponder and an identification beacon to satisfy Federal Communications Commission requirements. "Luxuries" such as telemetry and redundant sub-systems for reliability had to be eliminated. The TRW team did, however, decide that the spacecraft would be solar powered and designed the system with a one-year-lifetime goal. The crossband transponder received (uplink) on 144 MHz and transmitted (downlink) on 432 MHz. Its design borrowed a number of ideas from a standard ranging transponder NASA used at the time. Power was set at 3 W PEP and the bandwidth was 10 kHz.

On launch day, Dec 21, 1965, the fear that constantly haunts satellite builders came partially true—the top stage of the launch rocket failed and the spacecraft never reached the targeted orbit. Had it achieved the intended orbit, it would have hung directly over the equator, drifting slowly eastward at just under 30° per day. Instead, OSCAR IV entered a highly elliptical orbit inclined to the equator at 26°; the high point (apogee) was 21,000 miles above earth and the low point (perigee) about 100 miles.

The rocket failure presented amateurs with a number of serious problems. Consider tracking, for example. Because of the spacecraft's height and low-power transmitter, ground stations needed high-gain (narrow-beamwidth) antennas. With the planned orbit, antenna aiming would have been simple, but with the actual orbit, it was nearly impossible. Suitable techniques for tracking satellites in highly elliptical orbits hadn't been devised. Because the downlink signal strength fluctuated rapidly, even scanning the receive antenna to peak signals didn't work well. This latter problem was closely related to the rocket malfunction. The attitude stabilization scheme chosen for OSCAR IV and the antenna configuration were based on the spacecraft's being spun off correctly from the top stage of the launch vehicle. Rocket failure meant loss of planned spin stabilization, and consequently inadequate control over the antenna's orientation. Amateurs attempting to use the transponder encountered additional difficulties that some attributed to spacecraft electronics. It's probable, however, that many of these difficulties were from the lack of attitude stabilization and could have been overcome had stabilization been achieved.

Had OSCAR IV operated a sufficient length of time, the attitude probably would have stabilized naturally (though not necessarily in a preferred orientation) and ground stations would have devised methods to overcome the tracking and radio link problems by, for example,

OSCAR IV, the only amateur satellite to be designed around a tetrahedral frame. Intended for a 21,000-mile high circular orbit, OSCAR IV was doomed to a short operating life when the top stage of the launch rocket failed, leaving it in an elliptical orbit for which it had not been designed.

agreeing to specified uplink power levels and using circular polarization. But the transponder ceased operating after a few weeks. Since the spacecraft didn't have a telemetry system, we can only guess the cause: either battery failure from thermal and power supply stresses, or solar cell failure from the radiation levels encountered, both possibilities arising from the unexpected orbit.

Even with the enormous difficulties encountered, several amateurs completed two-way contacts through OSCAR IV. One contact, between a station in the United States and another in the Soviet Union, was the first direct two-way satellite communication between these two countries. The mission provided considerable information that would prove valuable in designing future spacecraft.

Although OSCAR IV was, in some respects, a major disappointment, it's important to keep in mind that the key failure occurred in the launch vehicle, a possibility that radio amateurs working with satellites must learn to accept. The amateurs who designed and built OSCAR IV did an extraordinary job, and the users whose ingenuity salvaged so much from the mission were a credit to Amateur Radio.[31-33]

OSCAR-RELATED EXPERIMENTS

An intriguing announcement appeared in April 1967 *QST* (p 87). A young German engineer, Karl Meinzer (DJ4ZC), was offering to loan, free of charge, a 1-W 2-meter single-band linear transponder to an American group interested in organizing balloon launches. The unit was a prototype for an OSCAR mission. During the previous two years the German group had 18 balloon flights up to 65,000 ft. Karl reported that "During the first two or three flights a lot of confusion was present, but in later flights skills improved and people in general are now quite capable of making the best of this way of communication. There is still some time until the [next OSCAR] launch, so I think it would be nice to give the American hams an opportunity to train themselves, too." You'll be seeing Karl's name often in the following chapters, as he went on to play, and still plays, a key role in OSCAR spacecraft design and construction. If you're currently on AMSAT-OSCAR 13, you're using a transponder developed by Karl's group.

When the Germans tried to begin their balloon-flown transponder program, government authorities refused permission to launch. Finally, Dutch amateurs volunteered to obtain the required permission and launch from the Netherlands. Once a balloon was up, there was no regulation preventing the Germans from using it. After several flights without incident the Germans were finally able to obtain their government's permission to launch from Hessel. Over the next several years more than 60 flights were undertaken under the direction of Fritz Herbst, DL3YBA.

Nearly 25 years later—the date is 1988 and the place is Great Britain—another group of amateurs runs into a stone wall when trying to obtain permission for a balloon launch of a prototype OSCAR transponder. Nico Janssen, PA0DLO, offers to apply for permission to fly in the Netherlands. This story is still in progress; tune in next week.

Notes

[1]G. S. Sponsler, "Sputniks Over Britain," *Physics Today*, Vol 11, no. 7, Jul 1958, pp 16-21. Reprinted in *Kinematics and Dynamics of Satellite Orbits*. American Association of Physics Teachers, 335 East 45th St, NY, NY 10017 ($2.50).

[2]F. L. Whipple and J. A. Hynek, "Observations of Satellite I," *Scientific American,* Vol 197, no. 6, Dec 1957, pp 37-43.

[3]R. Buchheim and Rand Corp Staff, *New Space Handbook* (New York: Vintage Books, 1963), pp 283-312.

[4]From a condensed translation of the June 1957 article in the Soviet journal *Radio:* V. Vakhnin, "Artificial Earth Satellites," *QST*, Nov 1957, pp 22-24, 188.

[5]From a presentation by Dr Van Allen at a special program commemorating the 20th anniversary of the launch of Explorer I. Held at the National Academy of Sciences, Washington, DC, Feb 1, 1978.

[6]A discussion of Dr Reber's work and original references can be found in J. D. Kraus, *Radio Astronomy* (New York: McGraw- Hill, 1966, Chapter 1). Dr Reber is currently living in Bothwell, Tasmania. Dr Kraus, director of the Ohio State University Radio Observatory, is W8JK. Also see: R. W. Miller, "The Listener," *QST*, Jan 1989, pp 58-59.

[7]E. P. Tilton, "Lunar DX on 144 Mc.," *QST*, Mar 1953, pp 11-12, 116.

[8]D. Stoner, "Semiconductors," *CQ*, Apr 1959, p 84.

[9]W. I. Orr, "Sixty Years of Radio Amateur Communications," *QST*, Feb 1962, pp 11-15, 130, 132.

[10]P. McKnight, "*QST* Profiles Chuck Towns, K6LFH: 'Let's Fly High'," *QST*, Jan 1987, pp 68-69.

[11]H. Gabrielson, "The OSCAR Satellite," *QST*, Feb 1962, pp 21-24, 132, 134. Technical description of OSCAR I.

[12]W. I. Orr, "OSCAR I: A Summary of the World's First Radio-Amateur Satellite," *QST*, Sep 1962, pp 46-52, 140.

[13]W. I. Orr, "OSCAR II: A Summation," *QST*, Apr 1963, pp 53-56, 148, 150.

[14]W. I. Orr, "OSCAR I: A Summary of the World's First Radio-Amateur Satellite," *QST*, Sep 1962, p 56.

[15]R. Soifer, "Space Communication and the Amateur," *QST*, Nov 1961, pp 47-50. The references in notes 14 and 15 treat basic concepts in a comprehensive manner, and the information contained is still of interest to experimenters involved in radio astronomy, direct reception from lunar and deep space probes, and reception of commercial satellite TV.

[16]R. Soifer, "The Mechanisms of Space Communication," *QST*, Dec 1961, pp 22-26, 168, 170.

[17]R. Soifer, "Amateur Participation in ECHO A-12," *QST*, Apr 1962, pp 32-36. Note: ECHO 2 was known as ECHO A-12 before launch. R. Soifer, "Project ECHO A-12," *QST*, June 1962, pp 22-24.

[18]R. Soifer, "Amateur Radio Satellite Experiments in the Pre-OSCAR Era," *Orbit*, Vol 2, no. 1, Jan/Feb 1981, pp 4-7.

[19]D. Blaschke, "The Evolution of the 'Mighty Big Antenna'," *QST*, Jan 1989, pp 15-19.

[20]For information on the history of EME see: W. Orr, "Project Moon Bounce," *QST*, Sep 1960, pp 62-64, 158; F. S. Harris, "Project Moon Bounce," *QST*, Sep 1960, pp 65-66; H. Brier and W. Orr, *VHF Handbook for Radio Amateurs* (Wilton, CT: Radio Publications, 1974); T. Clark, "How Diana Touched the Moon," *IEEE Spectrum*, May 1980, pp 44-48.

[21]J. D. Kraus, R. C. Higgy and W. R. Crone, "The Satellite Ionization Phenomenon," *Proc IRE*, Vol 48, Apr 1960, pp 672-678.

[22]C. Roberts, P. Kirchner and D. Bray, "Radio Detection of Silent Satellites," *QST*, Aug 1959, pp 34-35.

[23]R. Soifer, "High-Frequency Satellite Scatter," *QST*, Jul 1960, pp 36-37. R. Soifer, "Satellite Supported Communication at 21 Megacycles," *Proc IRE*, Vol 49, no. 9, Sep 1961.

[24]"The 1963 Geneva Space Conference," *QST*, Jan 1964, pp 60-61, 152.

[25]W. I. Orr, "The OSCAR III V.H.F. Translator Satellite," *QST*, Feb 1963, pp 42-44.

[26]A. M. Walters, "OSCAR III—Technical Description," *QST*, Jun 1964, pp 16-18.

[27]A. M. Walters, "Making Use of the OSCAR III Telemetry Signals," *QST*, Mar 1965, pp 16-18.

[28]W. I. Orr, "OSCAR III Orbits the Earth!," *QST*, May 1965, pp 56-59.

[29]D. M. Chapin, C. S. Fuller and G. L. Pearson, "A New Silicon p-n Junction Photocell for Converting Solar Radiation into Electrical Power," *J. Applied Physics*, Vol 25, May 1954, p 676.

[30]H. C. Gabrielson, "OSCAR III Report—Communications Results," *QST*, Dec 1965, pp 84-89.

[31]"OSCAR IV News," *QST*, Dec 1965, p 41.

[32]"OSCAR IV Due Dec 21," *QST*, Jan 1966, p 10.

[33]E. P. Tilton and S. Harris, "The World Above 50 Mc.," *QST*, Feb 1966, pp 80-82.

Chapter 3

The 1970s

After the disappointment over OSCAR IV, amateurs were to be treated to a string of amateur spacecraft that not only met, but exceeded, expectations. Australis-OSCAR 5, designed and built at the University of Melbourne, worked almost flawlessly, while AMSAT-OSCARs 6, 7 and 8 brought reliable two-way satellite communication to amateurs and students around the world.

OSCAR 5

The OSCAR 5 story begins in Australia. Late in 1965, several students at the University of Melbourne, mostly undergraduate members of the Astronautical Society and Radio Club, seriously began to consider building a satellite. Though none of them had any spacecraft construction experience, they were competent in electronics and mechanical design. When the California-based Project OSCAR agreed to take care of final environmental testing, locating a launch, and launch operations for Australis-OSCAR 5 (AO-5), the "down under" crew began the project in earnest. (Note: With the fifth amateur satellite being readied for flight, amateurs decided to acknowledge the advantage of Arabic numerals over their Roman counterparts—hence OSCAR 5, not OSCAR V.)

Members of the Melbourne group wanted to make a unique and significant contribution to the amateur space program, but they recognized that their isolation and lack of experience dictated a relatively simple spacecraft. The design, finalized in March 1966, showed that their desire and the constraints they were working under could be reconciled.

AO-5 would attempt to: (1) evaluate the suitability of the 10-meter band for a downlink on future transponders; (2) test a passive magnetic attitude stabilization scheme; and (3) demonstrate the feasibility of controlling an amateur spacecraft via uplink commands. The flight hardware to accomplish these goals included telemetry beacons at 144.050 MHz (50 mW) and 29.450 MHz (250 mW at launch), a command receiver and decoder, a seven-channel analog telemetry system and a simple manganese-alkaline battery power supply. The spacecraft did not contain a transponder or use solar cells.

Though technical aspects of the AO-5 project went smoothly, they turned out to be just the tip of the project's iceberg; administrative concerns were a constant frustration. Air-posting a special 50-cent part from the US to

Australia might cost $10, and clearing the part through customs often required pages of paperwork and several trips to government offices. You probably get the picture: Technical competence isn't enough. People who build satellites also need great perseverance. Step-by-step, Australian dollar by Australian dollar, AO-5 took shape. On June 1, 1967, 15 months after final plans were okayed, the completed spacecraft was delivered to Project OSCAR in California. A launch opportunity was targeted for early 1968. Delay followed delay, however, until the host mission was indefinitely postponed. No other suitable launch was immediately available.

So stood the situation in January 1969 when George Jacobs (W3ASK) spoke to the COMSAT Amateur Radio Club in downtown Washington, DC. Jacobs suggested that, with the space-related expertise and facilities in the area, the amateur space program might benefit from an East Coast analog of Project OSCAR. As a result, AMSAT (the Radio Amateur Satellite Corporation) was founded. Formal incorporation took place on March 3, 1969, in Washington, DC, and the first task of the new organization was arranging for an Australis-OSCAR 5 launch. (Excerpts from the bylaws appear in Table 3-1.)

Environmental and vibration tests of AO-5 showed that some minor changes were needed. AMSAT performed the modifications and identified a suitable host mission. Finally, on January 23, 1970, AO-5 was launched on a National Aeronautics and Space Administration (NASA) rocket (previous OSCARs had all flown with the US Air Force). Electronically, the satellite performed almost flawlessly. One small glitch prevented telemetry data from being sent over the 29-MHz beacon. Since the same telemetry information was available on 144 MHz, the problem had little impact on the overall success of the mission. The magnetic attitude stabilization system worked beautifully. The spacecraft's spin rate decreased by a factor of 40—from 4 revolutions per minute to 0.1 revolution per minute—over the first two weeks. A network of ground stations periodically transmitted commands to the satellite, turning the 29-MHz beacon on and off. Allowing the beacon to operate only on weekends helped to conserve the limited battery power. The first successful command of an amateur satellite took place on orbit 61, on January 28, 1970, when the 29-MHz beacon was turned off. The demonstration of command capabilities was to prove very important in obtaining FCC licenses for future missions.

Performance measurements of the 29-MHz beacon confirmed hopes that this band would prove suitable for transponder downlinks on future low-altitude spacecraft, and led to its use on OSCARs 6, 7 and 8. As the battery became depleted, the transmitters shut down: The 144-MHz beacon went dead 23 days into the mission, and the 29-MHz beacon, operating at greatly reduced power levels, was usable for propagation studies until day 46.[1-5]

At AMSAT, the project manager responsible for final testing, modification, and integration of AO-5 was a young engineer named Jan King (W3GEY). It's difficult for people not directly involved in a program of this scope to imagine the pressure on the project manager. But Jan must not have minded too much, as he went on to oversee the design and construction of several AMSAT spacecraft over the next quarter century.

AO-5 met its three primary mission objectives. In addition, careful analysis of reports submitted by ground stations that monitored the mission showed that such stations were capable of collecting reliable quantitative data from a relatively complex telemetry format. All in all, AO-5 was a solid success. But radio amateurs wanted a transponder they could use for two-way communication, and five years had passed since the last one had orbited.

Space Communication II

Deep space probes. While waiting for the next active relay satellite, radio amateurs experimented in related areas. A few constructed 2.3-GHz (S-band) microwave receiving stations to monitor the Apollo 10, 12, 14 and 15 lunar flights. During the Apollo 15 mission (August 1971), amateurs received voice transmissions from the Command Service Module as it circled the moon.[6] Although the S-band RF equipment needed for monitoring unmanned deep-space probes is similar to that needed for listening to manned flights, efforts have focused on the latter. Probably, this is because decoding the voice channels is much easier (and more exciting?) than extracting information from the sophisticated telemetry links. Though a long lull in lunar exploration began with the late 1970s it's inevitable that humans will one day again visit the moon and set up a camp on Mars. In the not-too-distant future, some amateurs will certainly attempt to monitor the first human expedition to the red planet. With a significant number of scientists in both the US and USSR pressing for a joint venture, the Mars trip might take place a lot sooner than many people realize.

ATS-1. The US government launched ATS-1 (Applications Technology Satellite) into a geostationary orbit on December 7, 1966. Satellites in such an orbit appear to remain fixed above a particular spot on the equator. Of interest to radio amateurs was an experimental 100-kHz-wide, hard-limiting transponder carried by ATS-1 that received near 149.22 MHz and retransmitted at 135.6 MHz. Professor Katashi Nose (KH6IJ) of the University of Hawaii was one of the scientists working with NASA to evaluate this system. By monitoring the transponder operation, radio amateurs were able to learn a great deal about the performance of radio links to geostationary satellites near the 144-MHz amateur band, and about the performance of hard-limiting transponders similar to those being considered for future amateur missions.[7]

ATS-1 served for 20 years before it was officially retired when it ran out of positioning fuel. However, as of mid-1990, it's still operational and frequently heard on 137.35 MHz.[8]

But this isn't a record. Relay I, launched in 1962 (see Table 2-1), is still often heard on 136.140 MHz and 136.620 MHz. Today's commercial satellites are being designed with projected 10-year lifetimes. Recent amateur spacecraft have shown that we should be able to obtain similar life

---

Table 3-1

The Introductory Section of AMSAT's Bylaws (as revised in 1989)

BYLAWS of the RADIO AMATEUR SATELLITE CORPORATION

\# These Bylaws have been adopted pursuant to the Articles of Incorporation, which provide, in part, as follows:

\# FIRST: The name of the corporation is Radio Amateur Satellite Corporation.

\# SECOND: The period of duration is perpetual.

\# THIRD: Said corporation is organized exclusively for scientific purposes, including, for such purposes, the making of distributions to organizations that qualify as exempt organizations under Section 501(c)(3) of the Internal Revenue Code of 1954, as amended (or the corresponding provision of any future United States Internal Revenue Law).

\#\# The scientific purposes for which said corporation is organized shall be the carrying on of scientific research in the public interest by the means of:

\#\# A. Developing and providing satellite and related equipment and technology used or useful for amateur radio communications throughout the world on a non-discriminatory basis.

\#\# B. Encouraging development of skills and the advancement of specialized knowledge in the art and practice of amateur radio communications and space science.

\#\# C. Fostering international goodwill and cooperation through joint experimentation and study, and through the wide participation in these activities on a non-commercial basis by radio amateurs of the world.

\#\# D. Facilitating communications by amateur satellites in times of emergency.

\#\# E. Encouraging the more effective and expanded use of the higher frequency amateur radio frequency bands.

\#\# F. Disseminating scientific and technical information derived from such communications and experimentation, and encouraging publication of such information in treatises, theses, publications, technical journals and other public means.

\#\# G. Conducting such lawful activities as may be properly incident to or aid in the accomplishment of provisions A-F hereinabove, and which are consistent with the maintenance of tax-exempt status pursuant to Section 501(c) of the Internal Revenue Code.

spans. When this occurs, it may be necessary to shut off still-functioning older spacecraft so the link frequencies can be used more effectively.

Modifying the ionosphere. Radio amateur interest in projects ECHO and West Ford (see Chapter 2) focused on reflecting radio signals off objects launched into space. A closely related class of experiments involves direct physical modification of the ionosphere to change its radio-reflecting characteristics. Two approaches that have received a great deal of attention involve (1) releasing chemicals, such as barium, from rockets directly into the ionosphere[9] and (2) employing very-high-power, ground-based radio transmitters, operating in the vicinity of 3-10 MHz, to produce an "artificial radio aurora" (raise the temperature of electrons in the ionosphere).[10,11] Both types of experiments are expected to continue through the '90s.[12] The main payload on the ill-fated AMSAT Phase 3-A mission, for example, was a barium-release experiment known as Firewheel.[13]

An artifical radio aurora has marked effects on propagation over the range of 20-450 MHz. The major facility for ionospheric heating, located in Platteville, Colorado, USA, began operating in 1970. Amateur experiments with this communications medium first took place in 1972 and are continuing. During the week of March 17, 1980, for example, the Platteville heater was scheduled for 20 hours of operation. Though articles suggest that one should be located within 800 miles of Platteville to take part in these experiments, the distance can be greatly extended by studying satellite links.

1971 Space Conference. In June 1971, the ITU convened a World Administrative Radio Conference (WARC) on Space Telecommunications to review and make necessary changes to the regulations concerning extraterrestrial radio communication and radio astronomy. This time around, radio amateurs were much better organized. They had prepared a modest set of proposals that would permit an orderly growth of amateur satellite activites over the next decade. Frequencies requested were all in existing amateur primary or shared allocations and had been carefully selected to minimize any possibilities of interference to other services. Reports explaining the amateur position had been circulated to and discussed with many ITU delegations before the conference. As a result, official proposals to the ITU on behalf of the Amateur Radio satellite program were presented by Argentina, Australia, Brazil, Canada, the Federal Republic of Germany, the Netherlands, New Zealand, the United Kingdom and the United States. With this type of support and preparation, you'd think the job would be easy. But France led intense opposition to any allocation in shared bands and proposed that the amateur-satellite service be prohibited from using any frequencies between 146 MHz and 24 GHz.

Although there was widespread support for amateur satellite activities at the conference, official delegations often had to work out compromises in order to attain their highest priority objectives, so the amateur satellite program was in serious jeopardy. As is often the case, many of the official delegates and advisors at the conference were radio amateurs. These individuals, and a group of observers representing the IARU, gathered together an informal team that kept attention focused on the needs of the amateur satellite program. One of these observers was Perry Klein (W3PK), the first president of AMSAT. A quote from another attendee, ZL2AZ, provides some insight into what the conference was like. "It is surprising how hoarse one can get even though he is supposed to be only an observer!"

As a result of the 1971 WARC, amateur satellite frequency allocations were expanded to include all exclusive worldwide amateur bands plus an extremely important segment at 435-438 MHz. In a practical sense, the victory was very modest, since many of these frequencies are poorly suited to satellite links and requests for key allocations in the 1.2, 5.6 and 10-GHz bands had been turned down. In addition to these frequency allocation

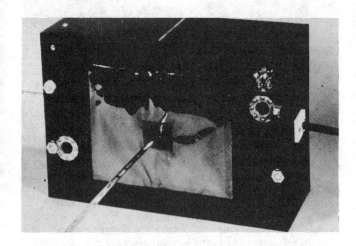

Australis-OSCAR 5 shown with its antennas deployed. Note that this satellite carried no solar cells and that actual steel carpenter rule was used for the antenna elements.

In this view of Australis-OSCAR 5, the flexible antennas have been tied back, where they will remain during launch. At the proper time, as the satellite is separated from its launch vehicle, these elements will spring out to their full pre-cut length.

actions, the ITU officially recognized and defined the Amateur Satellite Service in the International Radio Regulations. The implications of this recognition weren't clear. It could conceivably make it more difficult for an amateur group to obtain a special temporary authorization to experiment with certain frequencies on a satellite link. In any event, amateurs were no longer officially restricted to 144-146 MHz for space communications. In the near future they would also look into using 21.0-21.45, 28.0-29.7, and 435-438 MHz.[14]

AMSAT-OSCAR 6

Amateur Radio took a giant stride into the future on October 15, 1972, when AMSAT-OSCAR 6 (AO-6) was launched successfully. Although it was more complex than all previous OSCARs combined, ground stations interested in communicating through its transponder or studying its telemetry found AO-6 to be the easiest amateur satellite to work with. Phase 2 of the amateur satellite program, the age of long-lifetime satellites, was under way. While the aggregate operational time of all previous OSCARs amounted to considerably less than one year, AO-6 was to "do its thing" for more than four and a half years. From October 15, 1972 forward, the Amateur Radio community would have at least one transponder-equipped, low-altitude satellite in operation.[15]

To understand the significance of AO-6, we must go beyond the impressive facts and figures and look at the philosophy underlying its construction. Two ideas were central. First, the investment needed in both dollars and effort to produce spacecraft that could make significant new scientific, engineering or operational contributions to

Amateur Radio was such that only long-life (at least one-year duration) satellites could be justified. Second, constructing a reliable long-life satellite required much more than replacing a battery power supply with a power system consisting of solar cells, rechargeable batteries and related control electronics. Long lifetime could be reasonably assured only if the spacecraft contained (1) a sophisticated telemetry system permitting onboard systems to be monitored, (2) a flexible command system so that various

AMSAT-OSCAR 6, as shown on the plaque mounted inside the satellite structure, was dedicated to Capt Harry D. Helfrich, W3ZM, an active AMSAT participant in the OSCAR 6 project who became a Silent Key shortly before the satellite was launched.

AMSAT-OSCAR 6, mounted to its launch vehicle's attach fitting, only a few hours from launch. Note the flexible "carpenter-rule" 10-meter antenna elements on each side that are tied back for launch, and the tiny gold-plated piano-wire 70-cm whip antenna atop the spacecraft. After launch, upon separation from the launch vehicle, these pre-cut antennas were freed from their restraints and sprang to operating length. (The Plexiglas cover shown mounted over the front solar panel is there to protect the delicate solar cells while AMSAT and NASA personnel work around the spacecraft; it was removed prior to launch.)

Jan King, W3GEY, adjusts OSCAR 7 on its perch atop a "shake table." All OSCARs must undergo rigorous testing to prove that they will survive the rigors of launch and the hostile space environment without damaging or otherwise affecting the mission of the primary payload. The shake table, a distant relative of your local hardware store's paint shaker, is used for vibration tests in which the structure is subjected to the severe vibrations that will be experienced during launch. Secondary payloads, the "piggyback riders" of the aerospace world, aren't certified for flight until they have passed such tests.

spacecraft subsystems could be activated or deactivated as conditions warranted, and (3) redundancy in critical systems. In addition, the design strategy must attempt to prevent catastrophic failure by anticipating possible failure modes and incorporating facilities for isolating defective subsystems.

AMSAT-OSCAR 6 wasn't the first amateur spacecraft to use solar power or include command and telemetry systems. But it brought each of these subsystems a quantum jump forward. For example, the command system on AO-5 could only turn the 10-meter beacon on or off. AO-6 recognized 35 distinct commands, 21 of which were acted on. The most sophisticated telemetry system used previously (on OSCAR 5) included seven analog channels. The AO-6 downlink contained 24 telemetry channels. Furthermore, the spacecraft carried a newly designed processing system that greatly simplified telemetry decoding equipment requirements: Ground stations had only to copy numbers in Morse code and refer to a set of graphs.[16]

AMSAT-OSCAR 6 carried a 100-kHz-wide transponder running about 1 W output at 29 MHz, the frequency tested as a downlink on AO-5. The transponder was extremely sensitive. Ground stations running as little as 10 W to a groundplane antenna on the 146-MHz uplink would put through solid signals as long as the transponder wasn't fully loaded, or gain-compressed by users running excessive power. The ease and reliability of communication through the transponder were enhanced greatly by the spacecraft's magnetic attitude-stabilization system, another feature pioneered on AO-5. AO-6 also carried a unique digital store-and-forward message system called Codestore. Suitably equipped ground stations could load messages into Codestore using Morse code (or any other digital code conforming to FCC regulations) for later playback, either continuously or on command. Codestore was often used to relay messages between Canadian and Australian command stations, and many radio amateurs outside the USA depended on Codestore for pre- and post-launch information relating to AMSAT-OSCAR 7.

Although AMSAT-OSCAR 6 turned out to be an overwhelming success, it did have problems. The 435-MHz beacon failed after about three months. It lasted long enough, however, to test 435-MHz as a downlink for future low-altitude spacecraft and to enable John Fox (WØLER) and Ron Dunbar (WØPN) to discover an interesting Doppler anomaly.[17] A second problem, one of major importance, involved mode falsing. The satellite control system turned out to be very susceptible to internally generated noise. Noise would often be interpreted as a command and turn the transponder and other subsystems on or off. Working with AO-6 in those early months was very frustrating. It seemed as if the transponder would regularly choose the most inopportune times to shut down.

The solution to the falsing problem is an interesting story. Although the orbiting spacecraft couldn't be repaired, it was suggested that the difficulties would be minimized if a constant stream of commands directing it into the correct mode was sent to the satellite. To make

this idea work, automated ground command stations had to be developed and a number of stations around the world would have to volunteer, often at significant personal cost (in time and cash), to accept responsibility for building and operating these command stations. Larry Kayser (VE3QB) was among the first to feverishly attack the falsing problem. He quickly put a command station in operation, automating it, after a fashion, with a tape-recorder control loop connected to his telephone. By ringing up his home phone he would activate the system. As Kayser tells it:

> For the next few weeks, it was not uncommon (for me) to dash for a telephone, dial a number, and hang up. This went on several times in a 10-minute period for each pass, sometimes from Montreal, Toronto, a gas station on the highway, or wherever...Full automation was certainly a more desirable way to go.

Kayser went on to design a series of systems, each one buying a little more time so that the satellite could be kept operating while more reliable techniques were developed. By August 1973, a system capable of automatically generating 80,000 commands per day was in operation. This was quite a change from the twice-weekly commands used to control the 29-MHz beacon on AO-5.[18]

In addition to keeping AO-6 on a reliable schedule, command stations were largely responsible for the spacecraft's 4.5-year lifespan. Without their careful management, it's doubtful that even the original one-year intended design lifetime of the spacecraft could have been reached. AO-6 died when several battery cells failed (shorted) during its fifth year in orbit.

Perry Klein, W3PK, former AMSAT President, stands in front of OSCAR 7 before closing the door to the thermal-vacuum test chamber. One of the many tests that the OSCARs must pass, the thermal-vacuum test measures a structure's cleanliness in the harsh space environment. The spacecraft, in a high vacuum, is heated to the very high temperatures it will encounter in space (in a sense boiling off impurities into their gaseous state) for several days. Then a super-cold "cold finger" (a special thermal probe) is activated and the gaseous impurities condense on its surface, where they can be measured quantitatively. Other phases of the test include several days at "room temperature" and several days at the extreme cold temperatures that the satellite will experience in space.

Subsystems for the AMSAT-OSCAR 6 spacecraft were built in the US, Australia and West Germany. Ground command stations were activated in Australia, Canada, Great Britain, Hungary, Morocco, New Zealand, the US and West Germany. Users in well over 100 countries reported two-way communications.[19]

Though AO-6 was awarded a free ride into space for several reasons, its potential value as an educational tool was paramount. The introduction of long-lifetime radio amateur satellites made it feasible for science instructors at all levels to incorporate class demonstrations of satellite reception into regular course work. To assist teachers pioneering this path, AMSAT and the American Radio Relay League granted funds to the Talcott Mountain Science Center in Connecticut to produce an instruction manual aimed at educators working with grades 1 through 12. The result was the well-received *Space Science Involvement* manual, first published in 1974. Thousands of free copies were distributed to teachers over the following six years. In 1978 a follow-up publication geared to college level instruction was published. *Using Satellites in the Classroom: A Guide for Science Educators* was produced with the financial assistance of the National Science Foundation and the Smithsonian National Air and Space Museum. The OSCAR education program includes additional activities supported by the ARRL in a continuing program of local assistance referral, personalized educational bulletins via satellite, special satellite scheduling, the publication of newsletter updates for science educators and development of a slide show library.

The outstanding OSCAR education program is probably, by itself, sufficient justification for free launches. Our emphasis on it here, however, should in no way be construed as downgrading the significance of OSCAR contributions in other areas such as emergency communications, scientific exploration and public service, which are discussed elsewhere in this book.

AMSAT-OSCAR 7

November 15, 1974 marked the beginning of another success story. AMSAT-OSCAR 7 (AO-7) was launched on that date, and for the first time amateurs had two operating satellites in orbit. While AO-6 represented a quantum leap forward technically, AO-7 was more of an evolutionary step in technical improvement.[20] It contained two transponders, one similar to the unit flown on AO-6, using a 146-MHz uplink and a 29-MHz downlink (known as Mode A), and the second with an uplink at 432 MHz and a downlink at 146 MHz (known as Mode B).

The Mode B transponder was based on a unique design developed by Dr Karl Meinzer (DJ4ZC), whose mid '60s balloon-flown transponder experiments were described earlier. Running 8 W (PEP), it featured a highly efficient method of linear frequency translation. Built in West Germany under the sponsorship of AMSAT-Deutschland, the Mode B transponder (in concert with the frequencies used, the antenna system and the magnetic attitude control) provided outstanding performance. Whereas AO-6 demonstrated that simple ground stations could commu-

AMSAT-OSCAR 7 (lower left) is dwarfed by the primary payload, the ITOS-G satellite, as it sits attached to its Delta launch vehicle. The similar-looking spacecraft, shown opposite to and counterbalancing the weight of OSCAR 7, is the Spanish INTASAT. Note that the flexible elements of OSCAR 7's 2-meter canted-turnstile antenna protrude downward and will ride within the launch vehicle cowling.

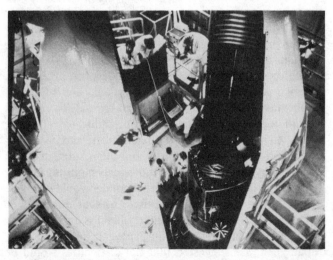

A last look at OSCAR 7 before the cowling is secured around the trio of fellow space travelers.

nicate via satellite, AO-7 showed that low-altitude satellites could, under many conditions, provide simple stations with communications capabilities over moderate distances (200-4500 miles) far exceeding any alternative mode.

The AMSAT-OSCAR 7 spacecraft carried Codestore and telemetry units nearly identical to those of AO-6. It also contained a new high-speed, high-accuracy telemetry encoder (designed by an Australian group) that transmitted radioteletype. Beacons at 146 MHz, 435 MHz (built in Canada) and 2304 MHz were also flown. The 100-mW, 2304-MHz beacon, contributed by members of the San Bernardino Microwave Society (California, USA), was potentially one of the most interesting technical experiments aboard AO-7. Much has been learned from this beacon, though not in the areas anticipated. Because of international treaty constraints, the FCC decided to deny amateurs permission to turn the 2304-MHz transmitter on. As a result it was never tested. In 1979, at the World Administrative Radio Conference, the Amateur Satellite Service received several important new frequency allocations in the microwave portion of the spectrum. Although the events at the 1979 WARC and the legal constraints on the 2304-MHz beacon appear, at first glance, unrelated— are they? Might the new allocations have resulted, in part, from the responsible, restrained manner in which radio amateurs handled the sensitive 2304-MHz beacon issue?

The transponder frequencies chosen for the AMSAT-OSCAR 7 Mode B transponder made it theoretically possible for two ground stations to communicate by transmitting to AO-7 on 432 MHz, having the signals relayed directly to AO-6 on 146 MHz, and then relayed back down to the ground on 29 MHz. Many such contacts were made when AO-7 was in Mode B and the two satellites were physically close.[21] Never before, in any radio service, had two terrestrial stations been linked by a direct satellite-to-satellite relay.

Launched in late 1974, AO-7 operated until mid 1981, a period covering more than six and a half years. The cessation of operation coincided with the beginning of a three-week eclipse period in which the satellite entered the earth's shadow for up to 20 minutes on each orbit. When the eclipse period began, the average spacecraft temperature dropped. It is believed that thermal stress caused a battery cell that had previously failed in the open mode to short out, placing a very large load across the solar panel output.

Flight hardware for the AMSAT-OSCAR 7 satellite was contributed by groups in Australia, Canada, the US and West Germany.

AMSAT-OSCAR 8

Each time a new amateur satellite is placed in orbit, launch-day radio networks provide information on countdown, liftoff, rocket staging and the first user reception. AMSAT-OSCAR 8 (AO-8) was orbited successfully at 1754 UTC on March 5, 1978. Barely minutes later, those monitoring the nets heard G2BVN report reception of the 435-MHz beacon. But the suspense didn't end. AO-8 carried a novel 10-meter antenna which had to be deployed

by ground command before the Mode A transponder could be turned on. There was a lot of pressure on the project managers to extend the antenna as soon as possible, since it was believed that the reliability of the mechanical deployment mechanism would rapidly decrease in the cold vacuum of space. However, trying to extend the antenna before the satellite spin decreased to an acceptable level could be catastrophic.

Pre-launch speculation was that it might take a week for the spacecraft to slow down sufficiently. Roughly seven hours after launch, however, as the satellite was passing over the East Coast of the US, the spin rate looked good and the decision was made to send the antenna deployment command. Hundreds of stations listening to the 80-meter AMSAT net breathed a simultaneous sigh of relief as the antenna extended and the Mode A transponder responded to the "on" command.

Previous launch descriptions would have ended here as the joyous bedlam of early two-way communication began. But this time amateurs were being asked to refrain from transmitting to the spacecraft until it was fully tested in orbit. The reason was twofold: (1) An empty transponder would permit orbit-determination measurements and engineering evaluation of the satellite to proceed as quickly and efficiently as possible, and (2) plans for the critical transfer-orbit stage of future missions depended on users exhibiting such self-control; AMSAT had to know if these plans were realistic. While the satellite was over North America, cooperation proved excellent. The waiting period was easier on the users than expected, because monitoring the engineering tests proved intrinsically interesting. During sensitivity tests, for example, a transmitting station would announce the power levels it was using: "...10 watts...1 watt...one-tenth watt..." and the hundreds of silent, monitoring stations would witness the results first-hand in real time. The unloaded sensitivity

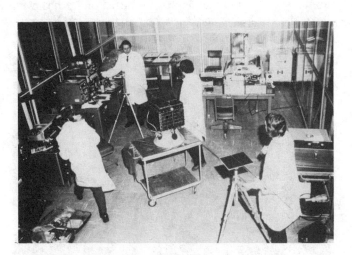

AMSAT engineers very carefully test the deployment of OSCAR 8's 10-meter dipole antenna in the lab at Vandenburg AFB in the days before launch. Each antenna element, on earth command, was deployed in flight by driving pre-cut rolls of thin copper-beryllium foil through a circular opening that formed it into a long tube of precisely the proper length.

of the transponder was remarkable. Two weeks after launch, AO-8 was officially opened for general operation, with all systems in excellent shape.[22]

Let's backtrack a bit to look at some of the events leading to the launch of AO-8. After AO-7 was placed in orbit late in 1974, the AMSAT design team focused on the next major step in the radio amateur space program—building a high-altitude, long-life (Phase 3) spacecraft. Early in 1977, however, when it became clear that AO-6 was nearing the end of its lifespan, the Phase 3 effort was interrupted. With the fear that AO-7 might not last until the first Phase 3 satellite was launched, a commitment was made to provide continuity of service to the thousands of amateurs and educators who had built Mode A ground stations and had financially supported the AMSAT satellite program. AMSAT had a serious problem; the resources (financial and volunteer) for building both spacecraft just weren't there.

To resolve this dilemma, the American Radio Relay League offered to donate $50,000 to AMSAT so that an interim Phase 2 satellite could be built, an offer AMSAT accepted. The initial plans for AO-8 called only for a Mode A transponder and a minimal telemetry and command system. When JAMSAT (the Japanese affiliate of AMSAT) learned of plans for AO-8, they offered to develop a second transponder for the mission. The JAMSAT transponder (Mode J) would use an uplink at 146 MHz and a downlink at 435 MHz. AMSAT agreed to provide antennas and interface circuitry so the Mode J transponder could be included if JAMSAT could deliver the transponder in time for launch integration. The time schedule for preparing the transponder and the spacecraft was extremely tight, but both groups met their deadlines and the satellite was launched with both Mode A and Mode J transponders. An interesting feature of the AO-8 spacecraft was that the transponders could be operated simultaneously, as long as the batteries maintained a sufficient charge. Since a single uplink signal could then be retransmitted on both downlinks, the two modes could easily be compared. AO-8 operated flawlessly in orbit from March 1978 through mid 1983.

Dick Daniels, W4PUJ, applies a cleaning swab to the OSCAR 8 spacecraft on the pool table in the family room of his Arlington, Virginia home. The solar panels have not yet been mounted.

Flight hardware for AO-8 was provided by AMSAT, JAMSAT and Project OSCAR. By prior agreement, the responsibility for operating AO-8 resided with the ARRL.

The '70s Draw to a Close

While most amateurs were using OSCAR 6 and 7 to communicate, a few individuals were engaged in more serious pursuits. One set of experiments focused on the feasibility of transmitting medical telemetry from isolated locations via linear transponders on low-altitude satellites.[23,24] Another set of experiments involved using OSCAR 6 and 7 to validate the concept and viability of the Search and Rescue Satellite system known as COSPAS/SARSAT.[25] Satellite-aided search and rescue works as follows. A low-power transmitter is placed aboard every civilian aircraft. In case of sudden impact (that is, crash), the transmitter is automatically activated. When the signal is relayed to a central processing center via a low-altitude spacecraft, the Doppler shift signature can be used to determine the location of the downed plane. The COSPAS/SARSAT project is a cooperative effort involving the US, the Soviet Union, Canada and France.[26] According to James Bailey, SARSAT program manager at the National Oceanic and Atmospheric Administration, more than 1150 lives had been saved by the system by mid 1988.

A new major contributor to the amateur satellite scene emerged in the late 1970s. The first two Soviet amateur satellites, RS-1 and RS-2, were successfully launched in October 1978. In the next chapter we'll look at these two spacecraft and the extensive program that was to follow. Meanwhile, AMSAT rushed to complete work on Phase 3-A, hoping to see it in space before WARC-79 took place. A series of delays at the European Space Agency, however, pushed the launch back into the '80s.

As the '70s ended, the Amateur Radio scorecard read: 10 launches, by the US and USSR, over 19 years. Things were to become a bit more hectic in the '80s. As of July 1990 the total has reached 33 satellite launches, an active ham-in-space program has begun, and several new series of spacecraft are either in space or on the construction table. The '70s closed with another WARC.

WARC-79. Most WARCs are specialized, taking into account one part of the radio spectrum or a limited class of users. However, every so often there's a grand general WARC where everything is laid on the table. WARC-79, which opened September 24, 1979, was the first general WARC since 1959.[27] IARU preparations for this conference were extensive and had taken place over many years. The underlying strategy was simple to express—get the IARU member societies to agree on a set of common objectives, and then work to generate maximum support for these objectives by the voting members of the ITU—but it required a tremendous effort to achieve. Four objectives were agreed upon in a series of regional IARU meetings in 1975 and 1976. One of these was to obtain authorization for amateur satellite radio links in sections of each of the amateur bands between 1 and 10 GHz. To an Amateur Radio operator, the social and economic benefits that a country gains from a healthy Amateur Radio service seem obvious. However, the value of having a large body of trained, proficient communications specialists wasn't necessarily known to many of the voting WARC delegates who came from countries where Amateur Radio hardly existed. And, as many as possible of these delegates had to be reached and personally briefed in the years preceding the WARC.

Most will agree that, on the whole, efforts put into preparing for WARC-79 paid off. The amateur satellite program obtained access to all bands between 1 and 10 GHz (inclusive). However, there were some restrictions on the links and we didn't always receive the specific allocations requested. Still, the new allocations will enable us to look past the Phase 3 series of spacecraft. In very simple terms, Phase 3 is designed to marginally improve upon, and provide an option to, currently existing long-distance communications mediums (that is, the HF bands). The new allocations will one day permit activities that would otherwise be impossible—fast-scan TV across continents and to amateurs aboard space stations, a worldwide digital data system capable of handling very high data rates in near-real-time, and so on. In the next chapter, we'll look at Phase 3 and follow-on programs in detail.

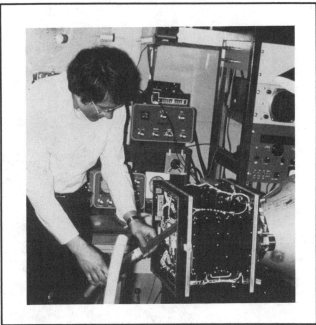

OSCAR 8 gets a thorough cleaning from W3GEY. At every step of the way, though the tools may not always be of the sophisticated laboratory type, AMSAT engineers pay meticulous attention to cleanliness. In the extreme temperatures and near-vacuum of space, minute debris from sloppy work habits could contaminate other satellites aboard the launch vehicle and even jeopardize the primary mission. AMSAT's "compulsive" care and attention have paid off well, as the record shows.

Notes

[1]D. T. Bellair and S.E. Howard, "Australis-Oscar," *QST*, Jul 1969, pp 58-61.

[2]D. T. Bellair and S. E. Howard, "Obtaining Data from Australis-Oscar 5," *QST*, Aug 1969, pp 70, 72, 82.

[3]J. A. King, "Proposed Experiments with Australis-Oscar 5," *QST*, Dec 1969, pp 54-55.

[4]R. Soifer, "Australis-Oscar 5 Ionospheric Propagation Results," *QST*, Oct 1970, pp 54-57.

[5]J. A. King, "Australis-Oscar 5 Spacecraft Performance," *QST*, Dec 1970, pp 64-69.

[6]P. M. Wilson and R. T. Knadle, "Houston, This is Apollo . . ." *QST*, Jun 1972, pp 60-65.

[7]K. Nose, "Using the ATS-1 Weather Satellite for Communications," *QST*, Dec 1971, pp 48-51.

[8]*Spacewarn Bulletin*, May 25, 1990. Prepared by World Data Center A for Rockets and Satellites, Goddard Space Flight Center.

[9]O. G. Villard, Jr and R. S. Rich, "Operation Smoke-Puff," *QST*, May 1957, pp 11-15.

[10]V. R. Frank, R. B. Fenwick and O. G. Villard, Jr, "Communicating at VHF via Artificial Radio Aurora," *QST*, Nov 1974, pp 27-31, 34.

[11]V. R. Frank, "Scattering Characteristics of Artificial Radio Aurora," *Ham Radio*, Nov 1974, pp 18-24. (Contains an extensive bibliography.)

[12]L. Van Prooyen, "Stimulating the Ionosphere in Alaska," *QST*, Jul 1989, pp 22-26.

[13]M. W. Browne, "June Space Test (Firewheel) to Outdo Moon in Brief Display," *New York Times*, Vol CXXIX, no. 44547, Tues 8 April 1980, pp C1, C2.

[14]J. Huntoon, "The 1971 Space Conference," *QST*, Sep 1971, pp 78-81.

[15]J. A. King, "The Sixth Amateur Satellite," Part 1, *QST*, Jul 1973, pp 66-71, 101; Part II, *QST*, Aug 1973, pp 69-74, 106. This article is highly recommended for anyone interested in satellite design.

[16]P. I. Klein, J. Goode, P. Hammer and D. Bellair, "Spacecraft Telemetry Systems for the Developing Nations," *1971 IEEE National Telemetering Conference Record*, Apr 1971, pp 118-129.

[17]J. C. Fox and R. R. Dunbar, "Preliminary Report on Inverted Doppler Anomaly," *ARRL Technical Symposium on Space Communications*, Reston, VA, Sep 1973, pp 1-30.

[18]L. Kayser, "SMART—System Multiplexing Amateur Radio Telecommands," *ARRL Technical Symposium on Space Communications*, Reston, VA, Sep 1973, pp 31-45.

[19]P. I. Klein and J. A. King, "Results of the AMSAT-OSCAR 6 Communications Satellite Experiment," *IEEE National Convention Record*, NYC, Mar 1974.

[20]J. Kasser and J. A. King, "OSCAR 7 and Its Capabilities," *QST*, Feb 1974, pp 56-60.

[21]P. Klein and R. Soifer, "Intersatellite Communication Using the AMSAT-OSCAR 6 and AMSAT-OSCAR 7 Radio Amateur Satellites," *Proceedings of the IEEE Letters*, Oct 1975, pp 1526-1527. M. Davidoff, "Predicting Close Encounters: OSCAR 7 and OSCAR 8," *Ham Radio*, Vol 12, no. 7, Jul 1979, pp 62-67.

[22]P. Klein and J. Kasser, "The AMSAT-OSCAR D [8] Spacecraft," *AMSAT Newsletter*, Vol IX, no. 4, Dec 1977, pp 4-10.

[23]J. P. Kleinman, "OSCAR Medical Data," *QST*, Oct 1976, pp 42-43.

[24]D. Nelson, "Medical Relay by Satellite," *Ham Radio*, Apr 1977, pp 67-73.

[25]D. L. Brandel, P. E. Schmidt and B. J. Trudell, "Improvements in Search and Rescue Distress Alerting and Locating Using Satellites," *IEEE WESCON*, Sep 1976.

[26]"Four Nations Sign 15-Year Sarsat Pact," *Aviation Week & Space Technology*, Oct 24, 1988, p 41.

[27]R. Baldwin and D. Sumner, "The Geneva Story," *QST*, Feb 1980, pp 52-61.

Chapter 4

1980 into the Future

The amateur satellite program grew rapidly in the 1980s. Growth areas included numbers of: radio amateurs using satellites, satellites being launched, distinct spacecraft series, countries providing launches, and core groups around the world working on construction of spacecraft and support facilities. And, an entirely new activity was opened to the amateur space program—hams in space. To describe the '80s we switch from chronological coverage of events to a focus on specific satellite series, programs and activities.

PHASES 1, 2, 3 AND 4

The whole concept of Phases in the amateur satellite program grew out of the need of those working on the Phase 3A spacecraft to describe its radically different nature. For discussion purposes it was (and remains) convenient to divide amateur satellites—past, existing and future—into groups. The criteria for including a spacecraft in a particular Phase were never formally specified. However, certain constellations of satellite traits relating to design goals, orbital characteristics, primary mission subsystem, launch date, and so on, can be identified. Table 4-1 lists several key classification criteria and catalogs most amateur spacecraft. Though some of the spacecraft assignments in the Table are not clear-cut, this doesn't cause any practical problems.

The short-lived Phase 1 satellites, designed to gather information on basic satellite system performance, appealed mainly to a relatively small number of hard-core radio amateur experimenters—perhaps several thousand. The communications capabilities of the long-lifetime Phase 2 satellites attracted a significantly larger, new group of operators to the space program, amateurs who shared a vision of the future communications capabilities that satellites would provide. They wanted to get started on the learning curve early and assist in working towards its success. Estimates are that by 1983, between 10,000 and 20,000 amateurs had communicated through a Phase 2 satellite. With Phase 3 spacecraft in operation since 1983, amateurs have gained access to reliable, predictable long-distance communication capabilities of a type never before available. As a result, the number of radio amateurs engaged in space communications continues to increase.

PHASE 3 PROGRAM

Phase 3 satellites spend most of their time at high altitudes, where they are in sight of more than 40 percent of the earth. For example, during the period 1989 to 1995 AMSAT-OSCAR 13 will be in view of Northern hemisphere

A visitor's eye-view of the Phase 3A structure. The room in which Phase 3A was built was affectionately known as the "Fishbowl" by the AMSAT personnel who assembled the satellite in full view of the public at the Goddard Space Flight Center Visitor's Center, in Greenbelt, Maryland. What appears to be the ghost of Dr Robert Goddard watching over the satellite is the reflection of his statue: a memorial to the late rocketry pioneer for whom the Space Flight Center was named. Other rocketry and astronautics displays surround the area. *(W4PUJ photo)*

ground stations, on average, more than 10 hours each day. By comparison, the Phase 2 satellite AMSAT-OSCAR 8 was only in view of about 9 percent of the earth at any given time, and accessible to most ground stations for less than two hours each day.

Modern Phase 3 satellites are designed to provide modestly equipped ground stations with reliable, predictable, long-distance communications capabilities of a quality unmatched by any other amateur mode. With the large satellite-earth distances involved, attaining this objective places stringent requirements on Phase 3 spacecraft: high-power transmitters, large solar power systems, high-gain directional antennas, attitude sensing/adjusting systems, sophisticated computer control, rocket motors, and so on.

One of the best ways of describing the communications performance of a Phase 3 satellite is to compare it to an HF band that almost all radio amateurs are familiar with—14 MHz. A modestly equipped 14-MHz station can communicate with any place on the earth by exploiting favorable conditions. But if the station is interested in scheduled communications, either over specific point-to-point paths or involving a multi-point network, reliability is not very high. The unpredictable nature of the 14-MHz band isn't necessarily a negative attribute. In fact, in many situations it's a feature that makes the band interesting and exciting.

In contrast, Phase 3 spacecraft provide highly reliable, predictable and consistent communications over long paths to stations modestly equipped for the VHF and UHF frequencies used by the satellites. Signals levels are far from overpowering, but they're sufficient to provide excellent readability. The predictability of Phase 3 spacecraft radio links makes them an invaluable asset when natural disasters occur and in other situations where getting a message through in a timely fashion is of paramount importance. Examples include general bulletins, code practice, phone patches, coordinating DXpeditions or arranging moon-bounce schedules.

Satellite links don't produce any "skip zone." As a result, nets run much more smoothly and efficiently, and accidental QRM is easily avoided since anyone listening to the downlink frequency will know if it is in use. Although the skip zone can cause problems to 14-MHz users, it can have positive effects. For example, it often allows a single frequency to accommodate several QSOs, with each group unaware of the presence of the others.

This brief comparison of the 14-MHz band and Phase 3 satellite communication links illustrates their complementary nature. Whatever your primary interests are, don't get caught in the trap of viewing the situation as a competition where one tries to decide whether satellite communication

is "better" than HF communication. Phase 3 is designed to provide a new dimension to Amateur Radio, not to replace existing options.

We now look at four spacecraft in the Phase 3 series. Three (Phase 3A, OSCAR 10 and OSCAR 13) have been to the launch pad and the fourth, Phase 3D, is tentatively scheduled for 1995. Note that satellites are not assigned a number until after they've been successfully deposited in orbit. For example, the Phase 3B spacecraft became known as AMSAT-OSCAR 10 once the Ariane rocket dropped it off in space.

AMSAT PHASE 3A

The launch window opened at 1130 UTC on Friday, May 23, 1980, with AMSAT Phase 3A perched atop an

ESA technicians mount the AMSAT-OSCAR Phase 3A spacecraft to the CAT (Application Technology Capsule). Above Phase 3A is Firewheel, the primary payload. Its cylindrical canisters contained lithium, barium, explosives and other compounds that when exploded would have provided a visible, "glowing," steam-like cloud enabling scientists to study the earth's magnetic-field patterning. Had the mission not crashed in the Atlantic a scant four minutes after launch (a launch vehicle problem, not an OSCAR malfunction), Phase 3A would have been separated from the CAT and been clear of the experiments long before the fireworks began. *(AMSAT-DL photo)*

Table 4-1

Classification Criteria: Amateur Satellite Phases

Stage	Characteristics	Satellites
Phase 1	experimental, short-lifetime, low-earth-orbit	OSCARs I, II and III, Australis-OSCAR 5, ISKRAs 2 and 3
Phase 2	developmental and operational, long-lifetime, low-earth-orbit	*General communications,* AMSAT-OSCARs 6, 7, 8 RS-1, 2, 3, 4, 5, 6, 7, 8, 10/11 Fuji-OSCARs 12, 20 *UoSAT series* UoSAT-OSCARs 9, 11, 14, 15 *Microsat series* OSCARs 16, 17, 18, 19
Phase 3	operational, long-lifetime, high altitude elliptical orbit	AMSAT-Phase 3-A, AMSAT-OSCAR 10, AMSAT-OSCAR 13, AMSAT-Phase 3-D*
Phase 4	operational, long-lifetime, geostationary or drifting geostationary orbit	AMSAT-Phase 4-A** AMSAT-OSCAR 4

Notes: Satellites have been classified in terms of mission objectives. Satellites marked * or ** have not yet been launched. * indicates launch commitment has been received.

Ariane rocket sitting on a launch pad in Kourou, French Guiana. Following nine years of planning, including four of intensive construction, AMSAT workers around the world could now only sit and listen as Dr Tom Clark (W3IWI), president of AMSAT, relayed the countdown as it occurred on the northern coast of South America. Would the tropical weather clear for a liftoff? Would the rocket systems remain "go"? We all listened to the continuous reports being broadcast on several HF bands as Phase 3A waited within the cowling of the newly developed European Space Agency (ESA) launch vehicle. The amateur spacecraft was awarded this prized position in a stiff international competition involving more than 80 applicants. Finally, at approximately 1430 UTC, the liftoff signal was given and the Ariane rose from its pad. For several minutes spirits soared.

Then disaster. A great many amateurs monitoring the net had their hopes dashed as they heard the words "non-nominal flying...problem in one engine...the rocket is going down...splashdown." The first stage of the Ariane rocket had failed and unceremoniously dumped Phase 3A in its final resting place several hundred feet under the Atlantic Ocean.

While many dejected AMSAT members wondered if they had just witnessed the end of the amateur space program, Clark drafted a statement objectively describing the situation. Later that evening it was read over AMSAT post-launch nets.

> What we lost on Black Friday was sheet metal, solar cells, batteries, transistors, a lot of sleep and a major portion of our lives for the last few years. What we gained over those same years was knowledge; knowledge that we could make a complicated spacecraft. Knowledge in areas of aerospace technology that none of us had before. Knowledge that we could work as a team, despite national boundaries, differences in our cultures, lifestyles and personalities. Knowledge that, from within the ranks of Amateur Radio, we could draw upon enough resources to attempt a project with a complexity rivaling commercial satellite endeavors funded at levels of tens of millions of dollars. The knowledge is still intact. We even had the forethought to purchase a duplicate set of sheet metal that constitutes the spaceframe. We have a second set of solar panels, batteries and sensors. We have on hand the documentation and art work necessary to replicate all the printed-circuit boards. We have in place and ready to go a network of ground telecommand stations.[1]

The situation was bleak but not completely hopeless.

At this point the AMSAT Board of Directors looked to the membership for guidance. Did members have the heart and confidence to continue? Being realists, the Board couldn't commit to a follow-on spacecraft without reasonable assurance of financial and moral support. Over the next several weeks the support was overwhelming. One by one, key volunteers, convinced that Phase 3B was possible, recommitted themselves to the project. Like the

pieces of a giant jigsaw puzzle, the elements fell into place with a picture of Phase 3B emerging. The amateur space program would continue.

We now take a brief look at some of the hardware that was lost. Phase 3A was the first amateur satellite to carry its own propulsion system, a rocket (kick motor) accounting for roughly half the launch weight of the spacecraft. The kick motor was required for this mission because the initial (transfer) orbit targeted by ESA had a low point (perigee) of only 125 miles. Left here, the orbit would rapidly decay and the spacecraft would re-enter the atmosphere and burn up within the year. The kick motor would have enabled AMSAT to (1) prolong the spacecraft's lifetime by increasing the perigee and (2) enhance the spacecraft's utility to Northern hemisphere ground stations by raising the inclination of the orbit.

It's fair to say that the Phase 3A project was more complex, required a larger financial investment and reflected a greater total effort than all previous OSCARs combined.[2,3] The spacecraft contained a 50-W Mode B transponder and an energy supply system capable of supporting the transponder; a computer for flexible control of command, telemetry, Codestore and housekeeping functions; a sophisticated attitude sensing and control system permitting the use of high-gain antennas; and two beacons sandwiching the 180-kHz-wide 146-MHz downlink. Building and testing a spacecraft was only part of the Phase 3A project. AMSAT also coordinated the development, construction and deployment of a series of ground telecommand stations and provided a crew to oversee the connection of the satellite to the launch rocket in Kourou. The telecommand stations, located around the world, were capable of loading the spacecraft computer, providing real-time orbit determination and attitude control data, and reducing the relatively sophisticated telemetry to meaningful values.

Flight hardware for the project was produced in Canada, Hungary, Japan, the US and West Germany. Primary responsibility for spacecraft and ground support systems resided with the US and West Germany. The project co-directors were Jan King (W3GEY) and Dr Karl Meinzer (DJ4ZC). Jan, acting as the senior spacecraft engineer, oversaw the entire construction project beginning with highly speculative feasibility studies in 1971. Karl was responsible for the overall technical design of the spacecraft and for the design and construction of many of the unique high-technology subsystems developed for it. In recognition of the ESA-sponsored launch and the major contributions to the spacecraft by AMSAT-Deutschland (AMSAT-DL) members, the satellite was licensed as DLØOS.

AMSAT-OSCAR 10

On 16 June 1983 AMSAT Phase 3B was successfully launched by a European Space Agency Ariane rocket along with the European Communications Satellite ECS-1. Upon separation from the launcher, Phase 3B became known as AMSAT-OSCAR 10. The usual worldwide launch net enabled a great many amateurs to share the anxiety of another liftoff. By now, even the launch nets were becoming a mammoth activity. Information was flowing into the net control from amateur stations at the launch site (Kourou, French Guiana), the AMSAT-DL technical crew in Marburg, Federal Rep of Germany and commercial communications circuits of NASA and ESA. Outgoing information flowed from several stations on a great many HF and VHF bands to listeners on six continents. At the announcement of each milestone—attaining orbit, separation of various payloads—you could almost feel a collective sigh of relief. The OSCAR 10 beacon was scheduled to spring to life about two and a half hours after liftoff. The delay was necessary to allow the satellite to outgas and stabilize in temperature. Until the beacon began operating, we wouldn't know whether OSCAR 10 was alive or simply a misplaced boat anchor. Needless to say, there was a feeling of great elation when the first report of telemetry reception, from New Zealand station ZL1AOX (Ian Ashley), arrived.

But the feeling of relief was short lived as analyses of the telemetry showed that the spacecraft was behaving oddly. The orientation and spin were far from nominal and, as a result, the spacecraft was capturing very little solar power and running dangerously cold.

To gauge the real competence of a technical staff you have to observe how they handle the unexpected. The AMSAT crew performed flawlessly under pressure. Problems were analyzed, corrective measures taken, and plans for orbit transfer revised before most radio amateurs realized the serious nature of the situation. Weeks later, AMSAT learned the cause of the problems. ESA was testing a new procedure for separating the AMSAT satellite

from the Ariane rocket's third stage and an unforeseen set of circumstances had caused the third stage to bump into OSCAR 10 twice shortly after separation.[4]

The collision affected the mission in several ways. At least one of the elements of the 2-meter antenna on OSCAR 10 appears to have been damaged. This caused a slight reduction in gain, on the order of 2 dB, and a small increase in spin modulation on the 2-meter downlink due to the resulting antenna asymmetry. The impact also affected the spacecraft's initial orientation and spin. As a result, OSCAR 10 was subjected to temperatures considerably below the range it was designed and tested for. This is believed to have caused a problem in the plumbing associated with the spacecraft's liquid fuel rocket. The rocket ignited for the first burn so that the perigee was raised. But a leak in a high pressure helium tank prevented operation of the valves needed to re-ignite the rocket. So the second burn, designed to adjust the perigee height and orbit inclination, never took place. OSCAR 10 ended up residing in a final orbit having an inclination of 26 degrees (the target was 57 degrees) and a perigee near 4000 km (the target was 1500 km). Superficially, the final orbit doesn't appear to differ too much from the target, but a careful analysis shows several major effects, including the following: Ground control stations had to adjust the satellite's attitude very often to ensure sufficient solar illumination and had to accept orientations that were far from optimal with respect to antenna pointing. It's believed that increased cumulative radiation exposure due to the resulting orbit reduced the operating life of the spacecraft's control computer. However, the picture wasn't too bleak—we didn't have to worry about early re-entry. OSCAR 10 would remain in orbit for thousands of years.

Despite the depressing aspects of the picture just painted, AMSAT-OSCAR 10 was a great success. The Mode B transponder performed superbly, providing amateurs with their first exposure to the communications capabilities of a high-altitude Phase 3 satellite. OSCAR 10 was available a great many hours each day, it was where you expected it to be, it was turned on when the schedule said it would be on, and with a little experience you could easily predict link performance before you even turned on your rig.

Technically, AMSAT-OSCAR 10 was closely patterned after Phase 3A, but there were two significant design changes involving the rocket motor and transponder complement. Phase 3A's solid-fuel kick motor was replaced with a liquid-fuel rocket. The new rocket was more powerful and its design permitted multiple re-ignition (when no breakdowns occurred). A solid rocket can only be ignited once. The new rocket enabled AMSAT to target a more desirable final orbit and plan a series of burns to get there, eliminating the dangers associated with certain single-burn orbit changes (see Chapter 15). The liquid fuel motor, valued at two million dollars, was donated by the manufacturer, Messerschmitt-Bolkow-Blohm (MBB). It had been a backup for the European Symphonie communications satellite.

AMSAT-OSCAR 10 rests atop its attach fitting on a laboratory bench in preparation for packaging and shipping to the launch-vehicle integration site. Antennas occupy the top face, solar panels the side faces, sensors and antenna reflectors protrude off the ends of the arms. At the proper time in the launch-separation sequence, the satellite is literally "sprung" free of the launcher, leaving the conical attach fitting behind. (The kick motor is hidden from view within the cone.)

The added payload capabilities of the rocket also permitted AMSAT to add a second transponder and additional radiation shielding to the main computer. The new transponder, an 800-kHz wide unit using a 1269-MHz uplink and a 435-MHz downlink, was called Mode L. The primary transponder remained Mode B. Plans were to schedule the Mode B transponder most of the time during the early part of the satellite's life and then, as the number of amateurs with the ability to access Mode L increased and as Mode B got crowded, to raise the percentage of the time allocated to the high-capacity (wide-bandwidth) Mode L unit. These plans had to be curtailed for several reasons. A biasing regulator in the Mode L transponder failed. As a result, users required approximately 10-15 dB more uplink power than anticipated. Also, the highly directional Mode L antenna on the spacecraft often had to be off pointed in order to adjust sun angles so that the satellite would have adequate power and maintain the proper temperature. Because of these factors, Mode L remained essentially experimental and the amount of operating time allocated to it was limited. However, when the spacecraft antennas were properly lined up toward earth and the Mode L transponder was turned on, the relatively small number of amateurs set up to access it found that it performed phenomenally. The total lack of spin modulation, the excellent linearity, and an exceptionally good signal-to-noise ratio produced a link that sounded better than most telephone connections.

OSCAR 10 had an Achilles heel. It was known before launch that the satellite control computer was susceptible to cumulative radiation damage and its projected lifetime, in the aimed-for orbit, was at most five to seven years. Everything possible had been done to maximize this figure using the best technology available to AMSAT in late 1982 when the spacecraft was buttoned up. If all other subsystems performed up to expectations this is where catastrophic failure would occur. In May 1986, after approximately three years of excellent service, serious radiation damage began to be observed, and by the end of the year the computer on OSCAR 10 was declared lost.

However, the damage to OSCAR 10 turned out not to be fatal. The satellite's attitude couldn't be controlled, the telemetry system no longer provided information, and command stations couldn't provide a reliable schedule. But OSCAR 10 refused to die. When the spacecraft orientation caused the solar cells to turn away from the sun, the battery power dropped nearly to zero and subsystem temperatures plummeted far out of the safe range. OSCAR 10 ceased operating and we thought it might never be heard from again. However, the condition turned out not to be permanent. When conditions improved, OSCAR 10 returned to life and continues to operate. Command stations turn on the Mode B transponder and the low-gain omni antennas regularly. To prolong the operational life of OSCAR 10, AMSAT requests that amateurs only use the transponder during certain parts of the orbit and at certain times of the year when the sun angles are providing adequate power. For updated information on when operation is permitted, check local AMSAT nets and packet bulletin boards. As this is being written, more than four years after OSCAR 10 was declared brain dead, the spacecraft continues to provide excellent communications links to the flexible experimenter who is willing to put up with somewhat erratic scheduling. Its longevity is an excellent testimony to the robustness of the basic design. In effect, OSCAR 10 now provides the same element of surprise as the 15-meter HF band.[5]

AMSAT-OSCAR 13

The AMSAT-OSCAR 13 launch on June 15, 1988 and the orbital transfer operations that followed went flawlessly. Two burns placed the spacecraft precisely in the desired orbit. (See Appendix B: Spacecraft Profiles.) For the first time, AMSAT had demonstrated its ability to take a spacecraft through the series of reorientations, spin-ups, spin-downs and rocket firings needed for a complex orbit change. At each stage the spacecraft was deposited in the exact location and orientation targeted. All of these maneuvers were accomplished using Amateur Radio communications equipment and personal computers for ranging, telemetry capture, spacecraft control and orbit and attitude determination. Of course, in those cases where commercial/government data sources were available they were used for confirmation. However, it's important to note that these sources were only used as a check and backup—amateurs have total command of the technology, equipment, techniques and expertise needed to control all aspects of orbital maneuvers.

OSCAR 13 was modeled on OSCAR 10, with small, evolutionary changes in many subsystems plus some major changes and additions. One major change was to the spacecraft computer. Thanks to a donation from Harris Semiconductor of special radiation-hardened ICs, radiation damage to key computer circuitry is not likely to affect

AMSAT-DL team members work on AMSAT-OSCAR 13 during integration in Golden, Colorado. (Left to right) Hanspeter Kuhlen, DK1YQ; Karl Meinzer, DJ4ZC. (AMSAT-DL photo)

satellite life. OSCAR 13 has no known Achilles heel. New systems on the spacecraft include an experimental Mode S transponder (435-MHz uplink, 2.4-GHz downlink—future Mode S uplinks will be 1.2 GHz) and a special digital transponder, known as RUDAK, which uses the Mode B set of frequencies. Several changes were made to the Mode L transponder that traded off some bandwidth for versatility and robustness. A narrow (50-kHz wide) 2-meter secondary input channel was added with the output superimposed on the normal Mode L downlink (250 kHz wide). When the secondary input channel is operating the mode is designated JL. To provide additional protection against surprise hazards like those that occurred with OSCAR 10, several key systems were designed and tested over an extended temperature range.

As this is written, OSCAR 13 has been in orbit for over two years. The Mode B transponder, which is activated about 60-75% of each orbit, is performing excellently. It is extremely difficult to make quantitative Mode B comparisons between OSCAR 13 and OSCAR 10, but subjective reports seem to agree that the new downlink is slightly stronger (the change is less than 3 dB) and that spin modulation is less obtrusive. The spin modulation performance is of special interest; since Mode L is used when spacecraft antenna orientation is optimal, Mode B is generally operative when the spacecraft is somewhat off-pointed.

The Mode L transponder is turned on during approximately 20% of each orbit, usually near apogee when the high-gain spacecraft 1.2-GHz antenna is pointed almost directly at the earth. With respect to audio quality this link is clearly superior. The downlink is providing excellent performance. A reasonably equipped ground station can hear the transponder noise floor. Unfortunately, the receiver aboard the spacecraft appears to be generating AGC signals in response to some spurious signal and this is limiting its sensitivity. While it's much easier to access Mode L aboard OSCAR 13 than it was with OSCAR 10, the improvement is still about 6 dB short of the objective. An uplink with 50 W to a 19-dBd antenna will produce a good SSB downlink under most conditions. (Details on operating Mode L will be found in the next chapter and in Appendix B.) A report by Bill McCaa (KØRZ) listed more than 200 stations known to have been active on Mode L during its first three months of operation. On GMT weekends the secondary 2-meter uplink is usually activated during the time scheduled for Mode L (now Mode JL). The 2-meter uplink is very sensitive, but please note that it was specifically added to the spacecraft to provide access to Mode L for amateurs in countries where operation on the 1.2-GHz uplink is not permitted or equipment is not available. Please do not use the 2-meter Mode JL uplink for general operation if you don't fall in one of these categories. Cooperation on this point can have an important impact on the position of the amateur-satellite service at future WARCs.

The Mode S transponder exhibited some initial problems related to sensitivity of the uplink receiver. But long-distance troubleshooting and software repairs have fixed the glitch and the transponder is now performing well. As an experimental system, it is only operated a small percentage of the time. Transponders are turned off about 5-20% of each orbit to allow for attitude adjustments via magnetorquing.

There is a serious problem with the RUDAK transponder which has so far prevented its general use. Some rather ingenious long-distance troubleshooting has enabled AMSAT-DL to identify the problem. Various approaches for reactivating or bypassing the malfunctioning circuitry are being tested, but the situation is not promising.

AMSAT PHASE 3D

AMSAT-DL has received a commitment for a launch of the Phase 3D spacecraft in approximately 1995 on the first test flight of the Ariane 5 rocket. As has been true for other members of the Phase 3 series, we're likely to see a spacecraft comprised of the best of the past plus some new experimental systems. Phase 3D will be considerably larger and more powerful than OSCAR 13 and three-axis stabilization is being considered. Since 2-meter spectrum space for satellite downlinks is already extremely crowded, the Mode B transponder on Phase 3D will be regarded as a backup for OSCAR 13 and the Mode L unit will be considered the primary payload. The aim is to improve downlink performance by 10 dB. Since a ground station using a small 6-element Yagi on a 2-foot boom now obtains marginal performance when receiving OSCAR 13, the same antenna should provide a 10-15 dB signal-to-noise ratio on Phase 3D. Ground stations may find that a simple ground-plane antenna will suffice for casual monitoring of the downlink. The uplink should also provide a 10-dB performance improvement over OSCAR 13. This will significantly simplify setting up transmitting equipment at a ground station. The Mode L transponder will implement the LEILA (*LEIstungsLimit Anzeige*) concept. LEILA consists of a spectrum analyzer which monitors the transponder passband. If a signal is observed using too much power, it is attenuated by a tunable notch, and a CW message indicating the problem is superimposed on the downlink. Several notches can be handled simultaneously.

Phase 3D design and construction is being handled by AMSAT-DL under the direction of Karl Meinzer, DJ4ZC. AMSAT-US is playing a support role while devoting attention to the Phase 4 and Microsat programs which we'll describe shortly. This is not a radical change, since AMSAT-DL was centrally involved, perhaps even the major contributor to, the Phase 3-A, -B and -C spacecraft.[6]

UoSAT PROGRAM

UoSAT spacecraft are built at the University of Surrey (England) by a group led by Dr Martin Sweeting, G3YJO. Each satellite carries a collection of experimental payloads of interest to radio amateurs, educators and scientists. Several payloads have been related to studies of radio propagation. These include beacons from 7 MHz to 10 GHz, magnetic field measuring instruments, charged particle detectors, and so on.[7] Others were designed to provide amateurs with information regarding the reliability of electronic components used on amateur spacecraft, especially the digital components used in computer control systems. Another major area of interest has been techniques for digital store-and-forward message systems. A synthesized voice telemetry unit has proven extremely popular for educational demonstrations. This voice beacon provided position information to a team of Canadian-Soviet arctic explorers during an early 1988 expedition. Schoolchildren around the world were able to monitor the progress of the expedition by listening to the telemetry link each day.[8,9]

Amateurs are often disappointed when they learn that many UoSATs do not carry open-access transponders. There are important reasons for this. All amateur spacecraft to date have received free, or nearly free, launches in recognition of their potential value for scientific studies, educational applications and disaster communications. Naturally, accomplishments are more important than vague promises. When Dr Sweeting suggested that it might be possible to obtain government and commercial financial support for satellite construction if he focused on payloads that were of interest to both the amateur community and the scientific/educational community, he was encouraged to do so by various national AMSAT groups. The rationale is that these spacecraft provide a test platform for developing techniques and testing new technology of interest to amateurs and that this work is, to a large extent, being subsidized by contributions that would not otherwise be available to the amateur satellite community.

The UoSAT program has provided numerous indirect benefits to the amateur space program. Our access to UoSAT means that valuable resources aboard operational AMSAT communications satellites can be focused on the communications transponders. Supporting a crew to oversee activities at a launch site is a big expense. For the 1990 multiple Microsat launch from Kourou, a group from UoSAT took care of the four AMSAT spacecraft, saving the amateur community a great deal of money.

Since UoSAT spacecraft employ frequencies allocated to the amateur-satellite service, they only include payloads that are of interest to radio amateurs and consistent with the rules and regulations of the amateur-satellite service. The amateur community is provided with complete information on each subsystem and the related telemetry available. Although UoSAT spacecraft may only be of direct interest to a small subset of the amateur satellite user group, they have had a very positive impact on the quality of service provided to users of other amateur spacecraft.

Two satellites in this series, UoSAT-OSCAR 9 and UoSAT-OSCAR 11, were launched in the early 1980s. UoSAT-C had been targeted for late 1988, but when its launch was canceled and a new opportunity arose, the subsystems were resurrected as UoSAT-D and UoSAT-E and launched, along with four Microsats, by ESA in early 1990. A brief review of the highlights of the UoSAT program follows. See Appendix B (Spacecraft Profiles) for more detailed information on these spacecraft.

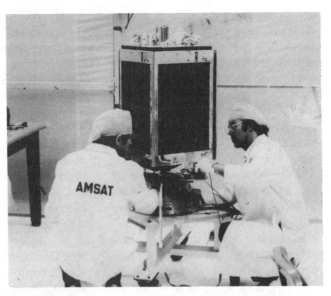

Jan King, W3GEY, and Martin Sweeting, G3YJO, apply the finishing touches to UoSAT-OSCAR 9. The spacecraft is roughly twice the size of AMSAT-OSCAR 8, having been built around the equivalent of two OSCAR 8 frames, one atop the other.

UoSAT-OSCAR 9

UoSAT-OSCAR 9 was launched into a low earth orbit (LEO) on October 6, 1981 with the Solar Mesosphere Explorer (SME) spacecraft. (UoSAT-OSCAR 9 is sometimes referred to as UoSAT-1 since it's the first spacecraft in the UoSAT series.) The payload included a General Data Beacon at 145.825 MHz compatible with standard amateur NBFM receivers, an Engineering Data Beacon at 435.025 MHz, phase-locked HF beacons at 7, 14, 21 and 28 MHz for propagation studies, and microwave beacons at 2.4 and 10.47 GHz, also for propagation observations. In addition, the spacecraft carried a CCD camera to send back pictures of the earth formatted to be viewed on a regular TV after minimal processing; a three-axis, wide-range, flux-gate magnetometer for measurement of the earth's magnetic field; and high-energy particle and radiation detectors. The 146-MHz telemetry beacon could be switched between digital data and a voice synthesizer. Voice-synthesized telemetry has a very low data rate, but it's excellent for educational demonstrations.

Commanding this complex spacecraft was a real challenge. For the first few months that UoSAT-OSCAR 9 was in orbit, a typical day at Surrey would go as follows. Last orbit of set: Collect telemetry. Between sets: Study telemetry, plan actions, write and load computer software, test software on spacecraft simulator. Next orbit (about 10 hours between sets): Collect telemetry and check status, transmit commands, verify correct receipt, instruct spacecraft to act on commands, collect telemetry. Between orbits: Study telemetry, plan actions and so on.

While the University of Surrey team perfected the commanding operation (techniques, hardware, software), checked out the spacecraft systems, and oriented the spacecraft using the new dynamic magnetorquing system to prepare for extending a 50-foot gravity gradient boom that would provide a measure of passive attitude control, radio amateurs around the world grew impatient at what appeared to be snail's pace progress. It was a frustrating situation for the Surrey crew, who were working till they dropped from exhaustion, and for the amateurs who wanted to put the experimental packages aboard UoSAT to practical use. Then disaster struck.

While uplinking commands on April 4, 1982, a software glitch caused both the 2-meter and 70-cm telemetry beacons to be turned on. As a result, satellite command receivers were desensed. When the uplink power at Surrey proved insufficient to overcome the desense, Dave Olean, K1WHS, offered the use of his powerful 2-meter EME station for the command link. The 0.5 megawatt EIRP available still proved insufficient, so a group of amateurs in northern California, under the leadership of Dr Robert Leonard, KD6DG, obtained permission to reactivate a 150-foot dish antenna owned by the Radio Physics Laboratory of Stanford Research Institute (SRI). This dish, when used in conjunction with an amateur 70-cm transmitter, could produce an EIRP of 15 megawatts. The beamwidth was only 0.6 degrees, so precise aiming was critical. This setup would either get through to the command receiver or fry the spacecraft. Returning the Stanford dish to service required several months of work

UoSAT-OSCAR 11, shown here undergoing final checkout, was the second amateur spacecraft designed and built at the University of Surrey, England. (photo courtesy University of Surrey)

resuscitating drive motors, hydraulic components and control computers. Then, on September 20, 1982, after operating out of control for nearly six months, UoSAT-OSCAR 9 was salvaged and found to be in perfect health. It went on to perform nearly flawlessly until late 1989, when it re-entered the earth's atmosphere and burned up. UoSAT-OSCAR 9 provided solid evidence that amateur projections of seven- to ten-year lifetimes for satellites currently being constructed were realistic.

UoSAT-OSCAR 11

In late 1983 the UoSAT team was unexpectedly offered another launch opportunity. But there was a catch attached—the date was six months away. Was it possible to design, construct and test a spacecraft in this limited amount of time? After evaluating the risk, Martin Sweeting committed the Surrey team to the project. UoSAT-B carried five experimental payloads and several new or modified spacecraft engineering systems. The experimental payloads focused on (1) particle-wave detection, (2) earth imaging, (3) synthesized speech, (4) packet communications and (5) space dust detection. The spacecraft systems experiments focused on (1) navigation, attitude control and stabilization, (2) computer hardware, software and memory technology and (3) communications systems.

The Surrey crew completed UoSAT-B on time and it was successfully launched on March 1, 1984. A serious problem quickly arose. On its third orbit of the earth, the spacecraft stopped accepting commands and went into hibernation. Nothing was heard on any of the radio links. The limited amount of data that had been received on the first three orbits did provide some clues. The prevailing theory was that lower than expected spacecraft temperatures following launch, combined with the way in which the spacecraft was tumbling, made it extremely difficult to transfer commands to the satellite. After attempting to access the spacecraft for several weeks without any success, the Surrey crew began to suspect that they might have misdiagnosed the problem and that the situation might be more severe than first thought, perhaps even fatal for OSCAR 11. They devised a plan which would enable them to determine whether the satellite was still alive and confirm the accuracy of their tracking data. The plan involved listening for the extremely weak signal generated by the 1.2-GHz local oscillator in the satellite's command receiver. In order to detect this feeble signal they needed access to a very big antenna, so they once again approached Dr Leonard, Director of the Radio Physics Laboratory. This time, Dr Leonard suggested using a 100-foot dish at Sondre Stromfjord in Greenland which already had a 1.2-GHz receiver in place. The station chief in Greenland, Finn Steenstrup, OX3FS, wasted no time. As soon as the proper equipment could be assembled, signals were heard from OSCAR 11's local oscillator and the information was relayed back to Surrey. The Surrey crew renewed their control efforts using the updated orbital information and a few days later, on 14 May 1984, they regained control of the bird. Barely six months later, Steenstrup died in an accident while working on the Greenland antenna. Thanks to his efforts, UoSAT-OSCAR 11 was able to provide many years of service to the amateur community.

Reviewing articles that appeared in various periodicals between 1981 and 1984, one can't help but notice how the UoSAT saga resembles a script for a serial adventure story—hopeless predicaments followed by last-minute rescues. But the radio amateur-satellite service is not immune to the laws of probability—Phase 3A met a wet early end, there was a serious launch vehicle failure with OSCAR 4, and so on. If there's a message here, it's that taking part in the amateur satellite program is not for the faint-hearted. Setbacks and barriers will always be part of the picture. And, the most rewarding successes will probably come from employing ingenuity and tenacity to overcome the "impossible" hurdles.

UoSAT-OSCARs 14 and 15

On Jan 22, 1990 (UTC), an ESA Ariane rocket launched from Kourou placed seven spacecraft in circular low earth orbits. The payload included six OSCAR

Jeff Ward, K8KA, takes thermal surfaces off the bottom of UoSAT-3 (which became UoSAT-OSCAR 14 after achieving orbit) at the launch site in Kourou, French Guiana. *(G3YJO photo)*

spacecraft (two UoSATs and four Microsats) and the commercial satellite SPOT 2. The OSCAR spacecraft provided ESA a chance to test the new *A*riane *S*tructure for *A*uxiliary *P*ayloads (ASAP).

Once they reached orbit, the University of Surrey satellites received the names UoSAT-OSCAR-14 and UoSAT-OSCAR-15. The two spacecraft are structurally similar, have a number of subsystems in common and are in nearly identical orbits.

UoSAT-OSCAR 14 is the first spacecraft in the UoSAT series to carry an open-access transponder, which was made possible by a grant from AMSAT-UK. It is a store-and-forward unit using Mode J frequencies and is available for use by all radio amateurs. Other payloads on OSCAR 14 focus on studying radiation damage due to high-energy particles, experimenting with onboard data handling, and improving previous spacecraft attitude control and stabilization systems.

UoSAT-OSCAR 15 carries an improved CCD Camera Imaging Experiment, and a set of microcomputers which can be connected together in various ways to study reliability, parallel processing and onboard data manipulation. The spacecraft also carries samples of several solar cell technologies so that the long-term behavior of each type of cell can be observed in a low earth orbit environment. Additional information on UoSAT spacecraft can be found in Appendix B (Spacecraft Profiles).

AMSAT MICROSAT PROGRAM

About 1987, several government agencies and private corporations began to realize that small satellites, the size of OSCAR 7, could serve some very practical needs. While AMSAT was pleased that everybody was "discovering" what we knew all along, the impact on our launch opportunities was serious. With a large launch backlog existing in the US, and all these groups willing to pay big bucks to get into space, it looked like the plug had been pulled on our hopes for placing a pacsat (small store-and-forward satellite for packet radio) in low earth orbit.

Nadia King, daughter of Microsat Chief Engineer Jan King, W3GEY, seems pleased with the way DOVE-OSCAR 17 came out. *(photo courtesy Rich Ensign, N8IWJ)*

Studying the payload envelope of ESA's Ariane Rocket, Jan King noted that there were several places where very small satellites could be mounted. One constraint placing a lower limit on the size of a useful spacecraft is that the surface area must be large enough to hold sufficient solar cells to power the mission. Rough calculations showed that if we took advantage of advances in solar cell efficiency that had occurred in recent years and took great care with power budgeting, a pacsat could be built in a cube roughly 9 inches on edge (a Microsat).

Jan presented these ideas to several key AMSAT technical volunteers at the annual meeting in November 1987. Everyone felt it was the right idea at the right time, and the pieces quickly fell into place. In the past, AMSAT satellites had used redundant systems to enhance reliability. With the premiums on space and power in a Microsat, the reliability approach would be redirected—AMSAT would build multiple spacecraft. In fact, it made sense to think about a satellite production line with spacecraft sharing a common structure, solar array architecture, telemetry and control unit, and power conditioning module. This approach would minimize the cost and construction time for each satellite. It would also provide new groups wishing to build amateur spacecraft an excellent entry point—they could design their own mission payload around the basic Microsat foundation.

Of course, the idea had to be sold to ESA. AMSAT offered to work with ESA to demonstrate how space that was currently being wasted could be marketed to commercial users. ESA liked the idea and went on to develop the Structure for Auxiliary Payloads. Meanwhile, AMSAT started working on Microsats A, B, C and D, and the University of Surrey began constructing UoSATs D and E. All six were dropped off in circular low earth orbits on January 22, 1990. AMSAT and UoSAT had to reimburse ESA for direct costs associated with interfacing to the ASAP platform. The charges amounted to approximately $150,000.

AMSAT-OSCAR 16 (Microsat-A) contains the store-and-forward digital transponder using mode J frequencies that stimulated the entire program. It was built by the North American AMSAT group (AMSAT-NA). Lusat-OSCAR 19 (Microsat-D) unit carries a similar payload. A group from AMSAT Argentina (AMSAT-LU), under the direction of Arturo Carou (LU1AHC), was responsible for its construction. Both of these spacecraft are often referred to as pacsats (*pac*ket radio *sat*ellites).

Led by Junior Torres DeCastro (PY2BJO), AMSAT Brazil (BRAMSAT) undertook the construction of DOVE-OSCAR 17 (Microsat-B). The payload, called DOVE, for *D*igital *O*rbiting *V*oice *E*xperiment, produces a strong downlink signal that can be received on the inexpensive scanner radios widely available. It's designed to support educational activities and encourage young people to become interested in the peaceful uses of space and Amateur Radio. When not devoted to time-critical projects, DOVE carries messages promoting peace and friendship in various languages produced by classes of young school children around the world. Finally, Webersat-OSCAR 18 (Microsat-C), sponsored by Weber State University in Utah, carries a TV camera. Pictures are encoded in normal AX.25 packet frames so that they can be received using packet-radio equipment. For details on the Microsats, see Appendix B.

THE JAPANESE AMATEUR SPACE PROGRAM

Japanese involvement in the spacecraft hardware end of the Amateur Radio satellite program began in earnest with the construction of the Mode J transponder for OSCAR 8. With the encouragement of Harry Yoneda, JA1ANG, a long-time member of the AMSAT Board of

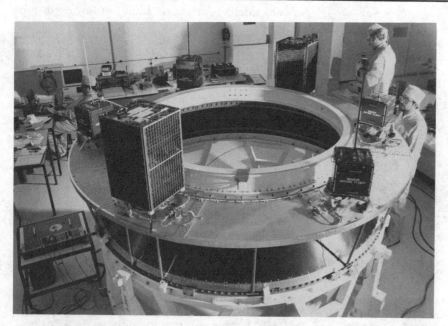

Four Microsats and two UoSATs mounted on the launch platform of the Ariane rocket that was to propel them into space in January 1990. *(photo courtesy Joe Kasser)*

Directors, Japan AMSAT (JAMSAT) went on to organize support for constructing a complete spacecraft to be launched on a Japanese rocket. The satellite would be called JAS-1. The construction of JAS-1 was jointly sponsored by four groups: JAMSAT, JARL (Japan Amateur Radio League), NEC Corporation and NASDA (National Aerospace Development Agency). The payload included two transponders, one analog and one digital. Since congestion makes the 2-meter band almost useless as a satellite downlink in Japan, the designers of JAS-1 elected to use Mode J (146-MHz uplink, 435-MHz downlink) for both transponders. The satellite was successfully placed in a low earth orbit on 12 August 1986 by a two-stage H-I rocket of the Japanese Space Development Agency. After launch, the name Fuji-OSCAR 12 (FO-12) was assigned.

Fuji-OSCAR 12 was a very sophisticated first satellite for JAMSAT and JARL. Most systems worked well, but there were some difficulties. These problems can be placed in three categories (1) spacecraft management, (2) uplinking digital data, (3) power budget. As the UoSAT crew had learned earlier, it's easy to underestimate the time it takes to learn how to manage a powerful low earth orbiting spacecraft. Because of the complexity of FO-12, writing and testing control programs took considerable time. When loading command programs into FO-12 proved more difficult than anticipated, new procedures had to be devised. The times at which the transponder was turned on had to be frequently revised because of the spacecraft's negative power budget. Since FO-12 didn't carry a real-time clock, it was difficult for users to keep track of the scheduled switching times between the digital and analog transponders and the recharge mode.

Back when satellite operators didn't have other options, they were willing to put up with a great deal of frustration. Readers who remember the early days of OSCAR 6, when it was continually shutting down at the most inopportune times, know what I mean. But in 1986,

amateurs had a choice of several spacecraft. So, even though the transponders aboard FO-12 provided first-class performance when they were available, lack of access to a stable, reliable operating schedule caused many amateurs to concentrate on other spacecraft. In spite of all these difficulties, JAMSAT reports that more than 300 stations sent packets through the digital transponder on FO-12. As FO-12 aged, the solar cells and storage batteries continued to deteriorate. On November 5, 1989, having exceeded its three-year design lifetime, FO-12 became the first amateur spacecraft to be withdrawn from service. This enabled JAMSAT to concentrate efforts on preparing JAS-1b for flight re-using FO-12 frequencies. JAS-1b, now known as Fuji-OSCAR 20 (FO-20), was successfully placed in orbit by the Japanese National Space Agency on February 7, 1990. JAMSAT mission planners requested that NASDA use fuel reserves to raise the apogee of the 900-km circular orbit. NASDA complied and the apogee was increased by about 800 km. As a result, the orbit has a pronounced elliptical component and the total amount of time that the spacecraft is eclipsed by the earth is greatly reduced during certain parts of the year. This significantly increases the average power available for transponder operation.

Based on experience gained with FO-12, FO-20 designers made several changes which should significantly

The second flight model of JAS-1, Japan's first Amateur Radio satellite, undergoes close scrutiny by technicians and some members of the Japan Amateur Radio League after its completion at an NEC factory in December 1985. *(photo courtesy JA1ANG)*

Fuji-2 (Fuji-OSCAR 20 after it achieved orbit) is dwarfed by other nosecone paraphernalia just before liftoff. *(photo by NASDA; QSL card courtesy Fujio Yamashita, JS1UKR)*

improve the user friendliness of the new spacecraft. Ground stations who perfected their satellite packet communication techniques on FO-12 now have access to digital transponders on OSCARs 14, 16, 19 and 20. For additional details on FO-20 see Appendix B.

SOVIET PROGRAM

RS Satellites

In the mid 1970s, during the planning stages of the US-USSR Apollo-Soyuz earth-orbiting mission, several Soviet engineers, some of whom were radio amateurs, visited NASA facilities in the US. Their itinerary included the Goddard Space Flight Center in Greenbelt, Maryland, an area where several AMSAT designers lived. During these visits to Goddard SFC, the Soviet radio amateurs and the AMSAT crew met, as hams are apt to do, and discussed the technical and operational aspects of the OSCAR program. At these meetings, it became clear that the Soviet amateurs were interested in producing their own radio amateur satellites. In fact, a satellite coordinating group had already been formed, and construction of prototype equipment was underway.

Though time passed—the Apollo-Soyuz linkup in space succeeded in July 1975—not much was heard of Soviet amateur satellite plans. Then, in the October 1975 issue of *RADIO*,[10] a very widely read Soviet electronics magazine, the awaited article appeared. It focused on experiments with terrestrial linear Mode A type transponders in Moscow and Kiev and discussed, for the first time in the Soviet press, the OSCAR program. Several US amateurs were able to directly monitor the 1-W 29-MHz output of the Moscow transponder whenever 10 meters was open. Although no mention was made of Soviet Amateur Radio satellites in the *RADIO* article, some Soviet hams were clearly laying the groundwork. Speculation as to Soviet plans wasn't officially confirmed until July 1977, when the USSR filed a notice with the International Frequency Registration Board (IFRB) of the International Telecommunication Union announcing that a series of satellites in the amateur-satellite service would be launched.[11] This was followed by a series of *RADIO* articles on the RS spacecraft.[12]

Finally, on October 26, 1978, a rocket lifted Cosmos 1045 and two radio amateur satellites, Radio-1 and Radio-2 (RS-1 and RS-2), into low earth orbit. Each spacecraft carried a Mode A transponder. Since Soviet radio amateurs are limited to 5 W in the 145-MHz uplink band, the transponders contained very sensitive receivers optimized for low-power terrestrial stations. Automatic shutdown circuitry protected them from excessive power drain. In operation, the protective circuits, acting like a time-delay fuse, would shut the transponder off when ground stations used too much uplink power for more than a few seconds. Unfortunately, the fuse could only be reset when the spacecraft passed near a Russian command station. Although the Soviets appeared to make every reasonable effort to keep the satellites on and available to the rest of the world, the transponders were often off over the Western Hemisphere because of the actions of a few inconsiderate high-power users.

The Soviet approach to improving reliability through redundancy merits note: When possible, launch two spacecraft at the same time. In addition to a transponder, each satellite contained a telemetry and command system, a Codestore-like device and a power system using solar cells. The primary telemetry system used Morse code letters and

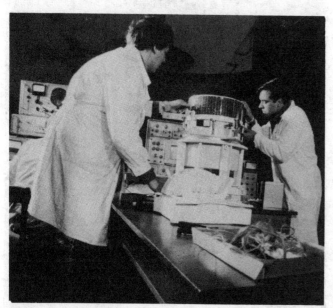

Two technicians assemble the Radio-2 artificial satellite. With Radio 1, this satellite provided communications for more than 700 Amateur Radio operators from 70 countries on all continents. Its communication range was over 8000 km (5000 miles). *(Novosti photo, provided by Embassy of the USSR)*

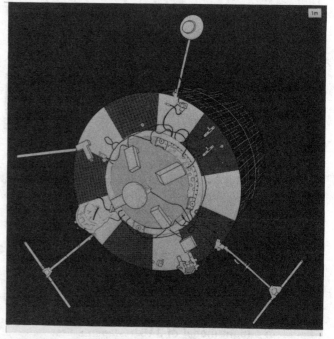

Artist's rendition of RS-14 (AMSAT-OSCAR 21) on a Soviet GEOS spacecraft. *(courtesy AMSAT-DL)*

numbers that identified the parameter being measured, encoded the most recent value of that parameter and indicated the status (on/off) of the transponder. Specific decoding information was provided by the Soviets about a month after launch. At least one of the spacecraft also contained an infrequently used high-speed digital telemetry system described as an experimental prototype.

Observations during the first few weeks that these spacecraft were in orbit revealed that changes in the operating status of RS-1 and RS-2 (except for transponder shutdown) took place only at certain times and from certain locations. This strongly suggested that command of the spacecraft was confined to Moscow and limited to normal working hours. Further observations indicated that two additional ground command stations were soon activated, one in eastern Asia and another in central Asia. The Soviets later announced that the primary command station was in Moscow, a secondary command station was in Arsen'yev (44.1 N, 133.1 E) near Vladivostok, and that a third command station, which was portable (Novosibirsk?), had been tested.

RS-1 and RS-2 were designed for passive temperature control using techniques similar to those employed on OSCARs 1-8 (see Chapter 15). RS-2 also included a quasi-active thermal regulating system employing a heat bridge connecting the interior of the spacecraft with a heat exchanger on the outer surface. The bridge would automatically turn on whenever the internal temperature exceeded a predetermined level.

The transponders aboard RS-1 and RS-2 could be kept operating for only a few months before power supply (battery) problems disabled both spacecraft. Reception of a weak telemetry signal on 29.401 MHz signing a number group, often 5015 or 5501, has been reported into 1988. The source is believed to be RS-1. RS-1 and RS-2 were certainly an impressive and successful first step.[13] As later RS satellites demonstrate, the Soviets used the results of the first mission to design spacecraft with significantly longer operational lifetimes. Insofar as is possible, the Soviets have attempted to provide information about RS spacecraft, coordinate frequencies with OSCAR satellites and make RS spacecraft available to radio amateurs around the world. The international radio amateur community sincerely appreciates this effort.

On December 17, 1981, the Soviets simultaneously launched six satellites, RS-3 through RS-8, into low earth orbit. The launch had been expected, but the number of spacecraft certainly was a pleasant surprise. During the previous year, the club station at the University of Moscow, RS3A, had tested many of the spacecraft RF subsystems on the 10-meter band. The equipment observed during the tests included a Mode A transponder, a Codestore device, a Morse code telemetry system and an autotransponder (called Robot), all of which were flown. In addition, an engineering prototype of one of the spacecraft was exhibited at TELECOM-79, a large international telecommunications conference held in Geneva in 1979. The longest-lived members of the group, RS-5 and RS-7, ceased

operating reliably in 1988 after providing service for nearly 7 years.

On June 23, 1987, RS-10/11 was successfully launched. The unusual designation is due to the fact that two smaller spacecraft were integrated into a single package to take advantage of a launch opportunity. The primary payload aboard RS-10/11 was a linear Mode A transponder (145-MHz uplink, 29-MHz downlink). However, the spacecraft was also able to receive on 21 MHz and transmit on 29 MHz, and the flight control unit was able to interconnect the available uplinks and downlinks in a variety of ways, making several modes available. As a result, a QSO using Mode KA might involve one station uplinking on 21 MHz and the other on 145 MHz, with both listening to the 29-MHz downlink. The 29-MHz downlink can run up to 5 W, so signals are relatively strong for a Phase 2 low earth orbit satellite. The only problem using RS-10/11 is keeping track of the operating schedule. A quick check of the 10-meter beacon when the spacecraft comes into range will provide the information needed. See Appendix B for additional technical information and a list of specific transponder frequencies.

RS-12/13 is another dual spacecraft integrated in a single package. Launch is tentatively scheduled for sometime in 1990 or '91. Its specifications are similar to those of RS-10/11. The primary payload consists of several linear transponders with uplinks at 21 and 145 MHz and powerful (up to 8 W) downlinks at 29 and 145 MHz. RS-12/13 was built at the Tsiolkovskiy Museum for the History of Cosmonautics in Kaluga, an industrial center 180 km southwest of Moscow. Tsiolkovskiy, a school teacher and amateur space enthusiast, is very famous in the USSR. In the early 1900s, he wrote a number of extremely influential articles discussing the future of rocketry, space satellites and manned space stations. The chief engineers for the RS-12/13 project were Aleksandr Papkov and Victor Samkov.

RS spacecraft designers have been experimenting with prototype Mode B and Mode J transponders for future low earth orbit missions. Indications are that at least one of these modes will appear on future RS spacecraft, but the decision as to which has not yet been made.

ISKRA Satellites

Soviet radio amateurs have produced another class of satellites that go by the name ISKRA (*spark* in Russian). Two ISKRA spacecraft have been operated. Both were placed in orbit from the Salyut 7 space station. They're simple spacecraft, of an experimental nature, designed to operate for a short time before they burn up on re-entry.

ISKRA-2 (the first in the series) was launched May 17, 1982. Real-time TV coverage of the event was provided to viewers in the USSR and adjacent countries. As a result, students at the Moscow Aviation Institute had a chance to watch the birth of the satellite they helped build. According to *TASS*, solar-cell-powered ISKRA-2 contained a transponder, beacon, command channel, telemetry system and bulletin board (Codestore) facility. The novel 21-MHz up,

29-MHz down transponder (mode K) was intended to increase radio amateurs' store of practical information on the performance of a new combination of link frequencies. Mode K was later included on RS-10/11 and has been announced as being on RS-12/13. The transponder bandwidth was 40 kHz centered at 21.250 and 29.600 MHz and a beacon was placed at 29.578 MHz. Because of a malfunction, apparently associated with the command system receiver/decoder, the transponder was never activated in range of the US. During the seven weeks that the spacecraft was in orbit, it provided interesting telemetry.[14] Re-entry occurred on July 9, 1982.

ISKRA-3 was placed in orbit November 18, 1982. Like ISKRA-2, it carried a Mode K transponder. The telemetry beacon was on 29.583 MHz. A serious overheating problem severely curtailed operation. ISKRA-3 was the last one of the series to be launched and no new ISKRAs have been announced for the future. Although it is simpler to design a spacecraft for this type of get-away-special launch, the short lifetimes of satellites in the resulting orbit provide a poor payoff for the effort involved in constructing a conventional communication spacecraft. However, this doesn't mean we've seen the end of the ISKRA series. Self-powered satellites and manned missions can be combined in other innovative ways. For example, how about magnetic-mount solar-powered HF or microwave beacons that could be clamped to an external bulkhead during an EVA by the cosmonauts?

SOVIET HAMS IN SPACE

The Soviet radio amateur space program gained two new participants in late 1988 when cosmonauts Musa Manarov and Vladimir Titov, who were in residence aboard the MIR space station, qualified for amateur licenses. Manarov, the primary operator and initiator of the amateur activity, had the call U2MIR. Titov, first in command, was U1MIR. Operation on 145.5-MHz FM voice began on November 6, 1988, with the initial test producing a solid QSO with UA3CR in Moscow. A few days later, U2MIR had his first US QSO with W4BIH (UA3CR at the mike) from the AMSAT annual meeting in Atlanta. The station aboard MIR consisted of an off-the-shelf Japanese 2-meter hand-held transceiver feeding a ¼-wavelength ground-plane mounted outside MIR. Stories of how informal channels were used to convey the transceiver to the cosmonauts are intriguing and entertaining, but the cosmonauts deserve a chance to present their version before others begin circulating third-hand accounts. Once the "official" story is released, you'll see it in QST or the AMSAT Journal.

About six weeks after operations commenced, Manarov and Titov returned to earth after having set a record for continuous time in orbit—just over one year. It seems reasonable to assume that this background would place them in prime contention for flight crew aboard a Mars expedition if such a mission were authorized in the next decade. When Manarov and Titov returned to earth, the replacement crew carried up a new 10-W 2-meter transceiver. In early 1989, U4MIR (Alexander Volkov) and

U5MIR (Sergei Krikalev) were active during a relatively short stay aboard the space station.

Prospects For The Future

While predictions are always risky, based on some clear-cut observations it's reasonable to assume that Soviet Ham-In-Space activities will continue. The program has been in existence a long time. Cosmonauts started tossing ISKRAs from the Salyut 7 space station in 1982. Operation from the MIR Space Station began in late 1988. Since Soviet astronauts are often in space for very long periods of time, mission planners pay careful attention to physiological and psychological needs of the crew. As a result, Soviet astronauts are allocated a significant amount of time to exercise and relax. Place yourself in the position of a Soviet mission planner trying to think up diverting activities to provide to the crew? Sally Ride reports in a February 1989 article in Scientific American that Soviet cosmonauts are routinely sent videotapes, books and other personal items on resupply missions. As we all know, Amateur Radio operations can be an excellent diversion. I think you get the picture. Amateur operation from MIR encourages young people in the USSR to become involved in science and engineering, serves as a backup communications medium, and provides an important recreational outlet for the crew. An amateur 10-W 2-meter FM transceiver was ferried up to MIR in late 1988 and a ground-plane antenna was installed on the exterior of the spacecraft by Manarov and Titov during an EVA in late 1988. If future crew members become interested in Amateur Radio, Soviet Ham-In-Space operation is likely to be extensive and of an informal nature. However, there are indications that in the near future the Soviets will be allocating more resources to unmanned scientific space exploration and less to support manned activities.

You might want to think about creative ways of taking advantage of the communications opportunities that may arise. Cosmonauts generally keep to Moscow time, which is 8 hours later than EST. This means that their waking free time usually coincides with morning school hours on the east coast of the US. A good portion of the year MIR passes will occur at this time. A science or language teacher might like to have the class "adopt" a cosmonaut. A series of contacts directly from the classroom would give the cosmonaut and the class members a chance to get to know each other. The equipment needed to do this is relatively simple. A 25-W 2-meter FM transceiver and a ground-plane antenna should work fine.

US HAMS IN SPACE (AND RELATED ACTIVITIES)

In 1972, when Dr Owen Garriott (W5LFL) was selected to fly aboard the experimental US space station known as Skylab, a proposal to allow limited Amateur Radio operation from space was presented to NASA. The proposal was turned down, but the text of the rejection, and the fact that the decision came from the highest levels of NASA, made it clear that serious consideration would be given to future proposals. So, Garriott's sojourn in space from July 28, 1973 to September 25, 1973 was the Ham-In-Space mission that might have been.[15]

Ten years later Garriott was again selected for a space mission, this time aboard the US Space Shuttle *Columbia*. A proposal calling for experiments involving Amateur Radio to take place during free time periods allocated to the crew was accepted by NASA. Shuttle mission STS-9, involving Spacelab-1, took place from November 28 to December 8, 1983. During the flight, Dr Garriott was able to contact in excess of 250 amateurs using 2-meter FM simplex. His sons, Robert and Richard, operating from W5RRR, the club station at Johnson Space Center (JSC), were at the far end of one of these QSOs. W5LFL used a modified Motorola hand-held transceiver operating on internal batteries and a dipole antenna taped to a window. Operation was entirely independent of Skylab's power system. The three important objectives of the Amateur Radio experiment—(1) demonstrating that Amateur Radio can operate on the Space Shuttle without interference to primary mission tasks, (2) directly involving segments of the general public in US space activities and (3) demonstrating Amateur Radio's capabilities as a potential backup means of communication—were successfully attained.[16]

Another opportunity for operation from space arose in 1985 when Dr Tony England (WØORE) Dr John Bartoe (W4NYZ) and Commander Gordon Fullerton (ex-WN7RQR) were assigned to Spacelab-2 on Shuttle *Challenger* mission STS-51F. NASA granted amateurs permission to undertake a more ambitious set of tests known as SAREX (Shuttle Amateur Radio Experiment). This time, the emphasis was on communicating with groups of school children and on demonstrating the capabilities of slow-scan TV for uplinking and downlinking pictorial information. All operation was on 2 meters.

The experiment was a solid success. Pictures from the spacecraft were sent directly to classrooms and, for the first time, pictures were transmitted from the ground to an orbiting spacecraft. Amateurs also demonstrated their ability to connect to the spacecraft power buss and other equipment without compromising safety. Jess Moore, Associate Administrator of NASA stated, "I was watching the video of the Mission Control Center in Houston and actually saw the [pictorial] data that was being transmitted up. I was very impressed that we were able to do this, using Amateur Radio, and that's fantastic!" Since W4NYZ was tied up repairing some important scientific equipment, WØORE did most of the operating. Commander Fullerton, calling on his experience as a Novice class radio amateur when he was younger, handled some of the operations under the supervision of England, in compliance with US amateur regulations.

Preparing the SAREX proposal and then overseeing the construction and testing of the equipment used was a

The Space Shuttle *Columbia* rollout Are you ready for STS-9?

November 1983 $2.50

QST

devoted entirely to Amateur Radio

WORLD COMMUNICATIONS YEAR AÑO MUNDIAL DE LAS COMUNICACIONES 1983

Shuttle Astronaut Owen Garriott, W5LFL, goes through a trial run aboard the Shuttle Trainer at Johnson Space Center, Houston, Texas, before his historic November-December 1983 SAREX operation. *(NASA photo)*

Shuttle *Columbia* takes a final "roll" to its launch pad in preparation for the first Amateur Radio operation from space. W5LFL contacted or heard about 300 stations during the 10-day mission. *(NASA photo)*

mammoth job. Most of the work was handled by a group of amateurs at Johnson Space Center. Key were Louis McFadin (W5DID), president of the JSC Amateur Radio Club, who led the efforts to design and test the equipment, and Chuck Biggs (KC5RG), chief public affairs officer at JSC, who saw to it that the SAREX proposal and activities received needed support and recognition.[17]

Barely three months after SAREX another Ham-In-Space mission took place. This time (October 30 to November 6, 1985), the Space Shuttle *Columbia* carried the European Space Agency's Spacelab mission D1. Three hams were aboard: Dr Reinhard Furrer (DD6CF), Dr Ernst Messerschmidt (DG2KM) and Dr Wubbo Ockels (PE1LFO). All used the special call sign DP0SL, which had been assigned for the mission. Equipment on the spacecraft consisted of a cross-band FM transceiver which required ground stations to use the Mode B frequency combination (435-MHz uplink, 145-MHz downlink).[18]

The Ham-In-Space program also includes experiments aboard the US Space Shuttles and MIR that utilize Amateur Radio frequencies even if these experiments do not require direct involvement of a radio amateur. One such experiment, known as MARCE (The Marshall Amateur Radio Club Experiment), involved a GAS Can (get-away-special canister) payload on the Shuttle. Early sponsors were the Alabama Space and Rocket Center and the Alabama Section of the Institute of Aeronautics and Astronautics. The Marshall Space Flight Center Amateur Radio Club later joined in to provide the downlink telemetry system and coordinate the launch. The GAS Can contained four experiments, three provided by University of Alabama students, plus the communications/telemetry system. The student experiments focused on alloy solidification, plant physiology and crystal growth. AMSAT was particularly interested in the communications experiment, because consideration was being given to flying a pacsat as a get-away-special. The downlink employed a 70-cm FM transmitter

running 6 W to a dipole mounted directly on the GAS Can. A voice synthesizer signing WA4NZD spoke the telemetry. Of special interest to AMSAT was the possibility of relaying the MARCE Shuttle downlink via the Mode B transponder on AO-10. This would model, in a modest way, the TDRS (Telemetry Data Relay Satellite) employed by NASA to relay shuttle data.

MARCE was flown on STS-41G, launched October 5, 1984, but nothing was heard. It was later learned that, due to a procedural error, the experiment had never been activated. It was reflown on STS-61C which was launched January 12, 1986.

GAS Can payloads are placed in the shuttle cargo bay and exposed to space by opening the bay doors. The Shuttle *Columbia* flight plan called for the cargo bay to be earth facing, a good orientation for direct reception of the 70-cm telemetry at the ground, but one that would make a relay by AO-10 highly unlikely. The first report of direct 70-cm telemetry reception was by Junior Torres DeCastro (PY2BJO) of Sao Paulo. Partway through the six-day mission, *Columbia* had to be reoriented to warm a cold Auxiliary Power Unit. The new attitude greatly improved the 70-cm link between the Shuttle and AO-10, enabling several ground stations to obtain relayed telemetry. KG6DX provided an excellent tape illustrating system capabilities under conditions where the Mode B transponder on AO-10 was lightly loaded. The project coordinator for MARCE, the person who brought it all together, was Ed Stluka, W4QAU.[19]

Prospects for the Future

On January 28, 1986, a great many young schoolchildren all over the United States gathered around TV sets to observe the launch of the Space Shuttle *Challenger* and the start of the Teacher in Space Program. As they watched, a critical system aboard the rocket failed, and the rocket and Shuttle disintegrated, killing all seven crew

Shuttle *Challenger* astronaut Tony England, W0ORE, became the second ham in space in August 1985. This slow-scan self-portrait was just one of the impressive images he sent to classrooms while in orbit. W4NYZ and ex-WN7RQR also were aboard.

members. The Shuttle program ground to a halt.

Two and a half years later, after extensive hardware and procedural changes, the Shuttle program resumed. But it's a new, slower, more deliberate program with an extensive backlog of projects. To assess the prospects for future US Ham-In-Space operation from the shuttle, let's consider the ingredients that led to past support.

One of NASA's key justifications for supporting Ham-In-Space activity is the belief that these operations lead many young people to consider a career in science and engineering, and the widely accepted viewpoint that it's economically advantageous to a modern country to have a significant number of citizens well trained in these areas. The second key ingredient was the existence of a number of enthusiastic and dedicated volunteers (including astronauts) willing to devote a great deal of time and effort to prepare for a couple of hours of actual operations from space. Many of these people explain their motivation by stating that they greatly value the impact ham radio has had on their own lives and career choices and they want to pass this opportunity on to future generations.

So what happens next? Support for the program, both by NASA and individual volunteers, appears to run deep. But launch opportunities are limited and there's a great deal of competition for the astronauts' time. As a result, continued amateur activity from the Shuttle is likely, but we can't expect it to be too frequent. Activities that focus on particular experiments such as packet radio, fast-scan TV, Phase 3 relays, and so on, and are targeted at specific groups of ground stations, especially those involving young children and school groups, are the kinds of projects that are most likely to be supported. Opportunities for informal hamming do not appear very likely in the near future.

Two SAREX missions are scheduled for 1990 and '91. The first involves Ron Parise, WA4SIR, payload specialist for STS-35, who will operate 2-meter packet and FM voice. STS-37 Pilot Ken Cameron, KB5AWP, will fly with a crew consisting of Mission Specialist Linda Godwin, N5RAX; Mission Specialist Jay Apt, N5QWL; and Commander Steven Nagel, N5RAW. Scheduled modes include slow-scan and fast-scan TV (uplink only).

Amateurs learned a great deal from planning and executing the W5LFL, W0ORE and later Shuttle missions. One lesson is that a NASA desire to accept a proposed experiment can be balked by small practical details like whether or not a bulkhead feedthrough for antenna lines exists or a power outlet and operating position that won't interfere with other experiments is available. With plans for a permanent US space station firming up, Bill Tynan (W3XO—AMSAT VP for Ham-In-Space activities), is coordinating efforts of amateurs and educators to provide input to space station designers so that needs of these groups can be met. A small change or inexpensive addition today can eliminate an impenetrable barrier down the road.

PHASE 4 PROGRAM

After OSCAR 10 had been in orbit for a few years, several amateurs began to seriously think about the direction the amateur satellite program should take over the next decade. This led to a consideration of the inherent limitations of current programs such as Phase 3 (including the evolutionary improvements possible) and future programs involving digital transponders in low earth orbit. When you've worked on an immensely difficult project like Phase 3 for roughly eight years and the taste of success is still fresh, it's difficult to change stride and focus on limitations, but that's what had to be done.

The topic was actively discussed for a year. Then, Jan King prepared a document called the Phase 4 Technical Study Plan, which summarized the growing consensus. Many of the key proposals presented in the study plan were later described in a two-part article in *QST*.[20,21] We now turn to the limitations identified and the proposed system designed to alleviate them.

Important limitations of current amateur satellite programs identified in the Phase 4 Study Plan include the following:

1) With respect to the entire worldwide community of radio amateurs:

All current satellite programs require some specialized knowledge of satellite systems, satellite tracking, satellite scheduling, and so on. It's been said of radio amateur satellites that the medium is the message; that is, if you use satellites for communicating, you're probably more interested in satellites than in communications. As a result, if we elect to pursue evolutionary improvements in current systems, the size of the amateur satellite user community will show only modest growth since the program will continue to appeal to those primarily interested in space activities.

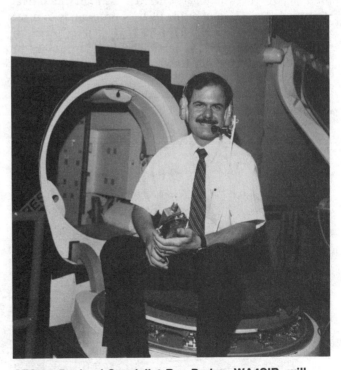

STS-35 Payload Specialist Ron Parise, WA4SIR, will operate packet radio during his flight. His SAREX operation will provide an Amateur Radio experience to schoolchildren around the country. *(NASA photo)*

2) With respect to the entire worldwide community of radio amateurs:

The UHF and microwave spectrum allocated to the amateur and amateur satellite services is an extremely valuable resource. Mode S transponders alone (1.2-GHz uplink, 2.4-GHz downlink) could provide 10 MHz of spectrum, more than three times the total HF spectrum currently available to radio amateurs. One can associate a monetary value with these frequency allocations by assuming that a commercial entity interested in acquiring a revenue-generating resource is generally willing to invest about six times the revenues the resource can return on an annual basis to purchase it. Viewed this way, our UHF and microwave allocations are worth hundreds of millions of dollars per megahertz. It's extremely unlikely that the amateur-satellite service, using Phase 3 technology and current communications techniques, can populate our UHF and microwave allocations to the extent needed to retain them at the next major WARC in the mid 1990s. In other words, if we want to keep this extremely valuable resource, something major must be done!

3) With respect to society at large:

Although Phase 3 satellites are very appealing to technically inclined radio amateurs, other groups, such as those who wish to use them in times of emergency, on the scene of natural disasters, or for educational activities, find that the limited access time, and the need to track antennas and calculate when the spacecraft will be accessible, pose an insurmountable barrier to such use.

The conclusion reached by the study team was that the only response option we have which directly addresses all of these key concerns is the construction of a constellation of high power, geostationary satellites, each carrying several transponders. The aim of the Phase 4 program is to place such a system in orbit. This leads to a long series of questions concerning the feasibility and desirability of AMSAT undertaking such a project. We'll consider a few

of the major concerns here.

Does AMSAT have the technical expertise to design and manage the construction of such spacecraft?

One way of responding to this question is to compare the state of the Phase 3 program in 1975 to the state of the Phase 4 program today (1990). In both cases, plans were being made to construct a new series of spacecraft that were many times more complex and costly than anything previously attempted. I think it's fair to say that, with respect to the state of the overall spacecraft system design, access to key proven technical people, experience in managing large projects, and availability of commercial equipment to the user community, the 1990 Phase 4 program is on a much firmer foundation than the Phase 3 program was in 1975.

What will Phase 4 cost?

The first Phase 4 spacecraft will cost approximately 2.5 million dollars, or roughly five times more than any previous satellite supported primarily by radio amateurs.

Is it possible to raise this amount of money?

It's not possible if we simply ask the current satellite user community to dig deeper in their pockets. It is possible if we can reach a significantly larger segment of the general amateur community and also reach the non-amateur groups who will benefit from the new services provided. To attain its financial goals, AMSAT will have to enlist the aid of individuals with appropriate fund-raising expertise. These people will have to identify the target audiences and devise and implement strategies for obtaining their backing. For example, for more than 15 years the NASA ATS Project Office used ATS-1 to support an active public service program. Now that ATS-1 has run out of fuel and been declared out of service (1986), and with prospects for a replacement spacecraft very poor, they may view support-

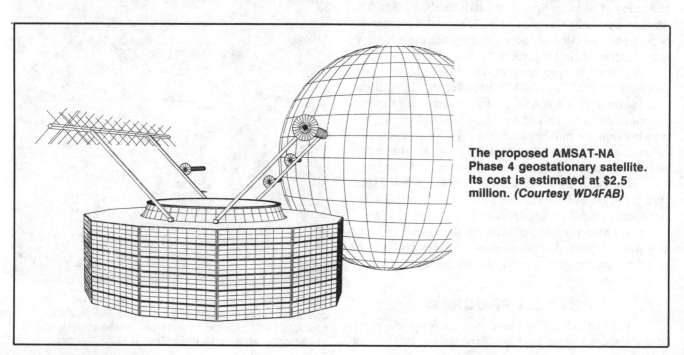

The proposed AMSAT-NA Phase 4 geostationary satellite. Its cost is estimated at $2.5 million. *(Courtesy WD4FAB)*

ing Phase 4 as being a cost-effective way of continuing NASA's public service activities. Other potential users, such as The Public Satellite Service Consortium, a group of educational and public safety organizations, have expressed interest in AMSAT's activities over the years, but we've never had anything to offer that met their needs. The point is, Phase 4 will provide an entirely new class of service and it's reasonable to believe that many of those who would benefit from these services will be willing to help support the program financially.

Will we be able to obtain a launch for a Phase 4 spacecraft?

This key concern has been carefully taken into account in the conceptual design of the Phase 4 spacecraft. Atop ESA's Ariane 4 rocket sits a payload compartment designed to carry two coaxially mounted large spacecraft into a geostationary transfer orbit. The two spacecraft are joined by a standard connecting ring, which is jettisoned in space. The Phase 4 satellite has been designed to incorporate this ring as an integral part of its structure. Launching our spacecraft will help ESA demonstrate the viability of this approach to satellite construction, an approach that may provide ESA with a new source of revenue. With US expendable launch vehicle activity growing, there may be launch possibilities here, too. Since ESA and US launchers use similar attach fittings, accommodating to a US launch shouldn't pose major problems.

We now take a look at the characteristics of a Phase 4 (AMSTAR) spacecraft, paying special attention to the user-oriented payloads it will carry and the ground station equipment required to access it. A geostationary satellite appears to sit at a fixed longitude above the earth's equator. The tentative positions envisioned for the first two Phase 4 spacecraft are longitude 47 degrees West (AMSTAR East) and longitude 145 degrees West (AMSTAR West). Coverage circles (footprints) for these two spacecraft are shown in Figs 4-1 and 4-2.

As an example of the type of service one of these spacecraft could provide, imagine the occurrence of a serious earthquake in Central or South America. AMSAT East could provide round-the-clock communications between workers and government officials on the scene, UN disaster relief agencies in New York and Red Cross Headquarters in Switzerland.

If you want to use one of these satellites, you only have to aim your antenna once, when you install it. The procedure is very simple; just point the antenna in the general direction of the spacecraft (from the US either southeast or southwest) and then peak it on a downlink beacon.

Operating schedules will be keyed to local time and will be very stable. For example, the AMSTAR East schedule might read: In 1994 the Mode XX transponder will be on daily from 2300 to 0300 GMT (6 to 10 PM EST). Contrast this to a Phase 3 satellite, where an announcement might state that for the next six weeks the Mode XX transponder will be on from MA (mean anomaly) 39 to MA 106, and from MA 146 to MA 250; of course, the spacecraft won't always be in view during these periods. The simplicity and day-to-day consistency of Phase 4 spacecraft schedules will greatly facilitate arranging nets, bulletins, point-to-point communications, classroom demonstrations, and so on.

Each Phase 4 spacecraft will carry high-power transponders and high-gain antennas. As a result, link margins will be considerably better than with AO-10 or AO-13. Therefore, ground stations wishing to directly access a Phase 4 spacecraft will find downlink signals stronger, and required uplink power lower.

Each AMSTAR will contain two linear transponders. One will use Mode JL (145-MHz and 1.2-GHz uplinks, 435-MHz downlink). The downlink will be 500 kHz wide and will run 120 W PEP. The second transponder will utilize Mode S (1.2-GHz uplink, 2.4-GHz downlink). It will be partitioned into several different IF channels, each with its own AGC and transfer function optimized for a different type of communications. In operation, Mode S will appear to be several independent transponders, one

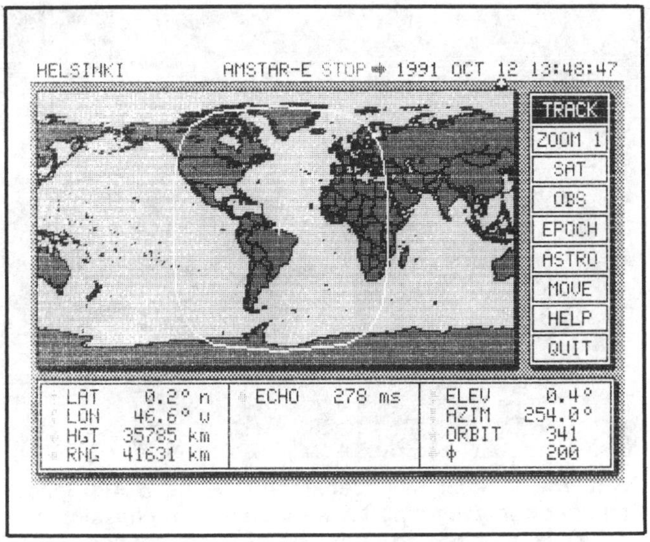

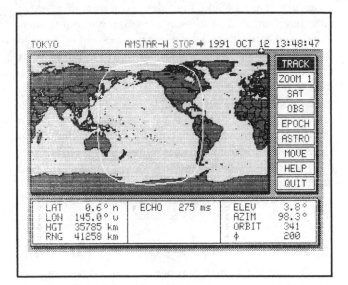

Fig 4-1—GRAFTRAK II map display showing footprint of AMSTAR East. *(photos courtesy Silicon Solutions)*

Fig 4-2—The footprint of AMSTAR West, a second proposed Phase 4 satellite.

for CW/SSB users, a second for those employing gateways, one for digital packet users, another for digital video, and so on. The Mode S video transponder will be capable of relaying real-time ATV signals if special video compression techniques are used to encode the video as a 500-kbit/s digital signal. Special ICs designed for video compression and decompression are just becoming available.

Many potential new users will find it much simpler to access the satellite indirectly through gateway ground stations. By connecting through a gateway, an amateur with a small hand-held VHF or UHF FM transceiver will be able to engage in continent-spanning QSOs. Gateways may become common adjuncts to terrestrial repeaters.

From a system viewpoint, the gateway operates as follows. One of the Mode S transponders aboard the spacecraft is optimized for gateway operation. The transponder can simultaneously handle approximately 20 communications channels and one or two command channels. A group of users pools their resources to build a standard gateway station (usually associated with a terrestrial repeater). Initiating a link involves several steps. A potential user calls up a local repeater using a VHF or UHF FM transceiver and indicates, via tone pad, that access to a specific distant gateway is desired. The transponder command links are used to determine whether the distant gateway is available and if there is an open channel. If the situation permits, a channel is assigned and the connection is opened. From the user point of view, most of this is totally transparent. Our potential user just calls up the local repeater and enters five tones, the first to indicate that access to the satellite is desired, the next four to identify the distant gateway. Within a few seconds the local gateway signals whether to "go ahead" or "wait." The signal might consist of synthesized speech, a CW "GA" or "AS" or a simple tone. If used effectively, the gateway system has tremendous potential, but it requires the user community to actively participate in establishing guidelines on how resources will be allocated.

Most of the year each Phase 4 satellite will be available to general users 24 hours per day. However, in cases of emergency or natural disaster (for example, hurricanes, floods and earthquakes), satellite resources will be devoted to relief efforts. Users will be informed in advance of disaster planning and procedures. Operations may also have to be slightly curtailed for short periods during certain parts of the year when the earth passes between the sun and the spacecraft or when the spacecraft passes between the earth and the sun. For example, we might shut down the 2.4-GHz downlink to conserve power when the spacecraft is in eclipse or when terrestrial antennas are aimed at the sun. However, these periods will be very brief, the Mode JL transponder will not be affected, and AMSAT will provide notice long in advance.

CONSIDERATIONS FOR THE FUTURE

The amateur satellite program is at an evolutionary junction point. We can continue to update the current Phase 2 and Phase 3 spacecraft or we can focus efforts

and resources on Phase 4.

There are still a great many improvements that can be made to Phase 2 and Phase 3 systems, enough to keep the technical workers challenged and happy. And, staying within the framework of Phase 2 and Phase 3, users can be provided with significant improvements in performance. If we stick with Phase 2 and Phase 3 systems, the amateur-satellite service is likely to see modest growth and a relatively healthy ongoing program over the next five to ten years. However, satellite communicators will remain a small minority segment of Amateur Radio. As a result, we might lose much of the 10-MHz-wide Mode S allocation and other important frequencies due to non-use and/or underpopulation. Many active satellite builders feel that focusing on Phase 2 and Phase 3 systems will lead to stagnation in the long term.

Devoting a major part of our resources to Phase 4 involves a calculated risk. If we succeed, (1) The profile of Amateur Radio in the eyes of government and private agencies involved in education, emergency relief and ongoing medical assistance will be greatly enhanced; (2) the number of radio amateurs involved in satellite communications will probably increase by a factor of three to five; it's even conceivable that the percentage of active amateurs equipped only for satellite communications will achieve parity with those equipped only for HF operation; (3) the amateur service will have access to wide bandwidth (multi-megahertz) transponders for worldwide real-time video and digital backbone networks; and (4) Phase 4 activities may provide financial support for Phase 2, Phase 3 and other serious microwave and space experimentation.

It can be argued that some of the activities proposed for Phase 4 are really not Amateur Radio. If we use the Mode S video transponder to relay pictures from the Space Shuttle directly to school classrooms on a regular basis, aren't we changing the nature of Amateur Radio and, in reality, giving up our frequencies? Yes and no. Yes, we are proposing a significant expansion in the type of public service Amateur Radio provides, and some of the rules and

Prototype of Solar Sail spacecraft that the World Space Federation is designing for a moon-orbiting mission. AMSAT is cooperating in design of telemetry and command electronics. *(photo courtesy World Space Foundation, PO Box Y, S Pasadena, CA 91031-1000)*

regulations governing the amateur service may have to be changed. But the world isn't a static place. Rules are always changing in response to new technologies. The question is whether we're to actively pursue changes we see as beneficial or sit back and let other groups impose their desires on us. I do not believe the proposed activities amount to giving up frequencies. Historically, amateurs have often been allocated spectrum that wasn't felt to be of much use. When the value was finally realized (often as a result of amateur experimentation), groups with financial clout would attempt to retrieve it. To a certain extent I believe that many governments view the UHF and microwave bands allocated to amateurs as strategic reserves. Such reserves can be reclaimed as needed with the explanation that amateurs aren't using them. Phase 4 is the only option I see for populating the 1.2- and 2.4-GHz bands before the next major WARC. If we don't populate them, they won't be there into the 21st century. If we elect to "share" them by making them available for educational and emergency activities, we retain access and control. In addition, we gain resources like a wideband transponder in space, and the support of major organizations in our dialogues with the FCC and ITU.

A successful amateur satellite program depends on four critical needs—our ability to obtain (1) volunteer technical and administrative workers, (2) launches, (3) access to appropriate spectrum and (4) financing.

Attracting Volunteer Support

In the past, the technical challenges inherent in the amateur satellite program have proved to be extremely successful lures in attracting excellent technical help to the project. As long as we continue to wrestle with interesting problems and investigate new innovative solutions, I think the program will continue to attract new blood. Administrative support is a different matter. Administrative tasks aren't as glamorous or as intrinsically satisfying as technical ones. As a satellite project becomes more complex, the administrative tasks appear to increase faster than the technical work. This means that one of the major challenges of Phase 4 is going to be to recruit and organize an administrative staff capable of managing a project of this magnitude. Without dedicated, competent people handling fund raising, parts acquisition, information dissemination, coordination with other groups, and so on, there won't be any Phase 4 spacecraft.

Obtaining Launches

Launch opportunities are always a key concern. It's very important that we position ourselves so that we can react very quickly to changing world situations. For example, in the early to mid 1980s, when it appeared that most US launches would be by the Space Shuttle, AMSAT actively investigated different types of rocket engines that could take us from Shuttle orbit to Phase 3 orbit. At about the time the US expendable launcher program was being reborn, several government agencies and large corporations became interested in small satellites similar (in size) to early AMSAT Phase 2 satellites. In a short time, a backlog

developed and it was clear that our chances of obtaining a launch were quickly approaching zero, so the Microsat program was born. The Phase 4 program is built on exploiting a launch niche that was previously overlooked. In sum, we don't have any guaranteed rides. We have to keep our eyes wide open and remain very flexible and innovative. Another possibility being closely watched is the new commercial launch vehicles which are approaching flight readiness (AMSAT might be able to trade some of the technology developed for OSCAR systems for launches). Finally, USSR-US cooperative efforts must be explored. During the past several years, the USSR has been launching roughly 75 percent of the payloads being placed in orbit. A precedent exists, as several US biological experiments were flown on COSMOS 782 and 936 in the late 1970s.[22]

Regulatory Constraints

As has been mentioned several times, frequency allocations made at international conferences have a very great impact on the amateur-satellite service. Where would we be today if the only band available for amateur satellite activities was 2 meters, as was officially the case from 1963 to 1971? The amateur-satellite service has planned for and participated in General and Space WARCs of 1963, 1971 and 1979. It's critically important that these activities continue in the future. Adequate planning is extremely important. Not many amateurs realize everything that goes into improving our WARC position. For example, Phase 4's real value to educational and disaster relief agencies is likely to result in these agencies providing support for our position at future WARCs. And, many of these groups have significant political clout in various countries around the world. The Microsat program will make it much easier for small groups of radio amateurs around the world to become actively engaged in satellite construction. This activity can have a major impact on the WARC delegations of these countries.

It should be apparent that WARC preparation requires more than writing papers and talking to delegates and other government officials. These activities are extremely important, but we must also design satellite systems not only for amateur communications, but to justify our existence as an internationally recognized service.

Financial Constraints

Last, and certainly not least, is the matter of financing the amateur satellite program. In the good old days of OSCARs I and II, when satellites were relatively simple, most flight hardware was donated. Out-of-pocket expenses incurred in building and launching a spacecraft were generally picked up by the same people doing the volunteer work. As amateur satellites grew more complex and expensive, it became necessary to seek additional donations to help finance the program. In 1962, the informal OSCAR Association incorporated as Project OSCAR and invited hams all over the world interested in the amateur space program to help support the program financially by signing on as members. Insofar as possible, dues were used to pay for flight hardware and an inexpensive newsletter supplying information about the satellite program to members. In the late 1960s, the hub of US amateur satellite construction activities shifted from southern California to Washington, DC, and AMSAT.

We'll now take a brief look at the factors contributing to the cost of a satellite and the actual cost figures for various amateur spacecraft. We'll then take a look at recent sources of support, and the new sources needed to make Phase 4 a reality.

COSTS

The direct expenses involved in placing a spacecraft in orbit include the following:

1) Launch fees
2) Technical expertise (engineering design)
3) Flight hardware (satellite parts)
4) Ground hardware (special test instruments, prototype subsystems, and so on)
5) Construction (salaries or contracted costs for machining, wiring, testing, and so on)
6) Administrative (parts procurement, required technical documentation, user documentation, bookkeeping, and so on)
7) Travel, shipping, customs, communication (telephone, telex, postage)
8) Miscellaneous (launch insurance, liability insurance, and so on)
9) Launch and post-launch operations

Note that this list contains only spacecraft-related expenses; costs of operating an organization, publishing a newsletter or providing membership services have not been included.

The largest single expense associated with placing a commercial satellite in position is the launch, which costs roughly 40 million dollars to geostationary orbit. Early OSCAR spacecraft were launched for free in recognition of their potential benefit to society in the areas of disaster communications, educational applications and scientific investigations. Later, as launching became more a commercial and less a governmental activity, we were asked to pay the additional costs the launch agency incurred in providing us with a ride. NASA and ESA will have to make some provision for non-commercial scientific payloads in the future, and we may be required to pay the going rate for this class of passenger unless we have something of value to barter in return or we volunteer for high-risk missions involving new rocket systems.

Technical expertise has been almost entirely provided by committed volunteers. Many companies in the aerospace industry knowingly contributed to the program in diverse ways such as donating hardware, authorizing computer time, granting access to facilities for vibration and environmental testing, and so on. Costs for ground stations used to control and monitor spacecraft have been borne by the person or group directly involved. However, as the amateur satellite program matures and gains momentum, more and

more expenses have to be picked up by the organization. There are bills for parts, plane tickets, telephone calls, insurance and launch integration costs that must be paid.

Spacecraft Cost

Trying to assign a cost to an OSCAR satellite is a little like playing catch with a glob of Jell-O—it helps to have a good imagination. Since it's nearly impossible to place a fair-market value on volunteer efforts or all contributed goods and services going into a spacecraft, the only thing we can focus on is out-of-pocket expenses that went through the books at AMSAT-NA and similar data from AMSAT-DL, the University of Surrey and JAMSAT. With this caveat, we present the best available figures in Table 4-2.

Past Sources of Support

Financial responsibility for the US amateur satellite program currently rests with AMSAT. It is important that radio amateurs understand that AMSAT and ARRL are separate organizations, each trying to serve the needs of their members as best they can.

The vitality of the amateur satellite program is critically dependent on the existence of a sound financial foundation. There are several important components to AMSAT's base. One component consists of a large number of individuals in the amateur community making modest contributions to AMSAT yearly. Some contribute the basic membership fee; many make significantly larger donations. Amateurs in this group serve a very important function. In addition to contributing to a stable financial base, they demonstrate to IARU member societies, to government agencies being approached for launches, to companies being asked for donations, and to the volunteers working on the project that there is widespread support for the amateur space program in the radio amateur community.

Table 4-2

Amateur Satellite Construction Costs

Satellite	Cost
OSCAR I	$26
Australis-OSCAR 5	$6,000
AMSAT-OSCAR 6	$15,000
AMSAT-OSCAR 7	$38,000
AMSAT-OSCAR 8	$50,000
AMSAT-Phase 3-A	$217,000
UoSAT-OSCAR 9	$100,000
AMSAT-OSCAR 10	$576,000
UoSAT-OSCAR 11	$200,000
AMSAT-OSCAR 13	$385,000
AMSAT-Phase 4-A	$2,500,000[1]

The cost figures listed represent direct cash outlays only. Several spacecraft are not included because figures were not available or only available in an incompatible format.

[1]Estimated in 1987.

Source: D. Jansson, "Spacecraft Technology Trends in the Amateur Satellite Service," *AMSAT-NA Technical Journal*, Vol 1, no. 2, Winter 1987-88.

As a result, these small donations are multiplied manifold. Early in the amateur satellite program, at critical junctures, a few individuals made very generous personal contributions. The ARRL has made several very significant contributions to the program in the past, including a $50,000 grant for the construction of OSCAR 8. Another major contributor has been the ARRL Foundation. Many amateurs are not familiar with the work of this group. The ARRL Foundation is a non-profit, tax-exempt corporation formed to complement the activities of the ARRL. It is engaged in raising funds for several important Amateur Radio projects such as college scholarships and the amateur satellite program.

While current sources of support appear to be able to provide for both Phase 3 and Microsat programs, new sources of support are needed to make Phase 4 a reality. AMSAT has to obtain the backing of a significant segment of the group of active Amateur Radio operators who are not particularly interested in satellites, but who are interested in the services Phase 4 will provide. While this might be easy once the spacecraft is in place, it's going to prove very difficult to do several years before the first Phase 4 satellite is launched. AMSAT also has to reach government agencies, service and professional organizations involved in education, emergency services, disaster relief, technical assistance, student exchange programs, and so on. The resources to make Phase 4 possible do exist. Whether we'll be able to tap them remains to be seen.

Notes

[1] T. Clark and J. Kasser, "Ariane Launch Vehicle Malfunctions, Phase III-A Spacecraft Lost!," *Orbit*, Vol 1, no. 2, Jun/Jul 1980, pp 5-9.

[2] J. A. King, "Phase III: Toward the Ultimate Amateur Satellite,"
Part I, *QST*, Jun 1977, pp 11-14;
Part II, *QST*, Jul 1977, pp 52-55;
Part III, *QST*, Aug 1977, pp 11-13.

[3] J. A. King, "The Third Generation,"
Part I, *Orbit*, Vol 1, no. 3, Sep/Oct 1980, pp 12-18;
Part II, *Orbit*, Vol 1, no. 4, Nov/Dec 1980, pp 12-18.

[4] J. Eberhart, "Satellite Hit By Its Own Rocket," *Science News*, Vol 124, Aug 6, 1983, p 87.

[5] K. Meinzer, "Three Years of Operation with AMSAT OSCAR-10," *AMSAT-DL Journal*, Sep/Oct 1986; English: *OSCAR News*, No. 62, Nov 1986, pp 15-19.

[6] K. Meinzer, "The Radio Links to Phase III-D: An Initial System Concept," *AMSAT-DL Journal*, Vol 14, No. 1; translation appears in *AMSAT-NA Technical Journal*, Vol 1, no. 2, Winter 1987-88, pp 23-26.

[7] M. Sweeting, "The AMSAT Amateur Scientific and Educational Spacecraft—UoSAT," *Orbit*, Vol 2, no. 2, Mar/Apr 1981, pp 13-17. Also see: *The Radio and Electronic Engineer*, Journal of the Institute of Electronic and Radio Engineers (England), Aug/Sep 1982, Vol 52, no. 8/9, Special issue on: "UoSAT—The University of Surrey's Satellite." This issue is highly recommended for anyone seriously interested in satellite design.

[8] T. Atkins, "USSR/Canada Polar Bridge Expedition," *QST*, Jun 1988, pp 62-63.

[9] A. Tropkin, "90 Days Over the Ice," *Soviet Life*, Oct 1988, No. 10(385), pp 2-7, 35.

[10] S. Budin and F. Fekhel, "Amateur VHF/UHF Repeaters," *RADIO*, no. 10, Oct 1975, pp 14-15.

[11] Special Section No. SPA-AA/159/1273 annexed to International Frequency Registration Board Circular No. 1273 dated 12 Jul 1977, submitted by USSR Ministry of Posts and Telecommunications.

[12] V. Dobrozhanskiy, "Radioamateur Satellites; The Repeater: How is it Used?" *RADIO*, no. 9, Sep 1977, pp 23-25. Also see July, Oct and Nov issues of *RADIO* for additional information.

[13] L. Labutin, "The USSR 'Radio' Satellites—Preliminary Results," *RADIO*, no. 5, May 1979, pp 7-8. For a summary of this article in English see: *Telecommunication Journal*, Vol 46, no. X, Oct 1979, pp 638-639.

[14] *ASR*, no. 34, May 31, 1982.

[15] Happenings of the Month, *QST*, Mar 1972, pp 75-76.

[16] P. O'Dell and B. Glassmeyer, "Well Done, W5LFL!," *QST*, Feb 1984, pp 11-14.

[17] P. Courson, "WØORE/*Challenger*: Picture Perfect from Space," *QST*, Oct 1985, pp 47-49.

[18] *ASR* 109, Sep 14, 1985, pp 3-4.

[19] *ASR* 113/114, Dec 12, 1985; *ASR* 115/116, Jan 16, 1986; *QST*, Mar 1984, p 91.

[20] J. King, V. Riportella, R. Wallio, "OSCAR at 25: The Amateur Space Program Comes of Age," *QST*, Dec 1986, pp 15-18. Reprinted in the *ARRL Satellite Anthology*.

[21] J. King, V. Riportella, and R. Wallio, "OSCAR at 25: Beginning of a New Era," *QST*, Jan 1987, pp 41-45. Reprinted in the *ARRL Satellite Anthology*.

[22] Souza, K. A., "The Joint U.S.-U.S.S.R. Biological Satellite Program," *Bio Science*, Vol 29, 1979, pp 160-166.

Chapter 5

Getting Started

If you're considering taking up satellite communications, your first question is, "Where do I start?". Since everyone has their own specific interests and a different equipment and experience base, no book is going to answer this question for you. What we can do here is explain the considerations that are important to every beginner and provide the information you'll need to make key decisions correctly the first time.

IS SATELLITE OPERATION FOR YOU?

Before you invest time and money in setting up a ground station, give some serious thought to whether satellite operation is really for you. Space communications has a certain "glamour" associated with it. But when you get down to the nitty-gritty, long-term appeal will depend on (1) your communications needs (do you really require the reliable and predictable long-distance capabilities of a Phase 3 or Phase 4 spacecraft, or are you attracted to the challenge of contest-style Phase 2 CW/SSB activity?), (2) your technical or scientific interests (do you especially enjoy the technical challenges or scientific aspects of space communications?) and (3) your time and financial resources.

It shouldn't be too difficult to locate a few local hams who have had some experience working with satellites. Ask them how they feel about it. Some will probably see satellites as the most exciting new dimension of Amateur Radio since the discovery, back in the early '20s, that "useless" short waves could propagate across the Atlantic. Others may firmly believe that satellite relays are as exciting as the telephone and expect them to have the same future in Amateur Radio as AM on 20 meters. Talk to advocates of both viewpoints and try to get at why they feel the way they do. If at all possible, visit an active satellite user during a satellite pass. Observing a ground station in operation firsthand is the best way to get a feel for what's involved. If the bug still has you in its clutches at this point, you may as well give in and start making plans to set up your own ground station.

Even though the International Telecommunication Union recognizes the amateur-satellite service as being distinct from the amateur service, you do *not* have to pass any new FCC test to use satellites. Your Amateur Radio license is your ticket to satellite operating.

FIRST STEPS

Assembling a ground station generally involves several steps.

1) Learn all you can about satellite communication.

2) Select your first goal: (A) CW/SSB or digital?, (B) high- or low-altitude?, and (C) mode and satellite(s). Keep in mind that any equipment acquired should be suitable for later objectives.

3) Set up a receive station. Use the receive station to familiarize yourself with satellite operating procedures and the basics of tracking.

4) Set up for transmission (CW, SSB or packet).

Parts I and II of this book are meant to provide you with most of the background information you need to make your selections. It's not necessary to stop and study everything at this time. You'll probably find it more helpful to skim large sections now so that later, as you need access to specific material, you'll know where to find it. In addition to the information here, you'll also need to know what satellites are currently available (new spacecraft may have been launched and others may have been put out to pasture since this book went to press) and you'll have to obtain some recent tracking data. The information on satellite availability will be found in monthly periodicals of ARRL and AMSAT. The tracking data can be obtained from newsletters, HF nets, computer bulletin boards and other sources listed in Chapter 10.

Having read about the history of the amateur satellite program, you already know a great deal about satellite communication systems, including the different capabilities of Phase 2 (low earth orbit) and Phase 3 (high altitude) spacecraft. Your initial ground station setup will probably be designed to give you access to a particular mode (set of uplink and downlink frequencies) or satellite. Mode and satellite selection are so closely related that they should be done together. Be sure to at least skim the remainder of Part II before committing yourself to a particular mode or spacecraft.

THE BASIC STATION

A satellite ground station and an HF or VHF station designed for regular terrestrial communication (either

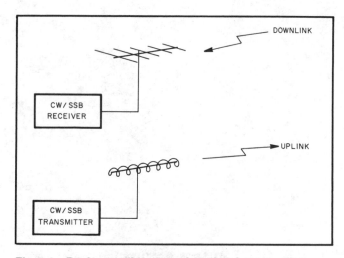

Fig 5-1—Basic satellite ground station for operating through a linear transponder.

direct or via repeater) have a lot in common. Both require a transmitter, a transmitting antenna, a receiver and a receiving antenna. (See Fig 5-1.) In the following sections we'll look at the frequencies used, the power levels required, the types of modulation employed, and so on. Naturally, there are also many differences between a satellite ground station and a station for terrestrial work, some subtle, some obvious. As we focus on these differences, don't forget the basic, underlying similarity.

Our analysis of the Amateur Radio satellite ground station looks at the uplink and downlink frequencies that will be in general use in the early and mid 1990s. Therefore, we'll be considering receiving equipment for 29, 145, 435 MHz and 2.4 GHz, transmitting equipment for 21, 146, 435 and 1269 MHz, and antennas for all of these frequencies.

As all current and planned satellite transponders are designed for cross-band operation, separate receive and transmit antennas will be required. For several reasons, which we'll discuss later, satellite operation also requires you to have a separate receiver and transmitter. If you currently own an HF or VHF transceiver, it can be used to satisfy (partially or completely) the receive or transmit requirements for your ground station, but not both.

CHOOSING: CW/SSB OR PACKET

The heart of a radio amateur satellite is the transponder. There are two types of transponders currently carried by amateur satellites: linear transponders and digital transponders. You need a brief understanding of what these transponders do before we go any further.

The Linear Transponder

A linear transponder takes a slice of the RF spectrum that is centered about a particular frequency, amplifies the entire slice and retransmits it centered about a different frequency. For example, the incoming slice might be a 100-kHz-wide segment centered about 145.950 MHz, containing dozens of SSB and CW signals. The slice is amplified a million-million times, and then retransmitted

as a 100-kHz-wide segment centered about 29.450 MHz. A linear transponder used on Amateur Radio satellites accepts any type of signal—CW, SSB, FM, digital, video, and so on. Each signal is retransmitted in its original format, but shifted in frequency and greatly amplified. Although a linear transponder will work with any type of signal, it should only be used for CW and SSB due to the limited power available on the spacecraft. (Other types of modulation may be used for special experiments.)

Linear transponders provide real-time communication. There is a short delay due to the distance between satellite and ground station, but even with high-altitude satellites this is only a fraction of a second.

A simple formula associated with each transponder enables one to predict the approximate downlink frequency corresponding to each uplink frequency. Actual downlink frequencies may vary by several kilohertz because of a phenomenon known as Doppler shift. Doppler shift refers to the fact that, on a link where a transmitter and receiver are in motion relative to one another, the received frequency is shifted from that transmitted. For more details, see the chapter on "Satellite Radio Links." The translation formula for each transponder is given in Appendix B.

There are two basic types of linear transponders: the noninverting transponder, which retransmits the entire slice as received (see Fig 5-2[A]), and the inverting transponder, which flips the slice (see Fig 5-2[B]). With a noninverting transponder, a signal with a frequency near the high end of the uplink passband comes back down near the *high*-frequency end of the downlink. With an inverting transponder, a signal with a frequency near the high end of the uplink passband comes back down near the *low*-frequency end of the downlink. Although frequency inversion may appear to be an unnecessary complication, it does have an important function—it reduces the magnitude of observed Doppler shifts. Noninverting transponders are used mainly for Mode A, since the Doppler shift on this mode is small. Other modes generally use inverting transponders. (Mode S on OSCAR 13 is an exception.)

When communicating through a linear transponder, you use standard CW/SSB transmitters and receivers for the uplink and the downlink. With a noninverting transponder, an upper-sideband uplink returns as an upper-sideband downlink. With an inverting transponder, a lower-sideband uplink returns as an upper-sideband downlink. The usual practice is to select the uplink sideband so that the downlink will be upper sideband.

The Digital Transponder

A digital transponder is a nonlinear device optimized to receive, process and retransmit digital data. Several types of digital transponders have been flown on amateur spacecraft. Our primary interest here is in the units used on OSCARs 16, 19 and 20, which were optimized for store-and-forward bulletin board and electronic mail activities. Insofar as possible, these transponders have been designed so that ground stations could use the same equipment and standards employed for terrestrial packet radio. However, certain characteristics of a communications system involv-

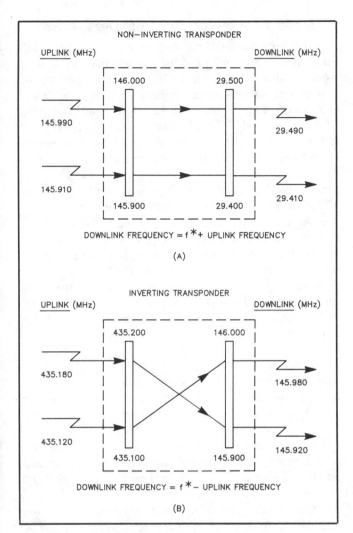

Fig 5-2—A noninverting transponder is shown in A. An inverting transponder is shown in B. Values of f* for each transponder currently in orbit are included in Appendix B.

ing low-altitude satellites make complete compatibility undesirable.

Current spacecraft digital transponders have several input channels and a single output channel. (The numbers are chosen before launch to balance data flow to and from the spacecraft.) The uplink channels and the downlink channels are characterized by their data rate and modulation scheme. These factors can be chosen to provide the best link performance, or to be compatible with equipment already used by amateurs, or some balance between these two criteria. It's even possible to change certain transponder operating characteristics such as data rate after launch by uplinking special command programs. The data rates and type of modulation may differ on the up- and downlinks.

A satellite digital transponder may be operated much like a terrestrial computer bulletin board with a "mailbox." You send messages up to the satellite. These messages are stored onboard. Ground stations can request that a message be read out over the downlink as the spacecraft passes near them. When a message is read out on the down-

link, it is not erased from the satellite's memory, so that many stations can request a readout. At some point, after a certain amount of time elapses or at the request of the originator or receiver, the message must be erased from the satellite's memory so that new messages can be added. A satellite with these features is said to have a store-and-forward communications system.

A digital transponder can also be optimized for digipeating. An example is the RUDAK I unit carried on OSCAR 13 and an RS spacecraft currently awaiting launch. We'll discuss this type of digital transponder later.

A satellite that carries a store-and-forward digital transponder designed so that it's relatively similar to terrestrial packet repeaters is frequently called a pacsat (*pac*ket radio *sat*ellite). Note that pacsat (lower-case p) is a generic term used to refer to any satellite using a digital store-and-forward message system. Pacsat (upper-case P) refers specifically to OSCAR 16, which was known as MicroSat-A before launch.

Several amateur satellites have carried experimental digital transponders which were not available for general use. The Codestore capability of various early OSCARs was a simple digital transponder. The DCE (Digital Communications Experiment) on UoSATs 9 and 11 was also a digital transponder. The RUDAK unit on OSCAR 13 (currently not operable) is a digital transponder optimized for real-time (digipeater) operation. To use any of these transponders, a ground station must have special uplink and downlink equipment.

In the discussion that follows we will focus on the equipment required to use the open-access pacsats. The first amateur satellite of this type was Fuji-OSCAR 12, launched in 1986. Because FO-12 had a very limited operating schedule (due to power budget problems), only a few hundred amateurs used it for two-way communication. This situation is changing rapidly now that OSCARs 16, 19 and 20 are in orbit.

The ground station equipment required to use a pacsat is discussed in detail in Chapter 9. Referring to Fig 5-1, we must replace the CW/SSB transmitter with an FM unit and add several items, including a standard terminal node controller (TNC), a special pacsat modem and a dumb terminal or computer. An Amateur Radio station that is already operating 2-meter packet will already possess the transmitter, TNC and terminal.

Suggestion: If you're currently operating 2-meter packet, then it's reasonable to make a move to packet satellite operations since you already have much of the equipment and operating experience required. If you've never operated packet and you want to operate satellite packet, you should first gain some experience operating terrestrial packet on 2 meters.

CHOOSING: HIGH- OR LOW-ALTITUDE SPACECRAFT

If you want to engage in real-time CW/SSB communications, then you're going to be interested in satellites carrying linear transponders, and you'll have a choice of working with high- or low-altitude satellites. If you wish to operate satellite packet radio, you need a spacecraft with the appropriate digital transponder. In 1990, the only satellites carrying this type of transponder are in low earth orbit.

Because satellite-to-ground-station distances are shorter with low-altitude satellites, you'll find that ground stations generally need less transmitter power and lower-gain antennas to access them. At 2 meters, 10 to 50 W of RF to an omnidirectional uplink antenna often provides good results. Low-altitude satellites have a number of limitations. Coverage distance for real-time CW/SSB operation is modest. And, since these satellites are only in range for about 4 to 6 brief passes (10 to 20 minutes each) per day, it's important to be able to predict when you'll be able to hear them. This is the main concern of tracking, a topic that we'll be covering in the next chapter.

High-altitude satellites are generally in view for many hours each day. For example, AO-13 will be in view of US stations for more than 10 hours per day until the mid 1990s. With this type of access, it's possible for the casual operator to dispense with tracking—just turn on the receiver and if weak signals are heard adjust the antenna by peaking the received signals. Since a high-altitude satellite is in view of a large region of the earth, communication over large terrestrial distances is possible. Because of the long satellite-to-ground-station distances involved, the ground station has to employ more power and/or higher-gain antennas than when using low-altitude spacecraft. At 435 MHz, for example, one might need 50 W and a 13-dBd antenna.

For various reasons, including ionospheric characteristics, path loss, satellite antenna requirements, Doppler shift, and so on, the frequencies used for low- and high-altitude satellites tend to be somewhat different (though there is considerable overlap).

Suggestions

In most cases, the easiest and cheapest way to get started in satellite communications is by working with CW/SSB and a low-altitude spacecraft. Keep in mind the fact that the communications capabilities you'll be acquiring are inherently limited. If you have to make a substantial investment in time or money to go this route, be sure to consider whether it may be more desirable to make a slightly larger investment to gain access to a high-altitude spacecraft. Also, if some of the equipment needed to work with a high-altitude spacecraft is already on hand, your decision as to where to begin should take this into account.

CHOOSING: MODES AND SATELLITES

All current and planned transponders on amateur satellites receive signals on one band and retransmit them on a different band. Each combination of frequency bands is called a *Mode*. The various modes used on amateur satellites are listed in Table 5-1. In the discussion that follows, assume that all satellite transponders mentioned are linear units suitable for CW and SSB unless they're specifically described as being for digital communications.

Mode A—2 Meters Up/10 Meters Down

If you own an HF receiver or transceiver, monitoring the 10-meter downlink of Mode A is an easy way to start working with satellites. In the 1972-1983 time period, the majority of satellite operators began this way. AMSAT-OSCARs 6, 7 and 8, and all RS satellites, had downlinks between 29.3 and 29.5 MHz. The popularity of this mode is due to the fact that it provided a quick and inexpensive route to satellite communications for amateurs migrating from the HF bands. If your HF receiver already tunes the correct frequencies and has a low-noise front end, you're all set to listen—the only question is when and to what precise frequencies. However, you may find that your receiver requires an accessory crystal to cover the upper end of the 10-meter band. A good low-noise preamp, especially one placed at the antenna, often produces a dramatic improvement in reception. If you're considering this approach, check *QST* to determine which currently active satellites have Mode A transponders. Mode A has only been used on low earth orbit satellites.

As this is written (mid 1990), the only spacecraft operating on Mode A is RS-10/11. See Table 5-2 for specific frequencies. Mode A is scheduled daily. RS-12/13 is due to be launched in 1990. So, if all goes well, as you read this there may be two Mode A transponders in orbit. Operating continuously, these spacecraft will provide users at mid latitudes a total of 3 to 4 hours of Mode A access each day. As you can see from Table 5-2, there are several additional spacecraft that may be heard between 29.3 and 29.5 MHz on an irregular basis.

Once you're able to hear the Mode A downlink fairly well, you'll no doubt want to contact some of the CW or SSB stations you've been listening to. To do so, you'll have to generate a CW or SSB signal on 2 meters. With low-altitude satellites, the power requirements are generally modest—on CW, 10 W often suffices. Several options for uplinking are discussed in the chapter on Receiving and Transmitting Equipment. Many amateurs made their first few Mode A QSOs by unplugging the mike on a 2-meter FM rig and keying the push-to-talk line. Please do not use FM on the uplink. This imposes an excessive power drain on the spacecraft and can be damaging.

Although Mode A is relatively easy to get on, it has several limitations which place severe constraints on what you'll be able to do once you get there. These constraints are due to two facts: the radio link employs an HF frequency, and the satellite is in a low-altitude orbit. The 29-MHz satellite downlink is very susceptible to ionospheric conditions. This is especially noticeable near peaks in the sunspot cycle. When the sun is active, the 10-meter downlink is very unreliable and unpredictable—signals may be excellent one minute and fade completely out of the picture a few minutes later. The same conditions that provide 10-meter HF operators with excellent worldwide skip can make the ionosphere nearly opaque to the satellite 10-meter downlink. Strange effects are often noted—one might hear solid downlink signals when the spacecraft is far out of normal range. Of course, many Mode A operators find the unpredictable nature of the 29-MHz downlink

Table 5-1

Transponder Mode Designations

Mode	Ground station transmit band (uplink)[1,2]	Ground station receive band (downlink)[1,2]	Current satellites[3] (8/90)	Future satellites (1990/91 only)
A	145 MHz	29 MHz	RS-10/11	RS-12/13
B	435 MHz	145 MHz	AO-10, AO-13	RS-14
J	145 MHz	435 MHz	FO-20	
JA[4]	145 MHz	435 MHz	FO-20	
JD[5]	145 MHz	435 MHz	FO-20 OSCARs 16, 19 OSCAR 18[6] UoSAT-OSCAR 14	
JL	1.2 GHz & 145 MHz	435 MHz	AO-13	
K	21 MHz	29 MHz	RS-10/11	RS-12/13
KA	21 MHz & 145 MHz	29 MHz	RS-10/11	RS-12/13
KT	21 MHz	29 MHz & 145 MHz	RS-10/11	RS-12/13
L	1.2 GHz	435 MHz	AO-13	
S	1.2 GHz	2.4 GHz	AO-13[7]	
T	21 MHz	145 MHz	RS-10/11	RS-12/13
U[8] (see Mode B)				

Notes

[1]Uplink is always specified first.

[2]Each spacecraft transponder uses a portion of the band. See Table 5-2 for specific frequency information. Bands are sometimes labeled in terms of wavelength as follows:

Band (frequency)	Frequency limits	Band (wavelength)
21 MHz	21.000- 21.450 MHz	15 meters
29 MHz	28.000- 29.700 MHz	10 meters
145 MHz	144.000-146.000 MHz	2 meters
435 MHz	435.000-438.000 MHz	70 cm
1.2 GHz	1.260- 1.270 GHz	24 cm
2.4 GHz	2.400- 2.450 GHz	13 cm

[3]This information should be updated by checking recent issues of *QST*.

[4]JA is short for J (A)nalog. Same as Mode J.

[5]JD is short for J (D)igital. Same input/output frequency bands as Mode J.

[6]The Mode JD transponder on Webersat-OSCAR 18 is a secondary payload.

[7]The transponder on AO-13 is *not* a true Mode S unit. It employs a 435-MHz uplink and is mainly designed to gather experience with the 2.4-GHz downlink.

[8]This is the same as Mode B. Since the West German amateurs who built the 435/145-MHz transponder often refer to it as the U-transponder, Mode B is sometimes (infrequently) referred to as Mode U.

one of the appealing attributes of Mode A. In sum, although Mode A is, in many cases, the easiest choice for the beginner, it does have several limitations. As a newcomer, you should at least consider the other modes before choosing your starting point. (See additional comments under Mode K.)

Mode K—15 Meters Up/10 Meters Down

Mode K is a relatively new mode that first appeared on the ISKRA spacecraft and then on RS-10/11 and RS-12/13. It's of interest both to newcomers and to serious experimenters involved in propagation studies. All of the limitations of HF satellite links and low-altitude satellites discussed earlier apply here also. If you already have a receiver and transmitter (not a transceiver) for HF, you're all set for Mode K. Mode K has generally been operated on Tuesdays through Fridays only, and in conjunction with

Mode A. It's assumed that the 15-meter uplink is being kept off on weekends when crowded HF conditions make unintentional QRM between those operating HF direct and those operating by satellite more likely. The combined Mode K, Mode A operation is called Mode KA. When Mode KA is activated, ground stations listen on 10 meters and have a choice of transmitting on either the 15-meter or 2-meter uplinks. In fact, a QSO can take place with one station transmitting on 2 meters and the other on 15 meters.

Stations who have had a chance to try it both ways report that the 2-meter uplink (Mode A) provides more reliable results than 15 meters (Mode K).

Mode KA—15 Meters and 2 Meters Up/10 Meters Down

See comments under Mode K.

Table 5-2

Radio Amateur Satellite Downlink Frequencies (as of 8/90)

Frequency (MHz)	S/C	(T)ransponder (B)eacon (R)obot	Comment
29.331	RS-5	R/B	not in regular operation
29.341	RS-7	R/B	not in regular operation
29.357	RS-10	B	Modes A/K/KA/KT
29.360-29.400	RS-10	T	Modes A/K/KA/KT
29.401	RS-1	B	not in regular operation
29.403	RS-10	R/B	Modes A/K/KA/KT
29.407	RS-11	B	Modes A/K/KA/KT
29.408	RS-12	B	Modes A/K/KA/KT (to be launched)
29.410-29.450	RS-5	T	not in regular operation
	RS-11	T	Modes A/K/KA/KT
	RS-12	T	Modes A/K/KA/KT (to be launched)
29.453	RS-11	R/B	Modes A/K/KA/KT
29.454	RS-12	R/B	Modes A/K/KA/KT (to be launched)
29.458	RS-13	B	Modes A/K/KA/KT (to be launched)
29.460-29.500	RS-7	T	not in regular operation
	RS-13	T	Modes A/K/KA/KT (to be launched)
29.504	RS-13	R/B	Modes A/K/KA/KT (to be launched)
145.810	AO-10	B	Mode B
145.812	AO-13	B	Mode B
145.822	RS-14	B	CW TTY
145.824	DO-17	B	synthesized speech
145.825	UO-11	B	may be synthesized speech
145.825-145.975	AO-10	T	Mode B
	AO-13	T	Mode B
145.852-145.932	RS-14	T	
145.857	RS-10	B	Mode T/KT
145.860-145.900	RS-10	T	Mode T/KT
145.862	RS-13	B	Mode T/KT (to be launched) Mode B (transponder 1)
145.866-145.946	RS-14	T	Mode B (transponder 2)
145.903	RS-10	R/B	Mode T/KT
145.907	RS-11	B	Mode T/KT
145.908	RS-13	R/B	Mode T/KT (to be launched)
145.910-145.950	RS-11	T	Mode T/KT
145.913	RS-12	B	Mode T/KT (to be launched)
145.952	RS-14	B	PSK TTY
145.953	RS-11	R/B	Mode T/KT
145.959	RS-12	R/B	Mode T/KT (to be launched)
145.983	RS-14	T	Multiple digital modes
145.985	AO-13	B	Mode B engineering beacon, usually off
145.987	AO-10	B	Mode B engineering beacon, usually off

Mode B—70 Centimeters Up/2 Meters Down

Mode B is suitable for low- and high-altitude satellites. It was first used on the low-altitude spacecraft OSCAR 7. During the six-plus years that OSCAR 7 was in operation, amateurs had a chance to compare the operation of Mode A and Mode B. Mode B was clearly superior. In mid-1990, two spacecraft, both in the high-altitude Phase 3 series, carried Mode B transponders—AO-10 and AO-13. Even though the control computer on AO-10 has been destroyed by radiation, it is still possible for command stations to activate the Mode B transponder and omnidirectional antenna. AO-10 is providing far less service than in its early life when it was fully operational, but it still provides better service than a low-altitude spacecraft like OSCAR 7 did, that is, averaged over a time span of several months AO-10 provides more access time and higher signal levels. Meanwhile, AO-13 yields excellent Mode B operation about 18 hours per day. Of course, a single ground station can't access it for the full 18 hours, but one can expect to be able to use it for eight or more hours per day for far into the future.

Mode B provides reliable, predictable communications. It's an excellent link and it's available a great many hours per day. For the next five to eight years, Mode B is likely to remain the mode of choice for anyone wanting first-class satellite CW/SSB communications. After that, it should remain popular and supported by satellite transponders for many more years, but other modes will take the lead. This is due to two serious problems with Mode B. First, Mode B is limited to a width of 200 kHz due to constraints on the 146-MHz downlink. This is already being nearly fully utilized. When the number of stations using Mode B becomes too large, which it eventually will, there will be very few options for expansion. Strategies to

Frequency (MHz)	S/C	(T)ransponder (B)eacon (R)obot	Comment
435.025	UO-11	B	
435.070	UO-14	B	
435.120	UO-15	B	(not operating 3/90)
435.651	AO-13	B	Mode L, JL
435.677	AO-13		RUDAK
435.715-436.005	AO-13	T	Mode L, JL
435.795	FO-20	B	Mode J Analog
435.800-435.900	FO-20	T	Mode J Analog
435.91	FO-20	T/B	Mode J Digital
436.028	AO-10	B	Mode L
436.048	AO-10	B	Mode L engineering beacon, usually off
436.150-436.950	AO-10	T	Mode L
437.026	AO-16	T/B	Mode J Digital (secondary)
437.051	AO-16	T/B	Mode J Digital (primary)
437.075	WO-18	T/B	Mode J Digital (secondary)
437.102	WO-18	T/B	Mode J Digital (primary)
437.125	LO-19	T/B	Mode J Digital (secondary)
437.127	LO-19	B	CW (TSFR module)
437.154	LO-19	T/B	Mode J Digital (primary)
2400.664	AO-13	B	Mode S
2400.711-2400.749	AO-13	T	Mode S
2401.143	AO-16	B	usually off
2401.221	DO-17	B	usually off
2401.500	UO-11	B	usually off

Notes

Transponders employing 29-MHz downlink are noninverting. Most other transponders are inverting.

RS-10 and RS-11 do not operate simultaneously

RS-12 and RS-13 do not operate simultaneously

RS-1, RS-5, RS-7 have been withdrawn from regular service, but may be heard occasionally

AO-10 is only in limited operation

get more use out of these frequencies are already being considered; for example, producing new Phase 3 spacecraft with 12-hour periods and precision orbital placement/control so we can keep satellites on opposite sides of the earth. Keep in mind the fact that the three MHz available to Mode L can support 15 times as many users as Mode B. The second serious problem with Mode B is that in some parts of the world QRM from terrestrial stations operating between 145.800 and 146.000 MHz makes the reception of weak satellite downlinks nearly impossible. This situation is likely to get worse rather than better.

Mode L—23 Centimeters Up/ 70 Centimeters Down

Mode L is especially well suited to high-altitude satellites. AO-13, the only satellite currently using Mode L on a regularly scheduled basis, operates this mode about 3 hours per day near the apogee of each orbit. Most northern hemisphere ground stations will have access to Mode L about 1.5 hours per day.

Listening to the AO-13 Mode L downlink provides a tantalizing glimpse into the future. If the transmitting station is producing a clean signal the received signal quality is superb—no flutter, no fading, no multipath distortion. This is due in large part to the spacecraft antennas. Because antennas for Mode L are physically much smaller than for Mode B, it is possible to achieve

better Mode L performance on a Phase 3 spacecraft through better antenna design and placement. This makes Mode L inherently superior to Mode B. When this fact is coupled with the 3-MHz bandwidth available, it becomes clear why Mode L will one day be the most popular satellite mode for CW/SSB communications. It's expected that by the late 1990s, after Phase 3-D is orbited, the number of Mode L users will exceed the number of Mode B users.

The Mode L receiver on AO-13 is not as sensitive as expected. Therefore, ground stations must produce a higher effective radiated power than pre-launch calculations suggested. This, and the modest amount of daily access time, tend to keep down the numbers of people currently operating Mode L. A survey of approximately 200 OSCAR 13 Mode L users during its first three months of operation produced the following information on 23-cm uplink equipment. With respect to transmit power: 30% were using less than 16 watts, 35% were using 16-64 watts, 34% were using more than 64 W. Quite a range of antennas were in use: 38% Yagi (or Yagi array), 30% parabolic dish, 24% loop Yagi (or loop-Yagi array), 7% helix.

Although there was only one operating Mode L transponder in orbit in the late 1980s, Mode L will probably be the prime mode on Phase 3-D and a primary mode on Phase 4 spacecraft. Mode L is not likely to be the first mode you choose to set up for, but it's important to understand Mode L's place in future plans, so that any major purchases you're considering can take into account the fact that you may one day want to gain access to it. Once Phase 3-D is in orbit, this is likely to be regarded as the most desirable mode for SSB/CW satellite communications. All the equipment needed to set up an SSB/CW station for Mode L operation can be purchased off the shelf in 1990.

Mode J (Also Called Mode JA)—2 Meters Up/ 70 Centimeters Down

Mode J is suitable for both low- and high-altitude spacecraft. It was first flown on OSCAR 8 using a transponder built by JAMSAT. More recently, Mode J has been used on FO-12 and FO-20. On JAMSAT spacecraft, this mode is often referred to as JA (J *Analog*) to distinguish it from JD (J *Digital*), which employs the same frequency combination. The Mode JA transponder currently in orbit on FO-20 is an excellent performer. Since modes JA and JD are usually alternated, Mode JA operating time is limited.

Radio amateurs also have limited access to a Mode J transponder on the high-altitude spacecraft OSCAR 13. The OSCAR 13 Mode L transponder has a secondary input at 2 meters which can be operated in conjunction with the normal Mode L input. This option was designed to provide Mode L access to radio amateurs in countries where 23-cm transmitting equipment is not available. When both inputs are active, the mode is referred to as JL. Mode JL operation only occurs part of the time that Mode L is active. The groups responsible for building and operating OSCAR 13 request that radio amateurs in developed countries refrain from using the 2-meter uplink on a regular basis (occasional experimental use is acceptable). Compliance with this request should strengthen support for the amateur-satellite service at future WARCs.

Because of the limited operating time currently (1990) available for Mode J operation, it does not appear appropriate to invest heavily in an analog Mode J station. However, if you already have equipment for the Mode J link frequencies (as many packet operators do), give FO-20 (Mode JA) or OSCAR 13 (Mode JL) a try—you'll be pleasantly surprised by the performance.

Mode JD—2 Meters Up/70 Centimeters Down (Digital Transponder)

Mode JD uses the same frequency bands as Mode J. The transponders on OSCARs 16, 19 and 20 employ four input channels and one output channel. Center frequencies are listed in Table 5-2. In early 1990, OSCARs 16 and 19 are providing round-the-clock mode JD operation and Fuji-OSCAR 20 operates in Mode JD more than 50% of the time.

Let's assume that you are already equipped to operate terrestrial packet on 2 meters and that you want to try satellite packet. What additional equipment is required? You'll need to build or purchase a special pacsat modem to connect to your TNC. This device will augment the modulator in the TNC modem on transmit, replace the demodulator in the TNC modem on receive, and provide

the signals needed to enable your receiver to track the satellite downlink frequencies as they shift due to Doppler. In the US, kit modems are available from TAPR and Radio-Kit (G3RUH design), and a ready-to-run model is available from PacComm. You'll also need SSB receive capabilities on 437 MHz. If you use a modern multimode single-band transceiver for the 70-cm downlink, automatic frequency tracking will be fairly easy to accomplish since these transceivers are designed to accept TTL-level signals from a computer for frequency control. Requirements are discussed in more detail in Chapter 9.

As you can tell from this brief introduction, communicating via packet satellite during the first few months after the launch of OSCARs 16 and 19 is challenging. However, now that the spacecraft are in orbit and operating, key pieces of equipment will quickly become commercially available and procedures will be standardized and publicized. When this occurs, it will be possible to purchase all the equipment needed to set up a ground station for packet satellite off the shelf. Those who are not savvy packet experimenters may avoid a great deal of frustration if they wait until pacsat's spacecraft software becomes more mature before setting up for this mode. When this occurs, detailed beginners' articles should appear in AMSAT publications and *QST*.

Mode JL

See comments under Mode J.

Mode S—24 Centimeters Up/13 Centimeters Down

(AO-13 only: 70 Centimeters Up/13 Centimeters Down)

The fact that Mode L will be needed one day, in the not-too-distant future, to alleviate overcrowding of Mode B has already been discussed. Since Mode L, the mode of the future, has 15 times the available bandwidth of Mode B, do we really have to worry about eventual overcrowding of Mode L? If we only consider CW and SSB communications, maybe a little slow-scan TV and point-to-point packet, the answer is probably no. Then why do we need Mode S? Mode S will permit a quantum jump in what we, as radio amateurs, can do. For example, the 10 MHz of spectrum available will permit fast-scan TV between continents, transfers of huge digital files in minutes instead of the days or weeks that would be required if they were handled piecemeal by Phase 2 pacsats, interlinking of Phase 2 pacsats, the construction of very small portable ground stations for transport to emergency areas, and so on. Many of the activities envisioned for Mode S do not require the individual radio amateur to set up a station for Mode S. The services will be provided by gateway stations associated with local terrestrial 2-meter and 70-cm FM repeaters. As a result, in the mid to late 1990s, the number of amateurs set up to directly access satellite Mode L transponders might be much larger than those similarly equipped for Mode S. However, Mode S, via a system of gateways, may be serving many more radio amateurs. For additional information on Mode S, see the discussion of the Phase 4 Program in Chapter 4.

At present, Mode S is certainly not for beginners, or amateurs primarily interested in communicating. Mode S is for the serious experimenter who wants to contribute to the future of Amateur Radio. If we want Mode S transponders to be a central component in Phase 4 spacecraft, some of us have to develop transponder and ground-station technology today. And, as discussed in Chapter 4, we have to start populating the frequencies involved and utilize them on satellites before they disappear at a future WARC. If we don't plan carefully and aggressively for future needs, these valuable frequencies are likely to be lost forever.

AO-13 is the first satellite to carry a Mode S transponder (actually a hybrid Mode S transponder since the uplink is not at 24 cm). This transponder is performing well and has already served one of its primary functions, which was to provide information on the performance of a 13-cm downlink from geostationary altitude. At present, Mode S is scheduled on a regular, but limited, basis. Note that the experimental Mode S transponder on AO-13 is noninverting.

Mode T—15 Meters Up/2 Meters Down

Mode T appears on RS-10/11 and RS-12/13. Because it's only been activated a very small percentage of the time, it can't be considered a possible starting point for beginners. However, many satellite operators have the equipment to operate this mode, so check current operating schedules to see if it's available and give it a shot if it's on. The results should be interesting. When the spacecraft is in range you may hear stations all around the world being accidentally relayed back down on 2 meters.

Mode KT—15 Meters Up/10 and 2 Meters Down

See comments under mode T.

Receive-Only Modes

Most amateurs are interested in two-way communications, so our focus to this point has been on transponders. There are, however, situations where one might be primarily interested in a downlink beacon. For example, some amateurs collect and analyze spacecraft telemetry; others study propagation; others might be involved in educational demonstrations. This section has been included so that some potentially interesting amateur satellite downlinks can be mentioned. Table 5-2 includes all amateur satellite beacons in current operation.

The UoSAT-11 145.825-MHz beacon is often in the synthesized-speech mode. It provides a strong signal to ground stations using simple receiving equipment. During the 1988 Skitrek Polar expedition, which involved a group of Soviet and Canadian explorers traveling from Cape Arktichesky in the Severnaya Zemlya Archipelago to Ward Hunt Island in Canada via the North Pole, this beacon was used to provide the expedition members with a daily report on their position as determined by the SARSAT/COSPAS Search and Rescue satellite system. By monitoring the UoSAT-11 downlink, amateurs and non-amateurs around the world were able to follow the expedition's daily

progress.

DOVE-OSCAR 17 contains a single voice-synthesized beacon that is operated continuously on 145.825 MHz. It produces a reliable, strong downlink which can be received on simple equipment. The better scanner radios will generally provide good reception if they can tune to the downlink frequency. The characteristics of DOVE make it excellent for demonstrations. Beacon messages will be provided in many languages and will focus on national events and holidays related to themes of peace, freedom and brotherhood among all peoples of the world. The beacon will also be used for special activities like Skitrek.

For information on other telemetry beacons see the Chapter on Satellite Systems and Appendix B: Spacecraft Profiles.

STARTING OUT: FOUR CASE STUDIES

Ham A

Subject A is an HF operator with broad operating interests that include a little DXing, some contesting and lots of casual rag chewing. He splits his time about 50/50 between SSB and CW and operates all bands from 80 to 10 meters using a 100-W transceiver and a collection of dipoles and ground planes. The shack also contains a 25-W 2-meter FM rig.

Subject A is interested in trying something new, but he doesn't want to spend much money. While he's not really into building equipment, he isn't afraid of taking the cover off a complex rig and making small modifications.

Subject A decides to try Mode A. He constructs a small Oscarlocator using the techniques described in the next chapter so that he can predict when RS-10/11 and RS-12/13 will be in range. He connects his regular 10-meter ground-plane antenna to the transceiver and tunes the appropriate 10-meter frequencies (see Table 5-2 or Appendix B: Spacecraft Profiles) when the satellite is predicted to be in range. Many solid signals, both CW and SSB, are heard. Some are engaged in brief contest-like QSOs, while others are rag chewing.

Liking what he hears, he unplugs the mike from his 2-meter rig, turns the mike gain down to zero and wires the push-to-talk line to his straight key using alligator clip leads. The usual 2-meter antenna is a large Yagi, but he knows that things get pretty hectic during a pass and that he won't have time to aim the beam, so he runs a spare cable out the window to the 2-meter ground-plane on his car sitting in the driveway. Everything is completed before the next pass. He picks a transmit frequency slightly below the middle of the uplink passband. As soon as the satellite comes in range, he starts transmitting dits and adjusts the frequency while listening for his own downlink. Sure enough, he finds it. A quick CQ and he makes his first contact. Three more QSOs follow before the end of the pass. He thinks to himself—46 more states and I've got WAS!

Ham B

Subject B is a 2-meter VHF/UHF packet operator who's addicted to the technical side of computers and ham radio. He owns the latest whiz-bang multimode transceivers for 2 meters and 70 cm and his shack is littered with terminals, computers and TNCs.

There's no question about what aspect of satellites appeals to Subject B. Having just seen a demonstration of packet satellite operation and learned that several low-altitude pacsats are in orbit and operating 24 hours per day, it's Mode JD here I come. He orders a software tracking disk direct from AMSAT to make sure he's got the latest version. Since he writes much of his own software, he appreciates the effort that goes into producing quality work, so he's pleased to make the requested contribution, especially since he knows it will go, in large part, to pay for future pacsats. He also purchases a special modem which connects to his TNC. For initial testing he disconnects his 2-meter and 70-cm beams and temporarily replaces them with simple ground planes. The special modem contains circuitry to adjust his receive frequency during a satellite pass, but the output polarities and waveforms don't match what his rig requires, so he grabs a couple of 555s and inverters and wire wraps an adapter. He then solders on a few connectors and plugs everything together. After it's checked out, there'll be plenty of time to neaten up the rat's nest.

It takes a little while to get everything working together. When things appear to be operating correctly, he gives a listen on the downlink and gets good copy. Switching between LSB and USB doesn't seem to have any effect. (It shouldn't if the receiver crystal filter is symmetric and things are working properly.) Next pass he tries transmitting and the results are somewhat erratic. Between passes, he reviews the information that came with the satellite modem and notes that it suggests that slight underdeviation of the transmitter, say ±3 kHz, often provides better results than normal deviation, so he adjusts his rig. Next pass, everything works perfectly.

Whereas most hams would be ready to sit down and pat themselves on the back at this point, Subject B sees his job as just beginning. He wants to write some software so his computer can send or retrieve his pacsat electronic mail automatically as the spacecraft flies by, even if he's not in the shack.

Ham C

Subject C enjoys RF construction projects and operating CW. Subject C hates spending money. In truth, money is actually a secondary issue—he really believes that one of the great pleasures of ham radio is the chance to solve problems by improvising, scrounging and using ingenuity. Most of his equipment, which covers the HF bands and 2-meter and 70-cm FM, is vacuum-tube vintage. He has nothing against solid-state equipment; it's just that "you can get your hands on a lot of excellent tube gear for next to nothing." After listening to a Mode B demonstration at his club's Field Day station, he decides that he wants to get on Mode B CW.

He sets up a 2-meter FM receiver RF deck for 145.89 MHz and then experiments with coupling RF from the

2-meter receiver's mixer and first IF over to his HF receiver. The 2-meter FM receiver is essentially being used as a converter. (See Chapter 9 for more information.) When everything appears to be working well, he connects the receive chain up to a ground plane and listens for Mode B on AO-13. The results are very disappointing—he can only hear a few very weak signals. He decides to splurge on a cheap preamp (on Mode B, there's not much to gain by trying to get below a 1.0-dB noise figure). He connects it in front of the receiver and it hardly makes any difference. Next he tries moving it to the antenna end of the feed line. This results in a noticeable improvement, but signals are still quite weak, so a beam antenna is clearly necessary. He knows his linearly polarized Yagi is not optimal, but he decides to give it a try. Since his wallet won't bear the shock of a second rotator for elevation control, he just raises the antenna boom to point about 25 degrees above the horizon. He now receives pretty decent signals. The strength isn't overwhelming, but most stations are solid R5. He notes that no one change had an overwhelming effect on performance, but that, in total, they produced a significant improvement. Now on to the transmitter.

Subject C checks his supplies (junk box) and finds a 10-W 430-MHz RF strip that he purchased as a backup for his old commercial 70-cm FM transceiver. He decides to swap it with the 20-W strip that's currently in the transceiver so he can use the higher-power unit for Mode B. He knows that the power rating is for CCS (continuous commercial service), so he decides to run it using ICAS (intermittent commercial/amateur service) ratings with an ac power supply, which will give him close to 40 W. For an antenna he uses a small Yagi mounted, like the 2-meter Yagi, at a fixed elevation. He has some higher-gain antennas around, but he figures that, with the fixed elevation mount and the 2-meter antenna in use, going to higher gain at 70 cm would likely reduce, rather than increase, his access time to AO-13. It takes a little experimentation to figure out how to obtain a clean signal when keying the transmitter. Finally, he adds a varactor across the capacitor which trims the crystal to frequency. This allows him to easily vary the frequency of the crystal-controlled transmitter ±15 kHz, which is adequate.

The setup works fine. Sometimes he wishes he had an elevation rotator or circularly polarized antennas, but he figures that these improvements would only increase his solid access time by about 20 percent. To him, these improvements aren't worth the effort. He looks over his station and feels rather pleased. The tube technology may be 30 years old and take up lots of room on the bench, but the equipment is reliable and it performs well. One change he does consider is replacing the new 2-meter preamp with an old nuvistor unit so he can print up some special satellite QSLs stating that his station is 100% tube, 0% solid state.

Subject C is not interested in tracking, but he quickly learns a couple of helpful operating tricks. If he catches AO-13 at a good high position one day, he can usually catch it at the same point on its orbit about 2 hours earlier the next day at an azimuth somewhat further east.

Ham D

Subject D is a high-school biology teacher. She's an active amateur, but she's never worked with the OSCAR satellites. It's January 15 and her school has just received a small grant, $300, to enable the science faculty, and students in the science, Amateur Radio and other clubs, to design and build a temporary exhibit for the city's Museum of Science and Technology. The exhibit will be on display during July and August. At a meeting called to set design objectives, she suggests the "Peaceful Uses Of Space" as a theme. The idea is accepted and the group decides to focus on earth satellites used for communications, environmental monitoring and special applications. To illustrate these classes of satellites, the group selects weather satellites, earth-resources satellites, SARSAT/COSPAS search-and-rescue satellites and Amateur Radio satellites.

Guidelines for the museum display suggest including both static and dynamic/interactive sections. The static section generally consists of posters, pictures, models, and so on. The dynamic/interactive component is the main area of concern. One student suggests a design focusing on four spacecraft that would require about 12 computer displays. Three displays would be allocated to each spacecraft: one showing the ground track, the second showing the view of the earth as seen from the spacecraft and the third having a running commentary on the satellite system or, in the case of weather satellites, showing a file of weather satellite images. While everyone agrees that this would make a very impressive exhibit and that the group has the expertise to build such a display, they feel it's a little too big to handle in the time available. However, elements of the suggestion are incorporated in the chosen plan.

The final design relies heavily on Amateur Radio for the dynamic/interactive element. Microcomputer software developed for tracking amateur spacecraft will be used to provide a graphic display of the real-time positions of the various satellites described in the static part of the exhibit. The computer-produced image will be projected on a large screen using an multicolor LCD shutter sitting on an overhead projector. This is the system used in the high school when a teacher wishes to display microcomputer output for a class.

Two receivers will be used. One, a scanner, will monitor the 145.825-MHz digitized-speech broadcasts of Dove and UoSAT-11. The second will be used to monitor SSB activity between 145.800 and 146.000 MHz. Subject D will be responsible for designing the RF sections of the display and obtaining equipment on loan when possible.

The RF design is complicated by several factors. Consider the 2-meter SSB receiver first. A modern multimode 2-meter transceiver that can be tuned by buttons on the mike is appealing because it probably would be easy to rig a remote tuning control on the front of the display using a couple of push switches. But money isn't available to purchase one, and if one were borrowed the possibility of damage would be a serious and constant concern. The solution is to invest $70 in a good 2-meter converter and borrow an old amateur-band-only tube receiver. With the help of the president of a local Amateur Radio club, a member with a spare receiver that's appropriate is located. The member readily agrees to loan it. In return, a notice about the club's fall Novice class will be prominently displayed at the science museum exhibit. Although the 1960-vintage 60-pound-plus receiver greatly reduces concern about theft, the group designing the display feels it's in the best interests of all concerned to agree on a value for the receiver ($90) ahead of time in case it is stolen or severely damaged.

The receiver will be placed behind a large sheet of plywood which forms the front of the display. Strategically placed cutouts will provide viewers with access to the volume control and main tuning dial only. The antenna will be a simple turnstile with the reflector pointing straight up. A high-quality preamp (borrowed) mounted directly at the antenna will be used.

The scanner receiver will be permanently set to 145.825 and used with a simple ground plane. Only one special piece of electronic equipment needs to be designed and constructed for the display. It will monitor the squelch line on the scanner receiver. When a spacecraft comes into range on 145.825 MHz, it will switch on a blinking red light and mute the SSB receiver.

The exhibit, finished on time and under budget, turns out to be very popular. The science museum asks the group to consider returning with a larger exhibit during the International Space Year (1992). The 2-meter converter, the only major piece of equipment purchased for the display, is donated to the high school radio club. Subject D, and several members of the high school radio club, become very interested in the OSCAR satellite program.

Chapter 6

Tracking Basics

This chapter focuses on basic satellite tracking: what it means, why it's usually necessary and how to do it. Our approach here is practical and applied. The mathematics and physics underlying the various techniques will be covered in later chapters. We'll be focusing on three types of orbits: low-altitude circular orbits as used by most Phase 2 satellites, elliptical orbits as used by Phase 3 spacecraft and geostationary orbits planned for Phase 4 satellites. Both graphic methods and computer methods are covered. If your main interest is in communicating through an amateur satellite this chapter will provide all the information needed to track any amateur spacecraft.

If you don't care much for mathematics and you know nothing about computer programming, don't worry. Wading through the details in this chapter does require patience, but the only skills you need are basic arithmetic and simple map reading.

TRACKING: WHAT, WHY, HOW?

To a scientist, tracking a satellite means being able to specify its position in space. To a radio amateur, tracking more likely refers to practical concerns: When will a satellite be *in range* (accessible to you) and where should the antenna be pointed? Satellites generally are moving targets, so when a ground station uses directional antennas, aiming information must be available. The ability to predict access times is also important because most satellites are in range of a specific ground station for only a part of each day. (Geostationary satellites, which remain over a fixed location on the equator, are an exception which we'll discuss later in this chapter.)

A low-altitude satellite (such as Fuji-OSCAR 20, RS-10/11 or a MicroSat) will generally be in range for less than 25 minutes each time it passes nearby (*satellite pass*). A ground station will usually see four to six passes per day for each satellite. As a result, a satellite that's operational 24 hours per day will be accessible only one to two hours each day at a specific ground station. Your average daily *access time* for a satellite is an important quantity in determining how useful the satellite will be to you.

A satellite in the high-altitude elliptical orbits used for Phase 3 spacecraft (such as OSCARs 10 and 13) behaves very differently. It will only provide one or two passes per day, but the total access time will be (very roughly) 12 hours for Northern Hemisphere stations. One way to look at this

is to say that one Phase 3 satellite will provide you with as much daily operating time as roughly eight Phase 2 satellites.

A geostationary satellite appears to hang motionless in the sky. If it's in range you'll have access to it 24 hours per day. If it's out of range you'll never see it.

Before we get down to the details of tracking, note that in several situations it can be ignored. For example, if you're in your shack frequently it's possible to periodically listen on 29.300 to 29.500 MHz using a dipole or ground-plane antenna on the receiver to see if any Mode A spacecraft are accessible. This can be frustrating, because the spacecraft may be out of range by the time your transmitter warms up. However, you can check Table 5-2 to identify the satellite and chances are roughly two out of three that there will be another pass in 1.5 to 2 hours.

For those who prefer to avoid tracking, the situation with a Phase 3 satellite is considerably more favorable. In fact, it's almost identical to checking an HF band to determine if it's open. Set your receiver to 145.900 MHz. If you're using a beam with azimuth and elevation control set the elevation to 25 degrees and scan in azimuth while tuning the receiver. If the *band is open* (satellite in range), weak signals will be observed. Just adjust the antenna to peak the receiver S meter. You usually won't have to peak the antenna again for at least a half hour.

Geostationary satellites which may become available in the late 1990s will make life even easier. A station in the US installing an antenna to access a Phase 4 spacecraft will point it roughly southeast (AMSTAR East) or southwest (AMSTAR West), peak it on received signal strength, tighten all the hardware and never touch it again. The satellite will be available 24 hours per day.

The knowledge that tracking can often be avoided might tempt you to skip the rest of this chapter. Try to resist! The ability to track will greatly add to your satellite operating pleasure by providing you the ability to arrange schedules with stations in specific locations, plan nets and demonstrations and schedule your operating efficiently. Sitting in front of a silent receiver can get very boring. We now turn to the details.

Radio amateurs wishing to track a satellite are interested in specific information. They want to know:

1) When the satellite will be in range; more specifically, times for *AOS* (*a*cquisition *of s*ignal) and *LOS* (*l*oss

*of s*ignal) for each pass;

2) Where to aim the antenna (azimuth and elevation) at any time;

3) The regions of the earth that have access to the spacecraft at any instant.

A good tracking device should be able to provide all this information. However, for many applications, only one or two of these features may be needed.

GRAPHIC METHODS VS COMPUTER METHODS

Most amateurs obtain their tracking information from either (1) a small map of the earth used in conjunction with a hand calculator and some transparent overlays (*Graphic Method*) or (2) a personal computer (*Computer Method*). It costs about $20 to purchase everything for the graphic method and about 10 to 100 times as much for the computer method. If the computer were only used for tracking most people would find the cost factor prohibitive. However, a great many hams have access to personal computers that were purchased for other applications. Once the computer is on hand, cost is no longer a major factor and we can pay attention to considerations like convenience.

In the 1960s and mid '70s Amateur Radio operators devised a number of ingenious and easy to use graphic methods for tracking OSCAR satellites in circular orbits. Many of these techniques are described in Chapter 12. In the late 1970s, with the elliptical orbit Phase 3-A spacecraft about to be launched, these methods were extended to enable amateurs to track spacecraft in elliptical orbits. The resulting devices were somewhat more cumbersome to use but they were cheap and they worked. Unfortunately, Phase 3-A ended up in the Atlantic Ocean so most satellite operators wouldn't need tracking devices for elliptical orbits until several years later.

Meanwhile, in the late 1970s Dr Tom Clark (W3IWI) had managed to compress the kernel of a professional tracking program which ran on a large mainframe computer into a program in BASIC that could be run on the small microcomputers being used by AMSAT development teams and control stations. In 1981 Tom's program was published in *ORBIT* and, over the next few years, dozens of amateurs translated it into several other languages and dialects of BASIC. By the time OSCAR 10 was launched (1983), small computers were becoming widely available and a great many amateurs were using them in conjunction with versions of the "IWI" program for tracking. However, some amateurs still preferred the graphic methods because the pictorial maps were more fun to work with than the long lists of numbers provided by the computer. Then, in the mid 1980s, Roy Welch (WØSL) and others added graphic output to the computer software and gave amateurs access to the best of both worlds—the convenience of the computer and the pictorial output of the graphic device (they could also have numerical output from the computer if desired.) I think it's fair to say that given the state of the software available by the mid 1980s, the

great majority of amateurs preferred computer tracking methods in most situations.

If you own a major-brand computer, there's probably a good tracking program available. However, keep in mind that though the graphic methods may be a little more tedious to use, they work fine. Since the pictorial output provided by computer programs is closely related to the graphic methods, an understanding of graphic methods can help you interpret what you see on the computer screen. Therefore, it's important that you at least skim the sections on graphic methods, even if you're planning to employ a computer for tracking.

No matter what approach you take to tracking, you're going to periodically need some updated information describing the orbit of the satellite you're interested in. Depending on which method you use, this data is either called *reference orbit data* or *orbital elements*. For now, just think of this information as a set of numbers that you plug into your tracking device. The meaning of each of these "numbers" will be discussed in the course of this chapter. For most satellites of interest this information only has to be updated every few months. However, for spacecraft at very low altitudes, like the Shuttle and Mir, it has to be updated once per week and, of course, after every orbital maneuver. Sources for this data are listed at the end of the chapter.

SOME BASIC DEFINITIONS

Most people find that the most irksome aspect of learning how to track is mastering the new jargon. But some familiarity with tracking terms is necessary, so take it slow and make sure you understand the informal, practical explanation given with each new italicized term presented. The definitions are summarized in Table 6-1. You're probably already familiar with several of the terms we'll be using.

The *subsatellite point (SSP)* is the point on the surface of the earth directly below the satellite. For most satellites (all but geostationary) the location of the SSP constantly changes as the spacecraft moves across the sky. If we were to watch the SSP as the satellite travels in space it would trace a curve on the surface of the earth, called the *ground track* or *subsatellite path*.

A satellite will be in range when the SSP is close to your ground station location, and out of range when the SSP is far away from your location. This seemingly obvious statement is the key to understanding how tracking devices work. Of course, we have to define how close "close" is. To do this we compute a critical *acquisition distance* associated with each satellite in a circular orbit. (Most low-altitude amateur satellites are in circular orbits.) When the SSP is closer to you than the acquisition distance, the satellite is in range of your station. See Table 6-2 for data on acquisition circles.

Fig 6-1 shows an *acquisition circle* for AMSAT-OSCAR 8 drawn about Washington, DC. Note how circles on the surface of the earth (roughly a sphere) appear distorted on most flat maps. Assume your station is located in Washington. If the ground track for a specific orbit of

Table 6-1
Summary of Tracking Terms

access range (acquisition distance)

acquisition circle: "Circle" drawn about a ground station and keyed to a specific satellite. When the SSP is inside the circle, the satellite is in range.

acquisition distance: Maximum distance between subsatellite point and ground station at which access to spacecraft is possible.

AOS (Acquisition Of Signal)

apogee: Point on orbit where satellite height is maximum.

argument of perigee: An angle that describes the location of a satellite's perigee. When the argument of perigee is between 0 and 180 degrees, perigee is over the northern hemisphere. When the argument of perigee is between 180 and 360 degrees, the perigee is over the southern hemisphere.

ascending node (EQX): Point where ground track crosses equator with satellite headed north.

ascending pass: Satellite pass during which spacecraft is headed in a northerly direction while in range.

azimuth: Angle in the horizontal plane measured clockwise with respect to North (North = 0°).

Bahn latitude and longitude (ALAT and ALON): angles which describe the orientation of a Phase 3 satellite in its orbital plane. When Bahn latitude is 0° and Bahn longitude is 180° the directional antennas on the satellite will be pointing directly at the SSP when the spacecraft is at apogee.

coverage circle: (With respect to a particular ground station) the region of earth that is eventually accessible for communication via a specific satellite; (With respect to a particular satellite) the region around the instantaneous SSP that is in view of the satellite.

decay rate: Short name for rate of change of mean motion. A parameter specifying how atmospheric drag affects a satellite's motion.

descending node: Point where ground track crosses equator with satellite headed south.

descending pass: Satellite pass during which satellite is headed in a southerly direction while in range.

eccentricity: A parameter used to describe the shape of the ellipse constituting a satellite orbit.

elevation: Angle above the horizontal plane.

elevation circle: The set of all subsatellite locations about a ground station where the elevation angle to a specified satellite is a fixed value.

epoch (epoch time): A reference time at which orbital elements are specified.

EQX (ascending node)

geostationary satellite: A satellite that appears to hang motionless over a fixed point on the equator.

ground track (subsatellite path): Path on surface of earth traced out by SSP as satellite moves through space.

inclination: An angle which specifies the orientation of a satellite's orbital plane with respect to the earth's equatorial plane.

increment (longitudinal increment)

longitudinal increment: Change in longitude of ascending node between two successive passes of specified satellite. Measured in degrees West per orbit (°W/orbit).

LOS (Loss of Signal)

mean anomaly (MA): A number between 0 and 256 that increases uniformly with time. Used to locate satellite position on orbital ellipse. When MA is 0 or 256, satellite is at perigee. When MA is 128, satellite is at apogee. When MA is between 0 and 128, satellite is headed up towards apogee. When MA is between 128 and 256, satellite is headed down towards perigee.

mean motion: Number of revolutions (perigee to perigee) completed by satellite in a solar day (1440 minutes).

node: Point where satellite ground track crosses the equator.

orbital elements: Set of six numbers specified at particular time (epoch) which completely describe size, shape, and orientation of satellite orbit.

OSCARLOCATOR: A tracking device designed for satellites in circular orbits.

pass (satellite pass)

PCA (Point of closest approach): Point on segment of satellite orbit, or ground track, at which satellite is closest to specific ground station.

perigee: Point on orbit where satellite height is minimum.

period: The amount of time it takes a satellite to complete one revolution, perigee to perigee, about the earth.

phase: See mean anomaly.

Phase3 Tracker: A tracking device related to the OSCARLOCATOR which is designed to be used with a satellite in an elliptical orbit.

point of closest approach (PCA)

RAAN (Right Ascension of Ascending Node): An angle that specifies the orientation of a satellite's orbital plane with respect to the fixed stars.

range circle: "Circle" of specific radius centered about ground station.

reference orbit: First orbit of UTC day for satellite specified.

satellite pass: Segment of orbit during which satellite passes in range of particular ground station.

semi-major axis (SMA): Half the long axis of an ellipse. Can be used to describe the size of an elliptical orbit in place of the orbital element mean motion.

SMA (semi-major axis)

spiderweb: Set of azimuth curves radiating outward from, and concentric elevation "circles" about, a particular terrestrial location.

SSP (SubSatellite Point)

stationary satellite (geostationary satellite)

subsatellite path (ground track)

subsatellite point (SSP): Point on surface of earth directly below satellite.

TCA (Time of Closest Approach): Time at which satellite passes closest to a specific ground station during orbit of interest.

window: For a specific satellite, the overlap region between acquisition "circles" of two ground stations. Communication between the two stations via the specified satellite is possible when SSP passes through window.

Expressions in parentheses are synonyms or acronyms. Note that true circles on the globe are often distorted when transferred to a map. Some minor differences will be found between the definitions in this table, which focus on the practical aspects of tracking, and those in the Glossary, where emphasis has been placed on technical precision.

Table 6-2

Distances Between SSP and Ground Station Corresponding to Specified Elevation Angles

Satellite name	Satellite height	Elevation circle radius		
		0°	30°	60°
OSCARs 14-19	800 km	3038 km	1078 km	403 km
	497 mi	1888 mi	670 mi	250 mi
		27.3°	9.7°	3.6°
UO-11	690 km	2840 km	958 km	354 km
	429 mi	1765 mi	595 mi	220 mi
		25.5°	8.6°	3.2°
FO-20				
apogee	1745 km	4257 km	1909 km	766 km
	1084 mi	2645 mi	1186 mi	476 mi
		38.3°	17.2°	6.9°
perigee	912 km	3223 km	1195 km	452 km
	567 mi	2003 mi	743 mi	281 mi
		29.0°	10.7°	4.1°
RS-10/11	1003 km	3362 km	1286 km	490 km
	623 mi	2089 mi	799 mi	305 mi
		30.2°	11.6°	4.4°
Mir	400 km	2201 km	603 km	215 km
	249 mi	1368 mi	375 mi	134 mi
		19.8°	5.4°	1.9°
AO-13 (1989)				
apogee	36,265 km	9052 km	5845 km	2859 km
	22,539 mi	5626 mi	3633 mi	1777 mi
		81.4°	52.6°	25.7°
perigee	2545 km	4936 km	2421 km	1008 km
	1581 mi	3068 mi	1505 mi	626 mi
		44.4°	21.8°	9.1°

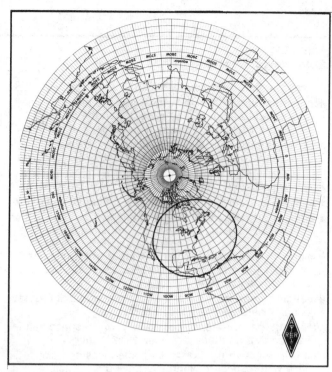

Figure 6-1—Typical acquisition circle for a Phase 2 satellite (OSCAR 8 in this case) drawn about Washington, DC.

a particular satellite passes inside this acquisition circle, the satellite will be accessible (*in range*) during the pass. AOS occurs when the SSP enters the acquisition circle; LOS occurs when the SSP leaves the acquisition circle. Determining when a satellite in an elliptical orbit is in range is a little more complicated. We'll consider this problem later.

For most orbits the ground track will cross the equator twice. The two points where the ground track and equator intersect are called *nodes*. The *ascending node* occurs when the SSP crosses the equator headed north; the *descending node* occurs when the SSP crosses the equator headed south. Some tracking methods use the time and longitude of the ascending node for *reference orbit* data. The ascending node is often abbreviated *EQX* (*eq*uator *cross*ing).

The amount of time it takes a satellite to go through one complete orbit (revolution of the earth) is called its *period*. The periods of amateur Phase 2 satellites range from about 98 minutes (UoSAT-OSCAR 11) to 120 minutes (RS-7). The Mir space station, which is slightly lower than OSCAR-11, has a period close to 90 minutes. The periods of Phase 3 satellites are usually between 11 and 12 hours.

If a satellite's period is known, it's easy to compute the number of orbits per day. Just divide the number of minutes per day, 1440, by the satellite period in minutes. The number of orbits per day is called the *mean motion*. Phase 2 spacecraft, the US Shuttle and Mir complete about 12-16 orbits per day. A Phase 3 satellite completes about two orbits per day. A geostationary satellite circles the earth once per day traveling exactly as fast as the earth rotates (it appears to remain fixed in space). Most tracking methods use either the period or the mean motion as an orbital element.

A GRAPHIC METHOD FOR LOW-ALTITUDE CIRCULAR ORBITS: THE OSCARLOCATOR

The OSCARLOCATOR is a graphic tracking device which is especially well suited for tracking satellites in low-altitude circular (or nearly circular) orbits. This category includes all OSCAR satellites launched to date except for OSCAR 4 and the Phase 3 spacecraft. To illustrate this method we focus on a single satellite—OSCAR 8.

The OSCARLOCATOR consists of two parts:

1) *Map board.* A map centered on the north pole like the one shown in Fig 6-1. A full-size map is presented in Appendix C and as a foldout from the back cover.

2) *Ground-track overlay.* An overlay, usually drawn on transparent material, is shown in Fig 6-2. The overlay is pinned to the map board so that it can be rotated about the pole.

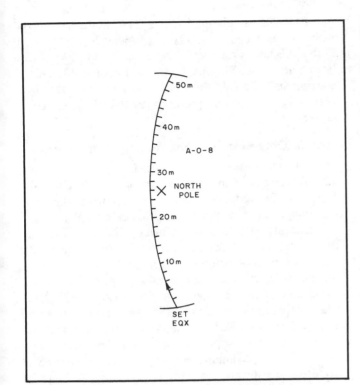

Figure 6-2—Typical ground track overlay (OSCAR 8).

Table 6-3
One Day From a Complete AMSAT-OSCAR 8 Orbit Calendar

AMSAT-OSCAR 8
21 June 1980 (173) Saturday

Orbit No.	TIME (UTC) HH:MM:SS	EQX °W
11695	00:53:37	66.0
11696	02:36:49	91.8
11697	04:20:01	117.6
11698	06:03:14	143.4
11699	07:46:26	169.2
11700	09:29:38	195.0
11701	11:12:50	220.8
11702	12:56:02	246.6
11703	14:39:15	272.4
11704	16:22:27	298.2
11705	18:05:39	324.0
11706	19:48:51	349.8
11707	21:32:03	15.6
11708	23:15:16	41.4

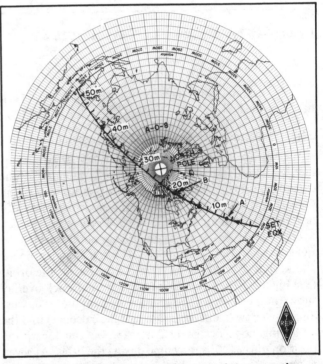

Figure 6-3—Ground track superimposed over a polar map with the ascending node set to 41° W longitude.

To set the overlay on the OSCARLOCATOR you need reference orbit data. This data is often presented in the form of an orbit calendar. For each date, the calendar provides information about all orbits that begin during that day. (For satellites in circular orbits the convention is that an orbit begins at the ascending node.) Table 6-3 shows how a calendar prepared by Project OSCAR presented data for OSCAR 8 on Saturday 21 June 1980. The 14 horizontal rows of information correspond to the 14 OSCAR 8 orbits that began this day. The first column contains a reference number that is meant to uniquely identify each orbit. For circular orbits the convention is to number orbits consecutively starting with orbit 1 at the first ascending node. This column is optional since, as we'll see shortly, orbit numbers are not needed to determine when the spacecraft will be in range. The next two columns in the calendar provide critical information—the time and longitude of each ascending node. Times in this particular calendar are given in UTC (Universal Coordinated Time) using Hour:Minute:Second (HH:MM:SS) notation. Longitudes are given in degrees West.

Although the calendar in Table 6-3 is dated June 21, keep in mind that this is a UTC date. A US East Coast station interested in determining whether any orbits were visible on Friday evening, June 20 EDT, would have to check the first two rows for June 21; a station on the US West Coast would note that the first four orbits begin June 20 PDT. The number "173" in parentheses following the date indicates that June 21 is the 173rd day of 1980. There's nothing special about the units chosen for the calendar; some calendars present time in EST, and/or provide the

longitude of ascending node in degrees East. When using a calendar be sure to note what units are used.

There are three steps to using the OSCARLOCATOR:
1) Set the ground track overlay. This will tell you where the SSP point will be at any time.
2) Determine AOS and LOS for your specific location.
3) Obtain antenna aiming information.

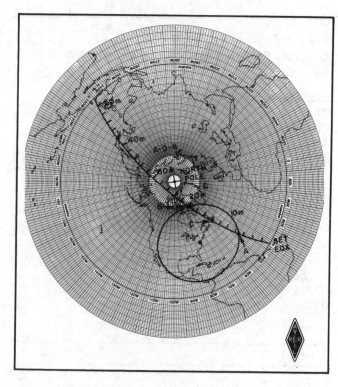

Figure 6-4—Typical ground track overlay (OSCAR 8).

An example will show how each of these steps is accomplished.

Step 1: Setting the Ground Track Overlay

Consider the last OSCAR 8 orbit of the day on Saturday, June 21, 1980 (orbit #11,708). The bottom row of the calendar (Table 6-3) contains the information describing this orbit. It tells us that the ascending node occurs at 41.4 degrees West longitude (°W), so rotate the ground track overlay until the ascending node (the point marked set EQX on the overlay) is over 41.4°W on the map board and keep it fixed at the position for the rest of the orbit. (See Fig 6-3.) The tick marks on the ground track overlay show the position of the satellite at 2-minute intervals. Since we know (from the last row of the calendar) that the ascending node occurs at 23:15:16, we can locate the position of the SSP at any time during the orbit. At seven minutes after the ascending node (about 23:22 UTC) the SSP will be at point A (Fig 6-3); at 20 minutes after EQX (23:35 UTC) the SSP will be at point B.

Step 2: Determining AOS and LOS

To determine AOS and LOS for your location you must add an acquisition circle for OSCAR 8, like the one shown in Fig 6-1, to the map board. Fig 6-4 includes an acquisition circle drawn about Washington, DC. The key to interpreting the OSCARLOCATOR is:

Whenever the OSCAR 8 SSP is inside your OSCAR 8 acquisition circle the satellite will be in range of your station.

AOS occurs at point A (23:22 UTC) as the SSP enters the acquisition circle. LOS occurs at point B (23:35 UTC) as the SSP leaves the acquisition circle. So OSCAR 8 will be in range of our Washington station for about 13 minutes starting at 23:22 UTC. Since Washington is on Daylight time in June, we might prefer to say that the pass begins at 19:22 EDT.

Step 3: Obtaining Antenna Aiming Information

To aim a beam antenna we need to know two angles: *azimuth* (measured in the local horizontal plane clockwise from North) and *elevation* (measured from the local horizontal plane, positive values indicate "up").

Consider elevation first. When the SSP coincides with the ground station location (satellite directly overhead), the antenna should be pointed straight up (90° elevation). At the positions where the SSP crosses the acquisition circle (AOS and LOS) the antenna is set horizontal (0° elevation). Between these extremes (SSP inside the acquisition circle), the elevation angle must be set somewhere between 0° and 90°. On most passes a satellite will not fly directly overhead. The maximum elevation will occur when the SSP is closest to your ground station. This is called the point of closest approach *(PCA)*.

Note that the acquisition circle consists of all points having 0° elevation. Circles corresponding to other elevation angles can also be drawn about your location on the map. In Fig 6-5 we've added elevation circles at 30° and 60° about the Washington, DC station. (The lines radiating out from the center may be ignored for now.) With the additional elevation circles on the map it's easy to estimate the elevation of the satellite if we know the location of the subsatellite point. Using "eyeball interpolation," accuracy is easily better than 10°, which is more than sufficient for most wide-beamwidth, low-gain beam antennas suitable for working with spacecraft in low earth orbits. Like the acquisition circle, the elevation circles are only valid for a circular orbit satellite at a specific altitude. Key distances for drawing acquisition and elevation circles for a number of popular spacecraft were listed in Table 6-2.

Returning to OSCAR 8's orbit number 11,708 once again (Fig 6-5), we can estimate the elevation at PCA to be approximately 20°. A reasonable strategy for a Washington, DC, station operating this pass would be to leave the antenna set at 10° elevation for the entire pass and only bother with azimuth tracking.

Now consider azimuth. Azimuthal directions radiating out from a ground station generally appear as curved lines on a map. Fig 6-5, shows a set of such curves centered on Washington. Noting the position of the SSP at any time in relation to these curves, we can again use "eyeball interpolation" to estimate the correct antenna azimuth angle. Returning to OSCAR 8 orbit 11,708, we see (Fig 6-5) that AOS occurs at an azimuth of roughly 110°, PCA at an azimuth of 50° and LOS at an azimuth of 5°.

Taken together, the set of concentric elevation circles and curved azimuth radials is often referred to as a *spiderweb*. Several methods for drawing spiderwebs will be covered later in this chapter and in Chapter 12.

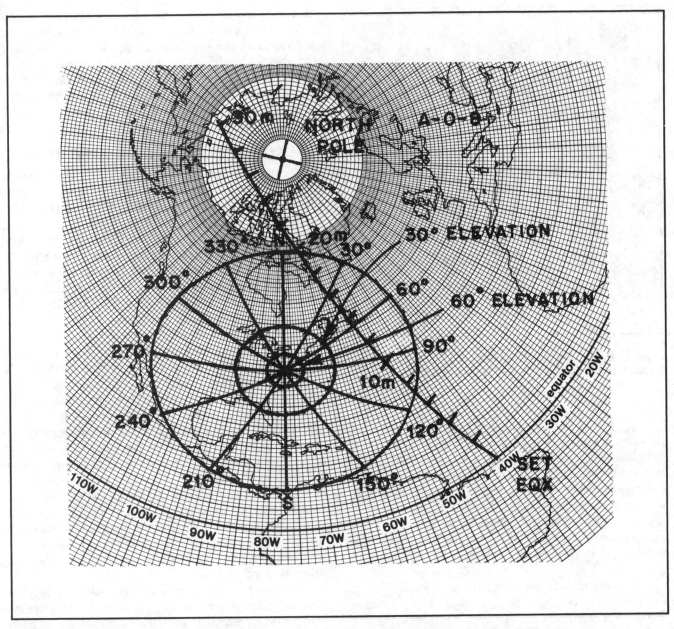

Figure 6-5—Added to the previous figure is the "spiderweb" of azimuth radials emanating from the station location, Washington, DC in this case, and concentric elevation circles. These are used for aiming antennas in azimuth (side to side) and elevation (up and down).

Simple extensions of the information just presented make it possible for us to determine (1) those regions of the earth eventually accessible to us via a specific satellite and (2) those orbits suitable for communicating with distant stations. Again using OSCAR 8 as an example, since the maximum acquisition distance for this satellite (Table 6-2) is 2000 miles, two stations separated by twice this distance (4000 miles) can communicate with one another, but only when the SSP is at the midpoint of the great circle path joining them. Just as we drew acquisition and elevation circles on the map board, we can draw a *coverage circle* around a ground station to show which regions of the earth will eventually be in range via a specific satellite. The radius of the coverage circle is twice the

acquisition distance. A station in New York who wanted to know if communication with London was possible through OSCAR 8 would only have to check his coverage circle to learn the answer (yes).

To select suitable orbits for communicating over the New York to London path, draw acquisition circles for both stations on the map board as in Fig 6-6. Whenever OSCAR 8's SSP is in the overlapping region (called the *window*), communications between the two stations is possible. To find the best pass, rotate the overlay so the ground track passes through the center of the window. As shown in the figure, when the equatorial crossing is 26°W the mutual window will open approximately 11 minutes after ascending node and last for about 8 minutes. Now rotate

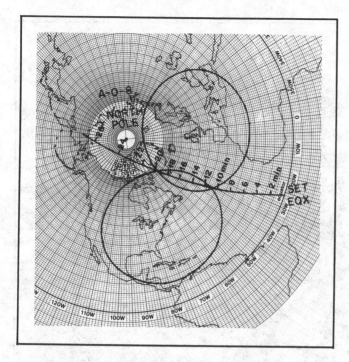

Figure 6-6—An AMSAT-OSCAR 8 orbit that passes through the center of the NY-London window.

the overlay slightly to determine the limits on equatorial crossings that permit communication. Passes with EQXs between 19°W and 31°W will permit New York to London communication. Descending passes, when the ascending node is between 217°W and 229°W, will also produce ground tracks passing through the window. Once you've got the longitude limits for ascending and descending passes check through an orbit calendar to locate appropriate orbits.

We now look at the orbit calendar in a little more detail. A careful analysis of the calendar entry shown in Table 6-3 reveals that each reference time differs from the preceding time by 1:43:12 (1 hour, 43 minutes, 12 seconds in HH:MM:SS notation), which is equivalent to 103.20 minutes in decimal minutes notation. Similarly, each reference longitude crossing is 25.8 degrees further west than the previous one. These two numbers characterize the OSCAR 8 orbit. Given these two numbers and data describing one northbound equatorial crossing time and longitude, we can generate data on future ascending nodes. For example, an abbreviated orbit calendar might just provide data for the first orbit each day (often called the *reference orbit*), leaving the user to fill in the remaining orbits on days of interest by successively adding 103.20 minutes to the time and 25.8 degrees to the longitude. The number 103.20 minutes is the *period* (time for one complete orbit) of OSCAR 8. The number 25.8 is the *longitudinal increment* (usually just called the *increment*) for OSCAR 8 and is given in degrees West per orbit (°W/orbit).

Though the various orbit calendars available use slightly different formats, heading abbreviations and time notation, you shouldn't have any trouble using them once

the basic ideas that have just been described are understood. Before you attempt to use an orbit calendar, make sure you know the format of the time notation being used and whether longitudes are being measured in degrees W or degrees E of Greenwich. Note that information on orbit numbers is not required for tracking. It's simply a convenience feature that makes it easier to refer to a specific orbit as we did with orbit 11,708.

COMPUTER METHODS: ANY ORBIT

In the great majority of situations, using a personal computer to track satellites is by far the most convenient method. Tracking software is available for many popular computers including the IBM® PC and clones, Commodore™ C-64, Commodore C-128, Commodore Amiga, Apple® II, Apple Macintosh and Radio Shack® TRS-80. An SASE to AMSAT will get you a list of what's currently available. (Software may be purchased from other sources, but when you order from AMSAT a significant portion of the purchase price will go towards supporting OSCAR satellite construction.)

Tracking software naturally answers the basic tracking questions: it tells you when the satellite of interest is in range and provides antenna pointing data. Most packages provide considerable additional information. For example, at each specified time the program may list *range* (the distance between your station and the spacecraft), the *Doppler shift* for the mode you specify (which helps you locate your downlink), the height of the satellite (for elliptical orbits this varies), the *Phase* or *Mean Anomaly* (a number that tells where a satellite is located in its orbit), *squint angle* (a parameter that tells how close to you the satellite antennas are currently aimed), predicted signal levels (on the downlink), path delay time (often labeled echo) and an orbit number (for reference purposes—no effect on tracking).

Because tracking programs provide so much information and often give you a choice as to how you'd like the information presented, they may at first appear overwhelmingly complex. But, when you start using a program, you'll find that you can initially just pay attention to elevation angle and azimuth. This will tell you when the satellite is in range (roughly, whenever the elevation angle is greater than 0 degrees) and where to point your antenna. Your radio horizon at the frequencies of interest will, of course, affect the actual elevation angle at which you can access a satellite. Later, once you know how to navigate through the tracking program and use the elevation and azimuth data, you can learn how to use the additional information provided.

A key feature of almost all tracking programs currently available is that they handle both elliptical orbits and circular orbits. (Since no orbit is perfectly circular, they do this by treating all orbits as being elliptical.) In this section we'll focus on how these programs operate from a user point of view—what information you (the user) must provide to the program and what information comes out. In later chapters we'll look at the underlying mathematics and physics. Although each tracking program has its own distinct appearance, all require a core set of input data to operate and they must provide certain key output data to the user. This makes it possible to discuss tracking programs in a general way. Once you know how a "generic" program operates, it's easy to pick up a new program and figure out how it works.

Let's look first at the input data the computer requires. Naturally it will need the location of your ground station in terms of latitude and longitude. Some sophisticated programs may even ask for your height above sea level. (This refinement has no observable effect for 99.99% of amateur applications, so even if you live in Denver and have a monster EME antenna, you can just enter zero or some approximate number if you don't know the correct value.) You may also be given the option of providing data for several secondary terrestrial locations. This information can be used for computing mutual windows or for providing antenna pointing data to those you are in contact with.

The computer also has to know the precise orbit of each satellite you're interested in. You must therefore provide the computer with information describing orbit size, shape, orientation with respect to the earth, orientation with respect to the stars, etc. This is done by entering a set of numbers called *orbital elements*. Since each satellite orbit has a different shape, size, orientation, and so on, it will have its own unique set of orbital elements. The elements consist of a set of six numbers specified at a given time (called the *epoch*). Many programs use an extended set of elements to provide additional and/or redundant information. After you enter orbital elements for each satellite of interest the program has all the basic information it needs to provide tracking data. You do not have to provide this data very often.

When you want to use the tracking program you just have to specify which satellite you're interested in, the dates and times you want covered and, perhaps, the format of the output information. To simplify operation most programs are menu driven. This means that you'll be presented with a series of choices of what you want done and requests (prompts) for information. To respond to a menu you generally must type a single letter or number followed by the enter key. When you're asked to provide information the prompt usually contains an example showing the exact format you should use to enter the requested data.

Once you have some understanding of what a tracking program does, the questions posed by the computer are usually clear. For example, when you first turn on the computer you may encounter a menu (the *main menu*) that asks:

(1) Do you want batch tracking data?
(2) Do you want real-time tracking data?
(3) Do you want to update or add orbital elements?
(4) Do you want to modify other parameters (ground station location, screen colors, and so on)?
(5) Do you want to exit the program?

Once you respond by typing a single number (perhaps followed by the enter key) you'll see a series of additional prompts. If you respond "1" to obtain batch tracking data,

the computer would need to know which satellite you're interested in, the date and time to start the calculations, the duration and time-step desired and so on. Exploring and familiarizing yourself with the capabilities of the powerful, flexible software currently available takes some time.

Computers can be very frustrating. If you're a computer novice, the following hints may prove helpful. When you're asked a question that requires a yes or no answer, most programs only require a one-letter response, but some may not accept a lower-case y or n, others may require you to press the enter key after you respond, and so on. When entering longitude (and other angles) make sure you know whether the computer expects degree-minute or decimal degree notation. This should be made clear by the question prompting you to provide the information. Also make sure you understand the units and sign conventions being used. For example, are longitudes specified in degrees East or West of Greenwich? Are Southern latitudes specified by a negative sign? Dates can also cause considerable trouble. Does the day or month appear first? Can November be abbreviated Nov or must you enter 11? (The number is almost always required.) Must you write 1990 or will 90 suffice? Should the parts be separated by colons or dashes or slashes? The list goes on and on. Once again, the prompt is your most important clue. For example, if the prompt reads "Enter date (DD:MM:YY)" one should enter Feb 9, 1992 as 09:02:92 following the format of the prompt as precisely as possible. (9:2:92 would probably work as well.) When entering numbers commas should never be used. For example, if a semi-major axis of 20,243.51 km must be entered type 20243.51 with the comma and units omitted. It takes a little time to get used to the quirks of each software package, but you'll soon find yourself responding automatically.

We now take a look at the batch output provided by a typical program. This type of output is especially helpful for planning future operations. For example, using it you can check to see which evenings next week will provide good access to OSCAR 13. Tracking data for these evenings can then be printed in advance so the computer can be free for other uses while you operate. Table 6-4 illustrates batch data output (modeled after ORBITS2, the WØSL version of the IWI program). The heading identifies the satellite (OSCAR 13), my ground station location (latitude, longitude and height) and the starting date. I asked the computer to print out data whenever the satellite would be in range (elevation greater than 0 degrees) for the 24-hour period beginning 0000 UTC Dec 23, 1988 and indicated that I wanted data every fifteen minutes (*time-step*). The first three columns in the body of the table present the basic information: time, azimuth and elevation. OSCAR 13 will come into range sometime between 1145 and 1200 UTC and remain in range for about 9.5 hours. Note that for six hours, between 1415 and 2015 UTC, the elevation remains between 45 and 55 degrees. During this time period most amateurs will be able to set their antenna to an elevation of 50 degrees and ignore elevation control. Of course, if you have a super-high-gain,

Table 6-4
Sample of Tracking Program Batch Output

AMSAT-OSCAR 13

Ground Station: Lat = 39°N, Long = 77°W, Ht = 0 km
Minimum Elevation = 0 degrees

Day # 358 - - - Friday, December 23 - - - 1988

UTC HHMM	AZ deg	EL deg	DOPPLER Hz	RANGE km	HEIGHT km	LAT deg	LONG deg	PHASE <256>
1145	167	5	—	23325	18353	−29.8	62.7	34
1200	166	11	−1867	25068	20664	−25.2	62.5	40
1215	165	16	−1733	26686	22773	−21.1	63.1	45
1230	166	21	−1596	28176	24694	−17.5	64.2	51
1245	166	26	−1462	29541	26438	−14.3	65.7	57
1300	168	30	−1332	30784	28017	−11.4	67.5	62
1315	170	33	−1206	31910	29438	−8.7	69.5	68
1330	173	37	−1083	32922	30711	−6.2	71.6	73
1345	175	40	−966	33823	31843	−3.9	73.9	79
1400	179	42	−851	34617	32838	−1.6	76.2	85
1415	183	45	−739	35308	33701	0.5	78.7	90
1430	187	47	−630	35896	34437	2.5	81.1	96
1445	192	49	−523	36384	35049	4.5	83.6	101
1500	197	51	−416	36773	35540	6.4	86.2	107
1515	202	52	−311	37063	35911	8.3	88.7	113
1530	208	53	−206	37255	36164	10.1	91.3	118
1545	213	54	−100	37348	36300	11.9	93.8	124
1600	219	55	7	37341	36320	13.7	96.4	129
1615	225	55	116	37233	36224	15.5	98.9	135
1630	231	55	226	37022	36011	17.3	101.4	141
1645	237	54	340	36705	35680	19.2	103.8	146
1700	243	54	456	36279	35231	21.0	106.2	152
1715	248	53	577	35741	34661	22.9	108.5	157
1730	254	53	703	35084	33967	24.8	110.8	163
1745	259	52	835	34305	33148	26.8	112.9	168
1800	264	51	973	33396	32198	28.9	114.9	174
1815	269	51	1119	32351	31114	31.0	116.7	180
1830	273	50	1276	31160	29890	33.3	118.3	185
1845	278	49	1442	29813	28520	35.6	119.6	191
1900	283	49	1621	28300	26996	38.1	120.5	196
1915	287	49	1815	26606	25310	40.8	120.9	202
1930	293	49	2027	24714	23451	43.7	120.6	208
1945	298	50	2258	22606	21407	46.8	119.2	213
2000	305	51	2511	20262	19166	50.0	116.2	219
2015	313	54	2784	17663	16715	53.3	110.6	224
2030	326	57	3066	14801	14045	56.3	100.8	230
2045	348	60	3300	11720	11160	57.5	84.5	236
2100	30	58	3243	8692	8115	53.9	61.2	241
2115	75	30	1933	6887	5126	39.9	36.4	247

narrow-beamwidth antenna, you might have to adjust it.

The next column (column 4) provides data on Doppler shift. At 1200 UTC a signal coming through the Mode B transponder will appear 1867 Hz lower than predicted using the transponder frequency translation constant. Because of the algorithm being used to compute Doppler shift, no value is provided for 1145 UTC, the first time the satellite comes into range. Looking at the trend in column 4 its easy to estimate that at 1145 the frequency will be about 2 kHz low. If you want the computer to provide Doppler data output you must first specify the frequency of interest. For an inverting transponder, use the difference between the uplink and downlink frequencies. For noninverting trans-

ponders such as Mode A specify the sum of the uplink and downlink frequencies. Since I was interested in Mode B, I entered 289 MHz (435 MHz − 146 MHz).

Column 5 gives the distance between your ground station and the satellite. When the satellite is closer signals are generally stronger, though this trend may be negated by the direction of the satellite antennas. Column 6 provides the height of the satellite above the earth (actually above mean sea level). Note that the highest point (called the apogee) occurs at about 1600 UTC. Columns 7 and 8 provide the latitude and longitude of the subsatellite point (longitudes are in degrees west and northern latitudes are positive).

Column 9, the phase, is an important parameter to anyone working with Phase 3 satellites such as OSCAR 13. Since OSCAR 13 has several transponders which cannot be operated simultaneously the control team must publish an operating schedule. This schedule is modified every few months to account for the fact that the satellite orientation must be adjusted several times a year to compensate for changes in the sun angle on the spacecraft. A typical schedule for OSCAR 13 might state:

Off: from MA 0 until MA 49
Mode B: on from MA 50 until MA 159
Mode JL: on from MA 160 until MA 199
Mode B: on from MA 200 until MA 255

The MA units refer to entries in the phase column. Checking column 9 we see that the Mode B transponder won't be turned on until just before 1230 UTC, so for the first 45 minutes OSCAR 13 is in view, all we can do is check on telemetry. The abbreviation MA stands for the term *mean anomaly*. The expression "anomaly" is just fancy term for angle. Astronomers have traditionally divided orbits into 360 mean anomaly units, each containing an equal time segment. However, because of the architecture of common microprocessors, it was much more efficient to design the computers controlling Phase 3 spacecraft to divide each orbit into 256 segments of equal time duration. Radio amateurs refer to these as *mean anomaly* or *phase* units. The duration of each segment is the satellite's period divided by 256. For example, a mean anomaly unit for OSCAR 13 is roughly 2.68 minutes. At MA 0 (beginning of orbit) and MA 256 (end of orbit) the satellite is at *perigee* (its lowest point). At MA 128 (half way through the orbit) the satellite is at *apogee* (its high point). If you check the height column and the phase column you can confirm that apogee occurs at approximately 1557 UTC. Referring to the schedule, when the mean anomaly reaches 160, at approximately 1720 UTC, the satellite will switch from Mode B to Mode JL.

Because radio amateurs and astronomers (texts on astrophysics generally follow the astronomical notation) use the term mean anomaly in a slightly different way, there's sometimes a question as to which system is being used. The following facts should clarify any minor confusion that may occur. The MA values transmitted on OSCAR 13 telemetry (including CW telemetry) refer to a 0-to-256 system. All operating schedules for OSCAR 13 produced by AMSAT and ARRL employ a 0-to-256 system. The term phase always refers to a 0-to-256 system. Some computer tracking programs use the traditional astronomical notation. Its easy to determine when this is the case because the mean anomaly column entries will vary between 0 and 360.

There are several additional output parameters (not illustrated in Table 6-4) sometimes provided by tracking programs which may be of interest. A column labeled *echo* or *delay* provides the time delay introduced by the up- and downlinks. At OSCAR 13's apogee this amounts to slightly more than a quarter second. A column labeled *squint angle* provides useful information on Phase 3 satellites. It tells

Table 6-5

Example of Real-Time Output in Tabular Form

ATLANTA
REALTIME SATELLITE TRACKING
Coordinates on 11-10-1988
at 15:52:47 UTC

NAME	AZ	EL	RNG km	HGT km	LAT	LONG	ORBIT	PHASE
AO13	144	40	34497	32557	−4.1	58.4	313	83
FO12	108	−16	6706	1495	5.1	30.5	10214	157
UO11	5	−44	9684	660	55.5	270.8	25100	113
RS5	192	−41	10606	1666	−60.5	106.2	30359	135
RS7	3	−26	8364	1652	78.5	271.3	30451	91
MIR	144	−40	8778	332	−39.4	32.9	5302	210

how the satellite antenna is oriented with respect to your ground station. Squint angle can vary between 0 and 180 degrees. A squint angle of 0 degrees means the antennas are pointed directly at you. When the squint angle becomes greater than approximately 20 degrees on OSCAR 13 Mode B, you're going to begin seeing some annoying spin modulation on up- and downlinks. For satisfactory Mode L operation, you may find that you have to stick to squint angles under 10 degrees. Not all programs include the algorithms needed to calculate squint angle. If yours does, it will need some information about the orientation of the satellite. To provide this information to the computer you have to know the satellite's *Bahn latitude* and *Bahn longitude*. These parameters are sometimes labeled BLAT and BLON or ALAT and ALON where the prefix "A" stands for attitude. This information is often provided along with the basic orbital elements. It can also be found on AO-13's CW telemetry. Programs that provide squint angle may also contain a column labeled *predicted signal level*. Values are usually computed using a simple prediction model which takes into account satellite antenna pattern, squint angle and spacecraft range. See Chapter 15 for details. For Phase 3 satellites the model assumes a 0-dB reference point with the satellite overhead, at apogee, and pointing directly at you. At any point on the orbit the predicted level may be several dB above (+) or below (−) this reference level.

In addition to batch output many computer tracking programs provide real-time output with positions of several satellites updated every minute or so. Real-time output may be in either tabular form (as illustrated in Table 6-5) or in graphic form. Since the output is constantly being updated, it almost always consists of a screen display (no provisions for hard copy). Note that real-time output does not mean current-time output. You can generally specify whatever starting time you wish and the display will be updated in real time. (This is generally accomplished by temporarily changing the current time registered in the computer's clock.) The column headings should be clear from our discussion of batch output. A quick check of the elevation column in Table 6-5 shows that the only satellite in range at 15:52:47 UTC was OSCAR 13.

Real-time graphic output takes the subsatellite location data from the longitude and latitude columns of

Table 6-6

Example of Orbital Elements Provided for Computer Tracking Programs

Parameter name	Value	Units
Satellite:[1]	OSCAR 13	
Catalog Number:	19216	
Epoch Year:	1988	
Epoch Time:	258.28144	days
Element Set:	RUH9-88	
Inclination:	57.57	deg
RAAN:[2]	239.56	deg
Eccentricity:	0.6563	
Arg of Perigee:[3]	190.53	deg
Mean Anomaly:	0.0	deg
Mean Motion:	2.09699369	rev/day
Decay Rate:[4]	0.	rev/day/day
Epoch Rev:[5]	193	
Semi-Major axis:	25783	km
Bahn Latitude[6]	0	deg
Bahn Longitude[7]	180	deg

Alternate parameter names
[1] Object
[2] Right Ascension of Ascending Node; RA of Node
[3] Argument of Perigee
[4] Drag factor; Rate of change of mean motion, first derivative of mean motion
[5] Revolution number, orbit number
[6] ALAT, BLAT
[7] ALON, BLON

Table 6-5 and plots this information on a map. A Cartesian map is usually used, but some programs will give you the option of using a polar projection. The graphic display will probably contain data on azimuth and elevation and include a coverage circle for a specific satellite of interest. This coverage circle shows which terrestrial stations have access to the satellite at the moment being pictured. Whenever I'm sitting in my radio room reading I usually leave the computer on with the graphic mode running. Every now and then I'll glance at the screen and if a satellite I'm interested in happens to be nearby I can quickly switch on the receiver.

We now look at how data on specific satellites is entered into the computer program. One of the choices provided by the main menu is switching to the part of the program where data on orbital elements is maintained. If you elect this option the first question you'll be asked is whether you want to add elements for a new satellite, change the elements of an old satellite, or delete a satellite and its elements. Suppose you choose to add a new satellite. The program will respond by asking you to enter a series of numbers. To respond you need current data listing the orbital elements of the satellite of interest. This data is provided by several sources. Although slightly different formats are used, the information you have will probably look something like Table 6-6. Note that since the published orbital data must work with a large variety of programs, the table will probably include several entries which your program does not require. It helps to have some

understanding of the meaning of each entry and to know if it's optional.

The first two entries identify the spacecraft. The first line is an informal satellite name. Since you'll be using this name when you run the program, select a label you're comfortable with and that's easy to type—you do *not* have to enter the name exactly as listed. The second entry, *Catalog number*, is just a formal ID assigned by NASA. Most programs will not ask you for this information.

The next two entries give the *epoch* (the time for which the elements were computed). Some programs ask for the year and date separately, others ask for both to be entered at the same time. For example, your program may require that the epoch data from Table 6-6 be entered as 88258.28144.

The next entry, *element set*, is used to identify the source of the information. For example, RUH9-88 refers to data provided by Jim Miller (G3RUH) in September 1988. Jim periodically processes data from other sources and provides elements for OSCAR 13 command stations. These elements produce excellent long-term (up to a year) results. The element set identity is not required for tracking. If your program asks for it you can either repeat the information provided or leave the entry blank.

The next six entries are the six key orbital elements. All programs require them. A brief description of what each one refers to follows. A more detailed description is given in Chapter 11. *Inclination* describes the orientation of the satellite's orbital plane with respect to the equatorial plane of the earth. *RAAN* (Right Ascension of Ascending Node) specifies the orientation of the satellite's orbital plane with respect to the "fixed" stars. *Eccentricity* refers to the shape of the orbital ellipse. *Argument of perigee* describes where the perigee of the satellite is located in the satellite orbital plane. When the argument of perigee is between 180 and 360 degrees the perigee will be over the southern hemisphere. Apogee will therefore occur above the northern hemisphere. *Mean anomaly* locates the satellite in the orbital plane at the epoch. When entering the orbital elements, all programs use the astronomical convention for mean anomaly units. The mean anomaly is 0 at perigee and 180 at apogee. Values between 0 and 180 indicate that the satellite is headed up towards apogee. Values between 180 and 360 indicate that the satellite is headed down towards perigee. *Mean motion* specifies the number of revolutions the satellite makes each day. This element indirectly provides information about the size of the elliptical orbit. Some programs require semi-major axis or period instead of mean motion.

Decay rate is a parameter used by some programs to take into account how the frictional drag produced by the earth's atmosphere affects a satellite's orbit. It may also be referred to as rate of change of mean motion, first derivative of mean motion or drag factor. This is an important parameter in retrospective scientific studies where information about solar activity can be used to model the earth's atmosphere. It has very little effect on producing good long-term predictions, even with satellites in very low orbits where drag is important. If your program has algorithms which take drag factor into account, enter the number provided. (If the element set does not contain this information, enter zero—you shouldn't discern any difference in predictions.) You usually have a choice of entering this number in either decimal form or scientific notation. For example, the number -0.0005 can be entered as $-5.\text{E-}4$. (If this is totally confusing and you're not sure what to do, just enter zero.)

Epoch revolution is just an orbit number that's used for reference purposes. The number provided here does not affect tracking data so don't worry if different element sets provide different numbers for the same day and time. In fact, it's often simplest to just enter a zero here and ignore information on epoch revolution provided by the program.

Semi-major axis is a parameter used to describe the size of the orbital ellipse. Since this information can be computed from the mean motion it is actually redundant and not requested by most programs. If you program asks for it, just provide the number given. Remember not to use commas when entering large numbers.

Since it's very easy for an error to creep in when tables of orbital elements are prepared, or when you're entering numbers into the computer, its best to use the following procedure when you want to update orbital elements for a specific satellite, say OSCAR 13. From the main menu enter the orbital element data base. Select "new satellite" instead of "change" or "update." If the old set is named OSCAR13a call the new set OSCAR13b. Your program will treat OSCAR13a and OSCAR13b as two separate satellites. Follow both spacecraft for several weeks. If the differences in position are small the new element set is probably okay and the old set can be deleted.

A Graphic Method for Geostationary Orbits

Tracking a stationary satellite (one that remains over a fixed spot on the earth's equator) involves determining whether the spacecraft is in range of your ground station and, if so, antenna azimuth and elevation angles. A chart like the one shown in Fig 6-7 makes it easy to obtain the needed information if the spacecraft's longitude is known. This approach works for all geostationary satellites including weather satellites, TV broadcasting satellites and future amateur Phase 4 spacecraft.

An example will illustrate how Fig 6-7 is used. Suppose a ground station in New Orleans (latitude = 30°N, longitude = 90°W) is interested in accessing a stationary satellite located at 40°W longitude.

Step 1. Compute a relative longitude by subtracting the ground station's longitude (90°W) from the satellite's longitude (40°W): 40°W − 90°W = −50°W. Plot the point consisting of the absolute values of (a) the relative longitude just computed and (b) the latitude of the ground station. See point A (50°, 30°) on Fig 6-7.

Step 2. If the point is inside the 0 degree elevation curve the satellite is in range of the ground station. The location of point A indicates that the satellite is in range of New Orleans.

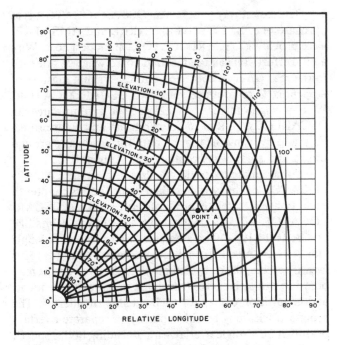

Figure 6-7—Chart for obtaining azimuth and elevation directions to geostationary satellites. See text for directions.

Step 3. Noting intersections of elevation curves and radial (azimuth) curves near point A, use "eyeball interpolation" to estimate the antenna elevation and radial values. In our example the elevation value is 27 degrees and the radial value is 113 degrees. Azimuth is determined from the radial value. For Northern Hemisphere stations the azimuth is equal to either (a) the radial value or (b) 360° − (radial value), depending on whether the satellite is east or west of the ground station. For Southern Hemisphere stations the azimuth is equal to either (a) 180° − (radial value) or (b) 180° + radial value, depending on whether the satellite is east or west of the ground station.

Step 4. The antenna should be aimed as per the results of Step 3 and then the direction varied slightly to peak received signals. If the satellite is being kept accurately on station the resulting antenna azimuth and elevation will remain fixed.

A full size chart of the type used for Fig 6-7 is contained in Appendix C. Computational methods for obtaining antenna pointing data for geostationary satellites are covered in Chapter 11.

Graphic Methods: Additional Information

Directions for using the OSCARLOCATOR were given earlier. We now look at some related practical concerns relating to where they can be obtained and how they can be built. We'll also consider some variations which will enable you to customize a graphic tracker to your particular needs.

The easiest way to obtain an OSCARLOCATOR is to purchase one. In 1990 both the ARRL and AMSAT-UK were providing OSCARLOCATORs at modest cost. With a little effort you can use the map board which is attached to the rear cover of this book and the tracing masters of Appendix C to produce your own OSCARLOCATOR. With a slight increase in effort you can produce an OSCARLOCATOR in conjunction with any polar projection map by using the tabular data for producing ground tracks and spiderwebs contained in Appendix C.

Some amateurs prefer to devote a single OSCAR-LOCATOR of the type described earlier in this chapter to each Phase 2 satellite orbit of interest. One OSCAR-LOCATOR will work for satellites in similar orbits like OSCARs 14-19. Other amateurs prefer to use a single OSCARLOCATOR for all active Phase 2 satellites. This is easily accomplished by limiting the spiderweb on the map board to azimuth curves (which are satellite independent) and acquisition curves for each satellite of interest. The ground tracks may be either placed on separate overlays or drawn on a single overlay with ascending nodes spaced out around the perimeter. Using a distinctive color for each satellite ground track and its accompanying acquisition circle aids readability. Since the multisatellite version of the OSCARLOCATOR omits elevation circles, elevation angle pointing data is not immediately available. This is of no concern if we're using an omnidirectional antenna. However, if a low-gain beam is in use some rough elevation angle data is desirable. A glance at the tracing masters in Appendix C shows that the radius of a 30-degree elevation circle is always less than half the radius of the associated acquisition circle. Using this fact we can usually guesstimate elevation angles to sufficient accuracy.

Large maps can be useful for demonstrations. They also can provide greater tracking accuracy if spiderwebs and overlays are carefully drawn. Some inexpensive map sources are listed in Table 6-7. Over the years I prepared several large OSCARLOCATORs for lectures; each took several hours to construct. Then, a few years ago, I discovered that copying machines with the ability to produce transparency masters (used in conjunction with overhead projectors) created excellent OSCARLOCATORs. The map board (with spiderweb) and the ground track are copied onto separate transparent overlays and combined in the usual manner. When the resulting OSCAR-LOCATOR is placed on an overhead projector the display is excellent.

If you decide to construct an OSCARLOCATOR, a trip to a drafting or art supply store will provided the needed materials: some transparent sheets or Plexiglas for the overlays, colored pens and nylon filament tape to reinforce the overlay and map board where they pivot at the pole. Since many transparent plastics are nearly impossible to draw on you'll find it convenient to first place a layer of Scotch® #810 tape in position. This tape is easily drawn on and it shows colors vividly. After the lines are drawn a second layer of tape will shield the OSCARLOCATOR from smudging.

Drawing the spiderweb is the most time-consuming

Table 6-7
Inexpensive Sources of Polar Maps

1) North pole stereographic projection, multicolor, extends to equator, "USAF Physical-Political Chart of the World"; GH-2A (41 cm diameter); GH-2 (82 cm diameter). Source: Department of Commerce, Distribution Division (C-44), National Ocean Survey, Riverdale, MD 20840.

2) North pole stereographic projection, 2-color, 105 cm/100 cm, stock no. DOD WPC xx032004. Source: same as 1.

3) South pole stereographic projection, 2-color, 105 cm/100 cm, stock no. DOD WPC xxp32007. Source: same as 1.

4) North pole azimuthal equidistant projection, black/white, extends to 30° South latitude, 61 cm diameter. Single copies available at no charge from APT Coordinator, Department of Commerce, NOAA, National Environmental Satellite Center, Suitland MD 20233. Request "APT Plotting Board" and reference this book.

5) Plain polar graph paper makes a very effective map board if political boundaries are not needed (azimuth equidistant projection).

Note: Azimuthal equidistant polar projection maps are characterized by equally spaced latitude circles. Stereographic polar projection maps are characterized by increasing spacing between latitude circles as they get farther from the center (see Chapter 12 for additional details).

part of producing a tracking device. In the "old days" many radio amateurs drew circles on a globe around their location and then transferred each circle, point by point, to a map—an extremely tedious process. We'll be discussing some shortcuts but the accurate methods still require considerable effort.

OSCARLOCATOR Spiderwebs: Several spiderweb tracing templates for satellites in circular low earth orbits are included in Appendix C. The templates are for stations at 30° N and 46° N latitude. If you're using the map in this book (rear cover flap) and live between 10° N and 60° N latitude, the template closest to your latitude should work well.

If you live further north, or are using a different polar map (one with equally spaced latitude circles) and you're willing to accept slightly lower accuracy, consider the following approach. By ignoring the distortions that occur when a circle is transferred from the globe to a flat map, we can approximate acquisition circles with true circles about your location and azimuth curves with straight lines radiating outward. Consider Mir as an example. Referring to Table 6-2 we see that the acquisition distance is 1368 miles, which corresponds to a 19.8° arc length measured along a great circle. The degree value makes it easy to determine where the acquisition circle will intersect the longitude line running through your ground station. A station in New Orleans (30°N, 90°W) would add 19.8° to the latitude to locate the point (49.8°N, 90°W) on the acquisition circle, which is then sketched in with a drawing compass. The degree values for the distances, contained in Table 6-2, can be used to draw elevation circles in a similar manner. The errors associated with this technique become worse as we get further from the pole and as the height of the satellite increases.

The tabular data of Appendix C can be used to produce more accurate spiderwebs. These tables will work with all polar maps (any size, any projection).

A less tedious method for drawing accurate spiderwebs on stereographic polar projection maps (characterized by increasing spacing between latitude circles as they get further away from the center) is covered in Chapter 12.

If you have access to a personal computer you can use the simple BASIC Spiderweb program in Appendix D to produce tables like those shown in Appendix C keyed to your specific location.

Methods for using the OSCARLOCATOR in conjunction with elliptical orbits are covered in Chapter 12.

COMPUTER METHODS: ADDITIONAL INFORMATION

The computer tracking programs currently available to radio amateurs contain a great many features and are considerably more complex than the ground-breaking W3IWI program presented in ORBIT in 1981. One that I've recently worked with has about 6000 lines of code; a listing runs 120 pages. Several authors have donated their work to AMSAT. When these programs are purchased from AMSAT, you can be assured that you're getting the latest version and that income in excess of costs goes toward satellite construction. A self-addressed stamped envelope to AMSAT (PO Box 27, Washington, DC 20044) will get you a list of what's available.

If you purchase a tracking program from AMSAT you can expect to receive a competently engineered program, but you won't see the glitzy packaging and multicolored manuals associated with many commercial packages. You should also be aware that these packages are not designed to teach you how to use your computer—they're aimed at those already familiar with fundamental computer operations such as copying a disk, loading an application program, setting up a self-booting disk, and so on.

Let's assume you've just received a tracking program on a disk from AMSAT. The first thing you should do is put a write-protect tab on this disk, make a copy (the working copy) and put the original in a safe place. For your convenience, AMSAT software is not copy protected—you're on your honor. The working copy should be used without a write-protect tab, since you'll be entering information on your ground station location, satellite orbital elements, etc. Place the working copy in your computer and obtain a list of files on the disk. You'll probably spot a file called read.me, or readme, or name.doc, or name.txt, where "name" is the name of your tracking program. The file you're searching for contains the instructions for using the tracking program. The information it contains has been stored in a plain text format so you can print it to the screen or printer. It's usually handy to have a hard copy of the

Table 6-8
Sidereal Time Constants

Greenwich Mean Sidereal Times (GMST) for January 00.00Z of each year

Year	Value	Year	Value
1989	0.27676777	1995	0.27552708
1990	0.27610467	1996	0.27486399
1991	0.27544157	1997	0.27693880
1992	0.27477847	1998	0.27627570
1993	0.27685328	1999	0.27561260
1994	0.27619018		

instructions, especially when you're first learning how to use the program, so print a copy if possible.

If you keep in mind the basic underlying organization common to most programs—a data base for satellite orbital elements, a data base containing ground station location information and a list of your computer resources, a batch tracking subprogram and a real-time tracking subprogram—you'll have a big head start on learning how to use a new software package. When working with a new program, if you get stuck at some point and you're not sure what to do, use the trial and error approach. It's often much quicker than wading through the directions. Remember that you always have your master disk as backup so nothing serious can go wrong.

In order to run quickly, most programs being distributed today have been compiled into machine language. This makes them difficult to modify. If you're interested in programming and you want to develop a tracking program for a special purpose the best place to start is by mastering the material in Chapters 11 and 12 and the sample programs in Appendix D.

If you have a tracking program that's been operating fine and it suddenly quits on January 1 or February 1 the most likely reason is that the software requires a new Sidereal Time Constant for each year and the table of values in your program has run out. Some programs treat January as the 13th month of the previous year, hence the February 1 date for appearance of the problem. Values that will take you through 1999 are listed in Table 6-8. For years after 1999, you'll have to check the *Nautical Almanac* (see Appendix E).

ACQUIRING UP-TO-DATE INFORMATION

How long does it take for tracking data to become out-of-date? It depends on the satellite orbit, the source of the data and your requirements. Most sources are disseminating good data. We can therefore focus on general guidelines related to satellite orbit and user needs. Elapsed times are measured from epoch, not from the day you obtain the elements.

With a Phase 3 satellite, updating elements every three months provides excellent results for most users. Using the special smoothed data provided by G3RUH one can often go for 6-12 months on one element set. Since the effects of changes in the elements are much more pronounced around perigee, you may want to update more often if you're involved in activities that require satellite access near perigee. If you're using squint angle information you should update the values for Bahn latitude and Bahn longitude whenever they're changed. (Changes usually coincide with a new operating schedule.)

Turning to satellites in low earth circular orbits, for those above 800 miles updating elements every six months is usually sufficient. For satellites between 500 and 800 miles in height, inserting new elements every three months is desirable. For spacecraft between 400 and 500 miles, every six weeks is suggested. And, for spacecraft under 400 miles (like the Shuttle and Mir), elements should be updated as often as possible.

Aside from orbital maneuvers, which are mainly of concern with the Shuttle and Mir, the main cause for changes in the orbital elements of low-altitude satellites is atmospheric drag. When the sun is inactive, the average status of the atmosphere can be fairly well predicted and this effect can be taken into account. When the sun is active, atmospheric composition changes radically over short time periods, making it impossible to take drag effects into account. As a result, when drag effects are small and relatively unimportant we can incorporate them into a tracking prediction model; when drag effects are large and important we have no reliable way of modeling them. When using the suggested time intervals for updating orbital elements, keep in mind that you might want to shorten the intervals near sunspot maxima and lengthen them near sunspot minima.

Several tracking programs currently in circulation use a prediction algorithm which requires that the orbital element epoch be of the current year. (During January, elements from the previous year are accepted.) If you're using one of these programs you must update your orbital elements sometime during January of each year.

Now that we've described what "up-to-date" means, we turn to sources for information, including publications, computer bulletin boards and nets.

Since most publications involve considerable time lag, they're not sufficient for providing information on all satel-

---

Table 6-9

Typical NASA Two-Line Orbital Element Set

```
ISSUE DATE: JUNE 6, 1989
BLTN     36 ELEM    36 OBJ  19216

          1         2         3         4         5         6
1234567890123456789012345678901234567890123456789012345678901234567890123456789

1 19216U           89147.06007421 0.00000137           10000-3 0    364

2 19216  57.2077 206.1830 6723768 204.7577  96.6993  2.09696457  7298
```

Explanation

Heading
   BLTN 36 = Bulletin number 36
   OBJ 19216 = NASA Catalog number (19216 is OSCAR 13)
   Column numbers (1-69) are not provided. They have been added here to clarify NASA format.

Line 1
   line number = 1
   19216 = NASA catalog number
   U = Unclassified
* 89147.06007421 = Epoch (1989, day 147 = May 27 at 01:26:30.41 UTC)
   0.00000137 = First derivative (rate of change) of mean motion
   364 = Element set number 36; checksum 4

Line 2
   line number = 2
   19216 = NASA catalog number
* 57.2077 = Inclination (degrees)
* 206.1830 = RAAN (degrees)
* 6723768 = eccentricity (note: leading decimal point is missing)
* 204.7577 = Argument of perigee (degrees)
* 96.6993 = Mean anomaly (degrees); multiply by 256/360 for phase units
* 2.09696457 = Mean motion (rev/day)
   7298 = Revolution 729 at epoch; checksum 8

*Fundamental orbital elements at epoch required by all tracking programs.

For information on entries not covered, request "Format Explanation of the Two-Line Orbital Elements" from:
   Project Operations Branch, Code 513, Goddard Space Flight Center, Greenbelt, MD 20771.

lites. However, you may wish to check *The AMSAT Journal*, *OSCAR News* (published by AMSAT-UK) or *OSCAR Satellite Report*.

Updates are also frequently reported on W1AW and on the various AMSAT nets, including the US nets on Tuesday evening (3.840 MHz, 9 PM local time) and international nets on weekends (14.282 MHz, 1900 UTC and 21.280 MHz, 1800 UTC on Sundays for North and South America, Europe, Africa; 14.282 MHz, 2200 UTC on Saturdays for US, South Pacific).

Check your local packet-radio bulletin board. Most keep a file of the latest orbital elements as distributed by AMSAT. If you don't have packet-radio equipment, you can probably locate a local amateur who'll download the data for you. Orbital element data is also available on several telephone bulletin boards and from CompuServe's HamNet. See Chapter 10 for phone numbers and modem settings.

Scientists, educators and those operating information distribution services can have elements mailed to them by NASA. Send your request on letterhead stationery to: NASA, Goddard Space Flight Center, Project Operations Branch Code 513, Greenbelt, MD 20771. You must specify the NASA Catalog number for each satellite of interest. The information is presented in the abbreviated two-line format shown in Table 6-9.

CONVERSIONS

Converting units between various systems and formats is a frequently encountered problem when using tracking devices. Techniques for performing several different types of conversions follow.

Time Zone Conversions

Table 6-10 enables ground stations in the United States to convert to UTC from local time, or to local time from UTC. The switch between daylight and standard time can introduce a lot of confusion when planing future events. The Uniform Time Act of 1966 (as modified in 1987) specifies that Daylight Savings Time will be observed for approximately six months each year, beginning the first Sunday in April and ending the last Sunday in October. However, Arizona, Hawaii, Michigan, parts of Indiana and possibly other regions do not conform.

Epoch Time and Standard Time Notation

When using tracking devices you will not normally be required to convert between month/day and day of year

---

Table 6-10

Time Zone Conversion Chart for USA

Time zone	EST	EDT	CST	CDT	MST	MDT	PST	PDT	AK/HI	AK/HI
Time difference	5	4	6	5	7	6	8	7	10	9

To convert from UTC to (time zone) subtract (time difference) hours.

To convert from (time zone) to UTC add (time difference) hours.

Table 6-11

Chart for Converting Between Day/Month and Day of Year Notation

Day of year = day of month + number listed

Month	Normal year	Leap year
January	0	0
February	31	31
March	59	60
April	90	91
May	120	121
June	151	152
July	181	182
August	212	213
September	243	244
October	273	274
November	304	305
December	334	335

Sample Problem 6-1

Convert day 91089.37166448 to standard date/time notation.

Standard date/time notation
  YY:MM:DD = Year:Month:Day
  HH:MM:SS = Hour:Minute:Second

Epoch time: 91089.37166448
  91089.37166448 = 1991, day of year 089, +0.37166448 day
  day of year 089 = 59 + 30 = March 30 (see Table 6-11)
  0.37166448 day = (0.37166448)(24) hours = 8.919948 hours
  0.9199475 hours = (0.9199475)(60) minutes = 55.1969 minutes
  0.1968512 minutes = (0.1968512)(60) seconds = 11.81 seconds
Day 91089.37166448 = 1991, March 30 at 08:55:11.81 UTC

Sample Problem 6-2

Convert from (YY:MM:DD/HH:MM:SS) date/time notation to Epoch time.

date/time = 1991, March 30 at 8 hours, 55 minutes, 11.81 seconds
  = 91:03:30/08:55:11.81 (YY:MM:DD/HH:MM:SS)

  Month 3 + day 30 = 59 + 30 = 89 day of year (see Table 6-11)
  55 minutes = (55)(60) seconds = 3300 seconds
  08 hours = (08)(60)(60) seconds = 28,800 seconds
  08:55:11.81 = (28,800 + 3,300 + 11.81) seconds = 32,111.81 seconds
  1 day = 86,400 seconds
  32,111.81 seconds = (32,111.81/86,400) days = 0.37166447 days
Epoch date/time: 91:03:30/08:55:11.81 = 91089.37166447

Sample Problem 6-3

Angle Conversions

To convert from degrees West of Greenwich to degrees
  East of Greenwich subtract from 360.
  Example (Washington, DC): 77°W = (360 − 77)°E
    = 283°E

To convert from degrees East of Greenwich to degrees
  West of Greenwich subtract from 360.
  Example (Moscow): 36°E = (360 − 37)°W = 324°W

To convert from degree-minute to decimal degree notation
  Example: 38 deg 41 min 16 sec
  38 + 41/60 + 16/3600 = 38 + 0.6833 + 0.0044
    = 38.6877 deg
  38 deg 41 min 16 sec = 38.6877 deg

To convert from decimal degree to degree-minute notation
  Example: 38.6877 degrees
    0.6877 deg = (0.6877)(60) = 41.262 minutes
    0.262 minutes = (0.262)(60) = 16 seconds
    38.6877 deg = 38 deg 41 min 16 sec

Sample Problem 6-4

Find the AMSAT day number corresponding to
10 September 1988

Let the date of interest be specified by:
  year = YY
  day of year = DOY (see Table 6-11)

The AMSAT day number (ADN) is given by

$ADN = (YY - 1978){*}365 + (DOY - 1) +$
      $INT[(YY - 1977)/4]$

where the last term adjusts for leap years by taking the
  integer part of the quotient.

From Table 6-11 we note that 10 Sep is the 254th day of
1988 (which is a leap year).

$ADN = (1988 - 1978){*}365 + (254 - 1) + INT[(1988 - 1977)/4]$
     $= 3650 + 253 + 2 = 3905$

notation or between epoch time and standard date/time
notation. However, these problems sometimes do arise, so
we have presented Table 6-11 and two examples (Sample
Problems 6-1 and 6-2) illustrating how the conversions are
accomplished.

Angle Units

When working with angles such as latitude and longitude you may have occasion to switch between decimal
degree and degree-minute notation. It's also sometimes
necessary to convert latitudes from degrees West of Greenwich to degrees East of Greenwich. Sample Problem 6-3
illustrates how these conversions are accomplished.

AMSAT Day Number

Shortly before the launch of the Phase 3-A spacecraft,
amateurs developing software for the project decided
it would be convenient to select a specific day to use as
a reference for orbit calculations. The day selected,
January 1, 1978, is referred to as day zero. AMSAT day
numbers appear on Phase 3 satellite telemetry. Sample
Problem 6-4 shows how to compute the AMSAT day
number when the calendar date is known.

Chapter 7

Antenna Basics

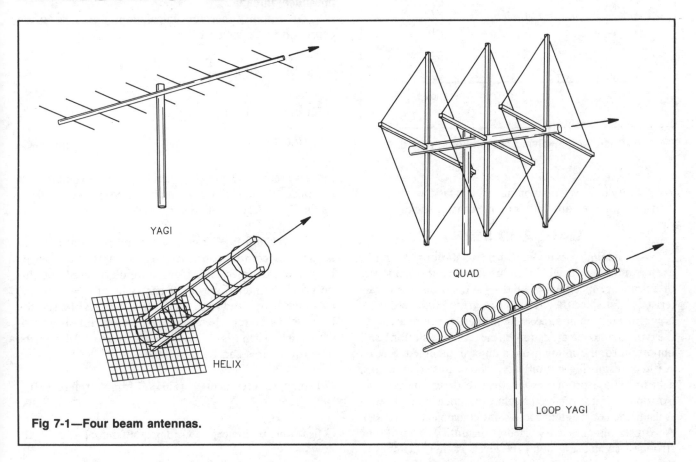

Fig 7-1—Four beam antennas.

G round station performance is affected by many factors, but one stands out as being critically important: the antenna system. Although there are no intrinsic differences between antennas for satellite use and those for terrestrial applications, some designs are clearly better suited to satellite work than others. Properties that make a certain type of antenna desirable for HF operation may make it a poor performer on a satellite link, and vice versa. In this chapter we'll consider the relation between basic antenna characteristics and satellite radio links. In the next chapter we'll look at several types of practical antennas useful for satellite communications.

Simply stated, an antenna for monitoring downlink signals should be chosen to provide an adequate signal-to-noise (S/N) ratio at the receiver output; an antenna for transmitting on the uplink should be chosen to provide the desired signal level at the satellite. While pursuing these goals we also try to keep costs down and minimize the complexity associated with large mechanical structures and high aiming accuracy.

The antenna system characteristics we'll focus on include:

1) Directional properties (gain and pattern)

2) Transmitting vs receiving properties
3) Efficiency
4) Polarization
5) Link effects (spin modulation, Faraday rotation)

One basic concept we'll refer to time after time is the *isotropic antenna*: an array that radiates power equally in all directions. No one has ever been able to build a practical isotropic antenna but the concept is still very useful as a "measuring stick" against which other antennas can be compared. Closely related is the *omnidirectional antenna*, one that radiates equally well in all directions in a specific plane. Practical omnidirectional antennas are common; the ground plane is one example. Any antenna that tends to radiate best in a specific direction (or directions) may be called a *beam antenna*. Several beams (the Yagi, quad, loop Yagi and helix) are shown in Fig 7-1. Even the common dipole can be regarded as a beam since it has favored directions. The "first law" of antennas is: You don't get something for nothing. A beam can only increase the power radiated in one direction by borrowing that power from someplace else. In other words, a beam acts by concentrating its radiated energy in a specific direction.

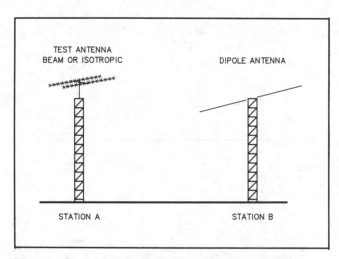

Fig 7-2—Radio link involving two stations, A and B.

To quantify how well it accomplishes this task, we compare it to the isotropic antenna, our measuring stick.

GAIN AND EIRP

An imaginary radio link with two stations, A and B, as shown in Fig 7-2, will help illustrate how the properties of a beam are specified. We'll discuss the transmit characteristics first since they're easier to grasp. Later we'll see how transmitting and receiving properties are related. As the type of antenna at Station B (the receiving station) isn't important for the comparison, a dipole is assumed. Station A (the transmitting station) has a choice of two antennas, a beam whose properties we wish to determine and an isotropic radiator. Our "thought experiment" begins with A using the beam antenna and some convenient power (P). A adjusts his antenna's orientation until B records the strongest signal, and notes the level. A then switches to the isotropic antenna and adjusts the power (P_i) until B reports the same signal level as noted earlier. The gain *(G)* of the beam is given by the formula

$$G = \frac{P_i}{P} \qquad \text{(Eq 7.1)}$$

For example, if 25 watts to the beam produced the same signal level at B as 500 watts to the isotropic, the gain of the beam would be

$$G = \frac{500}{25} = 20$$

This is roughly what would be expected from a well-designed Yagi with a boom length of 2 wavelengths.

Now suppose that B is the satellite and A is your uplink system. Aha! The satellite sees exactly the same signal whether you run 500 watts to an isotropic radiator or 25 watts to the beam. In either case we'd say the ground station *EIRP* (effective *i*sotropic *r*adiated *p*ower) is 500 watts. EIRP and the quantity P_i in our "thought experiment" are identical. We can rewrite Eq 7.1 as $P_i = GP$ (EIRP is equal to the product of "gain" and "power being fed into the antenna"). An EIRP of 500

watts can also be produced by a beam with a gain of 4 that is fed 125 watts, a beam with a gain of 10 that is fed 50 watts, and so on. The definition of EIRP we've been using just depends on gain and power fed to the antenna. Later we'll see how this can be generalized to include transmitter output power, feed-line losses and even the effects of a misaimed antenna.

To simplify certain calculations, gain is often expressed in decibels (dB).

$$G \text{ [in dB]} = 10 \log \frac{P_i}{P} \quad \text{or,} \qquad \text{(Eq 7.2A)}$$

$$G \text{ [in dB]} = 10 \log G \quad \text{or,} \qquad \text{(Eq 7.2B)}$$

$$G = 10^{G/10} \qquad \text{(Eq 7.2C)}$$

Since we refer to G and *G* as "gain," it's important to note the units. If gain is simply a number (a ratio), we're talking about *G* (Eq 7.1); if gain is given in decibels we're referring to G.

Eq 7.1 and Eq 7.2 clearly depend on what antenna is used for comparison (the reference antenna); it's the isotropic. At times, a half-wave dipole is used for this purpose. The half-wave dipole has a gain of 1.64 (2.14 dB) over an isotropic radiator. As a result, the gain of a specific beam looks better when the reference antenna is an isotropic than when it's a dipole. Eqs 7.3A and 7.3B describe how the figures can be translated.

$$G \text{ [isotropic reference]} = (1.64)(G \text{ [dipole reference]})$$
$$\text{(Eq 7.3A)}$$

$$G \text{ [isotropic reference]} = G \text{ [dipole reference]} + 2.14 \text{ dB}$$
$$\text{(Eq 7.3B)}$$

Obviously, it's very important to specify the nature of the reference. This is sometimes done by expressing gain in either dBi or dBd, where the last letter describes the reference antenna as *i*sotropic or *d*ipole. Note that so far we've looked only at the one direction in which the maximum signal is radiated.

GAIN PATTERNS

We've seen how one very important antenna characteristic, gain, is specified. Gain tells us nothing, however, about the three-dimensional radiation pattern of an antenna. A beam with a given gain might have one broad lobe as shown in Fig 7-3A, or several sharp lobes as shown in Fig 7-3B. A single broad lobe is generally more desirable because it makes the antenna easier to aim and is usually less susceptible to interfering signals. Because drawing quantitative three-dimensional pictures, like those in Fig 7-3, is difficult, the directional properties of an antenna are more often pictured using one or two two-dimensional cross-sections drawn to include the direction of maximum radiation. In Fig 7-4 we show two common cross-sections (gain patterns) used to describe a Yagi. When beams are installed for terrestrial communications the

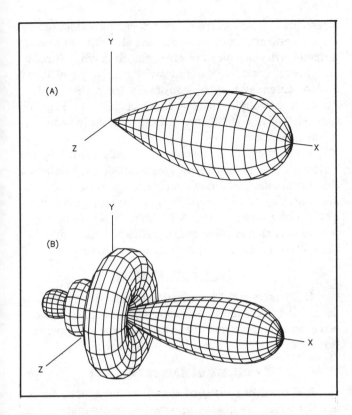

Fig 7-3—Three-dimensional illustrations of beam patterns. (A) single lobe and (B) multi lobe.

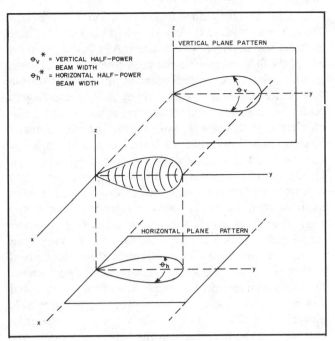

Fig 7-4—Relation between 3-D gain pattern and horizontal and vertical cross-sections.

cross-sections may conveniently be referred to as horizontal plane (azimuth plane) and vertical plane (elevation plane) patterns. When working with antennas that can be aimed upward or those using circular polarization, it's important to clearly specify the relation between any two-dimensional pattern pictured and physical orientation of the antenna. Before we continue, note that the gain pattern of an isotropic antenna is a circle in any cross-sectional plane, and the gain pattern of the omnidirectional antenna is a circle in one specific plane.

Gain patterns can be specified in terms of either power or field strength (field intensity). As power is directly proportional to the square of field intensity, translating back and forth between the two descriptions is relatively simple. For example, when field intensity drops to 0.707 of its maximum value, power will have dropped to 0.5 of its maximum since $(0.707)^2 = 0.5$.

One important characteristic of the gain pattern of an antenna is the *beamwidth*: the angle between the two straight lines that start at the origin of the pattern and that go through the points where the radiated power drops to one half its maximum value. See Fig 7-4. Since high antenna gains are obtained by concentrating the radiated power in a specific direction, it's clear that beamwidth and gain must be related. High gains can only be obtained by sacrificing beamwidth. It's possible, and sometimes desirable, to design an antenna so that two gain cross-sections taken at right angles are shaped significantly differently. For many familiar beam antennas, however, the two patterns are very similar. In cases where an efficiently designed antenna produces a symmetrical pattern (one

that's independent of cross-section orientation), the maximum beamwidth, θ^*, for a given gain is given approximately by Eq 7.4.

$$\theta^* \text{ [in degrees]} = 10 \sqrt{\frac{400}{G}} \qquad \text{(Eq 7.4)}$$

For example, an antenna with a gain of 20 would have a beamwidth of roughly 45°. Tracking a slowly moving target like an elliptical-orbit Phase 3 satellite near apogee with such an antenna would pose no problem. Tracking a speedy, low-altitude satellite, however, might be difficult.

Most discussions of antennas begin with a *free space* model that represents an antenna as if it were in outer space with no nearby objects to affect its performance. In the real world, RF reflections can have a large impact on an antenna's behavior. This is especially true for vertical plane patterns. At low elevation angles, for example, a receiving antenna will see two signals: a direct signal and a ground-reflected signal. Depending on the phase difference, these signals might add, giving up to 6 dB of ground reflection gain, or destructively interfere to produce a null. The real-world vertical pattern will therefore often consist of a series of peaks and nulls.

Ground reflection also has an impact on phase. A vertically polarized wave is reflected without any phase change, while a horizontally polarized wave undergoes a 180° phase change when reflected.

So far, we've been looking at the properties of antennas from a transmitting point of view. How do these properties relate to reception?

TRANSMITTING VS RECEIVING

A basic law of antenna theory, known as the *reciprocity principle*, states that the gain pattern of an antenna is the same for reception as for transmission. Let's see how this can be applied to the link shown in Fig 7-2. This time consider the situation where station A is at the receiving end of the link and assume that the incoming natural background noise at A is independent of direction. If station A measures the noise power arriving at the receiver with both a beam antenna and an isotropic antenna of the same efficiency, he'll obtain the same result. The beam actually picks up more noise than the isotropic from the primary direction and less noise than the isotropic from other directions, but the overall result is that both antennas capture the same total noise power. Now let's see what happens when station B transmits a reference signal at any convenient fixed power. The total amount of signal power reaching station A is fixed but the power is arriving from a particular direction. A beam antenna pointed toward station B will provide station A with more signal power than would an isotropic antenna. As a result, when signal and noise are present, the beam produces a better S/N power ratio at the input to A's receiver. For well-designed antennas, the improvement in the S/N power ratio over a communications link will be the same whether the antenna switch—beam for isotropic—is made at the transmitting end or the receiving end.

The reciprocity principle does *not* state that a particularly desirable receiving antenna is consequently also desirable as a transmitting antenna, or vice versa (though this is often the case). In transmitting, the objective is to produce the largest possible signal level at the receive point. High efficiency and gain are therefore very important. When receiving, the objective is to obtain the best possible S/N ratio. Though high efficiency and gain contribute to this goal, the shape of the gain pattern and the location of nulls may have a significant impact on S/N ratio by reducing noise and interfering signals. One reason for this is that, in the real world, background noise depends on direction. In sum,

> A good antenna for transmitting to a satellite is not necessarily a desirable antenna for receiving signals from a satellite.

Since separate antennas are required for the uplink and downlink at a satellite ground station, we can select each antenna independently.

EFFICIENCY

A transmitting antenna that is 100% efficient radiates all the power reaching its input terminals. Reduced efficiency causes an antenna to radiate less power in every direction; it has *no* effect on antenna pattern. A transmitting antenna that is 50% efficient only radiates half the power appearing at its input terminals. Since building high-efficiency VHF and UHF antennas is relatively easy, and producing transmit power at 146 MHz and higher frequencies is difficult, inefficient transmitting antennas should never be used at a satellite ground station.

The trade-offs are somewhat different for receive systems. A receiving antenna that's 50% efficient passes along only half the signal *and* half the noise power it

intercepts. If the receive chain's S/N ratio is limited by atmospheric or cosmic noise arriving at the antenna, poor antenna efficiency may not affect the observed S/N ratio. In practice, the only situation where relatively inefficient receive antennas may be considered is on the 29-MHz Mode A downlink where a compact half-size Yagi (or crossed-Yagi array) may provide the same performance as a full-size model.

Low antenna efficiencies are usually caused by (1) physically small (relative to design wavelength) elements that require inductive loading, (2) lossy power-distribution and matching systems, especially in multi-antenna arrays, or (3) poor ground systems (when the ground is an integral part of the antenna). Remember, efficiency does not affect the pattern of either transmit or receive antennas.

POLARIZATION

Our treatment of polarization begins with a technical description of the term polarization as it's applied to radio waves and antennas and goes on to examine how polarization affects satellite radio link performance.

Technical Description

Radio waves consist of electric and magnetic fields, both of which are always present and inseparable. Since most amateur antennas are designed to respond primarily to the electric field, it is possible to limit our discussion to it. When a radio wave passes a point in space, the electric field *at that point* varies cyclically at the frequency of the wave. When we discuss the *polarization of a radio wave* we're focusing on how the electric field varies.

The electric field can vary in *magnitude*, in *direction* or in both. If, at a particular point in space, the magnitude of the electric field remains constant while the direction changes, we have *circular polarization* (CP). [Note: All changes referred to here are cyclic ones at the frequency of the passing wave, and the direction of the electric field is confined to a plane perpendicular to the direction of propagation.] If, on the other hand, the direction of the field remains constant, while the magnitude changes, we have *linear polarization* (LP). If both magnitude and direction are varying we have *elliptical polarization*.

Imagine that we have a special camera set up at a fixed point in space which allows us to obtain a series of 12 time-lapse photographs of the electric field of a passing radio wave during one complete cycle. We begin our "thought experiment" by taking a set of pictures for each of three linearly polarized waves. See columns A, B and C of Fig 7-5. The three waves have the same frequency and maximum amplitude. The electric fields of waves B and C (columns B and C) are physically perpendicular to the field of wave A (column A). Wave B is in phase with wave A (both electric fields are a maximum at the same time), while wave C is 90° out of phase with wave A (the electric field of one is a maximum when the other is a minimum). If we combine (vectorially) linearly polarized waves A and B we obtain the linearly polarized wave shown in column D. Note its orientation and maximum amplitude (about 1.4 times the magnitude of the components). Observing

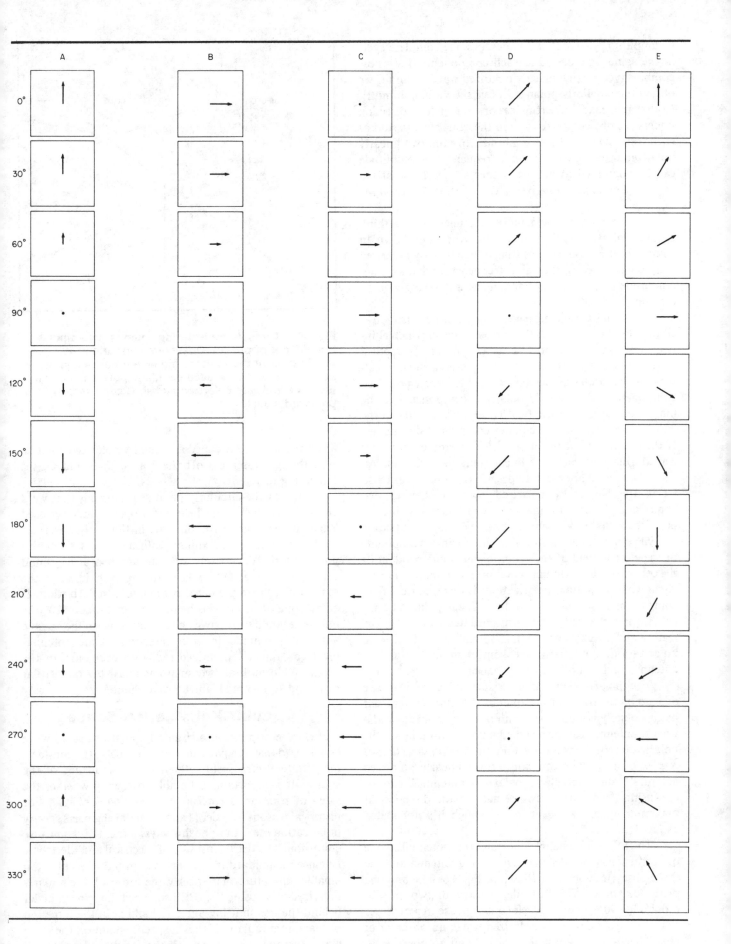

Fig 7-5—Electric field polarization at a point in space. See text for details.

its amplitude we see that it undergoes one complete cycle in the same time period as each component. If we now combine (vectorially) linearly polarized waves A and C we obtain the circularly polarized wave shown in column E. Note that its magnitude remains constant while its direction undergoes one complete cycle in the same time period as each component cycles in amplitude. In sum, two linearly polarized waves having the same frequency and amplitude can be combined to produce either a linearly polarized wave or a circularly polarized wave—*it all depends on the phasing*.

As a result of this imaginary experiment we can think of a circularly polarized wave, like the one shown in column E of Fig 7-5, as consisting of two linearly polarized components having the same frequency and maximum amplitude but oriented at right angles and having a phase difference of 90°.

Let's consider the most general type of polarization— elliptical polarization. Of the various ways in which elliptical polarization can be described, two are of practical interest to radio amateurs. The first pictures an elliptically polarized radio wave as consisting of a linearly polarized component and circularly polarized component. If the magnitude of the electric field varies only slightly in the course of each cycle, the circular component dominates. If the magnitude of the electric field decreases to nearly zero during each cycle, the linear component dominates.

The second approach to describing polarization also treats the elliptically polarized wave as having two components; but this time each component is linearly polarized with the two components at right angles physically and 90° out of phase electrically. The maximum amplitude of the electric field of each component is independent of the other. If the maximum values of both components are equal we have circular polarization. If one electrical field component is always zero we have linear polarization.

Both of the viewpoints mentioned treat circular polarization and linear polarization as special cases of elliptical polarization. And, both are helpful in understanding antennas and radio link performance.

The polarization characteristics of a radio wave depend on the transmitting antenna; the transmitter itself has absolutely nothing to do with polarization. Like radio waves, antennas can be assigned a polarization label: the polarization of the wave that they *transmit in direction of maximum gain*. The common terms "linearly polarized antenna" and "circularly polarized antenna" can be confusing if you aren't aware that "in the direction of maximum gain" is meant, even though it's not stated explicitly.

The following example illustrates this potential source of confusion. Consider the transmitting antenna and the three observing locations shown in Fig 7-6. The antenna pictured, known as a turnstile, consists of two crossed dipoles fed 90° (a quarter cycle) out of phase. An observer at A will see a circularly polarized wave; an observer at C will see a linearly polarized wave; and an observer at B will see an elliptically polarized wave. Since an observer at A, located in the direction of maximum gain, sees a CP

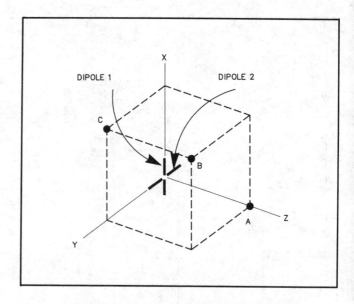

Fig 7-6 —Turnstile transmitting antenna (two dipoles fed 90° out of phase). Observers measuring the polarization of the radiated signal are positioned at points A, B and C. A sees circularly polarized signal. B sees elliptically polarized signal. C sees linearly polarized signal.

wave, the turnstile is called a circularly polarized antenna even though observers off the z axis, at B and C, see something different.

Most of the circularly polarized antennas that we'll be looking at, such as the helix and properly phased crossed Yagis, are similar to the turnstile in that only observers in the direction of maximum gain actually see circular polarization. When using such an antenna at a ground station it's important to keep the array pointed at the spacecraft to reap the benefits of circular polarization. If an antenna of this type is being used on a spacecraft, the only time you'll be receiving a CP wave is when the satellite antenna is pointed in your direction. Some antennas produce circularly polarized radiation over most of the beam. We'll look at two such antennas, the quadrifilar helix and Lindenblad, in the next chapter.

Circular Polarization Sense

Our description of a circularly polarized radio wave emphasized that at a particular point in space, the constant-magnitude electric field rotated at the frequency of the source. It's important to be able to specify whether the *sense* of rotation is *clockwise* or *counterclockwise*. For historical reasons, physicists and electrical engineers specify polarization sense in opposite ways, a fact that can cause confusion. The IEEE (Institute of Electrical and Electronic Engineers) standard is the one used in most recent radio amateur literature. To specify the sense of a circularly polarized wave using the IEEE standard, imagine yourself behind the transmitting antenna looking in the direction of maximum radiation. Pick a specific point on the main axis (any point will do) and note the position of the electric field at a particular instant and follow it through a cycle. If you observe the electric field rotating clockwise, the wave is *right-hand circularly polarized* (RHCP). If the electric field appears to be rotating counterclockwise, the wave is *left-hand circularly polarized* (LHCP). As we obviously cannot "see" the transmitted electric field, we'll discuss shortly how you can determine the sense of a helix or crossed-Yagi array by inspecting the antenna.

Although polarization-sense labels attached to an antenna depend entirely on its transmit properties, the same labels are applied when the antenna is used for reception. A circularly polarized receiving antenna responds best to circularly polarized radio waves of matching sense.

When a circularly polarized wave is reflected off an object (a metal screen, the ground or a house, for example), the sense of its polarization is changed. A RHCP signal aimed at the moon returns as an LHCP signal; a feed horn irradiating a parabolic reflector with LHCP produces a main beam that's RHCP.

Link Comparisons

We now look at how polarization affects a communications link involving two stations: T (the transmitting station) and R (the receiving station). Each station can choose from antennas that provide RHCP, LHCP or LP. All antennas are assumed to have the same gain and each is aimed at the other station. When we refer to the orientation of a linearly polarized antenna we're referring to the direction of the electric field it produces. For a Yagi this field is parallel to the elements. The orientation of an LP antenna on our test link can be varied by rotating the antenna about the line joining T and R.

Various possible link combinations can be characterized by the polarization at T, the polarization at R, and the relative orientation (one or both antennas linearly polarized) or sense (both antennas circularly polarized) of the antennas used. For example, (LP, CP, random) in Table 7-1 can mean either T has an LP antenna and R a CP antenna, or vice versa, and that the orientation of the LP antenna is random. The ambiguity is intentional; since the reciprocity relation previously mentioned states that system performance will be the same in both cases, there is no need to distinguish between them. When one antenna is circularly polarized, random orientation means the LP antenna can be vertical, horizontal or anywhere in between. When both antennas are linearly polarized, random orientation means the angle between the two antennas can be any value between 0 and 90°. Only five distinct combinations need be considered. (See Table 7-1.)

Arbitrarily choosing the Type 1 link as a reference, we compare the performance of the other four combinations.

Type 1 link (LP, LP matched). The received signal level is constant. This link is our reference.

Type 2 link (LP, LP random). The received signal strength varies monotonically from a maximum equal to the reference level when the two antennas are parallel down to zero (theoretically) when the two antennas are perpendicular. In practice, the attenuation is rarely more than 30 dB for the perpendicular situation.

Type 3 link (LP, CP, random). The received signal strength on this link is constant at 3 dB down from the

Table 7-1

Communications Links Categorized by Antenna Polarization

Type 1 link (LP, LP, matched)
Type 2 link (LP, LP, random)
Type 3 link (LP, CP, random)
Type 4 link (CP, CP, same sense)
Type 5 link (CP, CP, opposite sense)

reference level and is independent of the orientation of the LP antenna and the sense of the CP antenna.

Type 4 link (CP, CP, same sense). The received signal strength on this link is constant and equal to the reference level.

Type 5 link (CP, CP, opposite sense). A simple theoretical model predicts infinite attenuation compared to the reference signal link, but in practice attenuations greater than 30 dB are rare.

Having looked at the five basic links, we now compare the performance of various ground station antennas when operated in conjunction with a specific satellite antenna. If the satellite antenna is linearly polarized, our choice of ground station antenna is equivalent to choosing a Type 1, 2 or 3 link. Of the three, the Type 1 may appear best since it provides the strongest signals. From a practical viewpoint, however, in most cases it is nearly impossible to implement since the orientation of the incoming wave is continually changing. In reality our choice is limited to a Type 2 or Type 3 link. Although the Type 2 link will sometimes provide up to 3 dB stronger signals (matched orientation), the Type 3 link will equally likely provide up to 30 dB stronger signals than the Type 2 (perpendicular orientation). Of the two, the Type 3 link is clearly preferable.

We can perform a similar analysis for a satellite antenna that is circularly polarized. The choice of ground station antenna here is equivalent to choosing a Type 3, 4 or 5 link. A Type 4 link is clearly preferable. But, it should be noted that the Type 3 link results in signals that are only 3 dB weaker with none of the severe fading problems of Type 2 links. As a result, on links where the spacecraft is transmitting a CP wave and the S/N ratio is generally good, someone designing a ground station might elect to trade a little performance for the mechanical simplicity of LP antennas.

Signals arriving from most satellites are elliptically polarized. As we've noted, an elliptically polarized wave can be thought of as having linear and circular components. Since a CP ground station antenna produces the best performance whether the signal from a spacecraft is circularly polarized or linearly polarized, a CP receiving antenna at a ground station will provide the best results in the case where the downlink signal is elliptically polarized. A CP transmitting antenna at the ground station will also provide optimal results on the uplink.

Producing Circular Polarization

Numerous techniques for constructing circularly

Table 7-2

Methods for Producing Circular Polarization Using Linearly Polarized Antennas

1) Pair of similar antennas fed 90° out of phase
2) Pair of similar antennas fed in phase
3) Dual-mode horn
4) Combination of electric and magnetic antennas
5) Transmission-type polarizers
6) Reflection-type polarizers

For information on methods not covered in this text see: H. Jasik, *Antenna Engineering Handbook*, (New York: McGraw Hill, 1961), Chapter 17.

---

polarized antennas exist. One approach is to build an antenna like the helix (see Chapter 8) which, by its fundamental design, produces a CP wave. Another approach is to combine LP antennas in the proper manner. Several methods for producing a CP wave from LP antennas are listed in Table 7-2. The first two methods have been used widely by radio amateurs. Several EME operators have had success using the third method at frequencies above 1 GHz as a feed for parabolic antennas. The remaining approaches do not appear suitable for amateur applications at satellite ground stations, so they will not be covered here.

We'll look at Methods I and II in detail. Each requires a pair of matched LP antennas. Two identical 2-element Yagis, carefully adjusted to provide a 50-ohm resistive input impedance, will be used to illustrate each method. Two dipoles, two multielement Yagis, two Quagis, and so on, could also serve. (With adjustments in the phasing/matching harness, other-impedance antennas would also work.)

METHOD I

In this method the two antennas are mounted as shown in either Fig 7-7A (known as a single-boom or concentric boom array) or Fig 7-7B (known as a dual-boom or cross-boom array). The feed system is critical to the

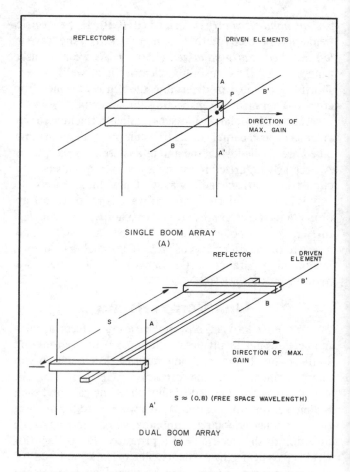

Fig 7-7—Yagi placement for production of circular polarization via Method I. Phasing/matching harness is shown in Fig 7-8.

operation of these arrays. A phasing/matching harness that produces the correct power division, matching and delay parameters is shown in Fig 7-8. Only when the two antennas are fed 90° *out of phase* with *equal power* will the array produce a circularly polarized wave. The effects of various errors in power division and phase difference are described in Table 7-3. For the feed system to perform its function, each antenna must be carefully adjusted to provide an unbalanced, 50-ohm, purely resistive input impedance before it's incorporated into the array. Small adjustment errors in each antenna, even though they may be identical, can have a large effect on power division and phasing and thereby produce an elliptical wave with a large linear component. This may occur even though the SWR in the main feed line remains acceptable.

Although each array in Fig 7-7 shows one Yagi mounted vertically and the other horizontally, this particular configuration needn't be employed as long as the Yagis are mounted at right angles to one another. The tilted arrangement shown in Fig 7-9 is commonly used so that interaction with the cross boom or rotators is balanced.

There's little difference in performance between the horizontal-vertical and skewed arrangements when a ground station antenna is aimed above about 20°. At low elevation angles, however, where ground reflections have

---

Table 7-3

The Effects of Feed Phase and/or Power Division Errors on the Performance of the Yagi Arrays shown in Fig 7-7A and Fig 7-7B

Phase Difference (θ)	Power Division	Resulting Wave Along Major Axis
90°	equal	circular polarization
90°	unequal	elliptical polarization
0°	equal	linear polarization in plane midway between planes of two Yagis
0°	unequal	linear polarization; plane depends on power division
0° < θ < 90°	equal	elliptical polarization
0° < θ < 90°	unequal	elliptical polarization

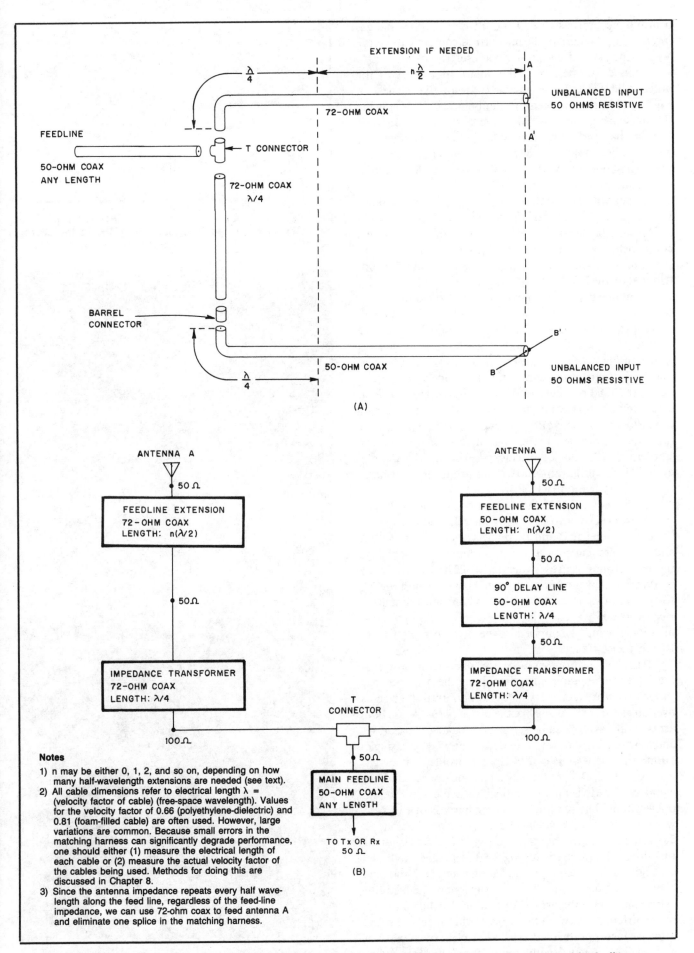

Notes

1) n may be either 0, 1, 2, and so on, depending on how many half-wavelength extensions are needed (see text).
2) All cable dimensions refer to electrical length λ = (velocity factor of cable) (free-space wavelength). Values for the velocity factor of 0.66 (polyethylene-dielectric) and 0.81 (foam-filled cable) are often used. However, large variations are common. Because small errors in the matching harness can significantly degrade performance, one should either (1) measure the electrical length of each cable or (2) measure the actual velocity factor of the cables being used. Methods for doing this are discussed in Chapter 8.
3) Since the antenna impedance repeats every half wavelength along the feed line, regardless of the feed-line impedance, we can use 72-ohm coax to feed antenna A and eliminate one splice in the matching harness.

Fig 7-8—Phasing/matching harness for arrays shown in Fig 7-7: (A) physical design, (B) function block diagram.

a pronounced effect due to the different phase changes imparted to the horizontal and vertical components of the signal, the skewed design may be preferable.

How do we determine the polarization sense of the antenna in Fig 7-7A when it's fed with the harness in Fig 7-8A? We could measure the polarization by testing the antenna on a link with CP antennas of known polarization at the other end or we could figure out the sense analytically as follows. Imagine yourself standing behind the single-boom array of Fig 7-7A looking in the direction of maximum gain.

Focus your attention on the electric field at the point P located at the center of the driven elements. The field at P results from the sum of two linear components: one component that is parallel to AA' (contributed by element AA'), and a second component that is parallel to BB' (contributed by the BB' element). Because of the 90° phasing, one component will be a maximum when the other one is zero. We wait until the field at P points toward 12 o'clock (parallel to AA', pointing in the direction of the element connected to the center conductor of the feed line, as shown in Fig 7-8A). Exactly one quarter cycle (90°) later, the RF currents at the end of the delay line will produce an electric field parallel to BB', pointing toward 3 o'clock since element B is connected to the center conductor of the delay line. From your observation position in back of the antenna, you see the electric field at P rotate from 12 o'clock to 3 o'clock (90° clockwise) during this quarter cycle. This configuration therefore produces right-hand circular polarization (RHCP).

How can we change the sense of polarization? We can switch from RHCP to LHCP by interchanging *either* (1) the connections at B and B' *or* (2) the connections at A and A'. Switching both sets of connections will *not* change the polarization sense. Switching to LHCP can also be accomplished by modifying the matching section of Fig 7-8 so that the two extensions differ by an odd number of electrical half wavelengths. Any of the techniques just mentioned can, of course, also be used to switch from LHCP to RHCP.

The polarization sense of the dual-boom array of Fig 7-7B can be predicted by imagining the two antennas slipped together (sideways motion only) until the booms overlap and form an array like the one in Fig 7-7A. As both arrays will have the same sense of circular polarization, and as we already know how to determine whether a single-boom array is RHCP or LHCP, the problem is solved.

As stated earlier, the 90° phase difference and the equal power division are critical to achieving circular polarization. The phasing/matching harness of Fig 7-8A was designed to work with antennas having an unbalanced, 50-ohm input impedance that's purely resistive. Using the functional block diagram of Fig 7-8B as a guide, we'll step through its operation starting at the antenna end and ending at the feed point. Temporarily ignore the feed-line extensions. Since the array's operation depends on a 90° phase difference between the two sets of elements, the first thing we incorporate is a delay line. A piece of coax that's electrically ¼-wavelength long does the job. (One

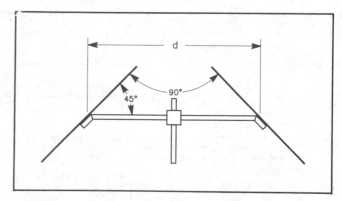

Fig 7-9—An end-on view of crossed Yagi antennas mounted at 90° to each other, and at 45° to the cross boom.

Crossed Yagis mounted in skewed orientation.

wavelength is 360°, one-quarter wavelength is 90°.) The coax delay line would also act as an impedance transformer if the characteristic impedance of the line didn't match the antenna. To obtain the proper power division we don't want any impedance transformation here. Therefore, we use 50-ohm cable. Next, we could connect the two branches in parallel at a coaxial T connector and have both equal power division and correct delay, but the feed-point impedance would be 25 ohms, a value that's awkward to match. Instead, we install two identical impedance transformers consisting of ¼-wavelength sections of 72-ohm coax to step up the impedance of each branch to 100 ohms. When the two 100-ohm lines are connected in parallel at a coaxial T connector, we obtain a good match to 50-ohm feed line. The two impedance transformers do, of course, also act as delay lines. But, since we've used a pair of equal lengths, the phase *difference* between the two Yagis isn't affected.

The harness is now complete except for one mundane consideration: The two branches may not be long enough to reach the antennas. Two identical pieces of 50-ohm coax (any length) would work as extensions; we can eliminate a coax connector and its consequent losses, however, by using electrical-half-wavelength sections cut from coax of different impedances as shown. (The input impedance of an electrical half-wavelength section is independent of the coax impedance—it depends only on the load.) Note that adding an extra half wavelength to one of the feed-line extensions will reverse the sense of polarization.

This completes our discussion of Method I. We now turn to the second technique for obtaining circular polarization.

METHOD II

To illustrate Method II we again use two 2-element Yagis. They can be mounted on a single boom as in Fig 7-10 or on two separate booms as in Fig 7-9. A 90° phase difference is again the key to the antenna's operation. This time it's obtained by physically offsetting one Yagi a quarter of a wavelength in the direction of propagation and feeding the two Yagis in phase. With this approach, no delay line is needed in the feed harness. The feed system need only take into account impedance matching and power splitting. An appropriate matching harness is shown in Fig 7-11. Starting at the antenna end, it operates as follows. The ¼-wavelength sections of 72-ohm coax step up the impedance of each Yagi to 100 ohms. When the two 100-ohm impedances are connected in parallel at the T connector, a good match to 50-ohm feed line results.

To analytically determine the polarization sense of the array shown in Fig 7-10, imagine yourself standing behind it looking in the direction of maximum gain. Focus your attention on the electric field at point P, the center of the front driven element. The field results from the sum of two contributions, one from element AA' in the vertical direction, and the second from element BB' in the horizontal direction. Note that, because of the time it takes to travel through space, the contribution from element BB' at point P was actually produced by BB' a quarter-cycle earlier.

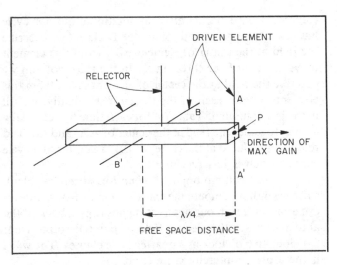

Fig 7-10—Array used for producing circular polarization from linearly polarized antennas via Method II. See Fig 7-11 for matching harness.

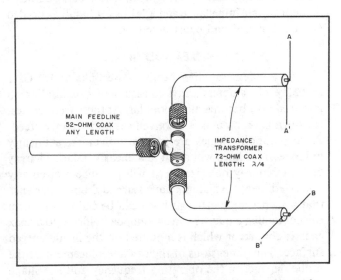

Fig 7-11—Matching harness used with array shown in Fig 7-10 to produce circular polarization from linearly polarized antennas via Method II.

Table 7-4

Data Used to Compute Polarization Sense of Antenna and Feed Shown in Figs 7-10 and 7-11

	Field at center of BB' (from element BB' only)	Field at center of AA' (from element AA' only)	Total field at P
Time 1	9 o'clock	12 o'clock	xxx
Time 2	zero magnitude	zero magnitude	9 o'clock
Time 3	3 o'clock	6 o'clock	6 o'clock

We're going to compute the direction of the electric field at P by combining the fields produced at the center of each element at three different times. Table 7-4 will help us keep track of all the needed information.

We start our observations at *Time 1* when the RF current in the feed line is producing a maximum field at each element. *Time 2* occurs after a quarter cycle has elapsed. *Time 3* occurs after an additional quarter cycle has passed. In the second column of Table 7-4 we describe the field at the center of element BB', from this element only, at each of the three times. In the third column we describe the field at the center of element AA', from this element only, at each of the three times. Finally, we fill in the last column of Table 7-4 for each time by vectorially adding together the field at the center of AA' and the field that was produced at the center of BB' a quarter of a cycle earlier which is just reaching P.

The last column opposite *Time 1* has been left blank since we didn't compute the contribution of BB' a quarter cycle earlier. From our observation position in back of the antenna we see the electric field at point P rotate from 9 o'clock to 6 o'clock as a quarter cycle elapses. The wave is therefore counterclockwise (LHCP).

Yagis using a balanced driven element such as a folded dipole are particularly well suited to Method II. An efficient matching harness using open wire balanced line of an appropriate impedance and a 1:1 or 4:1 balun as needed can be designed and constructed easily.

METHOD III

As amateur satellites begin to use links at 1.2 GHz and 2.4 GHz, antenna arrays consisting of a parabolic dish and feed may become more popular. At these frequencies, the horn antenna makes a convenient and effective feed. Since parabolic dishes are passive reflectors, a linearly polarized feed horn will result in a linearly polarized array and a circularly polarized feed will produce a circularly polarized array. Amateurs have learned from experience that a surprisingly efficient horn can be built from a tin can containing a quarter wave monopole soldered to a coax chassis connector which is mounted on the inside curved surface. The dimensions of the can and placement of the probe depend on the operating frequency and the shape of the dish. Circular polarization can be obtained by using a dual mode horn consisting of a single monopole and several strategically placed tuning screws. Information on parabolic dish antennas and feeds is contained in the next chapter.

COMPARING METHODS I AND II

The analytic procedures described for determining the polarization sense of a crossed Yagi or similar array may leave your brain feeling numb. Don't worry; you're in good company—the telecommunications engineers responsible for the first satellite transatlantic TV broadcast via TELSTAR set up a link with RHCP at one end and LHCP at the other! You can sidestep the calculation approach to determining antenna polarization sense by testing an unknown antenna on a link with a CP antenna of known sense at the other end. A small helix makes a good test antenna since its polarization is easily determined (see next chapter). If you have an array whose polarization sense can be switched from the operating position, you can even ignore the sense—just select the switch position that produces the strongest signals.

Using either method, two identical antennas having

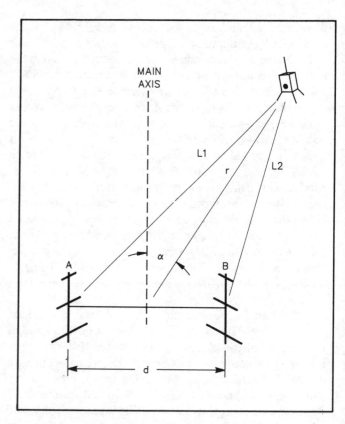

Fig 7-12—Geometry of dual boom antenna array used to compute off-axis polarization.

a gain of, for example, 10 dBi, can be combined to form an antenna array having a gain of 10 dBic. The "c" is sometimes used to indicate a CP antenna when there's a possibility of confusion. Note that combining the same two antennas with a phasing harness designed to produce linear polarization results in an array having a gain of 13 dBi.

One significant advantage of Method II over Method I is that the adjustment of each Yagi is not nearly as critical. As long as both Yagis are identical, small errors in the input impedance, or the presence of a reactive component, will not disturb the equal power split or phasing; the errors will only affect SWR. As long as the SWR is acceptable, the antenna will produce the desired circularly polarized pattern.

With either method one can mount the two antennas concentrically as in Fig 7-7A or Fig 7-10 or using the dual-boom arrangement illustrated in Fig 7-9. Both configurations produce the same results when aimed directly at the target.

To compare off-axis performance of the two approaches (antenna not aimed directly at target), we model the dual-boom array (using either Method I or II) in Fig 7-12. (Our analysis is very similar to the description of the operation of an interferometer using LP antennas.) Assume the figure represents a ground station transmitting an RHCP signal to the spacecraft. When the spacecraft is on-axis L1 = L2 corresponding to $\alpha = 0°$, and the satellite sees an RHCP signal. Now consider the situation where α, the squint angle, is such that L1 − L2 = ¼ wavelength. The spacecraft now sees the electric fields produced by the two antennas as differing by 180° (90° due to the feed harness plus 90° due to the squint angle). The resulting field will be linearly polarized. If the squint angle is such that L2 − L1 = ½ wavelength, the satellite will again see a circularly polarized signal but it will now be LHCP! When r is very much greater than d, as it is in all practical amateur installations, we obtain the following relation

$$L2 - L1 = \pm d \sin \alpha \qquad \text{(Eq 7.5)}$$

How large must the squint angle, α, be before these effects become noticeable? Consider a typical amateur antenna with d = one wavelength and select L2 − L1 = ¼ wavelength. Solving Eq 7.5 for the squint angle, we obtain the value 14.5°. This means that when the antenna is off pointed by 14.5° (in the direction shown), the satellite will see an LP signal. And, when the antenna is off pointed by 29°, the satellite will see a CP signal of the wrong sense! The conclusion is clear—the concentric boom arrangement is superior. However, the concentric boom method cannot be used in all cases. When combining antennas like the quad, Quagi and loop Yagi, interaction between elements of the two component antennas seriously compromises performance. Therefore one should use the dual-boom configuration with these antennas and make allowances for the fact that accurate antenna pointing will be a more critical concern.

Note that the interferometer-like effects we've been discussing also affect spacecraft antennas. Since the 2-meter and 70-cm antennas on OSCARs 10 and 13 are essentially arrays of linearly polarized antennas, they can be analyzed in a similar manner. This explains why, when the spacecraft squint angle becomes greater than about 10°, the links appear to have a significant linear component. When the squint angle becomes large enough, the circularly polarized component of the downlink may even have the "wrong" sense.

SPIN MODULATION

Since a satellite antenna and its gain pattern are firmly anchored to the spacecraft, a ground station's position relative to the pattern will change moment by moment. As we've noted, both the polarization and gain of an antenna vary with the observer's location. A ground station will therefore see gain and polarization changes on a downlink signal resulting from satellite rotation. These changes are called *spin modulation*. The spin-modulation frequency depends on the spacecraft's rotation which, in turn, depends on the attitude stabilization technique employed. After a few weeks in orbit, OSCARs 5, 6, 7 and 8 rotated at frequencies on the order of 0.01 Hz (about one revolution every four minutes). Spin modulation at 0.01 Hz sounds much like a slow fade. Its effect on intelligibility is minor unless the signal drops below the noise level.

The attitude stabilization scheme used on Phase 3 elliptical orbit missions differs considerably. The spacecraft is spun at roughly 20 revolutions per minute (RPM) about an axis that is ideally parallel to the line joining the apogee

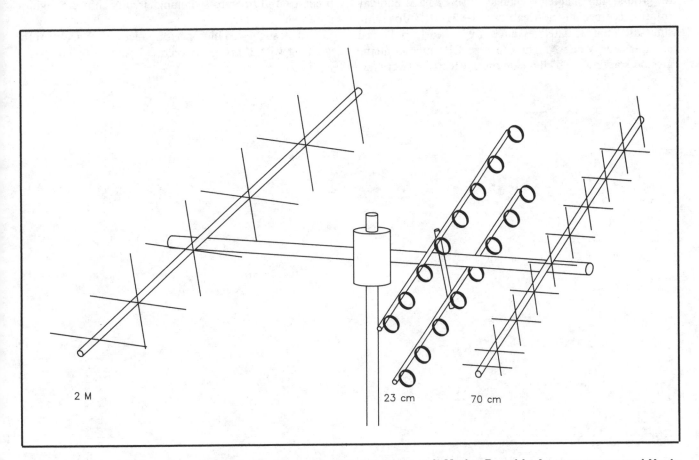

Fig 7-13—High performance antenna for working with Phase 3 spacecraft Modes B and L. Array uses crossed Yagis for 2 meters and 70 cm and a pair of loop Yagis on 23 cm, for circular polarization on all bands.

and the center of the earth. Because of the tri-star shape of Phase 3 satellites, gain and polarization variations on the links occur at three times the spin rate (60 RPM, or 1 Hz). When a ground station is located on the fringes of the satellite's antenna pattern, it may observe gain variations that exceed 10 dB. To a user, spin modulation at a frequency of 1 Hz resembles rapid airplane flutter. It can be very annoying and have a severe impact on intelligibility.

How serious a problem is spin modulation? It is mainly of concern with spin-stabilized spacecraft of the type used with Phase 3 elliptical orbits. Even with these spacecraft the effects become annoying only when the ground station is looking at the spacecraft from off to its side (large squint angle). A ground station can't do much to alleviate true gain variations due to asymmetries in the satellite's antenna gain pattern, but variations caused by polarization mismatch can be minimized by using a circularly polarized antenna.

FARADAY ROTATION

As a linearly polarized radio wave passes through the ionosphere, the direction of the electric field rotates slowly about the direction of propagation. This rotation, known as the *Faraday effect* (see Chapter 13), is most noticeable at lower frequencies, such as 29 MHz and 146 MHz. Its effects can be observed by ground stations that use linearly polarized antennas; slow fades will occur as the angle between the linear component of the incoming wave and the ground station antenna changes during a pass. Faraday rotation is especially noticeable on the 29-MHz downlink since all amateur satellites have used linearly polarized antennas for Mode A. The use of a CP antenna at the ground station would eliminate these effects, but very few ground stations employ CP at 29 MHz. It's important to note that circular polarization won't cure all Mode A fading, since much of it arises from the constantly changing orientation of the gain pattern of the antennas aboard the spacecraft as it spins. Other factors, such as absorption in the ionosphere, can also contribute to fading.

CONCLUSIONS

1) A circularly polarized ground station antenna will outperform a linearly polarized antenna most of the time. If you opt for the additional complexity of a circularly polarized antenna, it's definitely worthwhile to include provision for switching polarization sense since the off-axis signal produced by the spacecraft may, at times, have a sense opposite to the on-axis signal at 2 meters and 70 cm.

2) If you elect to use a circularly polarized array consisting of two linearly polarized antennas, the best approach is to select Yagis so they can be mounted using the concentric boom method. To simplify matching and phasing, and to assure equal power splitting, use the physical offset method to produce the 90° phase shift.

3) If you're interested in communicating by Phase 3 satellites and you're willing to restrict yourself to operating times when squint angles are small, a linearly polarized ground station antenna will provide excellent results.

4) Use different antennas for Phase 2 and Phase 3 spacecraft.

5) Read the sections in the next chapter concerning preamps and preamp switching circuits before designing an antenna system.

6) A suggested high-performance antenna system for working with Phase 3 spacecraft is shown in Fig 7-13.

Chapter 8

Practical Space-Communication Antennas

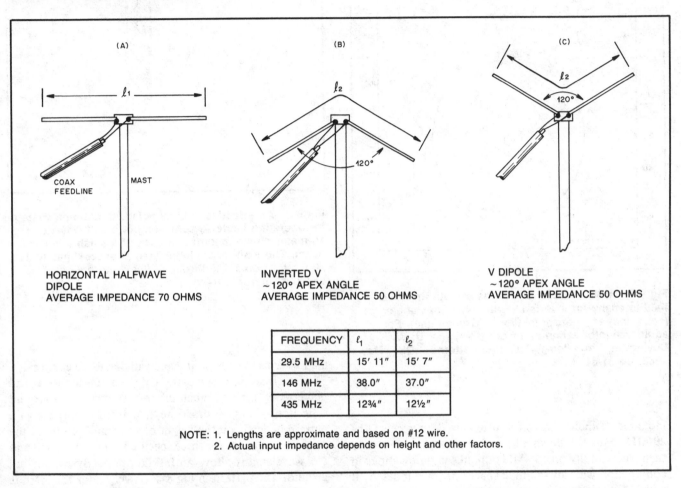

FREQUENCY	ℓ_1	ℓ_2
29.5 MHz	15' 11"	15' 7"
146 MHz	38.0"	37.0"
435 MHz	12¾"	12½"

NOTE: 1. Lengths are approximate and based on #12 wire.
2. Actual input impedance depends on height and other factors.

Fig 8-1—Three variations of the half-wave dipole.

This chapter focuses on several practical antennas that may be used at a satellite ground station. You'll no doubt recognize many, as they're also popular for terrestrial HF and VHF communication. We'll point out the advantages and disadvantages of each for accessing low- and high-altitude spacecraft, for construction difficulty, and for general utility as part of an overall antenna system. Construction details are provided for the more unusual models, and references to readily available sources of information are provided for the popular types.

THE DIPOLE AND ITS VARIANTS

The horizontal half-wave dipole (Fig 8-1A) is a familiar antenna that can be used at satellite ground stations. Two offshoots of the dipole, the inverted V (Fig 8-1B) and the somewhat less familiar V (Fig 8-1C), have also been used. Be sure not to confuse the V antennas discussed here with the V-beam, which is radically different in construction and operation. Our discussion will focus on the inverted V since it has been investigated thoroughly. Nonetheless, it's safe to assume similar characteristics for the V.

The dipole and Vs are usually mounted fixed in the same configuration for both satellite and terrestrial applications (as in Fig 8-1). It therefore makes sense to label patterns as vertical and horizontal. Gain patterns in the horizontal plane for the dipole and inverted V are shown in Fig 8-2. Note how the horizontal dipole has higher gain broadside and deeper nulls off the ends. Their low gain renders the dipole and V suitable mainly for use with low-altitude satellites. Their broad beamwidth provides reasonably good coverage when the antennas are fixed

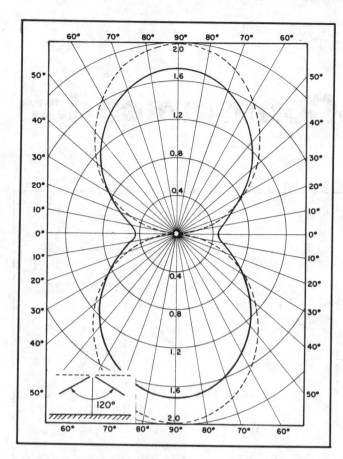

Fig 8-2—Horizontal-plane patterns showing relative field intensity for inverted V with 120° apex angle (solid line) and horizontal dipole (dashed line). For additional information on inverted Vs, see: D. W. Covington, ''Inverted-V Radiation Patterns,'' *QST*, May 1965, pp 81-84.

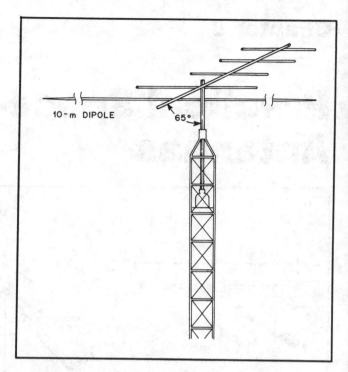

Fig 8-3—An effective linearly polarized antenna system for operating Mode A, consisting of a half-wave 10-meter dipole mounted in back of a small 2-meter beam. The main boom is inclined at approximately 25° above horizontal (65° from vertical) and only an azimuth rotator is used.

mounted. Dipoles are most often used to receive the 29-MHz Mode A downlink. A few amateurs have tried them successfully on 146-MHz uplinks and downlinks in conjunction with low-altitude spacecraft on Modes A, B and J, but this has mainly been for experimental, not for general, communication.

Let's look at some practical applications of dipoles and Vs at 29 MHz. Given the patterns in the horizontal plane, most amateurs who are constrained to using a single fixed antenna choose the V to reduce the effects of the deep nulls associated with the dipole. Slightly better overall performance can be obtained by using two totally independent dipoles mounted at right angles to one another. If feed lines for both are brought into the operating position, switching between them to find the dipole that produces the best received signals is a simple matter.

Another application, offering even better performance, consists of mounting a 10-m dipole behind a small 2-m beam, using a light-duty azimuth rotator to turn the whole array (Fig 8-3). Azimuth aiming requirements will be lax and, by inclining the 2-m beam at roughly 25° above the horizon, the elevation rotator can be eliminated. You'll note that for all three examples just presented, improved perfor-

mance seems to go hand-in-hand with increased complexity.

The free-space gain pattern of the dipole in the vertical plane really isn't of much interest to us because ground reflections change it drastically. As it turns out, the gain pattern depends on the height of the dipole. Look at the patterns in Fig 8-4 for three specific heights: 1/4, 3/8 and 1½ wavelengths above an infinite, perfectly conducting ground. The pattern in Fig 8-4C is very poor for satellite work since signals will fade sharply each time the satellite passes through one of the nulls. In reality, the nulls are not as severe as shown because the ground is not a perfect conductor and signals reflected off nearby objects often arrive at the ground station receiving antenna from several directions. The pattern in Fig 8-4B is most desirable since gain variations tend to balance out changes in signal level as the distance between spacecraft and ground station varies. In other words, the gain pattern of Fig 8-4B is high toward the horizon where signals are weak (large satellite to ground-station distances), and low in the overhead direction where signals are strong (small satellite to ground-station distances). The pattern in Fig 8-4A is acceptable, though not as good as the one in Fig 8-4B. Gain patterns for the V antennas are similar when height is measured from the feed point to the conducting surface.

As the effective *electrical* ground does not generally coincide with the *actual* ground surface, you can't simply measure height above ground to figure out which pattern

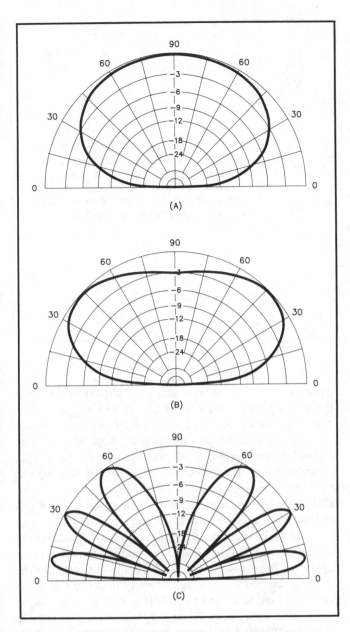

Fig 8-4—Vertical-plane gain patterns showing the relative field intensity for half-wave dipole above perfectly conducting ground. Pattern at right angles to dipole. (A) is for height of 1/4 wavelength, (B) is for height of 3/8 wavelength, and (C) is for height of 1.5 wavelengths.

applies to a given antenna. Many dipole users just orient the antenna with regard to the horizontal pattern and mount it as high and as clear of surrounding objects as possible. Although this does not always produce the best system performance, the results are usually adequate. Some users have tried to obtain the desired vertical patterns (Figs 8-4A or 8-4B) by simulating a ground with a grid of wires placed beneath the dipole as shown in Fig 8-5. Subjective reports suggest that even a single wire (the one labeled A) placed beneath a dipole or V may improve 29-MHz Mode A reception. At 146-MHz and higher frequencies, a reflecting screen can be used for the ground so that a vertical pattern similar to the one of Fig 8-4B can be achieved with

the antenna mounted in a desirably high location. Because the ground screen is finite, gain at take-off angles below about 15° is reduced.

The basic half-wave dipole can also be mounted vertically. In this orientation the horizontal plane pattern is omnidirectional while the actual vertical plane pattern, which depends on mounting height, is likely to have one or more nulls at high radiation angles. Although the characteristics of this antenna appear suitable for work with low-altitude satellites, there is a hitch: the feed line must be routed at right angles to the antenna for at least a half wavelength if one hopes to obtain the patterns described. As a result, it's usually easier to use a ground-plane antenna (see next section), which has similar characteristics. One novel configuration that has proved effective for working DX on Mode A consists of a vertical dipole for 29 MHz hung at the end of a tower-mounted 2-m beam. When the tower-to-dipole distance is set at roughly 6 feet, the tower will tend to act as a reflector and the resulting 29-MHz pattern will be similar to that obtained with a vertically mounted 2-element beam.

In truth, we've paid considerably more attention to the dipole and V than their actual use justifies. Nevertheless, they clearly illustrate many of the trade-offs between effective gain patterns and system complexity that a ground station operator is faced with.

THE GROUNDPLANE

The groundplane (GP) antenna, familiar to HF and VHF operators alike, is sometimes used at satellite ground stations. Physically, the GP consists of a 1/4- or 5/8-wavelength vertical element and three or four horizontal or drooping spokes that are roughly 0.3 wavelength or longer. At VHF and UHF, sheet metal or metal screening is often used in place of the horizontal spokes. The GP is a low-gain, linearly polarized antenna. The gain pattern in the horizontal plane is omnidirectional. Because of its low gain the GP is not generally suitable for operating with high-altitude satellites, though it may be used in special cases. We'll focus on its possibilities with respect to low-altitude spacecraft.

Gain patterns in the vertical plane for 1/4-wavelength GP antennas are shown in Fig 8-6. Although the vertical plane patterns suggest that performance will be poor when the satellite is overhead, stations using the GP report satisfactory results. The reasons are most easily explained in terms of reception. Downlink signals usually arrive at the ground station antenna from several directions after being reflected off nearby objects. These reflected signals can either help (when the direct signal falls within a pattern null) or hinder (when interference between the main and reflected signals results in fading). In practice, the good effects appear to far outweigh the bad; the GP is a good all-around performer for working with all low-altitude OSCARs (heights under 1000 miles) and the MIR and US Space Shuttle.

A GP may be useful for receiving signals from high-altitude satellites in certain situations. For example, although the downlink S/N ratio using a GP generally will

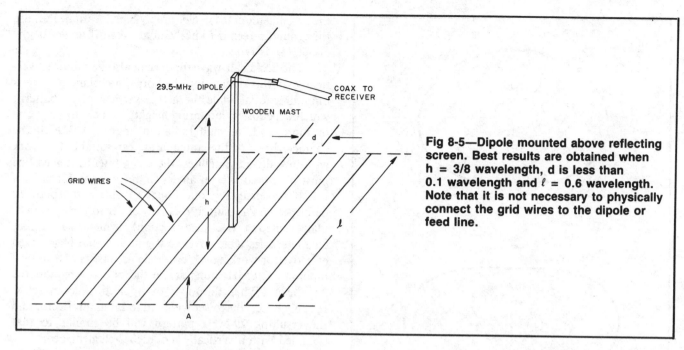

Fig 8-5—Dipole mounted above reflecting screen. Best results are obtained when h = 3/8 wavelength, d is less than 0.1 wavelength and ℓ = 0.6 wavelength. Note that it is not necessary to physically connect the grid wires to the dipole or feed line.

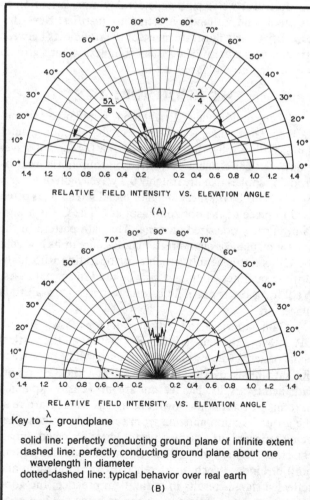

RELATIVE FIELD INTENSITY VS. ELEVATION ANGLE

(A)

RELATIVE FIELD INTENSITY VS. ELEVATION ANGLE

Key to $\frac{\lambda}{4}$ groundplane

solid line: perfectly conducting ground plane of infinite extent
dashed line: perfectly conducting ground plane about one wavelength in diameter
dotted-dashed line: typical behavior over real earth

(B)

Fig 8-6—(A) Vertical-plane gain patterns showing relative field intensity for 1/4 wavelength and 5/8 wavelength ground-plane antennas over ideal earth (perfect conductivity and infinite extent). (B) The effects of a limited ground plane and/or resistive ground on the 1/4-wavelength groundplane antenna.

not be adequate for communication, it should be sufficient for spotting (determining if the spacecraft is in range). The omnidirectional (horizontal plane) pattern of the GP makes it especially suitable for this purpose. Also, the GP may be useful near perigee of elliptical-orbit missions if spin modulation is not excessive. Some broad-beamwidth, circularly polarized antennas better suited to working with Phase 3 satellites near perigee are discussed later in this chapter. We now turn to some practical GP antennas.

GP antennas designed for the 27-MHz CB market are inexpensive and widely available. For Mode A downlink operation the 1/4-wavelength GP usually out performs the "bigger is better" 5/8-wavelength model because its vertical plane radiation pattern (Fig 8-6) is better suited to satellite operation. To modify a 1/4-wavelength CB antenna for 29.4-MHz the vertical element should be shortened about 9%. If a matching network is used it might also require a slight adjustment.

GP antennas designed for the 146-MHz and 435-MHz amateur bands are available commercially at moderate cost. Once again, the 1/4-wavelength models produce good results. Some users, however, prefer to use a 5/8 GP when the satellite is at low elevation angles, and a different type of antenna when the satellite is at higher elevation angles. A VHF or UHF 1/4-wavelength GP can be assembled at extremely low cost (see the illustration in Fig 8-7).

Tilting the vertical element of a 1/4-wavelength GP modifies the gain pattern in the vertical plane as shown in Fig 8-8B. Note how the overhead null has been eliminated. The horizontal pattern is slightly skewed, but remains essentially omnidirectional. Tilting also tends to reduce the already low input impedance of the GP. One way to compensate for this reduction is to use a folded element as shown in Fig 8-8A. As in the folded dipole, the folded 1/4-wave element in a tilted GP steps up the input impedance and gives a broader-bandwidth antenna. The

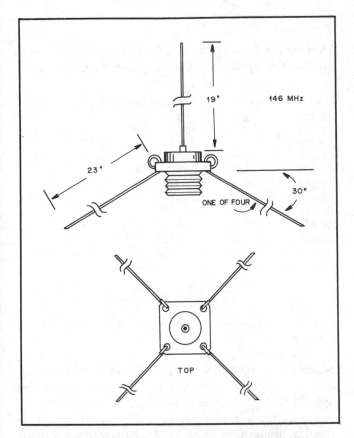

Fig 8-7—A groundplane antenna for 146 MHz is easily constructed using a chassis-mount coax connector. A type N connector is preferred, but a UHF type is acceptable at 146 MHz. Drooping the radials increases gain slightly at low elevation angles and raises input impedance to produce a better match to 50-ohm feed line.

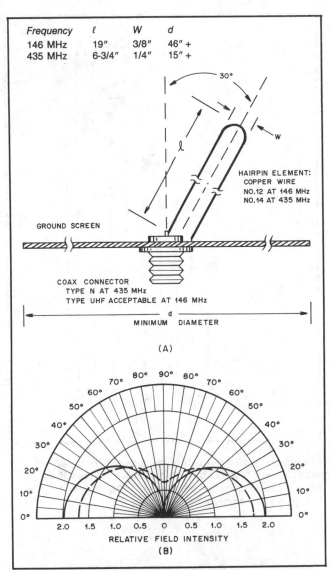

Frequency	ℓ	W	d
146 MHz	19″	3/8″	46″ +
435 MHz	6-3/4″	1/4″	15″ +

Fig 8-8—(A) ¼-wavelength ground plane antenna with tilted vertical element. Ground plane may be square or circular, solid or mesh. (B) Vertical-plane relative field intensity for ¼-wavelength groundplane; solid line—element vertical, dashed line—element tilted 30° from vertical.

dimensions shown should give a good match to 50-ohm coax.

Low gain omnidirectional antennas like the GP are especially useful with low altitude satellites. How does one choose which particular antenna is most suitable? One important consideration is the antennas vertical plane radiation pattern. This pattern should be matched to the daily average time that the satellite will appear at specific elevation angles. To analyze the situation we'll divide elevation angles into three sectors: 0 to 30°, 30 to 60°, and 60 to 90°. We'll then compute the ratio of the access time in a given sector to the total access time and express the result as a percentage.

For a satellite in a circular orbit the desired information depends on the ground station latitude and on the satellite's orbital inclination. However, a reasonably accurate estimate for mid-latitude ground stations and satellites in near polar orbits can be obtained by assuming that average daily access time in a given sector is proportional to the terrestrial area between the corresponding elevation circles. For example, the average daily time that RS-10/11 will appear between 30 and 60° in elevation is proportional to the area between its 30 and 60° elevation circles, and the average time it will be in range is proportional to the area between its 0 and 90° elevation circles

(area inside access circle). Details of the calculations may be found in Chapter 12. The results are shown in Fig 8-9. A more thorough analysis shows that, even for spacecraft like the US Space Shuttle where our assumptions are generally not valid, the results still hold.

From Fig 8-9 we see that a satellite at a height of 1200 miles will appear at elevation angles greater than 30° about 22% of the time it's available to us while a spacecraft at a height of 200 miles will appear at elevation angles greater than 30° less than 7% of the time it's available. Clearly, the lower the satellite height, the less important high elevation angle performance of a ground station antenna becomes. As a consequence, the ¼-wavelength GP is a good all around performer for working with low earth orbit spacecraft—the lower the orbit the better the performance.

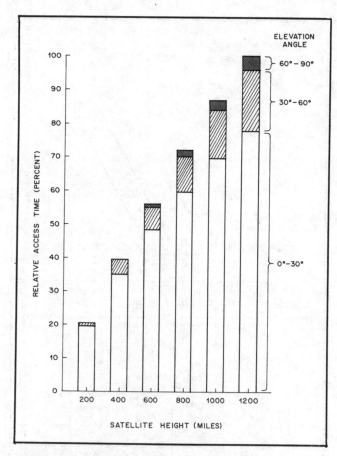

Fig 8-9—For circular orbits, as satellite height decreases a ground station's relative access time at high elevation angles decreases more rapidly than at low elevation angles.

BEAM ANTENNAS

Whether you work with low- or high-altitude satellites there are many situations in which a beam will be the preferred ground station antenna.

With low-altitude spacecraft, a beam could be used for (1) an uplink antenna when available ground station transmitter power is very low, (2) a downlink antenna when a very high downlink S/N ratio is required or (3) both link antennas when one is attempting to contact stations with the spacecraft near or below the normal radio horizon. Superior performance has its costs: the cash spent on rotator(s), and the inconvenience of having to "ride" the azimuth and elevation controls during a pass.

If several passes of a low-altitude satellite are previewed on an OSCARLOCATOR, you'll see that the satellite often just grazes the outskirts of your acquisition circle. During these horizon passes the satellite elevation angle will generally be between 0 and 15°, and azimuth changes, though larger, will usually be less than 90°. Readers who are already equipped for terrestrial VHF or UHF operation with a beam mounted on an azimuth rotator aimed at the horizon will find that their setup provides good satellite access on horizon-grazing passes. Before the pass begins, the antenna can be set to an azimuth about 20° past AOS.

Table 8-1

Half-power Beamwidth as a Function of Gain for Well Designed, Symmetric Pattern, Beam Antennas

G (gain)	G (gain)	θ^* (half-power beamwidth)
6 dBi	4.0	100°
8 dBi	6.3	80°
10 dBi	10.0	63°
12 dBi	16.0	50°
14 dBi	25.0	40°
16 dBi	40.0	32°
18 dBi	63.0	25°
20 dBi	100.0	20°
22 dBi	159.0	16°
24 dBi	251.0	13°

One or two azimuth updates will usually suffice for the entire pass.

For general operation with high-altitude satellites, beams are necessary for obtaining an adequate downlink S/N ratio and cost effective for obtaining the desired uplink EIRP. The burden of keeping the antennas properly aimed during a pass is not as severe with high-altitude elliptical orbits of the Phase 3 type as it is with low-altitude satellites because spacecraft motion near apogee appears very slow. With stationary satellites the antenna is simply aimed during the initial setup and then clamped in position. Once you commit to using a beam and rotators for the downlink, it's almost always least expensive to use a beam with similar gain on the uplink.

Free-space gain patterns for well-designed Yagis, quads, Quagis, loop Yagis and delta loops of equal boom-length are very similar. The patterns are roughly symmetrical (all cross sections look nearly the same) with a shape somewhat like the pencil beam pattern of Fig 7-4. The relation between half-power beamwidth and gain (Eq 7.4) has been used to prepare Table 8-1.

Ground reflections are of concern with all antennas, including beams. The vertical gain pattern of a beam mounted with its boom parallel to the surface of the earth does *not* look like the clean, free-space pattern shown in Fig 7-4. Instead, it breaks up into several lobes interspaced with nulls, the number and position depending on the antenna height (in wavelengths). An example can be seen in Fig 8-4C. These lobes and nulls result from constructive and destructive interference between the direct and ground-reflected signals, as discussed earlier in this chapter. In contrast, when the same beam is pointed significantly above the horizon, the ground-reflected signal contains only a relatively low proportion of the total power; interference effects (both constructive and destructive) become very small. As a result, the tilted beam does produce a clean pattern resembling that in free space.

To illustrate the practical implications of ground effects on vertical patterns, consider a typical ground station antenna for working with Phase 3 satellites. It gives

13 dBi gain and 45° beamwidth. Let's focus on the downlink and look at the satellite near apogee. Assume that both the satellite and the antenna are initially at an elevation angle of 40°. Suppose that one hour later the elevation rotator has not been touched, though the satellite has climbed to an elevation angle of 62.5°, a change of one half our antenna beamwidth. With the antenna set at 40° elevation, very little ground-reflected power reaches the antenna and the pattern can be thought of as a clean pencil-beam. When the satellite is at 62.5° elevation, it is at a point 3 dB down on the ground station antenna pattern; we'd expect the downlink signals to have decreased by 3 dB. Practical experience confirms these expectations.

Now consider a similar situation with the same satellite near apogee and same antenna, but this time let the initial elevations of both satellite and the antenna be 5°. Assume that one hour later the satellite elevation increases to 15° while the antenna elevation remains at 5°. What happens to the link? A prediction based on free-space patterns would yield an almost trivial 1- or 2-dB decrease in signal level since the 10° change in elevation is far less than the 22.5° (half-beamwidth) change it takes to reduce signals by 3 dB. But predictions based on the free-space model are totally inadequate at low antenna elevations where ground reflections play a very pronounced role. In reality, it's nearly impossible to predict the outcome, but changes in the downlink amounting to a decrease of 30 dB, an increase of 3 dB, or anything in between wouldn't be surprising. Even though the outcome can't be predicted, understanding the situation is important: At low satellite elevation angles, aiming the antenna in elevation becomes more critical. With a broad-beamwidth antenna it's very easy to ignore a small, seemingly insignificant change in satellite elevation. While this oversight is safe at high elevation angles, it can severely degrade performance at low angles.

Our discussion has focused on the downlink. The uplink is analogous except for one fact. Even if uplink and downlink antennas have identical free-space patterns and are mounted at the same physical height, their actual vertical patterns will not be the same since their electrical heights (measured in wavelengths) will be different.

When working at low elevation angles you may find, a small percentage of the time, that an elevation setting that results in good uplink signals is associated with a poor downlink and vice versa. The only solution, short of mounting each array on its own set of rotators, is to pick a compromise position. For reliable operation when a satellite is close to the horizon it's critically important to monitor your downlink and adjust antenna elevation as often as necessary.

Yagi, Quad and Related Beams

We now turn to some of the practical concerns involved in choosing among the Yagi, quad, Quagi, loop Yagi and delta loop. Since the performance of these antennas is similar in terms of pattern and ground effects we'll focus on difficulty of construction, mounting ease, commercial availability and suitability for later use as part of a circularly polarized system. Each type of antenna will be evaluated in terms of these criteria.

The Yagi has a number of positive attributes including its simple structure, light weight, and low wind load for a given gain. As a result, most commercial manufacturers favor it over other types of beams at 2 m and 70 cm. However, until recently, most published Yagi designs operated satisfactorily only over a very narrow bandwidth (often 1-2% of operating frequency). As a result, home brewers had to be extremely careful to replicate all dimensions and materials exactly as described to duplicate the performance of the original. The narrow bandwidth also made the antenna very susceptible to detuning effects from the mounting, nearby antennas, and rain, ice and snow. Back in the 1970s, attempts to conquer the bandwidth limitation focused on the use of log-periodic feeds. This approach was successful but the extra elements added to weight and wind load.

Until recently many engineers believed that the narrow-band nature of the Yagi was an intrinsic characteristic. But Steve Powlishen (K1FO) using computer design tools that became available in the late 1980s, demonstrated that it's possible to design high performance arrays with bandwidths on the order of 8% having extremely clean patterns (very low power in sidelobes). Computer analysis is ushering in a golden age in Yagi design. Although most of the cutting edge work is being done on very large antennas for EME operation, the smallest of these antennas make excellent building blocks for producing circularly polarized antennas. CP antennas formed from these new Yagi designs will be leading performers on 2 m, 70 cm and 23 cm satellite links over the next decade.

Quad antennas are easy to match and their dimensions are relatively uncritical. However, large quads have not been very popular at VHF and UHF because they're structurally cumbersome. Similar comments apply to the delta loop. Construction details for an easily duplicated 146-MHz, 3-el quad are given in Fig 8-10. I've had good results using this simple antenna for uplinking on Mode A and as part of a system designed to illustrate the minimal requirements for monitoring the OSCAR 13 Mode B downlink (a 1.0-dB preamp was mounted on the antenna). In addition, a set of these antennas used to construct an interferometer for a student experiment in orbit determination provided excellent performance.

The Quagi is a cross between quad and Yagi. It uses a quad reflector and driven element for easy, efficient matching, and Yagi directors for good gain, low wind load and simple structure. Since its introduction in 1977, the Quagi has quickly become popular with new VHF and UHF operators who want a simply constructed homemade antenna that, when put on the air without any specialized test equipment, performs up to expectations.

The loop Yagi (Fig 8-1) and the delta loop are close relatives of the quad and Yagi. Both have been used for satellite communication. Recently, the loop Yagi has received considerable attention. Since its structure is mechanically awkward at VHF and lower frequencies, it hasn't seen much use in this part of the radio spectrum.

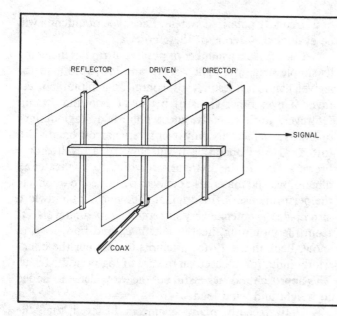

Element Dimensions

Element	Circumference	Note
Reflector	85 5/8 inches	closed loop soldered at bottom center
Driven	81 1/8 inches	feed at bottom center for horizontal polarization
Director	76 1/4 inches	closed loop soldered at bottom center

Element Spacing

Reflector—Driven Element: 18 inches
Director—Driven Element: 12 inches

Input Impedance: about 60 ohms. Feed directly with 50-ohm coax
Gain: about 9 dBd

Construction
Wood frame, size and materials not critical. Original used pine
molding 3/4 × 3/4 for boom, lighter grade for spreaders.
Elements: #12 copper wire

Originally described in: W. Overbeck, "A Small, Inexpensive
Moonbounce Antenna System for 144 MHz," EIMAC EME Notes,
AS-49-15.

Fig 8-10—A 3-el quad for 146 MHz.

Table 8-2

Sources for Construction Articles on Linearly Polarized Beam Antennas and Power Splitters Suitable for OSCAR Operation

Yagi

The ARRL Antenna Book, 15th Ed., 1988, Chapter 18.

S. Powlishen, "Improved High-Performance Yagis for 432 MHz," *Ham Radio*, May 1989, pp 9-10, 12, 17, 19-22, 25.

S. Powlishen, "An Optimum Design for 432-MHz Yagis, Part 1," *QST*, Dec 1987, pp 20-24; "Part II," Jan 1988, pp 24-30.

S. Powlishen, "High-Performance Yagis for 432 MHz," *Ham Radio,* Jul 1987, pp 8-9, 11-13, 15, 17-23, 25-27, 29-31. Correction, Oct 1987, p 97. Comment, Sep 1987, p 6.

S. Jaffin, "Applied Yagi Antenna Design, Part I: A 2-meter Classic Revisited," *Ham Radio*, May 1984, pp 14-15, 17-20, 23-25, 27-28.

J. Reisert, "Optimized 2- and 6-meter Yagis," *Ham Radio,* May 1987, pp 92-93, 95-97, 99-101. Correction, Aug 1987, p 41. Correction, Jul 1987, p 49.

R. J. Gorski, "Efficient Short Radiators," *QST*, Apr 1977, pp 37-39. Describes a 2-el Yagi design tested at 100 MHz. Should be excellent for Mode A reception when scaled to 29 MHz.

Quad

The ARRL Antenna Book, 15th Ed., 1988, Chapter 18. Description of 2-el quad for 144 MHz.

W. Overbeck, "A Small, Inexpensive Moonbounce Antenna System for 144 MHz," EIMAC EME Notes, AS-49-15. Describes an array of 16 3-el quads. The dimensions of the individual quads, scaled to 145.9 MHz, are given in Fig 8-10.

Quagi

W. Overbeck, "The VHF Quagi," *QST*, Apr 1977, pp 11-14. Includes designs for 144.5, 147 and 432 MHz.

W. Overbeck, "The Long-Boom Quagi," *QST*, Feb 1978, pp 20-21. Includes design for 432 MHz. Also see Technical Correspondence, *QST*, Apr 1978, p 34, for comments concerning scaling Quagis to other frequencies.

W. Overbeck, "Reproducible Quagi Antennas for 1296 MHz," *QST*, Aug 1981, pp 11-15.

The ARRL Antenna Book, 15th Ed., 1988, Chapter 18.

Loop Yagi

R. Harrison, "Loop-Yagi Antennas," *Ham Radio*, May 1976, pp 30-32. Includes designs for 28.5, 146 and 435 MHz.

B. Atkins, "The New Frontier," *QST*, Oct 1980, p 66. Includes two designs for 1296 MHz by G3JVL, a 38-element array on a 10-ft boom with about the same gain as a 4-ft dish, and a 27-element array on a 7.5-ft boom with about 1.5 dB less gain. Contains good construction diagrams.

The ARRL Antenna Book, 15th Ed., 1988, Chapter 18.

Delta Loop

A. A. Simpson, "A Two-Band Delta-Loop Array for OSCAR," *QST*, Nov 1974, pp 11-13. Includes designs for 146 and 435 MHz.

Power Splitters

J. Reisert, "VHF/UHF World," *Ham Radio*, May 1988, pp 80, 82-83, 85-86, 88-89.

It is gaining in popularity at 435 MHz, 1260 MHz and higher frequencies where a very straightforward mechanical design has evolved. Each loop is formed into a circle from a strip of flat, springy conductor. A single screw holds the loop in shape and secures it to an aluminum boom. Good loop Yagi designs appear at least to equal, and perhaps exceed, Yagis of the same boom length. As the bandwidth of a loop Yagi is several times that of currently existing Yagi designs, construction tolerances are considerably relaxed. Commercially made loop Yagis are one of the most popular uplink antennas for Mode L operation with OSCAR 13.

The quad, Quagi, loop Yagi and delta loop all suffer from the same shortcoming—if one wants to construct a circularly polarized array using two antennas they must be configured using the cross boom mounting to minimize interaction.

A list of relevant construction articles featuring the Yagi, quad, Quagi, loop Yagi and delta loop is contained in Table 8-2. It may take a little research to select the antenna that best meets your needs. Each year, at major VHF/UHF conferences around the US, test ranges are set up to compare antennas. The results are often presented in *QST*'s World Above 50 MHz column. Consistent top performers are quickly adopted by serious contesters and EME buffs, so you can also check recent contest results to see what antennas are being used. As a rule of thumb, if an array of eight brand-X Yagis is popular with EME operators, one brand-X Yagi will provide good performance to a Phase 3 satellite, and a pair, configured for circular polarization, will provide excellent results. Commercial sources for suitable antennas are listed in Table 8-3.

Circular Polarization from Linearly Polarized Antennas

The basic engineering concepts describing how two linearly polarized antennas (the component antennas) could be combined to produce a circularly polarized array were presented in Chapter 7. We now look at some of the practical aspects of implementing these ideas. There are two key, and totally independent, choices that have to be made: (1) whether to use a concentric boom or a cross-boom configuration, and (2) whether to use a delay line or to physically offset the antennas in the direction of propagation in order to achieve the required 90° phase difference.

The concentric boom configuration is preferable in that it eliminates off-axis circularity changes (the interferometer-like effects discussed in Chapter 7). However, of the beams discussed, it can only be used with the Yagi. With the Quagi, quad, loop Yagi or delta loop, the cross-boom configuration must be used. When using the cross-boom arrangement it's important to use the least separation possible to minimize off-axis circularity effects. A rule of thumb is to mount each component antenna at a 45° angle to the cross boom and to space the antennas so that there's a half wavelength between the tips of the closest elements. The cross boom should be non-conducting material (preferably fiberglass). If suitable material isn't available, a

Table 8-3

Commercial Manufacturers and Distributors of Satellite Antennas and Accessories*

Cushcraft Corp
48 Perimeter Rd
Manchester, NH 03108
603-627-7877

Down East Microwave
Box 2310 RR 1
Troy, ME 04987
207-948-3741 (9-2 Eastern Time)
Specialty: Loop Yagis for 902-3456 MHz

Mirage/KLM
PO Box 1000
Morgan Hill, CA 95037
800-538-2140

Rutland Arrays
1703 Warren St
New Cumberland, PA 17070
717-774-5298 (7-10 PM Eastern Time)
Specialty: K1FO Yagis

Spectrum International
PO Box 1084
Concord, MA 01742

Stridsberg Engineering Co
PO Box 7973
Shreveport, LA 71107
318-865-0523
Specialty: Power dividers for 144-1296 MHz

Telex/Hy-Gain Amateur Products Division
9600 Aldrich Ave S
Minneapolis, MN 55420
800-328-3771

Tonna
The PX Shack
53 Stonywyck Dr
Belle Mead, NJ 08502
Specialty: Tonna (F9FT) Yagis

*power dividers, VHF/UHF dividers, stacking frames, cross-booms, and so on, suitable for satellite communications

metal cross boom with nonconducting extensions fabricated from PVC plastic plumbing pipe can be used.

It's best to empirically check for interaction between the component antennas, and between each antenna and the cross boom. Set the array up at ground level and pointing at a high elevation angle. Disconnect any phasing/matching harness. Connect an SWR meter to one of the component antennas with a short length of coax and note the SWR. RF energy can be dangerous so use a low power level and keep clear of the antenna when it's energized. Move the other component antenna about and note whether the SWR is affected. Then remove the non-energized antenna from the cross boom. Rotate the energized antenna and again check for SWR changes. If the SWR is not affected by either operation the antenna configuration is acceptable.

Of the two methods described for obtaining the required 90° phase shift, our analysis in Chapter 7 showed that the physical offset method is clearly superior due to the simplicity and noncritical nature of the matching harness. It's not widely known that this is true even when a cross beam mounting is being used. Phasing/matching harnesses for the two methods are shown in Fig 7-8 and 7-11. These harnesses will only work with antennas having a 50-ohm resistive feed. The harness of Fig 7-8 provides a 90° phase shift. It will work for both the concentric boom and cross-boom arrays that require a shift. The harness of Fig 7-11 is designed for arrays using a physical offset to produce the needed 90° phase shift. It will work for both concentric boom and cross-boom arrays.

The phasing/matching harnesses just mentioned will also work with loop Yagis having a 50-ohm resistive input impedance. The loop Yagis must be mounted using the cross boom configuration with one feed at the bottom of the driven loop and the other fed at either 3 or 9 o'clock. The physical offset method of obtaining the required 90° phase shift is again preferred.

To obtain optimal performance from an antenna you should be able to switch polarization from the operating position. Choices might include circular or linear—sense if circular, orientation if linear. Using a pair of identical linearly polarized antennas mounted as in Figs 7-7 or 7-10, it's theoretically possible to obtain any polarization—linear (any orientation), circular (RH or LH) or elliptical (any combination of linear and circular)—by adjusting the power division between the two antennas and the relative phasing. In practice, systems providing a continuous range of choices are very complex and really unnecessary. When working with a satellite link where the polarization of the satellite signal is constantly changing, one only needs to select between RHCP and LHCP. Several simple switching systems requiring only a SPDT coax relay are illustrated in Fig 8-11. Although the illustrations in the figure employ a dual-boom configuration, all examples work fine with a concentric boom mounting.

Theoretically, the antenna switching shown in Fig 8-11C can actually be accomplished using a mechanical switch at the operating position if you run two feed-lines down to the shack. Even though the total feed line length is now twice as long, there's no additional loss. Although this may contradict intuition it's easily confirmed. Consider two identical 200-watt ground stations using crossed Yagi antennas. Let station A use a single 100-ft feed line and a power splitter at the antenna. Let station B use two feed lines, each 100 ft, and a power splitter at the operating position so that switchable polarization can be employed. Assume that 100 ft of feed line has 3 dB loss at the frequency of interest.

Station A

Transmitter output: 200 watts
Power at antenna end of feed line: 100 watts (3 dB loss)
Power reaching each Yagi: 50 watts (after power splitter)

Station B

Transmitter output: 200 watts

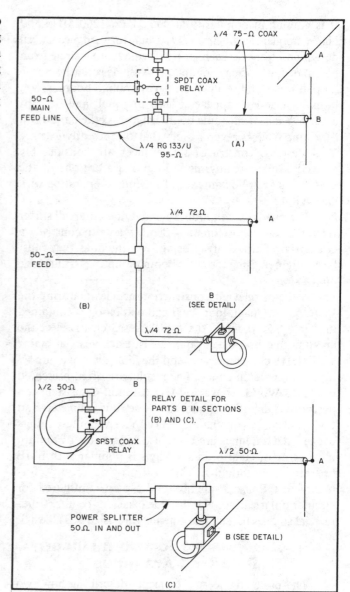

Fig 8-11—Systems for switching between LHCP and RHCP. Component antennas must be 50 ohms resistive. Component antennas may use either the concentric boom or cross-boom configuration. (A) system using 90° delay line (driven elements in same plane), (B) system using ¼ wavelength physical offset and (C) system using ¼ wavelength physical offset and power splitter (50 ohms in and out).

Power into each feed line: 100 watts (after power splitter)
Power reaching each Yagi: 50 watts (3 dB loss on each feed line)

Even though there's no additional loss with the two-feed-line system, it's definitely not recommended for several reasons. Fabricating cables of equal electrical length to the precision necessary is a difficult job and the cost of the

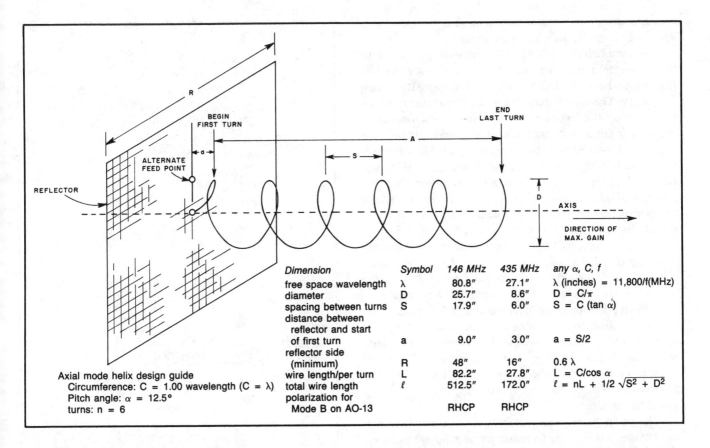

Axial mode helix design guide
Circumference: C = 1.00 wavelength (C = λ)
Pitch angle: α = 12.5°
turns: n = 6

Dimension	Symbol	146 MHz	435 MHz	any α, C, f
free space wavelength	λ	80.8″	27.1″	λ (inches) = 11,800/f(MHz)
diameter	D	25.7″	8.6″	D = C/π
spacing between turns	S	17.9″	6.0″	S = C (tan α)
distance between reflector and start of first turn	a	9.0″	3.0″	a = S/2
reflector side (minimum)	R	48″	16″	0.6 λ
wire length/per turn	L	82.2″	27.8″	L = C/cos α
total wire length	ℓ	512.5″	172.0″	$\ell = nL + 1/2 \sqrt{S^2 + D^2}$
polarization for Mode B on AO-13		RHCP	RHCP	

Fig 8-12—Dimensions for axial mode helix. For additional information see: J. D. Kraus, *Antennas* (New York: McGraw-Hill, 1950), Chapter 7; H. E. King and J. L. Wong, "Characteristics of 1-8 Wavelength Uniform Helical Antennas," *IEEE Transactions on Antennas and Propagation*, Vol AP-28, no. 2, Mar 1980, pp 291-296; *The ARRL Antenna Book*, 15th Edition, 1988, Chapter 19.

Table 8-4
Helix Characteristics†

No. of turns (n) [*1]	Gain (G) [*2]	Gain (G)	Half-power beamwidth [*3]	Approx boom length 146 MHz [*4]	435 MHz
3	10.0	10.0 dBi	64°	5.0 ft	2.0 ft
4	13.3	11.0 dBi	55°	6.5 ft	2.5 ft
5	16.6	12.2 dBi	49°	8.0 ft	3.0 ft
6	20.0	13.0 dBi	45°	9.5 ft	3.5 ft
7	23.3	13.7 dBi	42°	11.0 ft	4.0 ft
8	26.6	14.2 dBi	39°	12.5 ft	4.5 ft
9	30.0	14.8 dBi	37°	14.0 ft	5.0 ft
10	33.3	15.2 dBi	35°	15.5 ft	5.5 ft
11	36.6	15.6 dBi	33°	17.0 ft	6.0 ft
12	40.0	16.0 dBi	32°	18.5 ft	6.5 ft

[*1] For n less than 3 the helix pattern changes radically
[*2] Theoretical values: Measurements suggest these values are 1 or 2 dB too high. Gain (G) ~ 15 n tan α † = 12.5°)
[*3] Half-power beamwidth = $52°/\sqrt{n \tan \alpha}$ †
[*4] Boomlength = λ (n + 0.5) tan α

†Based on 1-wavelength circumference (C = λ) and 12.5° pitch angle (α = 12.5°)

extra coax usually eats up any savings on other components. In addition, the method does not lend itself to placing a preamp at the antenna.

The Helix

Imagine a beam antenna that (1) produces a circularly polarized wave without a complex feed harness, (2) operates over a wide bandwidth and (3) is very forgiving with respect to dimensions and construction techniques. Unlike the isotropic antenna, this one's for real. Called an axial-mode helix (*helix* for short), it's used by many satellite operators. It's also used on OSCAR 13 for the 1.2- and 2.4-GHz links. Before we get carried away describing the advantages of the helix, note that it does have several drawbacks which will be mentioned shortly.

A helix is characterized by three basic parameters:
- C, the circumference of the imaginary cylinder on which the helical element is wound (usually expressed in terms of wavelength so that it's frequency independent)
- α, the pitch angle, essentially a measure of how closely the turns of the helical element are spaced (also frequency independent)
- n, the total number of turns

When these parameters lie in these ranges,

$$0.8 \lambda \geq C \geq 1.2 \lambda$$
$$12° \leq \alpha \leq 14°$$
$$n \geq 3$$

the helix will produce a beam pattern similar to the Yagi and quad. Dimensions are given in Fig 8-12. A 6-turn helix suitable for use with AMSAT-OSCAR 13 is shown, but the number of turns may be scaled up or down (see Table 8-4) to change the gain and beamwidth.

When a helix is built with the circumference equal to the wavelength it is designed for, it will work well at frequencies between 20% below and 30% above the design frequency. The wide bandwidth is advantageous: It allows you to be a little less precise than usual when measuring the proper antenna dimensions. It also makes it possible to use the 146-MHz helix described in Fig 8-12 for monitoring scientific satellites that transmit near 137 MHz, and the 435-MHz model for listening to navigation satellites near 400 MHz. The bandwidth of the helix can contribute to receiver desensitization problems, however, if high-power commercial stations are located nearby. Unfortunately, megawatt EIRP TV and radar transmitters are common in the part of the spectrum that radio amateurs use for satellite links. A sharp band-pass filter at the receiver input may help if you encounter any trouble.

The input impedance of a helix that is fed at the center is usually close to 140 ohms. A matching transformer consisting of an electrical quarter wavelength of 75-ohm coax (RG-11) or 80-ohm coax (Belden no. 8221) will provide a decent SWR when 50-ohm feed line is used. The SWR improvement, however, exists only over a relatively small bandwidth.

In recent years a new matching approach with several advantages has become increasingly popular with professional space communication engineers. When the helix is fed at the alternate feed point on the periphery, as shown in Fig 8-12, the first turn may be thought of as an impedance transformer. To use this feed point, dimension a should be doubled (that is, set a equal to S, the spacing between turns). Displacing the first quarter turn toward the reflector tends to produce a better match to 50-ohm feed line. To bring the SWR down even closer to 1:1, increase the effective wire diameter of the first quarter turn by soldering a strip of thin brass shim stock or copper flashing (width roughly 5 times the wire diameter) to it. This technique is described in detail by J. D. Kraus in "A 50-ohm Input Impedance for Helical Beam Antennas," *IEEE Transactions on Antennas and Propagation*, Vol AP-25, No. 6, Nov 1977, p 913, and J. Cadwallader, "Easy 50-ohm Feed for a Helix," *QST*, June 1981, pp 28-29. With this matching technique the SWR remains below 2:1 over a range of about 40% of the center frequency.

The helical element must be supported by a nonconductive structure. Two common approaches to building such a frame are illustrated in Fig 8-13. Lightweight woods with good weathering properties, such as cedar or redwood, are preferred for large 146-MHz lattice structures, while varnished pine or oak dowels may be used for the smaller 435-MHz model. The construction of the reflector is not critical as long as it meets the minimal size requirements. Square or round sections of hardware cloth for 435-MHz helices, or 2″ × 4″ welded wire fencing for 146-MHz helices are suitable. A small aluminum hub with 18 or more evenly spaced spokes radiating outward can also be used. At 146 MHz the helical element may be wound from ¼-inch flexible copper tubing or from a

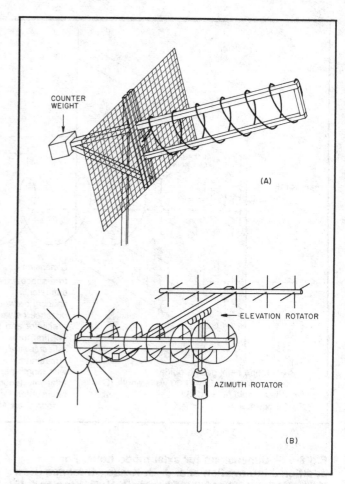

Fig 8-13—Frameworks that may be used for building helix antennas. A lattice structure, often used at 146 MHz, is shown at (A); the structure in (B) is popular at 435 MHz. For practical information on helix structures see: D. Jansson, "Helical Antenna Construction for 146 and 435 MHz," *Orbit*, Vol 2, no. 3, May/Jun 1981, pp 12-14.

length of old coaxial cable (impedance is not important) with the inner conductor and outer braid shorted together. At 435 MHz and higher frequencies, no. 12 wire is acceptable.

The main problem with using a helix is its cumbersome physical structure. Comparing a well-designed crossed-Yagi array and a helix with the same gain, you'll find that generally the Yagi array will be considerably shorter and have less than half the weight and windload (see Table 8-5). Several serious EME operators, experimenting with arrays of helices, have concluded that helices are not suitable for providing the very large gains required for EME communication.

A second problem with the helix is that there is no way to flip the polarization sense. Despite these problems you should consider the helix if you need an easily reproducible, moderate-gain, inexpensive antenna for satellite operation.

To determine the polarization sense of a helix, picture yourself standing in back of the reflector looking out in

Table 8-5

Comparison of Three Circularly Polarized Beam Antennas

	Crossed Yagis with delay Line	Crossed Yagis offset ¼ λ in direction of max gain	Single helix
Length for 12 dBi gain	1.0 λ	1.25 λ	1.4 λ (plus boom for counterweight if needed)
Bandwidth	~2% of center frequency	~2% of center frequency	From 20% below to 30% above center frequency
Matching/phasing system	Highly complex	Moderately complex	Relatively simple
Adjustment procedure	Complex	Complex	Simple
Are dimensions and construction materials critical?	Yes	Yes	No
Relative, size, weight, mounting complexity	Small, light, simple	Small, light, simple	Moderately large, heavy, complex
Can polarization sense be externally switched?	Yes	Yes	No

the direction of maximum gain. If you were to place your index finger on the feed point and slide it forward along the surface of the helical element, you would see it trace out either a clockwise pattern or a counterclockwise pattern. Clockwise corresponds to an RHCP helix; counterclockwise corresponds to an LHCP helix. As mentioned earlier, a 3-turn helix makes an excellent reference antenna for determining the polarization sense of a crossed Yagi or similar array.

THREE CIRCULARLY POLARIZED BROAD-BEAMWIDTH ANTENNAS

Three additional circularly polarized antennas are of

interest. All are low-gain, broad-beamwidth designs primarily suited for use with low-altitude spacecraft.

Lindenblad

The Lindenblad antenna, shown in Fig 8-14A, is not very well known to radio amateurs. However, it has been frequently used for VHF links at airport control towers. Its omnidirectional pattern and circular polarization are a near-perfect match for null-free reception from the linearly polarized, randomly oriented signals arriving from incoming and departing aircraft. The Lindenblad consists of four dipoles spaced equally around the perimeter of an imaginary horizontal circle about 0.3 wavelength in

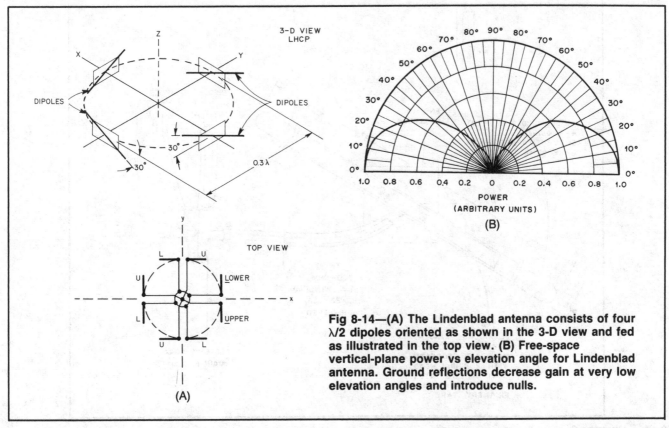

Fig 8-14—(A) The Lindenblad antenna consists of four λ/2 dipoles oriented as shown in the 3-D view and fed as illustrated in the top view. (B) Free-space vertical-plane power vs elevation angle for Lindenblad antenna. Ground reflections decrease gain at very low elevation angles and introduce nulls.

diameter. Each dipole is tilted 30° out of the horizontal plane; rotation (tilt) is about the axis joining the mid point of the dipole to the center of the circle. All four dipoles are tilted in the same direction: either clockwise (for RHCP) or counterclockwise (for LHCP) from the perspective of an observer located at the center of the array. Construction details for a 146-MHz version are given in Fig 8-15. Since all dipoles are fed in phase, power division and phasing are simple and the array can easily be dupli-

cated without test equipment. Furthermore, using folded dipole elements simplifies impedance matching.

Radiation from the Lindenblad is omnidirectional in the horizontal plane and favors low elevation angles in the vertical plane (see Fig 8-14B). When used with low-altitude, circular-orbit satellites, the increased power at low elevation angles compensates somewhat for increased satellite-ground station distance; signal levels therefore remain fairly constant over a considerable range of elevations. The

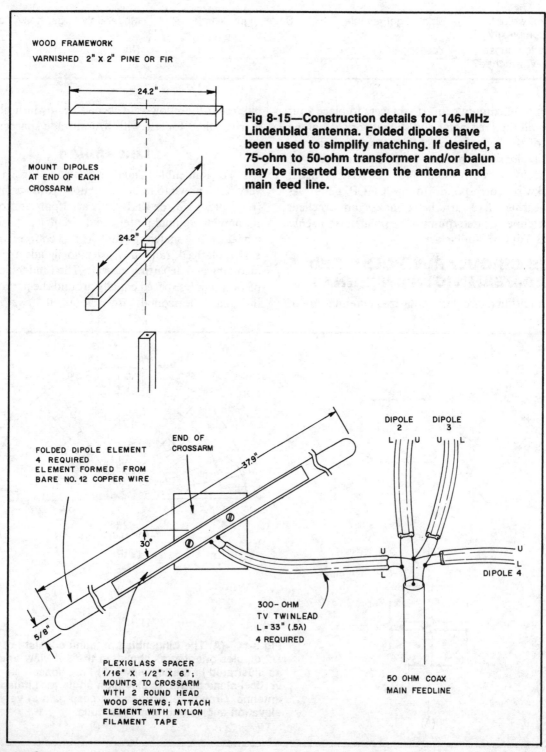

WOOD FRAMEWORK
VARNISHED 2" X 2" PINE OR FIR

24.2"

MOUNT DIPOLES
AT END OF EACH
CROSSARM

24.2"

Fig 8-15—Construction details for 146-MHz Lindenblad antenna. Folded dipoles have been used to simplify matching. If desired, a 75-ohm to 50-ohm transformer and/or balun may be inserted between the antenna and main feed line.

FOLDED DIPOLE ELEMENT
4 REQUIRED
ELEMENT FORMED FROM
BARE NO. 12 COPPER WIRE

END OF
CROSSARM

37.9"

DIPOLE 2   DIPOLE 3
L   U   U   L

30°

U
L   U
L
DIPOLE 4

300- OHM
TV TWINLEAD
L = 33" (.5λ)
4 REQUIRED

5/8"

PLEXIGLASS SPACER
1/16" X 1/2" X 6";
MOUNTS TO CROSSARM
WITH 2 ROUND HEAD
WOOD SCREWS; ATTACH
ELEMENT WITH NYLON
FILAMENT TAPE

50 OHM COAX
MAIN FEEDLINE

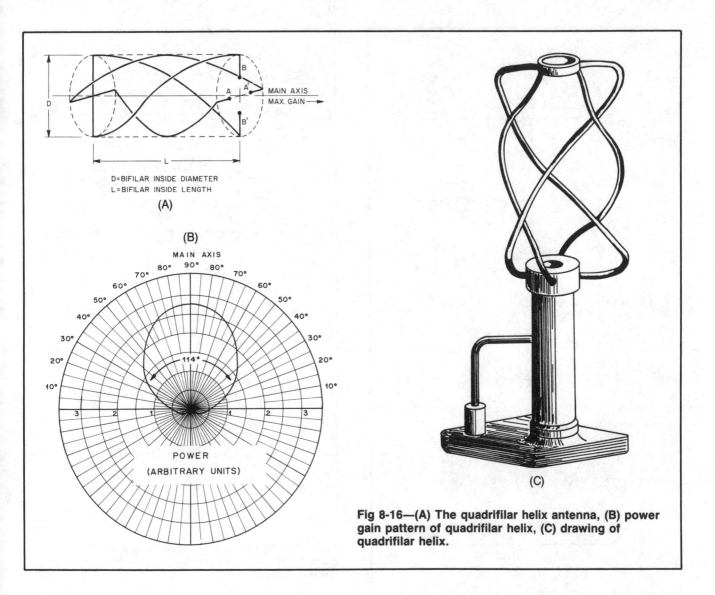

D=BIFILAR INSIDE DIAMETER
L=BIFILAR INSIDE LENGTH

(A)

(B)

MAIN AXIS

POWER
(ARBITRARY UNITS)

(C)

Fig 8-16—(A) The quadrifilar helix antenna, (B) power gain pattern of quadrifilar helix, (C) drawing of quadrifilar helix.

radiated signal is nearly circularly polarized in all directions, a very desirable characteristic. As mentioned earlier, the polarization sense is determined by the direction in which the dipoles are rotated (tilted) out of the horizontal plane. Polarization can't be reversed by modifying the feed harness; if you want to change from RHCP to LHCP, or vice versa, you must change the antenna structure.

Quadrifilar Helix

The quadrifilar helix (Fig 8-16A) consists of four ½-turn helices (A, A′, B, B′) equally spaced around the circumference of a common cylinder. Opposite elements (A and A′, B and B′) form a bifilar pair; the two bifilars must be fed equal amounts of power but 90° out of phase. As with other antennas requiring a 90° phase difference and equal power division, problems arise in designing an adequate feed system.

The solution favored by professional antenna engineers is to make one bifilar slightly undersize so it resonates above the operating frequency (input impedance has a capacitive component) and the other bifilar slightly oversize so it resonates below the operating frequency (input impedance has an inductive component).

If the diameters of the bifilars are adjusted so that the magnitudes of all resistive and reactive components of the two input impedances are equal, the current in the small bifilar will lead the input by 45° while the current in the large bifilar lags by 45°. This yields the desired 90° phase difference and a purely resistive input impedance of about 40 ohms when the two bifilars are fed in parallel. In effect, matching and phasing are built into the antenna itself. An "infinite balun" is conveniently used in conjunction with the self-phased quadrifilar.

The radiation pattern of a quadrifilar helix is omni-directional in the plane perpendicular to its main axis. In a plane containing the main axis (Fig 8-16B) the maximum gain is about 5 dBi, and the beamwidth 114°. Radiation is nearly circularly polarized over the entire hemisphere irradiated. In many situations an antenna with these characteristics is ideal for a ground station. For example, it could be used as part of an unattended automated command or data retrieval station. The quadrifilar helix also makes an excellent spacecraft antenna. One was used on AMSAT-OSCAR 7 for the 2.3-GHz beacon.

Because small changes in the dimensions and dielectric properties of the quadrifilar support structure, and the

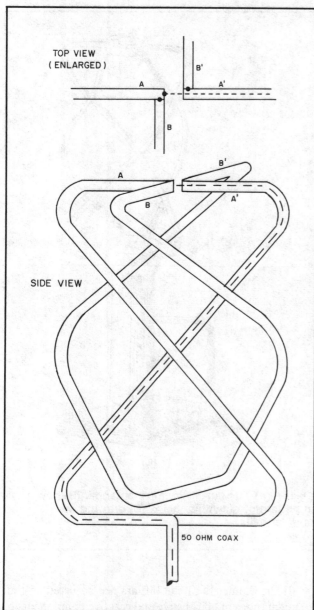

TOP VIEW
(ENLARGED)

SIDE VIEW

50 OHM COAX

Fig 8-17—Quadrifilar helix employing self-phasing and infinite balun. Note: Coax may be used to form all four helices. On three, the inner conductor and outer braid are shorted at cut ends. The fourth helix is part of the feed as shown. RG-58 and RG-8 may be used at 435 MHz and 146 MHz, respectively, but slight adjustments in the lengths of the helices will be needed since the element diameters will be smaller than specified in Table 8-6.

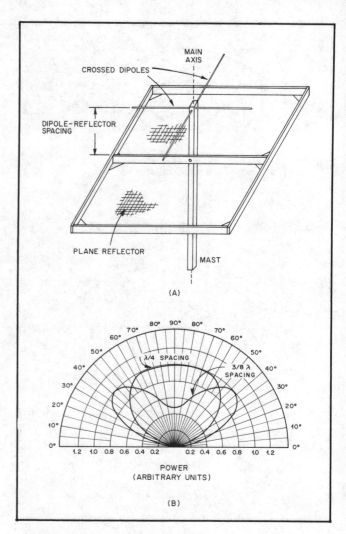

Fig 8-18—(A) Turnstile-Reflector array. (B) Vertical-plane power patterns for dipole-reflector spacings of λ/4 and 3λ/8.

presence of nearby objects, can have a large effect on power division and phasing, the radio amateur without sophisticated test equipment will have difficulty duplicating the desired performance. Nevertheless, the intrepid experimenter will find construction details for 146-MHz and 435-MHz quadrifilars in Fig 8-17 and Table 8-6. Dimensions, scaled from a 2-GHz model, should only be regarded as a guide. Phasing and balun details are also included.

Table 8-6
Design Data for Quadrifilar Helix

	Small Bifilar				Large Bifilar			
	D	L	Length A-A'		D	L	Length B-B'	Wire diam
146 MHz	12.62 in.	19.25 in.	82.19 in.		13.99 in.	21.03 in.	90.60 in.	0.71 in.
435 MHz	4.23 in.	6.46 in.	27.57 in.		4.69 in.	7.05 in.	30.39 in.	0.24 in.
Any frequency (λ)	0.156	0.238	1.016		0.173	0.260	1.120	0.0088

Note: Dimensions should be regarded only as a guide. Special thanks to Walter Maxwell, W2DU, for providing this information.

T-R Array

The T-R array (*t*urnstile-*r*eflector array) shown in Fig 8-18A consists of a pair of dipoles mounted above a reflecting screen and fed equal power, 90° out of phase. Performance is almost identical to the crossed 2-element Yagi array (Fig 7-7).

The T-R array produces a nearly omnidirectional horizontal-plane gain pattern. Vertical plane patterns, which depend on the dipole-to-reflector distance, are shown in Fig 8-18B for spacings of λ/4 and 3/8 λ. The 3/8-λ spacing produces an especially desirable pattern for a fixed ground station antenna. At high elevation angles, where this antenna is most useful, the changing gain tends to compensate for variations in ground station to satellite distance, yielding a relatively constant signal level. The T-R array produces a circularly polarized signal along the main axis. Off-axis circularity is fairly good at high elevation angles, but the Lindenblad and quadrifilar helix are superior in this regard.

The power division and phasing problems encountered with the crossed-Yagi array (Fig 7-7) are repeated with the T-R array. Fig 8-19A contains a matching/phasing harness for 3/8-λ spacing. Note that the impedance of the dipoles varies with dipole-reflector distance, so the matching network shown will not work with other spacings. An adjustment procedure, which requires only an SWR meter, should produce a 146-MHz version that yields optimal performance. Set up two slightly long dipoles 3/8 λ above the reflector. Feed one as in Fig 8-19B; let the other one float. Prune the active dipole for minimum SWR at 146 MHz. Don't worry about the actual SWR as long as it's below 1.5:1. Cut the second dipole to the same length. Reconfigure the feed system as in Fig 8-19A. Then increase the dipole-to-reflector spacing slightly until you obtain minimum SWR.

It is possible to "self-phase" the T-R array as was done with the quadrifilar helix by using one long dipole (resistive and inductive components of input impedance equal) and one short dipole (resistive and capacitive components of input impedance equal). Feeding these two dipoles in parallel will yield correct phasing, an approximately equal power split and a resistive input impedance. If you wish to experiment with the self-phasing approach, you'll have to determine dipole lengths empirically by using an impedance bridge or calculate values as explained in the article by M. F. Bolster (Table 8-7).

Summary

The properties of the three low-gain, circularly polarized antennas suitable for working with low-altitude satellites are summarized in Table 8-7. The table also includes references to literature discussing each of them.

If you've been using a GP antenna on a particular link and it has been yielding acceptable results, give the Lindenblad a try—it should result in a noticeable improvement. Amateurs interested in setting up an inexpensive Mode A station or an unattended station for Mode J packet radio should consider the T-R array. When making

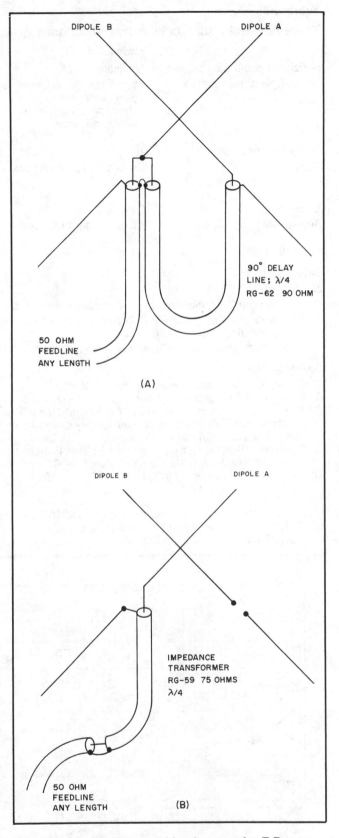

Fig 8-19—(A) Phasing/matching harness for T-R array. (B) Test harness for adjusting T-R array.

your decision keep in mind the fact that a given satellite will appear at elevation angles above 30° only a small percentage of the time it's in range (Fig 8-9).

Table 8-7

Three Low-Gain, Circularly Polarized Antennas

	Lindenblad	*Quadrifilar Helix*	*T-R Array*
Horizontal plane gain pattern	Omnidirectional	Omnidirectional	Omnidirectional
Vertical plane gain pattern	Favors low elevation angles, gain tends to compensate for changing satellite-ground-station distance	Favors main axis	Favors high elevation angles, gain tends to compensate for changing satellite-groundstation distance
Half-power beamwidth	NA	114°	140° (3/8 λ spacing)
Circularity	Excellent in all directions	Excellent in all directions	Falls off away from main axis, good over most of pattern
Construction	Easy to build	Moderately difficult to build	Easy to build
Adjustment	No adjustment required	Specialized test equipment for adjustment	Easy to adjust
Bandwidth	±8%	±2%	±4%

References

Lindenblad

G. H. Brown and O. M. Woodward, Jr, "Circularly Polarized Omni-directional Antenna," *RCA Review*, Vol 8, Jun 1947, pp 259-269.

R. Ott, "A Lindenblad Circularly Polarized Antenna for Amateur Satellite Communications," *The AMSAT Journal*, Vol 14, no. 6, Nov/Dec 1991, pp 10-12.

H. Sodja, "Lindenblad Serendipity and Enlightenment," *The AMSAT Journal*, Vol, 14, no. 6, Nov/Dec 1991, pp 15-18.

Quadrifilar Helix

C. C. Kilgus, "Resonant Quadrifilar Helix," *Microwave Journal*, Dec 1970, pp 49-54.

C. C. Kilgus, "Resonant Quadrifilar Helix," *IEEE Trans on Antennas and Propagation*, Vol 17, May 1969, pp 349-351.

R. W. Bricker, Jr and H. H. Rickert, "An S-Band Quadrifilar Antenna for Satellite Communications," Presented at 1974 International IEEE/P-S Symposium, Georgia Institute of Technology, Atlanta, GA. Authors are with RCA Astro-Electronics Div, Princeton, NJ 08540.

C. C. Kilgus, "Shaped-Conical Radiation Pattern Performance of the Backfire Quadrifilar Helix," *IEEE Trans on Antennas and Propagation*, Vol 23, May 1975, pp 392-397.

The ARRL Antenna Book, 15th Ed., 1988, Chapter 20.

T-R Array

M. Davidoff, "A Simple 146-MHz Antenna for OSCAR Ground Stations," *QST*, Sep 1974, pp 11-13.

M. F. Bolster, "A New Type of Circular Polarizer Using Crossed Dipoles," *IRE Trans on Microwave Theory and Techniques*, Sep 1961, pp 385-388.

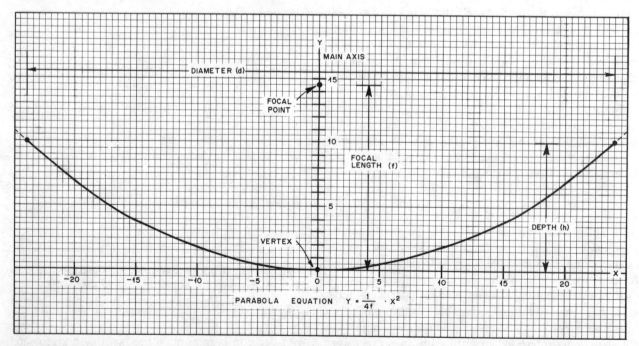

Fig 8-20—Parabola geometry and definitions. When a 2-D parabola is rotated about its main axis, a paraboloidal surface is formed. The parabola shown has an f/d ratio of 0.3.

REFLECTOR ANTENNAS

Reflector antennas consist of a feed antenna and a large (generally at least 6 wavelengths) passive focusing surface. The best known member of the family is the parabolic. The 8- to 10-foot dishes commonly seen in backyards and on rooftops are usually parabolics being used to receive 4-GHz TV transmissions from geostationary satellites. Two other members of this family which may be familiar are the spherical dish and corner reflector.

Parabolic Dish

To understand how the parabolic dish antenna operates, we have to look at dish geometry and feed systems, and the relationship linking these two factors. The shape of the reflecting surface is formally known as a paraboloid, but following common usage we'll refer to both the reflector and the entire antenna as a dish or parabolic dish. The three-dimensional dish surface is formed by rotating a parabolic curve (see Fig 8-20) about its main axis. The operation of the dish is based on the fact that incoming signals which arrive parallel to the main axis are concentrated at a point (the *focal point*) after being reflected off the dish. Similarly, a signal source located at the focal point that illuminates the dish will produce a beam parallel to the dish's main axis, in much the same way a flashlight focuses the light emitted by its bulb.

The location of the focal point depends on dish geometry. It's usually specified in terms of *focal length*, (f): the distance between the vertex (center) and the focal point. Note that modifying the diameter of a dish, by sawing off the outer rim or adding extensions, does *not* change the focal length. One key characteristic of a dish is its focal length to diameter ratio, f/d. The easiest dishes to feed properly are those having f/d ratios between 0.5 and 0.6.

The feed system is a critical element in the performance of a reflecting antenna. It's placed at the focal point and aimed at the center of the dish. A parabolic dish feed is usually designed so that its gain pattern is about 10 dB down at the edge of the dish (compared to the center). The pattern shape must therefore be selected to match the f/d ratio of the dish being used. As mentioned earlier, the horn is very effective when the f/d ratio is in the range 0.5-0.6. The NBS standard-gain antenna is often used when f/d is about 0.5. For lower values of f/d, a dipole or loop over a reflector can be employed. Details may be found in the references listed in Table 8-8.

The polarization of a parabolic dish antenna depends entirely on the feed. A linearly polarized feed antenna will result in a linearly polarized signal, a circularly polarized feed will result in a circularly polarized signal. With circularly polarized signals there is a sense change upon reflection. An LHCP feed will therefore produce an RHCP signal (and vice versa).

One of the most popular feeds is the cylindrical horn shown in Fig 8-21. Tin cans of various sizes make surprisingly efficient horns: 1-gallon motor oil cans (~7" diameter) work well at 1.2 GHz, and 1-pound coffee cans (4" diameter) are often used at 2.4 GHz. A horn feed at

---

Table 8-8

Sources of Information on Parabolic and Related Antennas and on Feed Design

Parabolic Antennas
General
D. S. Evans and G. R. Jessop, *RSGB VHF-UHF Manual*, 3rd Ed., 1976, pp 8.50-8.70.
The ARRL Antenna Book, 15th Ed., 1988, Chapters 18, 19.

Stressed Rib Design
R. T. Knadle, Jr, "A Twelve-Foot Stressed Parabolic Dish," *QST*, Aug 1972, pp 16-22. Reprinted in *The ARRL Antenna Book*, 15th Ed., 1988, Chapter 19.
A. Katz, "Simple Parabolic Antenna Design," *CQ*, Aug 1966, p 10.

Feed Design
N. J. Foot, "Cylindrical Feed Horn for Parabolic Reflectors," *Ham Radio*, May 1976, pp 16-20.
N. J. Foot, "Second Generation Cylindrical Feedhorns," *Ham Radio*, May 1982, pp 31-35.
N. J. Foot, "Cylindrical Feedhorns Revisited," *Ham Radio*, Feb 1986, pp 20-22.
M. Bachi, "XE1XA Circular Loop Dish Feed for 432 MHz," *VHF/UHF and Above Information Exchange*, Sep 1986, p 16. (For f/d = 0.45)
J. DuBois, "Mode L Feed Horn With Circular Polarization," *Orbit*, Mar/Apr 1983. (Reprinted in the 1989 *ARRL Handbook*, Chapter 23, pp 23, 24.)
B. Malowanchuk, "Scalar Feed-horn for 1296 MHz," *VHF/UHF and Above Information Exchange*, Mar 1989, p 18. (For f/d = 0.33 to 0.36)
B. Malowanchuk, "Deep Dish Feed Horns Revisited," *VHF/UHF and Above Information Exchange*, Jun 1989, pp 6-8.
Design Data for Resonant Ring Dipole Feed, *VHF/UHF and Above Information Exchange*, Aug 1986, p 19.
2.3-GHz dual-mode horn, *The ARRL Antenna Book*, 13th Ed, 1974, pp 259-260.

Spherical Reflector
General
A. W. Love, "Spherical Reflecting Antennas with Corrected Line Sources," *IRE Trans on Antennas and Propagation*, Vol AP-10, pp 529-539, Sep 1962.

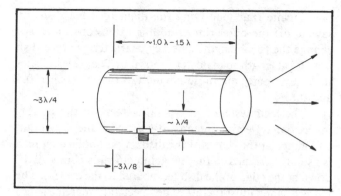

Fig 8-21—Feed horn suitable for illuminating a parabolic dish with an f/d ratio of 0.5 to 0.6. Horn has about 10 dBi gain.

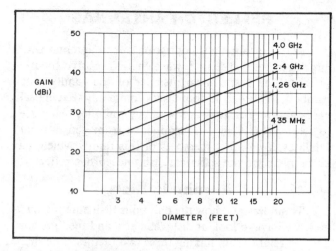

Fig 8-22—Parabolic dish gain vs diameter for several frequencies of interest.

435 MHz would be the size of a small garbage can. The diameter, not the original contents, is the important parameter. A quarter-wave monopole soldered to a coax connector typically is used to excite the horn. The arrangement shown in Fig 8-21 produces a linear wave, the dimensions are approximate. By varying the diameter and length of the feed horn (size of can), the spacing between the monopole and the closed end of the can, and the dimensions of the monopole element, we can shape the beamwidth and adjust matching. Brass screws protruding into the can and collars around the open end of the can are also frequently employed to improve performance. Optimizing performance often involves a trial-and-error approach. Working with a dish on a receiving station for a 2.1-GHz terrestrial TV link, I modified the horn, which was similar to the one shown in Fig 8-21, by adding three 8-40 brass bolts parallel to the monopole and spaced at ¼ wavelength intervals toward the open end. Nuts were soldered to the outside of the can so the bolts could be easily inserted to any length. The bolts were then adjusted while watching a received picture. A considerable improvement was noted. When setting up a parabolic dish on amateur frequencies the usual approach is to use the antenna on the receiving end of a link and adjust all parameters for maximum received signal level.

A circularly polarized horn-type feed can be produced in several ways. One method is to employ two monopoles at right angles fed 90° out of phase using an external delay line. A second method is to use the same monopole geometry in conjunction with a series of tuning screws to produce the phase shift (one plane retarded, the other advanced). The third method is to use a single monopole and a more complex arrangement of tuning screws to construct a device known as a dual-mode horn (references are in Table 8-8). Circular polarization can also be obtained by using a T-R array employing the self-phasing method (described in conjunction with the quadrifilar helix) or a 3-turn helix as the feed.

A dish's reflecting surface can be either solid or mesh. If mesh, the openings should be less than 1/10 wavelength in the largest dimension. Many people believe that mesh is desirable because it has a lower wind resistance. This is really a fallacy. At low wind speeds a mesh dish does present a much lower cross section than a comparably sized solid dish, but at high wind speeds, where problems are most likely to occur, any mesh smaller than 1″ chicken wire will produce nearly the same cross section as a solid dish. The real advantage of mesh is construction ease.

The gain and beamwidth of an efficiently fed parabolic dish are given by

$$G = 7.5 + 20 \log d + 20 \log F$$
(assumes 55% feed efficiency) (Eq 8.1)

and

$$\theta^* = \frac{70}{(d)(F)} \qquad \text{(Eq 8.2)}$$

where

G = gain in dBi
d = dish diameter in feet
F = frequency in GHz
θ^* = 3-dB beamwidth in degrees.

The gain (Eq 8-1) is plotted in Fig 8-22. Observed gains tend to be about 2 dB lower than the predicted values.

No dish is perfect. The surface will always depart slightly from a true paraboloid. If a mesh surface is used some signal will always leak through. The feed horn pattern will depart from the ideal, the horn and its mounting will block the beam path and the horn might not be positioned precisely at the focal point. How do these imperfections affect dish operation? Gain is surprisingly tolerant of sloppy construction and other imperfections. A home brew dish with a mesh surface having holes up to 1/10 wavelength in diameter and random surface inaccuracies on the order of 1/8 wavelength will probably exhibit a gain within 1 to 2 dB of a far more precise unit. However, the imperfections will degrade the pattern shape, often producing high-level sidelobes. As a result, the parabolic makes an effective transmitting antenna at frequencies above 1.2 GHz, where a 6-foot-diameter dish has a gain on the order of 25 dB. However, the same antenna that works well on transmission may be a poor performer on reception be-

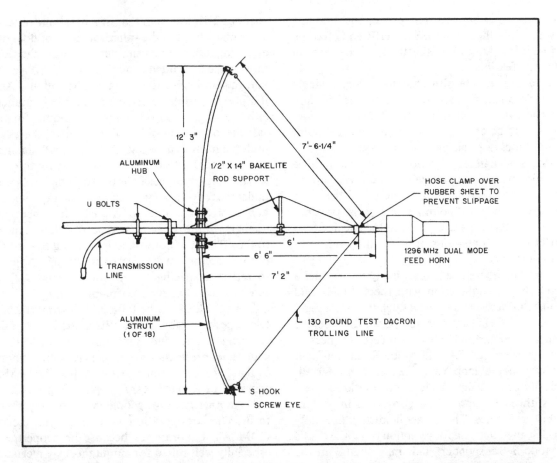

Fig 8-23—Side view of the stressed parabolic dish. See *The ARRL Antenna Book*, 15th Ed. (1988), Chapter 19, for details.

cause noise leaking in via sidelobes, spillover and feed-through may severely reduce the ultimate S/N ratio that can be obtained. An antenna with lower gain and a cleaner pattern could provide better reception.

In the past, amateurs have experimented with surplus dishes acquired from various commercial sources, modified parabolic TV antennas designed for terrestrial UHF TV reception, and dishes they've built from scratch. One construction method that has remained popular over the years involves the stressed rib design shown in Fig 8-23 and described by R. Knadle (K2RIW) and A. Katz (K2UYH) (see Table 8-8). It employs flexible ribs of wood or aluminum radiating out from a stiff central hub. Each rib is formed into a shape that's approximately parabolic by a string connecting the tip of the rib and the feed support. The ribs are covered by a flexible mesh reflecting surface. Antennas of this type, with diameters up to 16 feet, have been used for EME at 432 MHz and 1.2 GHz and for receiving 2.3-GHz transmissions from the Apollo Command module as it traveled to the moon. Dishes produced using this method tend to be lightweight and not very stiff. As a result, they don't stand up very well under severe weather and the pattern is generally not very clean. However, when the weather cooperates, they are suitable for temporary installations.

Direct satellite to home TV (4 GHz) has changed the dish availability situation radically. Secondhand dishes in the 6- to 12-foot-diameter range are now widely available. You don't even have to wait for a hamfest. Just check the classified ads in the Sunday papers under "Moving." Since these dishes were designed to operated at 4 GHz, even those marginally adequate at this frequency should provide decent performance at 1.2 and 2.4 GHz.

It's important to check the focal length of a dish before it's purchased in order to determine whether feeding it will be a problem and to determine its real value. The focal length can be computed from two dimensions that can be readily measured, the dish diameter (d) and the depth (h) (see Fig 8-20).

$$f = d^2/16h \qquad \text{(Eq 8.3)}$$

For example, suppose you locate an 8-foot diameter dish (d = 96″) with a depth of 16″ (h = 16″). Substituting in Eq 8.3 we find that f = 36″. Since the value of f/d is 0.375 (36″/96″), a feed horn does not appear appropriate. However, the dish could be used with a feed horn if one is willing to accept reduced performance. To estimate the effectiveness of the dish you have to determine how much of the dish a horn would illuminate. Let d′ be the effective diameter of the dish and solve the equation f/d′ = 0.5. Returning to our example, 36″/d′ = 0.5 and the effective diameter is therefore 6 feet. If the 8-foot dish is cheaper than 6-foot models, it might be a good buy.

If the thought of purchasing a small dish and enlarging

it by adding extensions has occurred to you, note that any extension will reduce the f/d ratio. Since efficiently feeding parabolics with low f/d ratios is difficult, this approach is generally not feasible.

One advantage of the dish antenna is that a single reflection surface can be used on both the uplink and downlink. The two feedhorns are mounted side by side. Though both may be offset slightly from the focal point and mutually block a small part of the main beam, the effect on gain is negligible (a fraction of a decibel). Once again, the main impact is on pattern sidelobes. When positioning the feeds for a satellite link it's best to either place the receive feed at the focal point and offset the transmit feed (best receive performance), or place the two feeds so they're offset by equal amounts in terms of wavelength, not in actual distance.

The situations where one is most likely to consider a dish antenna are in conjunction with the 1.2-GHz and 2.4-GHz links to Phase 3 and Phase 4 satellites. Since regulations restrict 1.2 GHz to use as an uplink, the sidelobe problem mentioned earlier is not of concern. Dishes are being used by several OSCAR 13 Mode L stations, but commercially available loop Yagis seem to be preferred. The complexity of the azimuth elevation rotation system associated with each approach is clearly an important factor in this preference. The parabolic dish is probably the most widely used antenna for reception of OSCAR 13's 2.4-GHz Mode S transponder. This may change in the future as new Yagi and loop-Yagi designs are introduced.

The 1.2-GHz antenna on OSCAR 13 is a single helix. A single helix is also used at 2.4 GHz. As a result, these antennas do not exhibit the off-axis interferometer-like circularity changes discussed earlier. Since this effect is one of the primary factors which makes circular polarization (especially with switchable sense) desirable, ground stations will find that using a linearly polarized feed only introduces a small performance penalty.

Professional antenna designers have investigated a number of variations of the simple parabolic we've been discussing. One can, for example, introduce a subreflector so that the feed horn can be placed at the vertex of the parabola or even below the dish. These variations can lead to increased gain, superior pattern shape, or a more convenient feed location (making it easier to switch feeds). Construction and performance testing of such antennas is a major undertaking—see the index to *IEEE Trans on Antennas and Propagation* for specific information.

Related Antennas

Several other reflector-type antennas may turn out to be useful at radio amateur satellite ground stations. We'll briefly mention two: the spherical reflector and the toroidal reflector.

Spherical reflectors are nothing more than sections of a sphere. A spherical reflector does not focus incoming signals at a point, so it makes little sense to talk about focal length. Instead, spherical reflectors are characterized by the radius (r) of the sphere they're cut from, rim-to-rim diameter (d) and r/d ratio. The gain of a spherical reflector

is about 2 dB less than that of a parabolic reflector of the same diameter, but the spherical reflector does offer other advantages beyond its simpler geometric shape. Notably, moving the feed antenna up to about 45° off axis is possible before gain begins to decrease substantially. As a result, it's possible to use a fixed reflector in conjunction with either a single feed on a movable mount for tracking a satellite over a considerable region of the sky, and/or multiple feeds to access several spacecraft simultaneously.

Although a feed designed specially for this application should be used, especially if the feed will be placed off axis, under the following conditions one can experiment by using a horn like the ones previously described. Select the diameter of the reflector so that (r/2)/d is in the range 0.5 to 0.6. Place the horn on axis pointing toward the center of the reflector at a distance of r/2 from it. Adjust the horn to reflector spacing for maximum gain. Adjust tuning screws and collar for maximum gain.

As a side note, the famous 1000-foot-diameter radio telescope at Arecibo, Puerto Rico, uses a spherical reflector covering 20 acres and having an accuracy of 1/8 inch! [A slide-tape presentation on the Arecibo "monster" is available for loan from the ARRL.] See D. DeMaw, "The Story of El Radar," *QST*, Sep 1965, pp 24-27.

Another reflector geometry of possible interest is the torus. This configuration has been examined carefully by COMSAT Laboratories because its properties make it especially well suited for simultaneously receiving signals from several stationary satellites that are spaced along the geosynchronous arc above the equator (Fig 8-24). Commercial toroidal reflectors and matched feeds for 4-GHz satellite TV downlinks were first marketed in 1980, and a large number are now used at cable TV earth stations. Although the torus does not have much to offer amateurs at this point in time, it may prove appealing once Phase 4 satellites are in orbit. One toroidal reflector could be used in conjunction with feeds for several transponders on each of the two planned Phase 4 satellites. As an added bonus (?) the same reflector could simultaneously provide satellite TV for the family.

ANTENNA SYSTEMS

An antenna system consists of several components in addition to the antenna itself. We'll look at these components briefly.

Feed Lines and Connectors

Satellite radio links generally use VHF and higher frequencies. In this part of the spectrum, RF power losses associated with both feed lines and connectors are a major concern. Since most ground stations use one or more coax feed lines and several connectors, it's important that selections be made to minimize losses.

All coaxial cable produces some attenuation. Typical losses for 100-foot runs of some common cables are shown in Table 8-9. The values quoted are for new, high-quality line; losses increase with age and exposure to the elements. The attenuation of bargain cable is often significantly greater. Measuring cable loss is relatively simple if you have

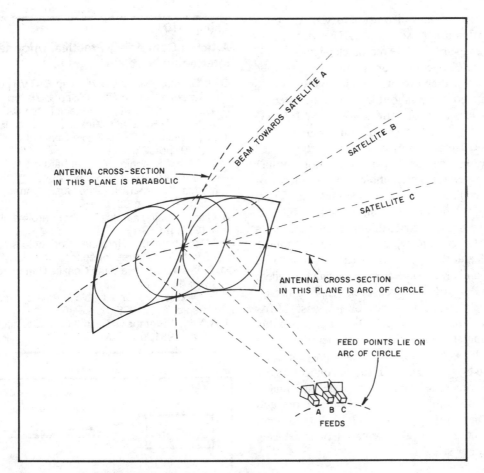

Fig 8-24—The multiple-beam torus antenna.

Table 8-9

Approximate Attenuation Values for Coaxial Lines

Power Loss Per 100 Feet (dB)

Cable	29.5 MHz	146 MHz	435 MHz	1260 MHz
RG-58 series	2.5	6.5	12	22
RG-58 foam	1.2	4.5	8	15
RG-8/M foam	1.3	3.2	7.2	13
RG-8 and				
RG-213	1.2	3.1	5.9	11
RG-8 foam	0.9	2.1	3.7	6.3
RG-17		1.0	2.3	
1/2″ Hardline	0.4	1.0	1.8	3.4
3/4″ Hardline	0.3	0.8	1.6	3.0
7/8″ Hardline	0.3	0.7	1.3	2.5

Note: Attenuation values for old or bargain cable may be much higher.

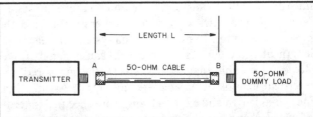

Method. Insert wattmeter at A (meter reads P_A) then adjust transmitter for proper operation. Now move meter to B and read P_S without touching transmitter adjustments. Cable attenuation, in dB, will equal 10 log(P_A/P_S) This value can be scaled to 100 feet as follows:

$$A_o = (100/L)(A_L)$$

where

A_o = attenuation per 100 feet at test frequency
A_L = measured attenuation

Compare the value obtained to the value listed in Table 8-9 to determine if cable is performing up to specifications. Although it's best to make measurements at the satellite link frequency you'll be using a test setup at 2 m will give a good indication of cable quality.

Fig 8-25—Experimental setup for measuring attenuation of a random length of cable. This approach is not designed for high accuracy, but it is useful for rough estimates and comparative measurements.

access to a wattmeter, and doing so is good insurance! (See Fig 8-25.)

Radio-frequency power attenuation is directly proportional to the coaxial cable's length. Doubling the length doubles the attenuation. To compute the loss expected from a given length of cable at a particular frequency use Table 8-9 (or your own measured attenuation value per hundred feed) and the formula

$$A_L = \frac{L}{100} A_o \qquad \text{(Eq 8.4)}$$

where

A_L = attenuation [in dB] of cable of length L
L = length [in feet] of cable
A_o = attenuation [in dB] of 100 feet of cable

Coaxial connectors may also cause losses. Amateurs working at HF often use the so-called UHF series of connectors (PL-259 plug and SO-239 receptacle) with RG-8/U and RG-213 cable. *UHF* connectors should never be used at UHF frequencies; they produce intolerable losses. In fact, this misnamed series shouldn't even be used at 146 MHz unless losses are of little concern. At 146, 435 and 1260 MHz, the Type N series of connectors (UG-21 plug and UG-58 receptacle) may be used with RG-8 sized cable. Though RG-58 can be used for very short jumper cables, there are several better choices, short pieces of which are often available through surplus channels. These include semi-rigid Uniform Tubing UT-141; RG-142, which features Teflon dielectric, double shielding, and a silver-plated center conductor; and the more common RG-141, RG-223 and RG-55. All these cables can be used with BNC, TNC and SMA connectors, which give excellent results up to 4 GHz at low power levels. E. F. Johnson produces a widely available series of low-cost SMA-compatible connectors (JCM type) which are justifiably popular with amateur microwave experimenters. Most Hardline cables have matching low-loss connectors that mate to the Type N series. Since Hardline connectors are relatively expensive, some amateurs have devised makeshift connectors by combining Type N connectors and standard plumbing fittings. Table 8-10 lists several references that contain practical information on interfacing Hardline at amateur stations.

Delay and Phasing Lines

Short sections of coax cable are often used as delay lines or matching transformers in antenna feed harnesses. Numerous examples were given earlier in this and the preceding chapter. In many antenna systems the electrical length of these devices is critical. Because signals travel slower in a cable than in free space, the measured and electrical lengths of a cable are not equal. Assume that measured length and electrical length are specified in terms of wavelength at a specific frequency. The velocity factor of a cable is equal to the ratio of its measured length to its electrical length:

$$\text{velocity factor} = \frac{\text{measured length}}{\text{electrical length}} \qquad \text{(Eq 8.5)}$$

Published values for velocity factor are generally in the range 0.66 (regular cable) to 0.80 (foam dielectric cable). Random measurements, however, show that these values vary by as much as 10% from cable to cable, or up to a few percent along the length of a given spool of cable. Although an error of a few percent may not be important, a 10% error can have a significant effect on antenna system performance. Therefore, it's best to cut all delay and matching lines about 10% long and then prune them to frequency using the dip-meter approach (Fig 8-26).

As an illustration, suppose we need a half-wavelength (electrical length) section of foam-dielectric coax line for 146 MHz. Assuming that our foam dielectric coax has a

Table 8-10

Articles Containing Practical Information on Interfacing Hardline

C. J. Carroll, "Matching 75-Ohm CATV Hardline to 50-Ohm Systems," *Ham Radio*, Sep 1978, pp 31-33.

D. DeMaw, P. O'Dell, "Connectors for CATV 'Hardline' and Heliax," Hints and Kinks, *QST*, Sep 1980, pp 43-44.

J. H. Ferguson, "CATV Cable Connectors," *Ham Radio*, Oct 1979, pp 52-55.

L. T. Fitch, "Matching 75-ohm hardline to 50-ohm systems," *Ham Radio*, Oct 1982, pp 43-45.

J. Mathis, "7/8-inch Hardline Coax connectors," *Ham Radio*, Sep 1988, pp 95-97.

B. Olson, "Using surplus 75-ohm Hardline at VHF," *QEX*, Mar 1988, pp 12-13.

D. Pochmerski, "Hardline Connectors and Corrosion," Technical Correspondence, *QST*, May 1981, p 43.

M. D. Weisberg, "Hardline Coaxial Connectors," *Ham Radio*, Apr 1980, pp 32-33.

G. K. Woods, "75-Ohm Cable in Amateur Installations," *Ham Radio*, Sep 1978, pp 28-30.

J. R. Yost, "Plumber's Delight Coax Connector," *Ham Radio*, May 1981, pp 50-51.

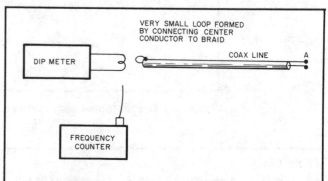

Method I. End A open

Lowest resonant frequency seen on dip meter corresponds to four times electrical length of line. Small pieces of coax are usually cut from end A until desired frequency is reached.

To cut a λ/4 line trim until lowest dip meter frequency = antenna design frequency.

To cut a λ/2 line trim until lowest dip meter frequency = 1/2 antenna design frequency.

Method II. End A shorted

Lowest resonant frequency seen on dip meter corresponds to two times electrical length of line. A pin may be repeatedly inserted near end A until the desired frequency is reached.

To cut a λ/4 line adjust until dip meter frequency = 2 times antenna design frequency.

To cut a λ/2 line adjust until dip meter frequency = antenna design frequency.

Reference formulas

$$\text{free space wavelength in inches} = \frac{11{,}810}{\text{frequency (MHz)}}$$

physical length of coax = (velocity factor) (electrical length)
velocity factor of regular coax = 0.66 (approx)
velocity factor of foam coax = 0.80 (approx)

Fig 8-26—Two methods for using a dip meter to prune a section of coaxial line to a specific electrical length. For additional details see: G. Downs, "Measuring Transmission-Line Velocity Factor," *QST*, Jun 1979, pp 27-28; A. E. Popodi, "Measuring Transmission Line Parameters," *Ham Radio*, Sep 1988, pp 22-25.

velocity of 0.80 as advertised, we can compute the measured length using Eq 8.4:

Measured length = (½) (0.80) = 0.40 wavelength

The free space wavelength (in inches) associated with a specific frequency, f, is given by 11,810/f. At 146 MHz this is 80.9 inches. The measured length will therefore be (0.40) (80.9) = 32.4 inches. A piece about 10% larger than this value should be cut and then trimmed to length using one of the methods shown in Fig 8-26.

Rotators (Azimuth and Elevation Control)

If you're using a beam to access a Phase 3 satellite, you generally have to reorient the beam as the satellite moves across the sky. The systems being used by amateurs to accomplish this task can be categorized as follows:

1) Azimuth rotator, fixed elevation angle
2) Azimuth rotator, manual elevation control
3) Azimuth and Elevation rotation
   a) Combination rotators
   b) Separate rotators,
   c) Azimuth rotator and elevation drive.

System (1) is essentially the normal HF/VHF/UHF terrestrial setup with the antenna tilted up at roughly 25°. If you're operating on a tight budget and using a beam with modest gain, this is a reasonable way to go. Of course, when the satellite elevation is less than 10° or more than 40°, performance may be poor.

Since the elevation angle of a Phase 3 satellite often stays within a narrow range (say 20°) for a long time interval, it's possible to devise an antenna mount incorporating a heavy-duty door or gate hinge making it possible to set the antenna elevation before operation begins. A visit to a sailboat supply house and/or a hardware store will provide the rope, pulleys, cleats and other hardware needed. A flagpole rigging kit will contain many useful parts.

Most amateurs prefer to control both elevation and azimuth from the operating position. Three rotators currently available combine the functions of azimuth and elevation control into a single housing—the Kenpro KR5400, the Dynetic Systems DR10 and the Yaesu G-5400B. Rotators of this type greatly simplify system assembly since the user only has to provide a mast and a cross boom. Another approach is to use separate rotators for each function. Azimuth rotators are widely available in models varying from light to heavy duty. Because the weatherproofing, lubrication and internal thrust bearings are designed for a vertical mast, you're likely to run into trouble if you try to use them for elevation control. However, the Alliance U110, the design of which allows the mast to go completely through the rotator, has proven to be reliable as an elevation rotator if it's crudely sheltered from the elements by an umbrella type arrangement and if the antenna load is arranged so that all static twisting forces are minimized. The light-duty Blonder-Tongue PM-2 can also be used to control elevation. There are several rotators specifically designed for elevation control, including the Kenpro KR500 and the Yaesu G-500A. When using separate azimuth and elevation rotators, you'll need a fitting that will allow you to mount the elevation rotator directly to the top of the azimuth rotator or mast. If your rotators come from different manufacturers you'll probably have to jury rig something from U bolts and a plate of heavy plywood, aluminum or steel.

Elevation control can also be accomplished by employing a hinged plate and a positioning controller like those used on 4-GHz satellite TV dishes. These positioners can often be acquired at modest cost. As they're designed to operate outdoors, reliability should be high. Fig 8-27 shows a possible arrangement. Some amateurs have built positioners from stock threaded rod driven by a motor designed to operate an automobile windshield wiper or power window. If you go this route it helps to have access to welding equipment, an automobile graveyard, a tractor parts dealer and a good hardware store.

Since most amateurs use small computers to obtain tracking information and then manually adjust their antenna controllers, the idea of eliminating the middle man and having the computer directly control the antennas has occurred to many. It's certainly possible to have a small computer take care of antenna pointing chores, but it's questionable whether the result is worth the effort. With slowly moving Phase 3 satellites tracking is a minor chore. And, with Phase 2 satellites it's probably simpler and cheaper to use an omnidirectional antenna and invest part of the money saved in a good preamp and/or a 6-8 dB amp for the transmitter. Nevertheless, having the computer take care of antenna pointing is the ultimate convenience. An automated azimuth/elevation aiming system requires (1) appropriate software for your computer (both tracking software and control software), (2) a computer and hardware interface board, or a dedicated hardware controller and (3) compatible rotators. Getting everything to work together is a major project, as a review of the literature in Table 8-11 will show. Unless you're a software and hardware design expert with lots of time to devote to development, it's best to acquire a set of components (software, hardware, and rotators) that are known to work together. In a 1988 article in *QEX*, Peter Prendergast (KC2PH) discussed the merits of six automated antenna aiming systems he had tested. Systems covered included two for the IBM PC and clones (*The Kansas City Tracker* and *The Mirage Tracking Interface*), two for the Commodore C64 (*Phase IV Systems Controller* and *Encomm KR-001*), *Autotrak* for the C64 or Timex 1000, and *The ARRL Automatic Antenna Controller*. Information can be obtained from the manufacturers listed in Table 8-11.

Radome Material

Most amateurs will not be constructing radomes to house their antenna systems. However, there are situations where you might wish to weatherproof part of an antenna system. For example, you may wish to seal the front of a coffee can dish feed, keep ice from building up on a gamma match, or cover a preamp. When choosing materials for these applications it's helpful to know how some common substances affect radio waves. Transmission attenuation

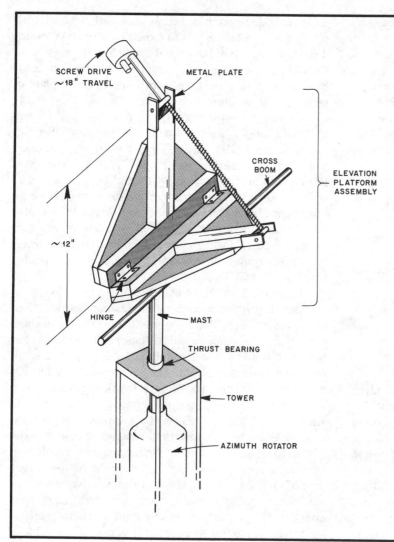

SCREW DRIVE ~18" TRAVEL

METAL PLATE

CROSS BOOM

ELEVATION PLATFORM ASSEMBLY

~ 12"

HINGE

MAST

THRUST BEARING

TOWER

AZIMUTH ROTATOR

Fig 8-27—Az-el mount with homemade elevation platform assembly using "dish" positioner, wood and stock hardware items. This unit is suitable for the antenna shown in Fig 7-13. Mast and crossboom are connected to platform assembly with U bolts and shear pins. Hardware for connecting screw drive to elevation platform should be chosen to match mechanical design of drive.

Table 8-11

Automated Antenna Controller Information Sources

Manufacturers and distributors

Kansas City Tracker
L. L. Grace Communications Products
41 Acadia Dr
Voorhees, NJ 08043
609-751-1018 (evenings and weekends)

ARRL Controller
A & A Engineering
2521 W LaPalma, Unit K
Anaheim, CA 92801
704-952-2114

Mirage Tracking Interface
Mirage/KLM
PO Box 1000
Morgan Hill, CA 95037
800-538-2140

Phase IV Systems Controller for the C64
Phase IV Systems, Inc
3405 Triano Blvd
Huntsville, AL 35805-4695
205-535-2100

Encomm KR-001 Controller
Encomm, Inc
1506 Capital Ave
Plano, TX 75074
214-423-0024

Autotrak
N. Hill
22104 66th Ave W
Mountlake Terrace, WA 98043

Articles

P. Prendergast, "Automatic Antenna Controllers," *QEX*, Aug 1988, pp 8, 14.

J. Bloom, "An Automatic Rotator Controller," *QST*, Sep 1986, pp 40-46.

N. Hill, "A Simple Rotor Interface Board for the C-64 and the VIC-20," *Ham Radio*, Dec 1987, pp 10-12, 16-19, 21-23, 25-27.

Table 8-12
Measured Transmission Attenuation Characteristics of Common Materials

Material	dB attenuation
0.005 in. plastic film bag	<0.1
0.05 in. Tupperware (cloudy)	<0.1
0.06 in. Rubbermaid (opaque)	<0.1
0.0625 in. Vector perfboard	<0.1
0.125 in. Plexiglas	<0.1
2.0 in. styrofoam	<0.1
20 lb bond paper	<0.1
cotton cloth	<0.1
0.07 in. Rubbermaid (cloudy)	0.2
0.125 in. ABS plastic	0.2
bubble pack, 0.25 in. bubbles	0.2
0.125 in. linen phenolic	0.5
0.375 in. dry plywood	1.0
0.70 in. plexiglas	1.2
0.25 in. solid phenolic	1.5
0.750 in. dry plywood	3.0
0.01 in. brass shim stock	>50
0.050 in. rubber sheet	2

See: J. Dubois, "Radome Materials for 1500-1800 MHz," *Journal of the Environmental Satellite Amateur Users' Group,* Vol 89, no. 1, Spring 1989.

is a good measure of how "invisible" a material is to RF radiation. J. DuBois (W1HDX) has measured the attenuation of a number of substances of interest. Though the measurements were made near 1.7 GHz they should be valid across the VHF/UHF spectrum. The results are listed in Table 8-12.

Calculating EIRP

To design the uplink side of a ground station, you usually aim at a specific EIRP (effective isotropic radiated power). This is the power that, if fed to an isotropic antenna, would provide the desired signal level at the spacecraft. Let's consider Mode L on OSCAR 13 for an example. The recommended EIRP is 3 kW. Assume our ground station transmitting system has the following characteristics:

Power output = 50 watts (P_o)

Antenna gain = 23.5 dBi (5-foot-diameter parabolic dish—see Fig 8-22)

Feed-line loss = 3 dB (50 feet of RG-8 foam —see Table 8-9)

Coax connector loss = 0.5 dB (two sets of Type N connectors)

Our calculation would proceed as follows:

Step I: Find G (gain, or loss, of entire feed and antenna system expressed in dBi)

$G = 23.5 \text{ dBi} - 3 \text{ dB} - 0.5 \text{ dB} = 20 \text{ dBi}$

Step II: Convert gain G (in dBi) to gain G (pure number)

$G = 10^{G/10} = 10^{20/10} = 10^2 = 100$

Step III: Calculate EIRP

$EIRP = G\,P_o = (100)\,(50 \text{ watts}) = 5000 \text{ watts}$

The results indicate that we have a slightly higher EIRP than is necessary. However, it's likely that our estimate of the parabolic antenna gain may be 1 or 2 dB high, so we're probably close to the desired level. In any event, keep in mind that published EIRP targets are only guidelines relating to performance under good conditions. To allow for poor squint angles and other factors that may degrade link performance, it's best to have a transmitting setup having more power than required and to include provisions for quick and easy reduction of the power level. If your station is in this category, it's critically important that you periodically compare your downlink signal level to that of the spacecraft general beacon. If you're louder than the OSCAR 13 beacon, you're running too much power! This is also true for most other amateur satellites; beacon power usually serves as a good reference level for checking uplink power.

Closing Hints

Having come this far, some brief final suggestions focusing on antenna systems seem in order. First, start simple and then make improvements where they most affect your operating needs. For example, with Phase 2 satellites try a groundplane or a Lindenblad before you decide that a circularly polarized beam with azimuth and elevation rotators is necessary. With Phase 3 satellites, listen to the 146-MHz downlink with a homebuilt, linearly polarized Quagi before deciding that you need full circular polarization on both links.

Second, don't get caught in the trap of thinking that you need one ultimate array. Often, it's more convenient and effective to have access to several simple antennas set up so that you can quickly switch to the one that produces the best results. Consider Phase 2 satellites again. The multiple antenna approach is most effective when the antennas are complementary in either (1) azimuth response (for example, two horizontal 29-MHz dipoles at right angles), (2) elevation response (for example, a 2-meter T-R array for high elevation angles and a beam aimed at the horizon for low elevation angles) or (3) polarization (for example, dipoles and a ground plane for 29 MHz).

Third, be sure to consider how the satellite of interest and your particular location affect antenna selection. A station at 50° N latitude that is interested in working with Mode B on AMSAT-OSCAR 10 might, after studying typical passes on the computer, decide that rotators are an unnecessary expense. A fixed-elevation array set at 20° and a manually adjustable azimuth control might be perfectly satisfactory. In many cases the operator could set azimuth prior to a pass and not need to adjust it any further.

The well-equipped station working with Phase 3 satellites will eventually desire the benefits of circular polarization. Over the next few years I believe the best antennas for Phase 3 operation will be: (1) the crossed Yagi (concentric boom, offset ¼ wavelength) with a circularity sense switcher at 146 and 435 MHz; and (2) a set of loop Yagis on a cross boom phased for circular polarization at 1269 MHz.

Chapter 9

Receiving and Transmitting

This chapter focuses on the transmitting and receiving equipment that radio amateurs need to communicate via satellite. Coverage includes equipment for CW/SSB communications via the linear transponders that have been in orbit since 1965, and gear for packet communications using digital transponders of the type that first appeared on Fuji-OSCAR 12 in 1986. We'll be looking at receiving equipment for 29.5, 146, 435 and 2400 MHz, and transmitting equipment for 146, 435 and 1269 MHz. The special accessories required for communicating via packet radio satellites are also covered.

If you're interested in telemetry reception, the information on receivers in this chapter will be helpful, but you may also need special modems and software. Information on telemetry reception can be found in Chapter 15 and Appendix B.

As recently as 1983, when OSCAR 10 was launched, little commercial CW/SSB equipment was available for 2 meters, 70 cm and higher frequencies. Even if an amateur used commercial equipment to set up a station, key building blocks were often missing, so assembling a Mode B or L station involved a considerable amount of effort and expertise. The situation has changed considerably over the past few years. Today, amateurs interested mainly in operating can purchase everything needed—transmitters, receivers, antennas and accessories—to put a station on any analog or digital mode currently used on amateur satellites. This approach is expensive. If you're willing to invest some time in shopping and some effort in modifying and/or converting used equipment, it's possible to substantially cut the cost. We begin by considering SSB and CW communications. The equipment needed to access satellites is essentially indistinguishable from gear built for terrestrial use at the same frequencies and power levels. Thus, the existing amateur literature on VHF and UHF construction is directly applicable to our needs. References to useful articles in various Amateur Radio publications are given throughout this chapter. Our main concern will be to describe several practical approaches to assembling a ground station by systematically evaluating the trade-offs involved in various choices.

RECEIVING

A CW/SSB receiver is a central component of the satellite ground station set up for CW/SSB communications via a linear transponder and for packet communicating via most digital transponders. The receiver

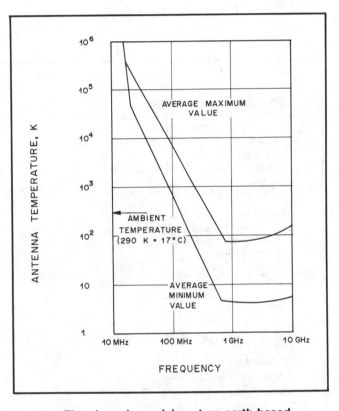

Fig 9-1—The sky noise arriving at an earth-based antenna depends on several factors, including (1) the portion of the galaxy being observed, (2) the elevation angle of the antenna and (3) to a lesser extent, the water-vapor content of the atmosphere. Average values of the upper and lower limits on sky noise are shown in the graph. For details, see J. D. Kraus, *Radio Astronomy* (New York: McGraw-Hill, 1966), p 237.

must meet certain minimal criteria with respect to sensitivity, stability, selectivity and freedom from overload or spurious responses. Anyone with hands-on HF experience knows roughly what these terms mean, and for our purposes we won't need to quantify most of them. Sensitivity, however, deserves special attention because downlink reception is often the limiting factor in the satellite communications chain.

Receiver Sensitivity

At all radio frequencies, noise arriving by way of the antenna ultimately limits our ability to receive weak signals. Over a considerable range of the VHF, UHF and micro-

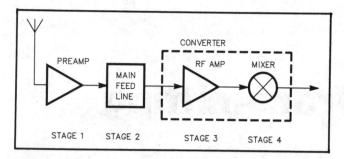

Fig 9-2—To analyze weak-signal performance, a receiver is pictured as consisting of a series of individual stages in the signal path.

wave spectrum, the dominant source of external noise is cosmic in origin. Fig 9-1 shows the background noise levels observed at various frequencies. The dip in the central section shows that the absolute level of this noise is very low at 146 and 435 MHz, making it possible, theoretically, to discern very weak signals. In practice, noise generated in the receiving system itself often masks these weak sources. Our ultimate goal in designing a ground-station receiver is to reduce the internally generated noise to a level below that of the incoming cosmic noise. In reality we usually don't reach this goal, but new receiver technology continually makes it easier and less expensive for us to approach it.

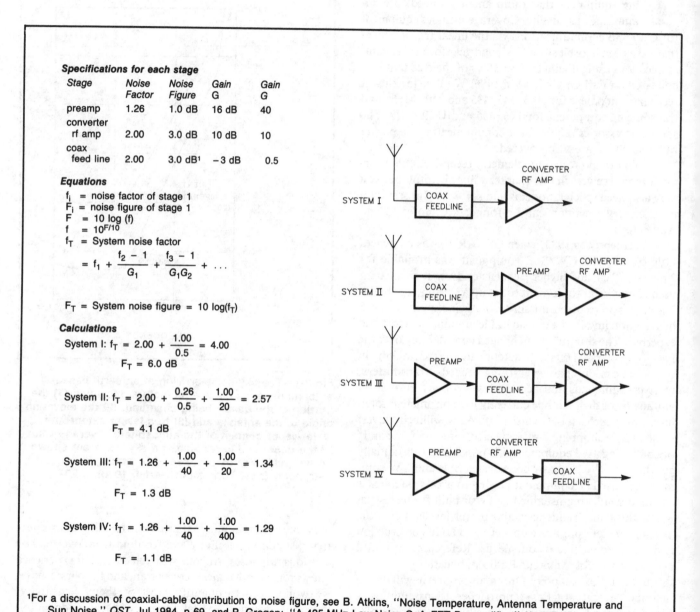

Specifications for each stage

Stage	Noise Factor	Noise Figure	Gain G	Gain G
preamp	1.26	1.0 dB	16 dB	40
converter				
rf amp	2.00	3.0 dB	10 dB	10
coax				
feed line	2.00	3.0 dB[1]	−3 dB	0.5

Equations

f_i = noise factor of stage 1
F_i = noise figure of stage 1
F = 10 log (f)
f = $10^{F/10}$
f_T = System noise factor

$$= f_1 + \frac{f_2 - 1}{G_1} + \frac{f_3 - 1}{G_1 G_2} + \ldots$$

F_T = System noise figure = $10 \log(f_T)$

Calculations

System I: $f_T = 2.00 + \dfrac{1.00}{0.5} = 4.00$

$F_T = 6.0$ dB

System II: $f_T = 2.00 + \dfrac{0.26}{0.5} + \dfrac{1.00}{20} = 2.57$

$F_T = 4.1$ dB

System III: $f_T = 1.26 + \dfrac{1.00}{40} + \dfrac{1.00}{20} = 1.34$

$F_T = 1.3$ dB

System IV: $f_T = 1.26 + \dfrac{1.00}{40} + \dfrac{1.00}{400} = 1.29$

$F_T = 1.1$ dB

[1]For a discussion of coaxial-cable contribution to noise figure, see B. Atkins, "Noise Temperature, Antenna Temperature and Sun Noise," *QST*, Jul 1984, p 69, and P. Gregory, "A 435-MHz Low-Noise GaAsFET Preamplifier," *Ham Radio*, Jul 1989, pp 9-12, 17, 19.

Fig 9-3—Comparing noise figures of four systems. Reference: J. R. Fisk, "Receiver Noise Figure and Sensitivity and Dynamic Range," *Ham Radio*, Oct 1975, pp 8-25. This article also contains a good discussion of noise temperature. Also see: B. Atkins, "Calculating System Noise Temperature," *QST*, Jan 1982, p 80.

A receiver can be depicted as a chain of individual stages (Fig 9-2), each characterized by two properties: *gain* and *noise factor* (or noise figure), a quantity related to the amount of noise the stage introduces. Noise factor is a dimensionless number greater than or equal to one (noise figure is given in decibels). The lower the noise factor, the better the performance. In a receiver, the noise contribution of each stage acts to reduce the overall system signal-to-noise ratio. The impact of a particular stage on total receiver noise factor depends on the gain prior to the stage and the noise factor of the stage. An analysis of the mathematics (Fig 9-3) shows that the first few stages in a receiver dominate the overall receiver noise factor. Therefore, using very-low-noise devices in the first stage or two of a receiver and avoiding runs of lossy feed line in front of active devices is very important. Once you grasp this basic point, you're well on your way to designing an effective receive system. Though many readers may prefer to skip the computations at this point, anyone interested in putting together a high-performance system should look at the sample calculations in Fig 9-3 to see the consequences of the various trade-offs.

Receiving Systems for 29, 146, 435 and 2400 MHz

29 MHz

Any HF communications receiver covering the 29.0- to 29.5-MHz range can be used to monitor Mode A downlinks (Fig 9-4). A good low-noise preamp will, in many cases, improve reception significantly. This is true even with expensive receivers, because they've been designed to satisfy the less-stringent sensitivity requirements (noise factor) for terrestrial communication. Mounting the preamp at the antenna always provides the best performance. At 29 MHz, however, the improvement is small when using less than 100 feet of RG-8 feed line; many operators give in to convenience and place the preamp at the receiver. Some amateurs with receivers having "hot" 10-meter front ends find reception satisfactory without a preamp. You may too, so give it a try without the preamp —you can always add one later.

146, 435 AND 2400 MHz

For receiving at VHF and UHF, we generally have a choice of using either a converter in conjunction with an HF receiver or acquiring a single-band multimode transceiver. If you already own an HF receiver, the converter approach is substantially cheaper. However, if you have a need for the transmitting capabilities provided by the transceiver, its real cost is substantially reduced. For example, a 435-MHz transceiver acquired for Pacsat reception (Mode JD) will give you the ability to transmit on Mode B. Multimode transceivers are available for the 2-meter, 70-cm and 23-cm bands. The receive sensitivity of older units is generally marginal as far as the needs of the satellite operator are concerned. It is therefore important to plan on using an external preamp with them.

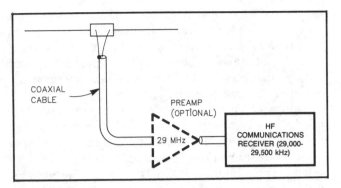

Fig 9-4—Basic receive system for Mode A reception.

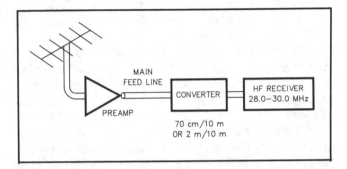

Fig 9-5—A crystal-controlled converter used in conjunction with an HF receiver acting as a tunable IF can provide state-of-the-art VHF/UHF receive capabilities.

Crystal-controlled converters using an HF receiver as a tunable IF are used at a great many satellite ground stations for reception at 2 meters, 70 cm and 13 cm (Fig 9-5). A good HF receiver coupled with a well-designed converter will provide state-of-the-art receive capabilities. Most modern converters for 2 meters and 70 cm use 28.000 to 30.000 MHz as the IF range. If you have a 70-cm converter that covers 435.000 to 437.000 MHz and you wish to monitor 435.300 to 435.500 MHz, you just tune the HF receiver between 28.300 and 28.500 MHz. Well-designed converters for 13 cm generally use 2 meters as an IF. This reduces problems related to image rejection and spurious responses, both of which can degrade the converter's effective sensitivity.

Characteristics that distinguish a good converter from a mediocre performer include low noise figure, freedom from spurious responses, low susceptibility to IMD (intermodulation distortion) and gain compression, high frequency stability and low susceptibility to burnout. Although noise figure is important, it shouldn't be of overriding concern. Once you commit to placing a good 1.0-dB-noise-figure, 16-dB-gain preamp at the antenna, whether the converter noise figure is 3 dB or 4 dB makes little difference. The next characteristic, spurious responses, often results from poor interstage filtering in the RF and local-oscillator circuits or instability (undesired oscillations) anywhere in the converter. IMD and gain compression can arise from poor choices of mixer injection power, bias levels or gain distribution. Poor stability is usually the result of cost-cutting shortcuts such as using cheap crystals and heavily loaded oscillators. To overcome the problems just mentioned, you may need to acquire special test equipment, develop good diagnostic skills and redesign the circuitry—in other words, these shortcomings are often difficult to cure.

Burnout is primarily associated with the first RF stage in solid-state converters or preamps. High-power transmitters and transients related to lightning, relay switching or other sources can burn out expensive RF devices even when extensive precautions are taken. If you build a solid-state preamp, don't skimp by omitting the recommended protection circuits. For a detailed analysis of converter design and hints for improving performance, see J. Reisert, ''Low-Noise Receiver Update,'' Part I, *Ham Radio*, Nov 1987; Part II, Dec 1987.

If you don't want to invest in a modern, well-engineered, low-noise-figure converter, you can often obtain excellent performance using an older, well-engineered one, even if it falls short in the noise-figure department; you need only add a good preamp. Several well-designed converters manufactured in the mid and late 1960s used Nuvistors®, miniature tube-type devices typified by the 6CW4, in the front end. The better units had noise figures in the 3.0- to 3.5-dB range at 2 meters and in the 4- to 5-dB range at 70 cm. You can often find these converters very inexpensively at hamfests. A number of operators actually prefer these older converters in situations where IM distortion, gain compression or burnout have been problems. In any event, avoid poorly engineered units, no matter how impressive the noise-figure specifications may seem.

Many operators report that their receiver performance degrades (desenses) whenever their transmitters are keyed. This problem is especially prevalent when operating Mode J, since the third harmonic of the transmitter is very close to the receive frequency. Spotting the downlink frequency and evaluating uplink performance become difficult, if not impossible, under these conditions. To alleviate the problem, amateurs have tried (1) filters between the transmitter and its feed line, and between the receiver and its feed line, (2) separating the transmit and receive antennas and feed lines physically and (3) using older-type converters in place of modern, solid-state units. Several users have reported cases where serious overload, intermodulation or other receive problems didn't respond to filters or physical displacement, but were cured by switching to a tube-type converter. Nonetheless, good filtering and adequate antenna and feed-line separation should also be pursued. One receive filter is shown in Fig 9-6. Recent editions of *The ARRL Handbook* contain other suitable designs.

Another feature that you may want in a converter is coverage of more than a 2-MHz segment of a band, for example, 144 to 148 MHz on 2 meters, or 432 to 434, 435 to 437 and 436 to 438 MHz on 70 cm. If so, look for provisions for switching crystals in the injection chain.

Many modern converters have good noise figures (often under 2 dB), suggesting the possibility of eliminating the preamp. To take advantage of the converter's noise

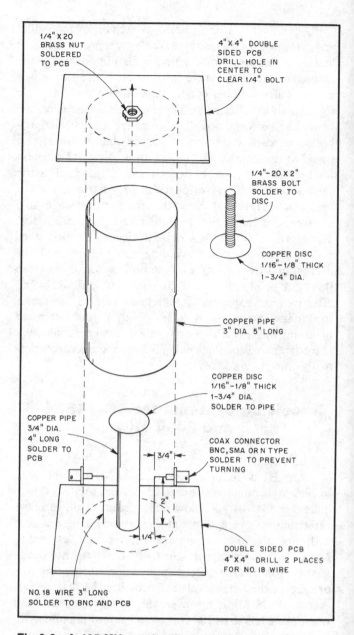

Fig 9-6—A 435-MHz cavity filter can significantly improve reception when front-end overload by out-of-band signals is a problem. Don't expect this filter to have much effect on the third harmonic of a 2-meter transmitter when operating Mode J, since the 3-dB bandwidth is generally about 15 MHz. The third harmonic should be removed at the transmitter by using a band-pass filter.

figure, however, it should be mounted remotely, at or very close to the antenna. Most feed-line losses will then occur after the point in the system where the noise figure is established. This approach generates a great many problems relating to weatherproofing, environmental temperature extremes, oscillator drift, switching the frequency range and adjusting the converter gain. Some of these problems are difficult to overcome. If you're considering remote converter mounting, be sure to check whether the converter is suited to the planned environment. In most cases, you'll have fewer headaches if you keep the converter in the shack and place a good preamp at the antenna.

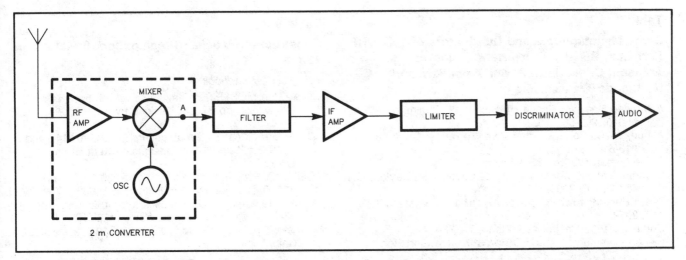

Fig 9-7—A block diagram of a typical 2-meter FM receiver. The section enclosed in the dashed lines forms a 2-meter converter. It's usually possible to pick off a portion of the RF signal at point A without impairing 2-meter FM operation. The frequency at point A varies from receiver to receiver, for example, ICOM 211 (10.7 MHz), Clegg FM-28 (16.9 MHz) and G E Progline (8.7 MHz).

Many HF operators don't realize that they already have a good 2-meter or 70-cm converter at the operating position: the receiver front end in an FM transceiver (Fig 9-7). It's usually a simple matter to "steal" a little of the signal from the FM transceiver and feed it into the communications receiver that is then used as a tunable IF. This gives full CW/SSB capabilities. If you're considering this approach, dig out the instruction manual for the FM transceiver and find the frequency of the first IF (point A in Fig 9-7). Next, determine whether the HF receiver following the converter can be adjusted to tune this range. If you have a modern continuous-coverage receiver (0.5 to 30 MHz), this shouldn't be a problem since the FM transceiver IF is almost always in this range. With older receivers using plug-in crystals, like the Drake R4 series, one can order an appropriate crystal or, in many cases, a standard crystal can be utilized by tuning the preselector to an image frequency. Modifying some receivers to cover the proper range may involve more work or expense than is justified. To pick off the signal, check the transceiver schematic for an easily accessible low-impedance point between the mixer and filter (or between the first mixer and second mixer in double-conversion units) and then try either a capacitive or inductive probe (Fig 9-8). Use a heavily shielded cable and keep the exposed ends short to minimize IF feed-through. A little experimentation with the point where the pick-off probe is attached may lead to significantly improved performance. If IF feed-through is a problem, you might try attaching the probe after the transceiver IF filter. This will make tuning somewhat less convenient since the HF receiver can only be used to cover a 20-kHz segment of the band. Additional segments are covered by switching channels on the transceiver. Instead of using a probe to steal some signal from an operating VHF or UHF FM transceiver, it may be preferable to salvage the front end of an inoperative unit obtained at low cost.

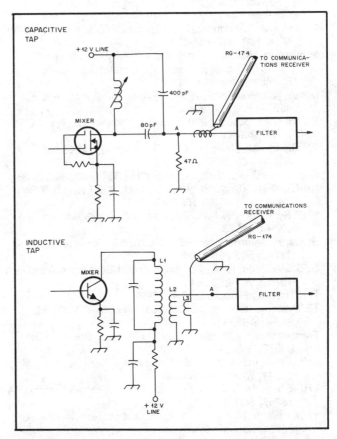

Fig 9-8—Techniques for tapping off a little RF from a 2-meter FM receiver.

Preamps

At 146 MHz and above, a good preamp mounted at the antenna almost always significantly improves down-link reception. Even at 29 MHz, an antenna-mounted preamp often provides a considerable improvement. The required preamp gain and noise figure depend on sky noise

Table 9-1

Some Manufacturers and Distributors of VHF/UHF Preamps, Receive Converters, Transverters, Transmit Converters, Power Amplifiers and Multimode Transceivers

Advanced Receiver Research, Box 1242, Burlington, CT 06013, 203-582-9409

Alinco, 20705 South Western Ave, Torrance, CA 90501, 213-618-8616

Angle Linear, PO Box 35, Lomita, CA 90717

Communications Concepts, Inc, 121 Brown St, Dayton, OH 45402, 513-220-9677

Down East Microwave, Box 2310, RR 1, Troy, ME 04987, 207-948-3741

Encomm, 1506 Capitol Ave, Plano, TX 75074

Falcon Communications, PO Box 8979, Newport Beach, CA 92658

GLB Electronics, Inc, 151 Commerce Pkwy, Buffalo, NY 14224, 716-675-6740

Hamtronics, Inc, 65 Moul Rd, Hilton, NY 14468, 716-392-9430

Hi-Spec, Box 387, Jupiter, FL 33468, 407-746-5031

ICOM America, Inc, 2380 116th Ave, Bellevue, WA 98004

Kenwood USA Corp, 2201 E Dominguez St, Long Beach, CA 90810

Microwave Components of Michigan, PO Box 1697, Taylor, MI 48180, 313-753-4581

Microwave Modules, Brookfield Dr, Aintree, Liverpool L9 7AN, England. In the US, contact PX Shack or Spectrum International.

Mirage Communications, PO Box 1000, Morgan Hill, CA 95037, 800-538-2140

Pauldon Associates, 210 Utica St, Tonawanda, NY 14150, 716-692-5451

PX Shack, 52 Stonewyck Dr, Belle Mead, NJ 08502, 201-874-6013

Radiokit, PO Box 973, Pelham, NH 03076, 603-635-2235

RF Concepts (Division of Kantronics), 2000 Humbolt St, Reno, NV 89509, 702-827-0133

SHF Systems, PO Box 666, Nashua, NH 03061, 603-673-1573

Spectrum International, PO Box 10845, Concord, MA 01742, 508-263-2145

SSB Electronic USA, 124 Cherrywood Dr, Mountaintop, PA 18707, 717-868-5643

Ten-Tec Inc, Sevierville, TN 37862

TE Systems, PO Box 25845, Los Angeles, CA 90025, 213-478-0591

Transverters Unlimited, PO Box 178, New Boston, NH 03070, 603-547-2213

VHF Communications, 280 Tiffany Ave, Jamestown, NY 14701, 716-664-6345

VHF Shop, 16 S Mountain Blvd, Rte 309, Mountaintop, PA 18707, 800-HAM-7373

Yaesu Electronics Corp, 17210 Edwards Rd, Cerritos, CA 90701, 213-404-2700

Table 9-2

Construction Articles: Preamps for 10 Meters to 13 cm

1991 ARRL Handbook

''Dual-Gate MOSFET Preamplifiers for 28, 50, 144 and 220 MHz,'' pp 31-1,2. Designs by Kent Britain, WA5VJB

''Dual-Gate GaAsFET Preamplifiers for 28, 50, 144 and 220 MHz,'' pp 31-3,4. Designs by Kent Britain, WA5VJB

''GaAsFET Preamplifiers for 144 and 220 MHz,'' pp 31-5,6. Designs by Kent Britain, WA5VJB

''Dual-Gate GaAsFET Preamplifiers for 432 MHz,'' pp 32-1,2. Designs by Kent Britain, WA5VJB

''GaAsFET Preamplifier for 70 cm,'' pp 32-3,4. Design by Chip Angle, N6CA

Periodicals

G. Barbari, ''UHF Preamplifier Centers on Budget Dual-Gate FET,'' *Microwaves and RF*, Feb 1984, pp 141, 143-145. (GaAsFET design for 70 cm)

K. Britain, ''A 2-Meter Preamplifier Using the TI S3030 Dual-Gate GaAsFET,'' *QEX*, Dec 1984, pp 8-9.

G. Ehrler and J. DuBois, ''A Very High Performance LNA for 1500-1750 MHz,'' *Journal of the Amateur Satellite User's Group*, Vol 89-1, Spring 1989.

P. Gregory, ''A 435-MHz Low-Noise GaAsFET Pre-amplifier,'' *Ham Radio*, July 1989, pp 9-12, 17, 18.

G. Krauss, ''VHF and UHF Low-Noise Preamplifiers,'' *QEX*, Dec 1981, pp 3-8

J. Reisert, ''Low-Noise GaAsFET technology,'' *Ham Radio*, Dec 1984, pp 99, 101, 103-106, 108, 111-112

J. Reisert, ''Low-Noise Receiver Update: Part 1,'' *Ham Radio*, Nov 1987, p 77; Part 2, Dec 1987, pp 72-76, 79-81

H. P. Shuch, ''A Low-Noise Preamp for Weather Satellite VISSR Reception,'' *QEX*, Feb 1989, pp 3-9. (An excellent design for 1.7 GHz that can be adapted for 13 cm.)

A. Ward, ''Simple Low-Noise Microwave Preamplifiers,'' *QST*, May 1989 pp 31-36, 75. (Includes designs for 13 cm.)

---

and at 146 MHz, about 1.5 dB. At 435 MHz, it's desirable to aim as low as you can go—0.7 dB is not too difficult to achieve. Since system noise figure will always be greater than preamp noise figure (see Fig 9-3), preamp noise figure targets should be slightly below the levels specified.

Ready-made preamps are available from a number of manufacturers (see Table 9-1). One manufacturer (Hamtronics) sells a bare-bones kit (no case, no connectors) for under $30 that meets our requirements at 29 MHz and 146 MHz, and is close to our objectives at 435 MHz.

A number of good preamp designs are available. One excellent source is the 1991 *ARRL Handbook*, which contains several suggested circuits with complete construction details, including lists of component suppliers. See Table 9-2 for additional sources. If you enjoy home brewing, a simple preamp is a good project.

At 29 MHz, a dual-gate MOSFET will provide about 20 dB gain with a noise figure under 1.5 dB. A 10-meter

seen by the antenna, feed-line length, noise figure of following stages, and so on. For most installations a gain of 15 dB is reasonable. If you keep lowering the noise figure of a receiving system, you'll eventually reach a point where further reduction provides no discernible improvement in performance. This is because noise arriving by the antenna has become the limiting factor. The target noise figure depends on frequency. At 29 MHz, it's about 2.0 dB,

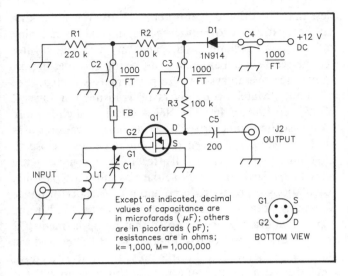

Fig 9-9—28-MHz dual-gate MOSFET preamplifier. Resistors are ¼-W carbon types. (See 1991 *ARRL Handbook*, p 31-1).

C1—15- to 60-pF ceramic trimmer, Erie 538-002F.
C2, C3—500- to 1000-pF feedthrough capacitor, solder-in type preferred.
C4—500- to 1000-pF feedthrough capacitor.
C5—100- to 200-pF silver-mica or ceramic capacitor.
J1, J2—Female chassis-mount RF connectors, BNC preferred.
L1—17 turns no. 28 enam on Amidon T-50-6 core. Tap at 6 turns from ground end.
Q1—Dual-gate MOSFET such as 3SK51 or 3N204.

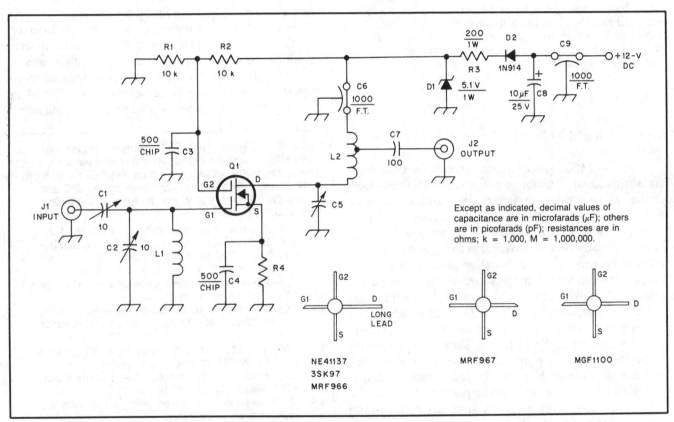

Fig 9-10—144-MHz dual-gate GaAsFET preamplifier. Resistors are ¼-W carbon-composition types unless otherwise noted. (See 1991 *ARRL Handbook*, p 31-3.)

C1, C2—10-pF (max) ceramic or piston trimmer capacitor.
C3, C4—200- to 1000-pF ceramic chip capacitor or leadless disc-ceramic capacitor.
C5—For 144 MHz: 20-pF trimmer capacitor. For 220 MHz: 10-pF trimmer capacitor.
C6, C9—400- to 1000-pF feedthrough capacitor.
C7—50- to 100-pF silver-mica capacitor.
C8—1- to 25-µF, 25-V electrolytic capacitor.
D1—5.1-V, 1-W Zener diode (1N4733 or equiv).
D2—1N914, 1N4148 or any diode with ratings of 25 PIV and 50 mA or greater.
J1, J2—Female chassis-mount BNC or type-N connector.
L1—See Note.
L2—See Note.
Q1—NEC NE41137, Motorola MRF966/967, Mitsubishi MGF 1100 or 3SK97 dual-gate GaAsFET.

R1—For NE41137, MGF1100, 3SK97: 10 kΩ; for MRF966/967: 4.7 kΩ.
R2—10 kΩ for all devices.
R3—150- to 250-Ω, 1- or 2-W resistor.
R4—For NE41137, MGF1100, 3SK97: 47 Ω; for MRF966/967: 100 Ω.

Note:
L1 and L2 Values

Device	L1	L2
NE41137	7 turns	7 turns, tap at 2 turns
3SK97 and MGF1100	5 turns	7 turns, tap at 2 turns
MRF966/967	12 turns	8 turns, tap at 2 turns

All coils are no. 20 to 24 wire, 3/16-inch ID, spaced one wire diameter.

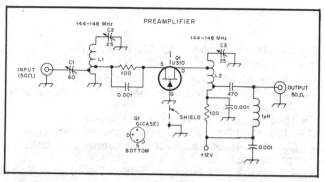

Fig 9-11—144-MHz JFET preamp. Noise figure: ~1.5 dB. Gain: ~10 dB. Q1: Siliconix U310 or 2N5397. See 1985 *ARRL Handbook*, p 18-25, for additional information. A similar design by C. E. Scheideler with detailed construction information appeared in "A Preamplifier for 144-MHz EME," *EME Notes*, AS-49-9, Eimac Division of Varian Associates, 301 Industrial Way, San Carlos, CA. L1 has 5 turns of no. 20 wire, ¾ inch (19 mm) long, with an ID of ¼ inch (6.3 mm). C1 tap approx ½ turn from ground, Q1 tap approx 1 turn from ground. L2 has same dimensions except for Q1 tap, which is approx 1 turn from C3 end.

preamp designed by Kent Britain (WA5VJB) is shown in Fig 9-9.

At 146 MHz, suitable preamps may be built from transistors based on silicon or gallium-arsenide technology. In the 1970s and earlier, it was extremely difficult to obtain noise figures under 2.0 dB at frequencies above 144 MHz. In about 1978, when GaAsFETs (gallium-arsenide field-effect transistors) became available to radio amateurs, it quickly became apparent that they would revolutionize preamp performance. Initially, these devices were extremely expensive, but today you can purchase a dual-gate GaAsFET that will provide 20- to 24-dB gain with a noise figure under 1 dB at 146 MHz for less than $5. A typical circuit, again by WA5VJB, is shown in Fig 9-10. GaAsFETs are hardy once they're installed in a preamp, but they are static sensitive and they can be easily destroyed if not handled properly while the preamp is under construction. For this reason, some amateurs prefer to use silicon devices. At 146 MHz, this is possible. The preamp shown in Fig 9-11, modeled after a design by C. E. Scheideler, W2AZL, uses an inexpensive silicon JFET in a grounded-gate configuration. The noise figure is 1.5 dB. The gain is only about 10 dB, so this circuit is suitable only if feed-line losses are relatively low and the following converter has a noise figure under 3.5 dB.

At 435 MHz, GaAs technology is the way to go. The main choice is whether to opt for a preamp using low-cost dual-gate devices or for one using a slightly more expensive, and better-performing, single-gate device. The 146-MHz dual-gate GaAsFET preamp (Fig 9-10) also works at 435 MHz if the tuned circuits are changed. At 435 MHz it provides a noise figure under 0.9 dB with 22+ dB of gain. The single-gate GaAsFET preamp shown in Fig 9-12 was designed by Chip Angle, N6CA. The noise figure is

less than 0.7 dB and gain is about 13 dB. When working at these extremely low noise figures, impedance matching of antenna, preamp input, and any band-pass filter connected between antenna and preamp becomes important if the overall system is to perform as expected. (See Z. Lau, "Matching Receivers to Transmission Lines," *QST*, Sep 1988, p 46.) The N6CA circuit is slightly more complex than some of the other designs, because it uses source feedback to keep the input impedance close to 50 ohms. As a result, the full effect of the low noise figure can be realized. In addition, the possibility of parasitic oscillations is greatly reduced and excellent performance is readily obtained. When a test unit was tweaked in the ARRL laboratory, it was noted that no adjustment (C2 and/or expansion/contraction of L3 and L5) caused the gain to dip below 12 dB or the noise figure to rise above 0.65 dB. As a result, you can build this preamp without any test equipment, simply tune it for maximum gain, and be reasonably sure of obtaining excellent performance.

One additional preamp design for 435 MHz deserves mention. The circuit shown in Fig 9-13, which is based on silicon bipolar transistor technology, was designed by Joe Reisert (W1JR). It has a noise figure of 1.5 to 2.0 dB and a gain of 12-14 dB. Note that there are *no* tuning adjust-

Fig 9-12—435-MHz GaAsFET preamplifier. Resistors are carbon-composition types. Resistor values are given in ohms; capacitor values are given in pF. This design, by E. R. "Chip" Angle, first appeared in the *432 and Above EME News*, Vol 9, no. 7, June 1981. Noise figure: 0.5 dB. Gain: ~16 dB.

C1—5.6-pF silver-mica capacitor, or same as C2.

C2—0.6- to 6-pF ceramic piston trimmer capacitor (Johnson 5700 series or equiv).

C3, C4, C5—200-pF ceramic chip capacitor.

C6, C7—0.1-μF disc-ceramic capacitor, 50 V or greater.

C8—15-pF silver-mica capacitor.

C9—500- to 1000-pF feedthrough capacitor.

D1—16- to 30-V, 500-mW Zener diode (1N966B or equiv).

D2—1N914, 1N4148 or any diode with ratings of at least 25 PIV at 50 mA or greater.

J1, J2—Female chassis-mount type-N connectors, PTFE dielectric (UG-58 or equiv).

L1, L2—3 turns no. 24 tinned wire, 0.110 inch ID, spaced 1 wire diameter.

L3—5 turns no. 24 tinned wire, 3/16 inch ID, spaced 1 wire diameter or closer. Slightly larger diameter (0.010 inch) may be required with some FETs.

L4—1 turn no. 24 tinned wire, 1/8 inch ID.

L5—4 turns no. 24 tinned wire, 1/8 inch ID, spaced 1 wire diameter.

L6—1 turns no. 24 tinned wire, 1/8 inch ID.

Q1—Mitsubishi MGF 1402.

R1—200- or 500-Ω cermet potentiometer set to mid-range initially.

R2—62-Ω, 1/4 W resistor.

R3—51-Ω, 1/8 W carbon-composition resistor, 5% tolerance.

RFC1—5 turns no. 26 enam wire on a ferrite bead.

U1—5-V, 100-mA 3-terminal regulator (LM78L05 or equiv; TO-92 package).

For reference see 1990 *ARRL Handbook*, pp 32-3 and 32-4, and etching patterns p 16.

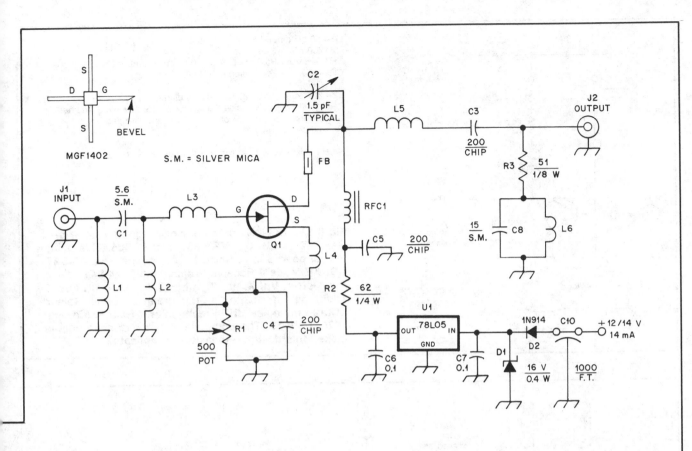

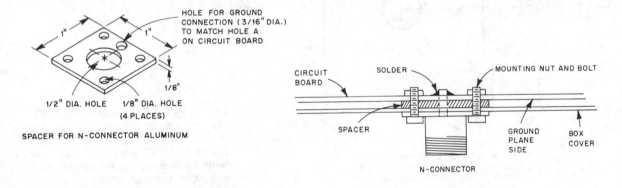

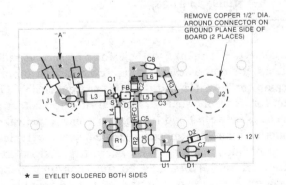

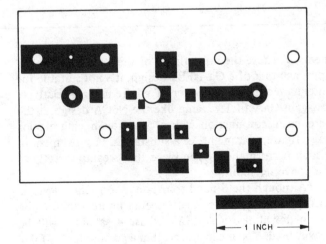

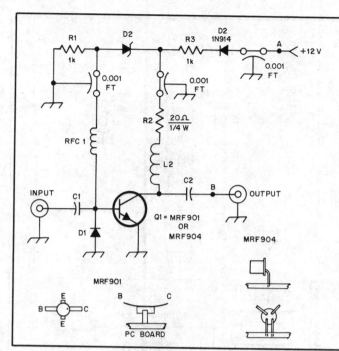

C1—50 pF dipped Mica.
C2—5.0 pF dipped Mica.
CR1—Hewlett Packard 5082-2810 or equivalent hot carrier diode.
CR2—6.2 volt Zener diode, 1N4735 or equivalent.
L1—Deleted.
L2—3t #24 on 1/10" ID space wire diameter.
RFC 1—0.47 μH Nytronics deciductor on 15t #32 AWG enamel
    covered copper wire on 1/10" ID space wire diameter.
R2—20 ohms, 1/4 watt.

Note: Mount transistor as shown with leads just touching PC board.

Fig 9-13—435-MHz preamp. Noise figure: 1.5-2.0 dB. Gain: 12-14 dB. Q1: MRF 901 or MRF 904. This design first appeared in J. Reisert, ''An Inexpensive AMSAT-OSCAR Mode J Receive Preamplifier,'' *AMSAT Newsletter*, Vol X, no. 2, June 1978, pp 10-11. For additional information on this preamp, see: J. Reisert, ''Ultra Low-noise UHF Preamplifier,'' *Ham Radio*, Mar 1975, pp 8-19. The complete absence of adjustments makes this design a pleasure to replicate.

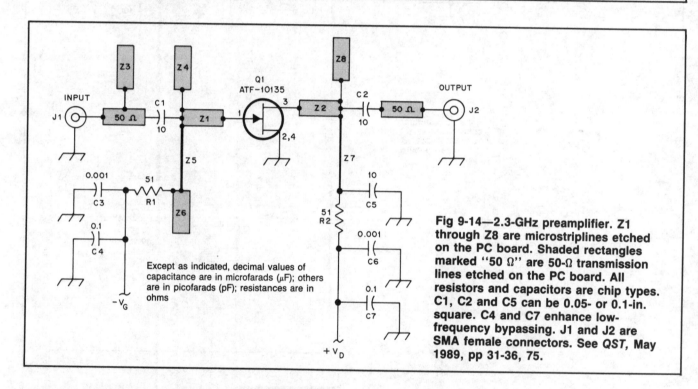

Fig 9-14—2.3-GHz preamplifier. Z1 through Z8 are microstriplines etched on the PC board. Shaded rectangles marked ''50 Ω'' are 50-Ω transmission lines etched on the PC board. All resistors and capacitors are chip types. C1, C2 and C5 can be 0.05- or 0.1-in. square. C4 and C7 enhance low-frequency bypassing. J1 and J2 are SMA female connectors. See *QST*, May 1989, pp 31-36, 75.

Except as indicated, decimal values of capacitance are in microfarads (μF); others are in picofarads (pF); resistances are in ohms

ments. Because the noise figure of this preamp does not approach that of a GaAsFET design, it's not suitable for antenna mounting. However, if you're using a relatively low-gain GaAsFET preamp like the N6CA design at the antenna in conjunction with a converter having a poor noise figure, you might see a substantial improvement in system performance if you place this preamp directly at the converter input.

Although the S-band transponder on OSCAR 13 is an experimental device that's seeing limited action, the 13-cm downlink on AMSAT Phase 4 satellites will be heavily used. You might suspect that a preamp for 2.4 GHz would be difficult to build. Prior to 1988 you would have

been correct, but thanks to the efforts of several radio amateurs, it's now as easy (perhaps even easier) to build a preamp for 2.4 GHz as it is for lower frequencies. The reason for this somewhat paradoxical situation is that at frequencies between 1 and 10 GHz it becomes practical to use microstriplines for impedance matching and to replace lumped circuit inductors. With microstrip design, construction of prototypes may be time consuming, but duplication of a successful model is relatively simple. A typical circuit, designed by Al Ward (WB5LUA) is shown in Fig 9-14. Note that the circuit has no tuning adjustments. At this frequency, extra gain is often needed to overcome feed-line losses. Paul Shuch (N6TX) has demonstrated that bipolar monolithic microwave integrated circuits (MMICs) are ideal for this application since they have a reasonably low noise figure, are extremely stable and are inexpensive (see Table 9-2). Since it's also possible to produce bandpass filters using stripline technology at these frequencies (R. Campbell, "A Clean, Low-Cost Microwave Local Oscillator," QST, Jul 1989, pp 15-21), one can build an entire 2.4-GHz converter that does not have any RF adjustments.

If you compare prices of commercial preamps, you'll see a very wide range. To a large extent, this reflects what's being included. The preamp itself just consists of a few components mounted on a small circuit board in a box with connectors. Quality connectors and a good box make a big impact on cost. So does the inclusion of a couple of coax relays for transmit/receive switching. After you take these factors into account, most of the remaining cost differential reflects the price of the active device being used.

Mounting Preamps at the Antenna

Three practical problems are encountered when mounting a preamp at the antenna: supplying power to the unit, weatherproofing the installation and switching the preamp out of the line if it's desired to use the antenna for transmitting.

Let's look at power first. Most solid-state units require a single positive supply of about +12 V dc. Batteries are inappropriate because they won't last long and the location may be difficult to access. The simplest method of supplying power to the preamp is to run a separate lead up to the antenna from a +12-volt power supply in the shack and use the outer braid of the coaxial cable as the power-supply ground return. A more elegant solution eliminates the need for any extra wires running to the antenna by using the coax feed line to carry both dc power and RF signals. Measurements reveal no discernible RF losses when this technique is used. Fig 9-15 shows two junction boxes which can be used with most preamps to accomplish this goal. One junction box mounts at the preamp, the other at the converter in the shack. In many cases, it's possible to eliminate one or both junction boxes by incorporating the components directly into the preamp and converter. This has been done in the preamp pictured in Fig 9-11. If this feature is incorporated in a preamp it can, at the user's discretion, be utilized or ignored. The capacitors C1 and C3 shown in Fig 9-15 can be omitted if neither the preamp-exit or converter-entry point is at ground potential. Note that the output circuit for Fig 9-12 does not satisfy this requirement. To protect against possible accidents, it's best to include these capacitors in all designs.

Weatherproofing a preamp is not a serious problem. One common technique is to enclose the preamp in a double plastic bag and mount it under a cover that protects against direct exposure to the elements. The rain cover may be plastic, glass or aluminum. Most plastics tend to deteriorate when exposed to sunlight, but enclosures cut from large plastic soda bottles seem to hold up well. (This is why their large-scale production causes serious environmental problems.) Plastic bags may be gathered together

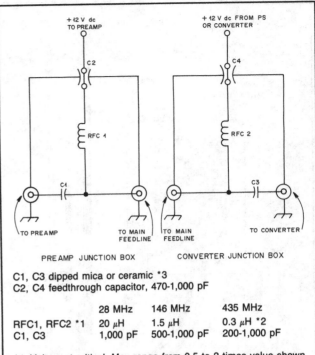

PREAMP JUNCTION BOX   CONVERTER JUNCTION BOX

C1, C3 dipped mica or ceramic *3
C2, C4 feedthrough capacitor, 470-1,000 pF

	28 MHz	146 MHz	435 MHz
RFC1, RFC2 *1	20 µH	1.5 µH	0.3 µH *2
C1, C3	1,000 pF	500-1,000 pF	200-1,000 pF

*1 Value not critical. May range from 0.5 to 2 times value shown.
*2 May use 15t #32 enameled copper wire on 1/10″ ID space wire diameter.
*3 C1 may be eliminated if preamp RF exit point is above dc ground. C3 may be eliminated if converter entry point is above dc ground.

Fig 9-15—The main feed line can be used to carry dc for the preamp as well as RF signals when the junction boxes shown are used.

using a twist tie; cables should be routed as shown (Fig 9-16) to discourage seepage. No attempt should be made to seal the unit. Seals eventually fail and lead to problems with condensation. How well does this approach to weatherproofing work? Many years ago I placed a microwave converter in a double plastic bag and set it out on the deck of my house where it was fully exposed to the weather. Several years later the deck had to be rebuilt so the experiment was terminated. The converter was working fine. Dick Jansson (WD4FAB) and John DuBois (W1HDX) report similar results.

If you wish to use the antenna for transmitting, the preamp must be switched off-line. The recommended switching circuit requires the installation of two SPDT coax switches between the antenna and preamp. The first one toggles the antenna between a transmission line to the transmitter and a jumper to the second relay. The second relay switches the preamp input between the jumper and a 50-ohm resistor to ground. This switching circuit requires two transmission lines, one for the receiver and one for the transmitter. Some commercial preamps use PIN-diode switching to automatically bypass the preamp when transmitter power is sensed. PIN-diode switching systems appear to be reliable when transmitter power limitations (generally on the order of 25 to 100 watts) are adhered to.

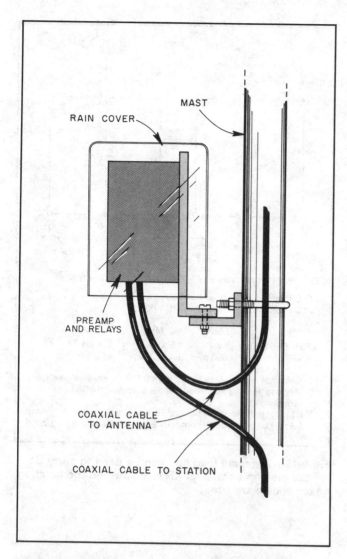

RAIN COVER

MAST

PREAMP
AND RELAYS

COAXIAL CABLE
TO ANTENNA

COAXIAL CABLE TO STATION

Fig 9-16—Protection for tower-mounted equipment need not be elaborate. Be sure to dress the cables as shown so that water drips off the cable jacket before it reaches the enclosure.

Picture a crossed Yagi for 70 cm that is used for both transmitting (Mode B) and receiving (Mode L). The antenna system includes a phasing harness, polarity switcher, preamp and relays for bypassing the preamp. Because of weight, wind load and possible antenna pattern distortion, it's best to mount all these components in a weatherproof enclosure attached to the mast, and to run a short length of coax from the enclosure to the antenna. All these relays, connectors and lengths of coax will add some loss to the receiving system and degrade overall performance. One way of avoiding a significant part of this loss is to use a dedicated receiving antenna, but the expense of this approach is generally prohibitive. 70 cm is the worst-case situation. Antennas for 23 cm and 13 cm are only used to uplink or downlink, so preamp-bypassing circuitry is not required. At 13 and 23 cm, it's also reasonable to omit the polarity switcher since the helix antennas on OSCAR 13 do not exhibit the off-axis sense reversals discussed in Chapter 7. At 2 meters, the preamp-bypassing

system can often be omitted since it is generally preferable to use a small beam or omnidirectional antenna to transmit to low-altitude spacecraft employing Mode A or Mode J rather than a high-gain 2-meter antenna designed for receiving Mode B from a Phase-3 spacecraft.

TRANSMITTING

All radio amateurs using the transponders on OSCAR satellites must share the available power and bandwidth. Cooperation is essential. Stations employing too high an EIRP will use more than their share of spacecraft energy and may even activate the transponder automatic-gain-control circuitry, making it impossible for low-power ground stations to be heard. For general communications, CW and SSB are recommended. The high peak-to-average power characteristic of SSB, and the low duty cycle of CW, use satellite energy effectively and efficiently. Users should generally not use FM, SSTV (slow-scan television), AM, or SSB with speech processing, because these forms of modulation drain an excessive amount of satellite energy. These modes may be used for special experimental purposes, however.

Cooperation

Recommended EIRP levels are listed in Appendix B: Spacecraft Profiles, for each transponder aboard each spacecraft. AMSAT has calculated these levels to ensure that the transponders provide reliable communications when *all* users cooperate. Though there's little chance of overloading the satellite when you abide by the EIRP guidelines, it can happen. In certain circumstances modestly higher EIRP levels may be used, particularly when path losses are exceptionally high (on passes below your radio horizon, when beaming through a thick stand of trees or when the ionosphere is disturbed). Courteous, experienced users monitor their downlink signals to ensure that they are not overloading or monopolizing the spacecraft transponder.

Cooperation also means taking care to produce a *clean* signal. Key clicks and SSB splatter will degrade transponder performance. In fact, SSB splatter can raise the noise floor of the transponder, making it impossible for any users to copy medium-strength stations that would otherwise be perfectly readable. This problem is potentially a serious one, because users of the wide, underpopulated terrestrial (non-satellite) 70-cm and 23-cm bands haven't had much incentive to worry about spectral purity. After all, it's difficult enough to put an SSB signal on 1.26 GHz; if you're only 10 dB above the noise at your closest neighbor's shack, no one will know whether your splatter is down 15 dB or 30 dB. As a result, not enough attention has been paid to the spectral purity of SSB signals at 435 and 1260 MHz in the past, but amateurs are finally beginning to focus on this problem. See, for example, "Solid-State VHF Linear Amplifiers," by C. F. Clark, *Ham Radio*, Jan 1980, pp 48-50, in which VHF solid-state amplifier design is discussed from the viewpoint of assuring linearity. Before turning to transmitting equipment, we'll consider a very important aspect of generating RF energy at VHF and UHF: RF power hazards.

RF Power Safety

Amateur Radio is basically a safe activity, but accidents can always occur if we don't use common sense. Most of us know enough not to place an antenna where it can fall on a power line, insert our hand into an energized linear amplifier, or climb a tower on a windy day. RF energy is also a potential hazard. Large amounts can cause damage in people by heating tissues. The magnitude of the effect depends on the wavelength, incident energy density of the RF field, exposure duration and other factors such as polarization. It is therefore important to take steps to prevent overexposure.

At the frequencies of interest to satellite users—145, 435 and 1269 MHz—large power densities may be accessible. The most susceptible parts of the body, the tissues of the eyes and gonads, don't have heat-sensitive receptors to warn us of the danger before the damage occurs. Symptoms of overexposure may not appear until after irreversible damage has been done. The potential danger should be taken seriously. With reasonable precautions, however, operation at 145, 435 and 1269 MHz can be safe.

The primary aim of this section is to show where protection from RF energy may be necessary. Our emphasis is on the practical problems encountered by satellite operators; for a more comprehensive technical treatment, see the references that follow this section. The RF-protection problem consists of two parts: (1) determining safe exposure levels and (2) measuring or estimating the local RF levels produced by a given power and antenna at a particular location. If the actual RF power density levels are greater than or even roughly equal to the safe levels, protection or precaution is required by limiting access or some other means. We begin with the question of safe exposure levels.

Safe Exposure Levels: In recent years, scientists have devoted a great deal of effort to determining safe RF-exposure limits. As the problem is very complex, it's not surprising that some changes in the recommended levels have occurred as more information has become available. The American Radio Relay League believes that the latest "Radio Protection Guide of the American National Standards Institute (ANSI)" is the best available protection standard; it took nearly five years to formulate and had undergone repeated critical review by the scientific community. This 1982 guide recognizes the phenomenon of whole-body or geometric resonance and establishes a frequency-dependent maximum permissible RF exposure level.

Resonance occurs at frequencies for which a body's long axis, if parallel to the ambient field, is about 0.4 wavelength long. Because of the range of human heights, the resonant region spans a broad range of frequencies. The most stringent maximum permissible exposure level, the bottom of the "valley" (see Fig 9-17), is 1 mW/cm^2 for frequencies between 30 and 300 MHz. On either side of those "corner" frequencies the rise is gradual. At 3 MHz,

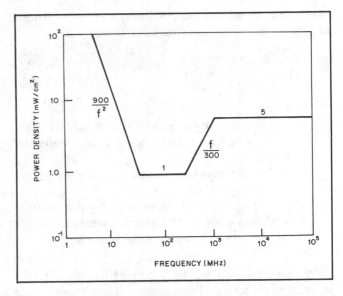

Fig 9-17—1982 ANSI RF Protection Guide for whole-body exposure of human beings.

the maximum permissible exposure level is 100 mW/cm^2; at 1500 MHz and above, 5 mW/cm^2. The valley region includes some active amateur bands (10, 6, 2 and 1¼ meters), as well as all FM, and some TV, broadcasting. The rationale for specifying a constant 5 mW/cm^2 above 1500 takes into consideration that with the extremely short wavelengths there is very little penetration into tissue. The levels specified refer to average power density allowed over any six-minute period. Until the Environmental Protection Agency (EPA) promulgates a general population standard, the ANSI Guide will likely be the most commonly accepted one.

Estimating Power Density: Our task is to determine the power density levels that could be produced at a given location by a specified antenna and power. Since most amateurs do not have the special equipment needed to measure RF electric fields accurately, power density will have to be estimated by calculations which involve approximations. While our estimates will always tend toward the conservative side, keep in mind that the results should be used only as a guideline for pointing out situations to avoid. The results should *never* be taken as *proving* that a particular setup is safe. For example, we consider only radiation from an antenna. Radiation can also take place directly from a power amplifier (if operated without proper shielding), from transmission lines (if poorly shielded, or if connectors are improperly installed) and in other situations. Take care to see that the only radiation from your station is at the antenna. Also, we'll be using a free-space propagation model to get a first estimate of power density. You may and should allow up to a 4- or 6-dB margin to provide for cases where a reinforcing reflection might occur.

Generally, antenna engineers divide the region around an antenna into a *far field* and a *near field*. At large distances from an antenna located in free space, power density can be estimated by applying Eq 9.1.

$$\rho = \frac{P\,G}{4\,\pi\,R^2} \qquad\qquad\qquad \text{(Eq 9.1)}$$

where

ρ = estimated power density at distance R from antenna (units will be W/m^2 if P is in watts and R is in m)

R = distance from observation point to closest point on antenna (in m)

P = *average* power at antenna feed point (watts)

G = gain as a power ratio, that is, the numerical gain (do *not* use gain expressed in decibels)

Note that Eq 9.1 holds only if both of the following requirements are met: (1) The free-space model is appropriate and (2) we're at a sufficiently large distance from the antenna to be in the far field. The far field is, in fact, defined as the region where Eq 9.1 is valid.

How far away from the antenna must the observer be for Eq 9.1 to hold? The answer depends on the type of antenna and on our accuracy requirements.

For several types of antennas likely to be used in space communication, we list the minimum distance at which Eq 9.1 may be applied with some confidence to provide an upper-bound (conservative) estimate.

To assure an upper-bound estimate, the free-space antenna gain should be used if the actual value is not known. A textbook value will generally be useful.

Antenna type	Dimension	Minimum distance, R_{min}, for Eq 9.1
Parabolic dish	diameter, D	$(1/2)(D^2/\lambda)$
Broadside array	max linear, L	$(1/2)(L^2/\lambda)$
End-fire types:		
Yagi	L = λ/2	$2L^2/\lambda$
Loop (quad) Yagi	L = max width of loop (quad)	$2L^2/\lambda$
Axial-mode helix	L = diameter of turn	$2L^2/\lambda$

Example: Consider a station with the following characteristics. A Yagi antenna with 13 dBi gain (G = 20) is mounted atop a 33-ft (10-meter) tower. Wavelength at 430 MHz is approximately 0.7 meter. Average power reaching the antenna terminals is 50 watts.

To determine the minimum R at which the far-field power density formula (Eq 9.1) can be applied, note that for a Yagi the maximum linear dimension "L" is roughly the length of an element, or λ/2.

$$R_{min} = 2L^2/\lambda = \frac{2(0.35\text{ m})^2}{0.35\text{ m}} = 0.35\text{ m}$$

Assume the interest is in the power density likely to occur at a point 50 feet on axis (15 meters) away. Since this distance is greater than R_{min}, we can use Eq 9.1.

Table 9-3

RF Awareness Guidelines

These guidelines were developed by the ARRL Bio Effects Committee, based on FCC/EPA measurements.

- Although antennas on towers (well away from people) pose no exposure problem, make certain that the RF radiation is confined to the antenna radiating elements themselves. Provide a single, good station ground (earth), and eliminate radiation from transmission lines. Use good coaxial cable, not open wire lines or end-fed antennas that come directly into the transmitter area.

- No person should ever be near any transmitting antenna while it is in use. This is especially true for mobile or ground-mounted vertical antennas. Avoid transmitting with more than 25 watts in a VHF mobile installation unless it is possible to first measure the RF fields inside the vehicle. At the 1-kilowatt level, both HF and VHF directional antennas should be at least 35 feet above inhabited areas. Avoid using indoor and attic-mounted antennas if at all possible.

- Don't operate RF power amplifiers with the covers removed, especially at VHF/UHF.

- In the UHF/SHF region, never look into the open end of an activated length of waveguide or point it toward anyone. Never point a high-gain, narrow-beamwidth antenna (a paraboloid, for instance) toward people. Use caution in aiming an EME (moonbounce) array toward the horizon; EME arrays may deliver an effective radiated power of 250,000 watts or more.

- With handheld transceivers, keep the antenna away from your head and use the lowest power possible to maintain communications. Use a separate microphone and hold the rig as far away from you as possible.

- Don't work on antennas that have RF power applied.

- Don't stand or sit close to a power supply or linear amplifier when the ac power is turned on. Stay at least 24 inches away from power transformers, electrical fans and other sources of high-level 60-Hz magnetic fields.

Further discussion of RF safety issues appears in both *The ARRL Antenna Book* (16th Ed.) and *The ARRL Handbook* (1992 and later editions). Also see I. A. Shulman, MD, "Is Amateur Radio Hazardous to Our Health?" *QST*, Oct 1989, pp 31-33, 37. Additional information appears in these sources:

ANSI C95.1-(1982). *Safety Levels with Respect to Human Exposure to Radio Frequency Electromagnetic Fields (300 kHz to 100 GHz)* (New York: American National Standards Institute).

Balzano, Q., O. Garay, K. Siwiak, "The Near Field of Dipole Antennas, Part I: Theory." *IEEE Trans Vehicular Technology* (VT) 30, p 161, Nov 1981). Also "Part II; Experimental Results," same issue, p 175.

Guy, A. W., C. K. Chou, "Thermographic Determination of SAR in Human Models Exposed to UHF Mobile Antenna Fields," Paper F-6, Third Annual Conference, Bioelectromagnetics Society, Washington, DC, Aug 9-12, 1981.

Lambdin, D. L. "An Investigation of Energy Densities in the Vicinity of Vehicles with Mobile Communications Equipment and Near a Hand-Held Walkie Talkie," *EPA Report ORP/EAD* 79-2, Mar 1979.

R. J. Spiegel, "The Thermal Response of a Human in the Near-Zone of a Resonant Thin-Wire Antenna," *IEEE Trans Microwave Theory and Technology* (MTT) 30(2), pp 177-185, Feb 1982.

$$\rho = \frac{P\,G}{4\,\pi\,R^2} = \frac{50 \times 20}{4\pi(15)(15)} = 0.35\text{ watt/m}^2$$

Since 1.0 watt/m^2 equals 0.1 milliwatt/cm^2, this yields

$$\rho = 0.035\text{ milliwatt/cm}^2$$

ANSI 1982 has a protection level of (430/300) milliwatts/cm^2 at 430 MHz, or 1.43 mW/cm^2, for far-field exposure. Our result in the example is less than 3% of the ANSI level.

Comments: The ARRL, through its Committee on the Biological Effects of RF Energy, will continue to keep the

Amateur Radio community informed on current protection issues and knowledge of electromagnetic bioeffects. Even though the preponderance of Amateur Radio operation, because of its fundamentally intermittent nature and relatively low power, poses little RF-protection requirement on antenna proximity, hams should keep abreast of developments and always follow recommended rules of good practice on the management and uses of RF energy.

Dr David Davidson, W1GKM (SK), a member of the ARRL Committee on the Bio-Effects of RF Energy, assisted with this section.

Transmitting Equipment

Amateurs use several methods for obtaining CW/SSB RF power at 145, 435 and 1269 MHz. The deluxe approach is to purchase a multimode transceiver for the desired uplink frequency. This option has not always existed, but today it's possible to buy a unit for all satellite uplink frequencies in use and/or planned for the next decade. Output powers are generally in the range of 10 to 40 watts.

Other approaches to producing an uplink signal include modifying FM equipment (commercial, amateur or military surplus), using a transverter in conjunction with an existing lower-frequency transmitter, exciting a varactor multiplier with an appropriate source, and so on. Power amplifiers are often used to boost the output power of these devices.

Transverter. A transmit converter works very much like a receive converter. Two RF sources are injected into a mixer, which produces sum and difference frequencies. One source is a CW/SSB transmitter and the other a fixed-frequency local oscillator. See Fig 9-18 for an example of a unit designed to generate a 435-MHz CW/SSB signal using a 28-30 MHz transmitter and a 407-MHz fixed frequency oscillator. Almost all commercial transmit converters use a mixer that operates at a very low (a fraction of a watt) power level. A unit will generally include one or more stages of linear amplification to bring the output signal up to the 0.5- to 10-watt range.

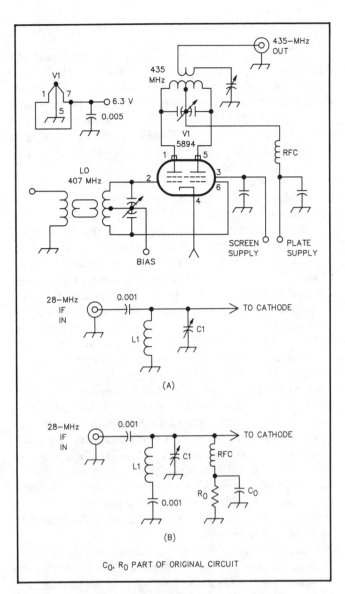

Fig 9-19—Partial schematic diagram of a 70-cm mixer, built from a converted FM transmitter. The original oscillator-multiplier-driver stages of the unit now provide LO injection. A filter should be used at the output of the mixer to prevent radiation of spurious products. See R. Stevenson, *QST*, Hints and Kinks, Mar 1976, p 40; and R. Stevenson, "SSB on Mode B, Using Modified FM Equipment," *AMSAT Newsletter*, Dec 1975, p 10. Reference: See 1990 *ARRL Handbook*, p 11-7.

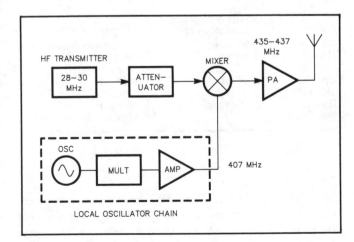

Fig 9-18—Block diagram of typical 10-meter-to-70-cm transmit converter.

Adding a receive converter to a transmit converter takes a minimal number of parts, since the local-oscillator signal is already available. A combined transmit/receive converter sharing a common local-oscillator chain is often called a *transverter*. However, many amateurs use the term transverter to refer to a transmit converter. When the distinction is important (like when you're purchasing one), make sure you know how the term is being used. Transverters designed for 145 and 435 MHz usually employ a 28-30 MHz IF. Units for 1269 MHz generally are designed to operate with a 145-MHz IF to minimize the need for filtering image and local-oscillator frequencies from the output.

High-level mixing can also be used in a transmit converter. This is not a very efficient approach if one is designing a unit from scratch, but it does lead to a relatively simple method for converting older tube-type high-band commercial FM (420- or 460-MHz) transmitter strips to CW/SSB transverters for 435 MHz. With 20 watts of drive from an HF transmitter, such a transverter will provide about 10 watts output.

An FM unit can be converted to a transmit converter by modifying the final amplifier to act as a mixer. A typical final amplifier circuit is shown in Fig 9-19. Injecting the 28-MHz signal into the cathode is usually the simplest approach. Fig 9-19 suggests two possible circuits for accomplishing this. If the cathode is initially grounded try (A). (B) can be used if the unmodified transmitter includes R_0 and C_0 in the cathode circuit. The tank circuit L1, C1 is resonant at 10 meters. The RFC prevents C_0 from detuning the tank circuit. It should have a broad peak at 10 meters (about 15 μH). Before any modifications are attempted, the FM strip should be tuned up and checked out at either 465 or 407 MHz. (If these frequencies are out of the tuning range, you can select frequencies that require 21-MHz drive from the HF transmitter.) Once everything is working correctly, modify the 5894 cathode circuit, activate the transverter and the HF transmitter (use CW), and retune the output circuit to 435 MHz. Switch the HF transmitter to SSB and adjust the drive level for the cleanest-sounding signal. Since the 5894 output tank circuit provides only minimal attenuation at the image and local-oscillator frequencies, you should include additional filtering before the antenna.

Varactor Multiplier. A power varactor is a type of semiconductor diode whose properties make it an efficient frequency multiplier in the 1- to 100-watt range. Although varactors can be used as doublers, triplers, quintuplers and higher-order multipliers, their most common application is tripling. Amateurs have used them to triple from 145 MHz to 435 MHz and from 420 MHz to 1260 MHz. The efficiency (output RF power × 100%/input RF power) is generally in the 50% to 60% range. A varactor multiplier does *not* require any dc power for operation, so it's possible to mount one remotely at the antenna. One serious limitation of varactor multipliers is that they produce severe distortion and spurious signals if used on SSB. Therefore, they are only suitable for CW. In recent years varactor multipliers have not been very popular, but situations may arise where their unique features make them an appropriate choice for producing an uplink signal. For example, a relatively inexpensive way of generating a 1269-MHz CW signal for Mode L is to use a varactor tripler in conjunction with the transmitter strip from a commercial 420-MHz FM transceiver. At least one manufacturer, Microwave Modules, has produced units for 145/435 MHz and 420/1260 MHz in the recent past.

The remarkable efficiencies exhibited by varactor multipliers are the result of an interesting design concept. The tripler, for example, can be thought of as simultaneously

Table 9-4
Varactor Multipliers

H. H. Cross, "Frequency Multiplication with Power Varactors at U.H.F.," *QST*, Oct 1962, pp 60-62. This pioneering article describes a 144/432-MHz tripler that uses conventional inductors. The unit yields 40% efficiency at 20 watts input with a Microwave Associates MA-4060A diode. The article gives lots of good practical advice.

D. Blakeslee, "Practical Tripler Circuits," *QST*, Mar 1966, pp 14-19. Contains a practical tripler that incorporates a strip-line output filter. The unit yields 60% efficiency at 20 watts input with an Amperex H4A (1N4885) diode. The basic design was reprinted in several editions of *The Radio Amateur's Handbook* in the late '60s and early '70s.

D. DeMaw, "Varactor Diodes in Theory and Practice," *QST*, Mar 1966, pp 11-14. Contains a thorough and understandable discussion of basic varactor doubler and tripler design considerations.

The Radio Amateur's VHF Manual, ARRL, 3rd edition, 1972. 144/432 tripler using H4A, pp 289-290; 432/1296 tripler using MA4062D, pp 292-293 (out of print).

D. S. Evans and G. R. Jessop, *VHF-UHF Manual*, 3rd edition, RSGB, London, 1976 (available from ARRL). Contains general information (pp 5.20-5.21), a 145/435-MHz tripler (pp 5.21- 5.23) that uses 1N4387 (40 W in/25 W out) or BAY 96 (15 W in/9 W out) and a 384/1152-MHz tripler (pp 5.70-5.71) using BXY 35A (30 W in).

FM and Repeaters, 2nd edition, ARRL, 1978. Contains a practical design for a 145/435-MHz tripler that uses an H4A (pp 49-50). (Out of print.)

D. R. Pacholok, "Microwave-frequency Converter for UHF Counters," *Ham Radio*, Jul 1980, pp 40-47. Describes how transistor collector-base junction can be used as varactor. As a result, a bipolar transistor can be used simultaneously to amplify at the input frequency and multiply using efficient varactor effect.

Complete triplers for 145/435 MHz and 420/1260 MHz are available from Microwave Modules at several power levels.

Note: In most cases, the varactors specified are interchangeable as long as power dissipation is taken into account. A summary of the varactors used in various amateur construction projects and the maximum RF power input follows.

Device	Max recommended input power	Manufacturer
MA-4060A	20 watts	Microwave Associates
H4A (1N4885)	20 watts	Amperex
BAY 66	12 watts	Mullard
BAY 96	40 watts	Mullard
1N4387	40 watts	Motorola
BXY35A	30 watts	Mullard
MA4062D		Microwave Associates

operating as a doubler and a high-level mixer. The mixer combines the fundamental frequency and the doubled frequency to produce a signal at 3 × the input frequency. Table 9-4 contains an extensive list of articles on varactor operation and construction.

Converting Commercial FM or Military Surplus Equipment: Old tube-type commercial FM gear designed for the land mobile service (130 to 160 MHz and 420 to 460 MHz) is widely available at modest cost at hamfests. Amateurs have successfully converted FM transmitter strips into CW transmitters, transmit converters for CW/SSB operation, and linear amplifiers for 146 and 435 MHz. Conversion may involve: (1) constructing an appropriate ac power supply, (2) retuning resonant circuits and cavities to the correct frequencies, (3) adding provisions for keying (producing a stable, chirp-free signal at 435 MHz can be challenging) and (4) changing power-amplifier biasing to AB_1 or AB_2 linear operation. Transmitter strips that are rated at 15-60 watts output in commercial service can safely provide 50% more power for amateur operation. A crystal-controlled 435-MHz transmitter can usually be pulled enough to give a 75-kHz tuning range. An annotated list of conversion articles is contained in Table 9-5.

In recent years, the military surplus market has been a very minor source for equipment. The technically astute amateur with access to a nearby surplus dealer warehouse will sometimes come across a desirable piece of equipment, but gear is generally not available in the quantities needed to stimulate the dissemination of conversion information. One desirable piece of gear that is at times available on the surplus market is the AN/UPX-6, which can be turned into a linear amplifier that will provide about 50 watts output at 1269 MHz when driven with a few watts. For information on the AN/UPX-6, see the article by R. Stein referenced in Table 9-6.

Modifying an Amateur 2-Meter FM Transceiver. This approach to producing an uplink signal will mainly appeal to beginners who are interested in gaining temporary access to Mode A. Most amateur 2-meter FM transceivers can be modified easily for CW operation on the Mode A uplink frequencies currently used by the RS satellites. Modification may be as simple as tuning to the correct frequency, removing the mike element and keying the push-to-talk switch. Of course, it's far better to change the push-to-talk circuitry so the unit can be left in transmit while only the driver and final amplifier are keyed. Amateur FM equipment produced in the mid '70s is often available at very low cost at hamfests. If the price is right, it may pay to cannibalize a rig to acquire a 10-watt CW transmitter for Mode A. The receiver front end can often be put to use as a converter for listening to Mode B.

Construction of Transmitting Equipment from Scratch: Collecting components and building and debugging equipment can involve a relatively large amount of time and expense. However, the educational rewards and the intrinsic satisfaction derived by constructing one's own equipment continue to appeal to many amateurs. Excellent plans exist for most any piece of equipment required at a ground station. See Tables 9-4 and 9-6 for a list.

Purchasing New or Used Amateur Equipment: Several major Japanese manufacturers supplying the amateur community now offer multimode transceivers for 2 meters, 70 cm and 23 cm. Some of these transceivers will operate on two or more bands. Earlier, we mentioned that it was

Table 9-5

Sources of Information on Converting Commercial FM Transmitting Equipment for Satellite Ground Station Use

FM and Repeaters, 2nd edition, ARRL, Newington, 1978. The chapter on surplus FM equipment contains a great deal of useful general information. (Out of print.)

D. P. Clement, "Using the Motorola TU-110 Series Transmitters on 420 MHz," *QST*, Sep 1971, pp 39-41, 45. Contains detailed information on converting the TU-110 to a 20-watt-output CW transmitter. Treats such topics as obtaining a stable, chirp-free signal.

F. R. McLeod, Jr, "ATV with the Motorola T44 UHF Transmitter," Part I, *QST*, Dec 1972, pp 28-32, Part II, *QST*, Feb 1973, pp 36-43. These articles are very useful to anyone wishing to put the widely available T44 on 435 MHz.

R. Stevenson, "SSB on Mode B, Using Modified FM Equipment," *AMSAT Newsletter*, Dec 1975, p 10. Shows how an RCA CMU-15 designed for 460 MHz can be converted to a 435-MHz transverter. Conversion involves modifying the 5894 power amplifier to operate as a high-level mixer as shown in Fig 9-19. This information was also published in *QST*, Hints & Kinks, Mar 1976, p 40.

W. R. Gabriel, "A 70-cm Linear Amplifier from a Motorola T44," *AMSAT Newsletter*, March 1977, pp 4-5. Illustrates how the 2C39 output stage of a Motorola T44 can be used as a 435-MHz linear amplifier. Specific power levels aren't given, but the design should provide 6-10 dB of gain at up to 40 watts output.

Sources for Manuals and Schematics

General Electric Co.
   Marketing Communications Production
   Box 4197 Lynchburg, VA 24502

Motorola Communications and Electronics, Inc
   1301 E Algonquin Rd
   Schaumburg, IL 60172

RCA Corporation
   Customer Technical Information Service
   Meadow Lands, PA 15347

extremely important for a CW or SSB ground station to be able to transmit and listen simultaneously so that the downlink can be monitored. Most multiband, multimode transceivers have this capability, but it's important to check.

Several manufacturers are producing solid-state power amplifiers that can raise the typical transceiver power output (10 to 30 watts) to up to 160 watts. If you plan to operate SSB, you'll need an amplifier that operates in a linear mode. Many older models were designed for class C operation, which is fine for CW or FM, but not for SSB. Some power amplifiers contain a preamp and automatic RF switching. Naturally, the cost is higher, but this feature can be a great convenience if you're setting up a ground station to operate several satellite modes.

A few smaller specialty manufacturers are producing transverters (see Table 9-2). One manufacturer, Ten-Tec,

Table 9-6

List of Articles: Constructing VHF and UHF Transmitting Equipment

Amplifiers over 100 watts are tube designs; others are solid state. All units are linear unless specified otherwise.

General

B. Olson, W3HQT, "RF Hybrid Modules: Building with Bricks," *QEX*, Jul 1988, pp 13-14.

J. Reisert, W1JR, "VHF/UHF Exciters," *Ham Radio*, Apr 1984, pp 84-88.

J. Reisert, W1JR, "Medium Power Amplifiers," *Ham Radio*, Aug 1985, pp 39-42, 45-46, 51-54.

2 Meters (145 MHz)

D. DeMaw, W1FB, "Some Basics of VHF Design and Layout," *QST*, Aug 1984, pp 18-22. (146 MHz, Class C, 2 W/15 W). Feedback, Oct 1984, p 42.

L. Leighton, WB6BPI, "Two-Meter Transverter Using Power FETs," *Ham Radio*, Sep 1976, pp 10-15. (Contains linear amplifier: 2 W/10 W).

B. Lombardi, WB4EHS, "A High-Performance 2-meter Transverter," *Ham Radio*, Jul 1989, pp 68-72, 75, 77. (Contains linear amplifier: 0.25 W/4 W)

D. Mascaro, WA3JUF "25-Watt Linear Amplifiers for 144 and 220 MHz," *QST*, Aug 1988, pp 15-21. (2 W/25 W)

R. S. Stein, W6NBI, "Solid-State Transmitting Converter for 144-MHz SSB," *Ham Radio*, Feb 1974, pp 6-18. (contains two linear amplifiers: 0.5 W/6 W, 6 W/30 W)

"A Medium-Power 144-MHz Amplifier," 1992 *ARRL Handbook*, pp 31-39 to 31-45. Design by Clarke Greene, K1JX, construction by Mark Wilson, AA2Z. (10 W/300 W)

"Linear Transverters for 144 and 220 MHz," 1992 *ARRL Handbook*, pp 31-17 to 31-28. Design and construction by P. Drexler, WB3JYO. Includes design for 1 W/10 W linear amplifier.

70 cm (435 MHz)

J. Buscemi, K2OVS, "A 60-Watt Solid-State UHF Linear Amplifier," *QST*, Jul 1977, pp 42-45. (Contains two-stage amplifier: 2 W/60 W).

R. K. Olson, WA7CNP, "100-Watt Solid-State Power Amplifier for 432 MHz," *Ham Radio*, Sep 1975, pp 36-43. (10 W/100 W)

J. C. Reed, W6IOJ, "A UHF Amplifier—From Scratch," *QST*, Aug 1987, pp 24-27. (2 W/45 W, class C).

J. C. Reed, W6IOJ, "A Simple 435-MHz Transmitter," *QST*, May 1985, pp 14-18, 45. (15-W, VXO-controlled CW transmitter.)

F. Telewski, WA2FSQ, "A Practical Approach to 432-MHz SSB," *Ham Radio*, Jun 1971, pp 6-21. Contains an extensive review of tube-type mixers and linear amplifiers at all power levels for 432 MHz. Since vacuum-tube techniques have remained relatively stagnant over the last two decades, the information here is still valuable for anyone working with 6939 mixers and the 2C39 family of amplifiers.

L. Wilson, WB6QXF, "Solid-State Linear Power Amplifier for 432 MHz," *Ham Radio*, Aug 1975, pp 30-35. (1 W/10 W)

23 cm (1.2 GHz)

Though most of the units referenced were built for 1296 MHz, they will work equally well at 1269 MHz.

E. R. Angle, N6CA, "A Quarter Kilowatt 23-cm Amplifier," Part I, *QST*, Mar 1985, pp 14-20; Part II, *QST*, Apr 1985, pp 32-37. Reprinted in 1992 *ARRL Handbook*, pp 32-29 to 32-42.

B. Atkins, KA1GT, "The New Frontier: 1296-MHz Bibliography," *QST*, Aug 1985, p 68.

B. Olson, W3HQT, "Focus on Technology above 50 MHz," *QEX*, Jun 1987, pp 11-15. 23-cm amplifier, 8 W/33 W.

R. S. Stein, W6NBI, "Converting Surplus AN/UPX-6 Cavities," *Ham Radio*, Mar 1981, pp 12-17. Describes a three-stage amplifier that produces 40 W output for 100 mW drive using 2C39 tubes.

A. Ward, WB5LUA, "1296-MHz Solid-State Power Amplifiers," *QST*, Dec 1985, pp 41-44. (2 units: 1.5 W/6 W and 5 W/18 W). Reprinted in 1992 *ARRL Handbook*, pp 32-19 to 32-21.

"A 1296-MHz Linear Power Amplifier," 1992 *ARRL Handbook*, p 32-18.

"1296- to 144-MHz Transverter," 1990 *ARRL Handbook*, pp 32-24 to 32-37. Design and construction by A. Ward, WB5LUA. Complete transmit and receive converter; 250-mW transmitter.

"1296-MHz Transverter," 1992 *ARRL Handbook*, pp 32-5 to 32-14. Design and construction by D. Eckhardt, W6LEV. Complete transmit and receive converter; 28-MHz IF, 250-mW transmitter.

produces a unit designed for Mode B users that contains a complete 10-watt CW/SSB 435-MHz transmitter and a 145/28-MHz receive converter.

Several excellent, though long discontinued, 2-meter transmitters and transverters sometimes appear on the used-equipment market at very reasonable prices. Many of the items use tube technology, so replacement parts may be a problem. Some of the older gear is useless, but age is not always a reliable indicator of quality. Many old tube-type transmitters (including some designed for AM/CW operation) and transverters are still providing good service. For example, the AMECO TX-62 40-watt transmitter, and gear by Clegg, can be used on CW. And transverters by Drake and Collins have a reputation for being reliable. Old receive converters by Parks and Tapetone are often seen going for peanuts at hamfests. With a good preamp these

units can provide excellent service. Table 9-7 contains a list of *QST* equipment reviews that may prove helpful if you're considering purchasing used equipment.

Transmitting Power Requirements

It's difficult to specify absolute power requirements for each satellite mode for several reasons. First, a mode may be used on several spacecraft operating at different altitudes. Second, the significant parameter is EIRP, not power, so it's important to consider the entire transmitting station, including antennas and transmission-line losses, when selecting an appropriate uplink power. Third, though a given power level may be sufficient 98% of the time, there will always be instances where the satellite is just outside your access circle or someone is desensing the spacecraft transponder and a little extra power would be useful.

Table 9-7

QST Equipment Reviews

The following product reviews may prove useful if you're considering purchasing used equipment for your satellite ground station. Much of the equipment listed is no longer manufactured.

(Manufacturer, item, initial page, month, year)

Ameco: CN-144 2-meter converter, 42, Sep 1962
Angle Linear: VHF/UHF preamps, 36, Aug 1978
ARCOS: 432-MHz transmit converter and PA, 40, Aug 1976
Advanced Receiver Research: MML 144VDG preamp and
    TRS04VD TR sequencer, 39, Feb 1987
Braun: TTV 1270 144/435-MHz varactor tripler and
    435/144-MHz receiver converter, 52, Jul 1971
Clegg: 22'er CW/AM transceiver, 38, Apr 1965
    AB-144 HF to 2-meter receive converter, 44, Oct 1980
    Zeus VHF transmitter, 55, Sep 1961
Collins: 62S-1 VHF converter, 52, Nov 1963
Cushcraft: 32-19 "Boomer" and 324-QK Stacking Kit (2
    meters), 48, Nov 1980
Drake: VHF converters, 51, Feb 1968
Gonset: Sidewinder 2-meter transceiver, 64, Mar 1965
    903A and 913A VHF PAs, 74, Aug 1965
Hallicrafters: HA-2 2-meter transverter, 43, Sep 1962
Hamtronics: VX-4 70-cm transmit converter, 43, Jan 1982
    XV-2 2-meter transmit converter, 46, Feb 1979
    70-cm converter kit, 29, Jul 1978
    P8 VHF preamp, 47, May 1977
Heath: VL-1180, 2-meter PA, 38, May 1982
    VL-2280 2-meter PA, 48, Jun 1982
    SB-500 2-meter transverter, 43, Sep 1970
Hy-Gain: V-2 2-Meter Antenna, 40, May 1982
    214 2-Meter Yagi, 32, Oct 1978
ICOM: IC-275A 2-meter multimode transceiver, 32, Oct 1987
    IC-290H 2-meter multimode transceiver, 36, May 1983
    IC-471A 70-cm multimode transceiver, 38, Aug 1985
    IC-271A 2-meter multimode transceiver, 40, May 1985
    IC-211 2-meter multimode transceiver, 30, Dec 1978
Janel Laboratories: 432CA 70-cm converter, 40, Dec 1975

Johnson: Viking 6N2 transmitter (2 meters), 46, Mar 1957
    6N2 Converter, 45, Nov 1959
    6N2 VFO, 43, Oct 1959
    6N2 Thunderbolt PA, 46, Jan 1960
Kenwood: TR-9000 2-meter multimode transceiver, 49, Dec 1981
    TS-700A 2-meter multimode transceiver, 38, Mar 1976
    TR-751A 2-meter multimode transceiver, 41, Mar 1987
    TS-700S 2-meter multimode transceiver, 31, Feb 1978
Klitzing: 70CM10W60 70-cm PA, 38, Jun 1979
KLM: 2M-22C and 435-40 crossed Yagis, 43, Oct 1985
    2M-16LBX 2-Meter Beam, 41, Mar 1985
    144-148-13 LBA 2-meter Yagi, 41, Feb 1985
    16-Element 2-Meter Yagi, 47, Aug 1979
    PA 15-80BL 2-meter PA, 43, Sep 1979
LMW Electronics: 1296TRV1k 23-cm transverter kit, 39, Dec 1987
    2304TRV2, 13-cm transverter, 37, Dec 1987
Microwave Modules, Ltd: MMT432 70-cm transverter, 43, Sep 1977
    MMT1296 435/1296-MHz varactor tripler, 41, Dec 1977
Mirage Communications: B215 2-meter PA, 40, Feb 1985
    D1010 70-cm PA, 42, Jan 1984
    B108 2-meter PA, 41, May 1979
Parks: 144-1 2-meter converter, 85, Jul 1964
    432-3 70-cm converter, 44, Oct 1966
RF Concepts: RFC-2-23 2-meter PA, 37, Mar 1988
    RFC 2-317 2-meter PA, 34, Oct 1987
RIW Products: 432-19 Yagi, 34, Dec 1978
Spectrum International: 1296-MHz Loop Yagi, 33, May 1978
Tapetone: 2-meter converter, 42, Jul 1957
    TC-432 70-cm converter, 46, Feb 1961
Telco: 125 2-meter PA, 40, March 1978
Ten-Tec: 2510 70-cm transmitter and 2-meter converter,
    41, Oct 1985
Tonna: F9FT 144/16 2-Meter Yagi Antenna, 48, Jul 1979
Trio-Kenwood: see Kenwood
VHF Engineering: BLE 10/40 70-cm PA, 33, Sep 1978
Yaesu: FTV-901R VHF/UHF transverter, 48, Feb 1983
    FT-736R VHF/UHF multimode transceiver, 30, May 1990
    FT-726R VHF/UHF multimode transceiver, 40, May 1984
    FT-480R 2-meter multimode transceiver, 46, Oct 1981

Despite these problems, it is possible to suggest power levels which will provide good performance 90 + % of the time when used in conjunction with a typical antenna system. Keep in mind that the levels specified are guidelines, taking into account several satellites, and may have to be modified to allow for your special needs. Specific EIRP recommendations for each mode aboard each satellite are contained in Appendix B: Spacecraft Profiles.

145 MHz: This band is used as an uplink with several low-altitude spacecraft. The relevant modes are A, J and JD. 40 to 80 watts to an omnidirectional antenna should provide excellent performance. Twenty watts to a rotatable 3-el beam set at a fixed elevation angle of 20 degrees will also work well. When transponder loading is light, many amateurs have communicated via Mode A using less than a watt to an omnidirectional antenna. 145 MHz is also used as an uplink on OSCAR 13 Mode JL. The 2-meter uplink was included for amateurs in countries which do not permit 23 cm transmission, and for amateurs in parts of the world where it's impossible to obtain equipment for 23 cm (or even 70 cm). If you fall in this category, you'll find that 40 watts to a beam having a gain of 13 dBic will provide good results.

435 MHz. This band is currently used as an uplink for Mode B on OSCARs 10 and 13. A transmitter putting out 40 watts in conjunction with an antenna having about 13 dBic gain will produce good downlink signals.

1269 MHz. This band is currently used as an uplink on OSCAR 13 Mode L. A survey of stations operating Mode L in 1988 showed that the great majority were using from 20 to 100 watts with beams averaging 20 dBi gain. Solid-state devices are generally used to produce powers up to the 20-30 watt level. Above this point, cavity amplifiers employing the 2C39 tube are popular. Future uplinks at this frequency may require 3 to 6 dB less power.

DIGITAL COMMUNICATIONS

Fig 9-20 shows a ground station designed for Mode JD (packet radio) operation. The station uses the same uplink and downlink frequencies as Mode J, but there are

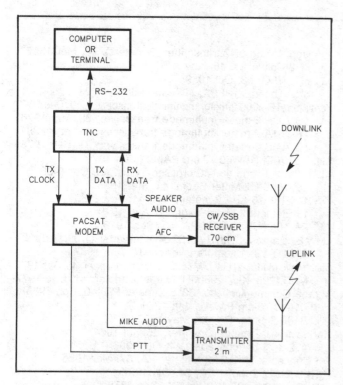

Fig 9-20—Ground station for Mode J packet operation with AO-16, AO-19 and FO-20.

Table 9-8

Packet Modems for Satellite Operation

Construction Information

J. Miller (G3RUH), "A Packet Radio PSK Modem for JAS-1/FO-12," *Ham Radio*, Feb 1987, pp 8-12, 14, 16-18, 20-22.

F. Yamashita (JS1UKR), "A PSK Demodulator for the JAS-1 Satellite," *QEX*, Aug 1986, pp 3-7.

L. Johnson (WA7GXD), "The TAPR PSK Modem," *QEX*, Sep 1987, pp 3-8.

B. McLarnon (VE3JF), "A 1200 Bit/s Manchester/PSK Encoder Circuit for TAPR TNC Units," *QEX*, Nov 1987, pp 10-11. (Contains helpful information for those using the TAPR *and* G3RUH modems.)

Sources for Kits

G3RUH Modem: In the US, contact Radiokit, PO Box 973, Pelham, NH 03076. Elsewhere contact AMSAT-UK, 94 Herongate Rd, London, E12 5EQ, England

JS1UKR/TAPR Modem: TAPR (Tucson Amateur Packet Radio Corporation), PO Box 12925, Tucson, AZ 85732

Sources for Assembled/Tested Units

PacComm, 3652 W Cypress St, Tampa, FL 33607

significant differences in the equipment. We'll look at the packet station in detail.

The packet radio uplink uses a 2-meter *FM* transmitter, and the downlink uses a 70-cm *SSB* receiver. A transmitter normally used for terrestrial 2-meter FM voice communications will work fine. So will the transmitter section of the TAPR (Tucson Area Packet Radio) packet transceiver and similar units. A standard TNC (terminal node controller) is used, but a special external modem is required. The modem connects to the TNC "Modem Disconnect Jack" on the TAPR TNC-1 or TNC-2. (Most clones of the TNC-1 and TNC-2 also have an appropriate jack.) Note that four lines (two signal and two control) connect the packet equipment to the radio gear. The two signal lines are transmitter audio input and receive audio output. The two control lines are transmitter PTT (push to talk) and receiver automatic frequency control (AFC).

Ground station power and antenna requirements are modest since the spacecraft is in a low earth orbit. Forty watts to an omnidirectional antenna should work fine for the uplink. Ten watts to a small rotatable 3-el beam at a fixed elevation angle of 20 degrees should also provide good performance.

The packet radio equipment consists of three major items: (1) A standard TNC (one with provisions for connecting an external modem), (2) the computer or terminal normally used with the TNC and (3) the special modem required for satellite packet operations. Current sources for information related to modem acquisition and construction are contained in Table 9-8. These sources are likely to increase rapidly now that OSCARs 16, 19 and 20 are in orbit, so check AMSAT publications and the

Amateur Satellite Communications column in *QST* for up-to-date details.

Of the four lines connecting the packet equipment to the RF equipment, the only one likely to be unfamiliar to packeteers is the AFC line. To understand its function, you need some familiarity with a phenomenon called Doppler shift. Briefly, if a transmitting station and a receiving station are in relative motion, the frequency observed by the receiving station and the frequency being transmitted will differ slightly. The effect, known as Doppler shift (see Chapter 13 for details), acts on the downlink channel and on the four uplink channels on each spacecraft.

Consider an uplink channel first. If you transmit on the designated uplink frequency, the spacecraft will see your signal as being a few kilohertz high at the beginning of a pass and a few kilohertz low at the end of the pass due to the Doppler effect. By using a relatively wide (about 20 kHz) receiving channel on the spacecraft, an uplink signal deviating less than 5 kHz will remain within the satellite receiver passband during the entire pass. The ground station therefore does not have to change the transmitting frequency as the spacecraft flies by. The wide bandwidth does degrade the spacecraft receiver's signal-to-noise ratio slightly, but this is easily overcome if necessary by a small increase in the ground station's power.

It would be great if the satellite downlink could use a system as simple as the one on the uplink, but calculations showed that doing so was impossible because of the low power available to the satellite transmitter. Using FM modulation on the downlink and an FM receiver at the ground station would result in inadequate signal levels. The spacecraft designers therefore had to select a more sophisticated modulation scheme for the downlink. The system chosen requires the use of a standard SSB receiver at the

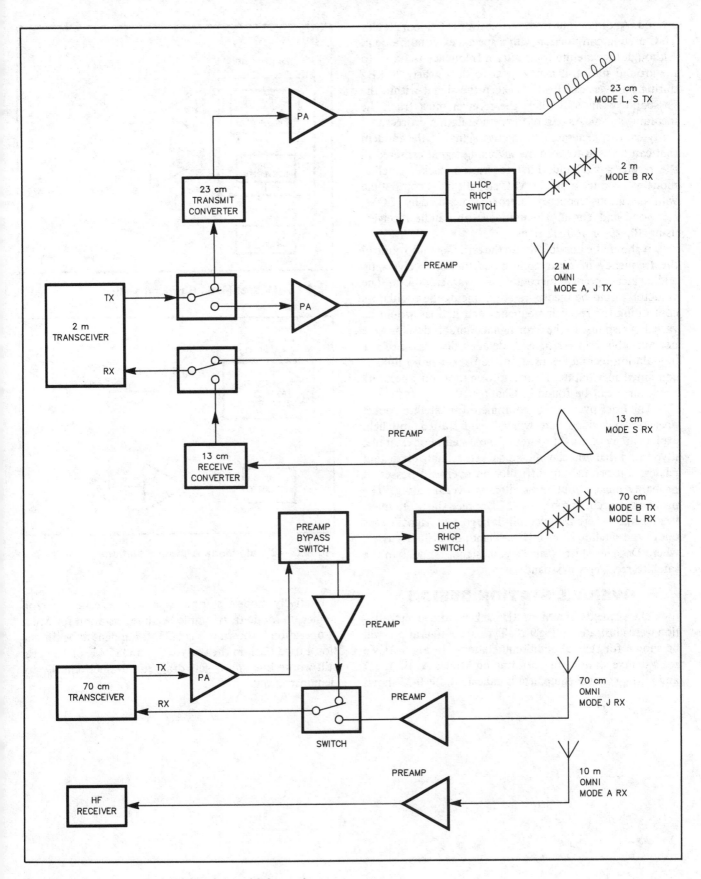

Fig 9-21—Deluxe ground station for multiple modes.

ground. Audio output from the receiver is fed to a regular TNC used in conjunction with a special external modem. Although the satellite transmitting frequency is fixed, to the ground station it may appear to shift nearly 20 kHz during an overhead pass. To keep the signal within the passband of an SSB filter, the receiver must track the incoming signal. As part of the demodulation process, it's fairly easy to generate a corrective signal in the modem that can be used to keep the incoming signal centered in the receiver passband. First-generation packet satellite modems produce a digital AFC signal that is compatible with the various frequency control schemes used by ICOM, Kenwood and Yaesu. The modems can handle receivers using 10, 20 or 100-Hz steps.

It should be possible to use the error signal to control the frequency of older continuous tuning receivers by adding an adapter circuit containing a varactor diode. The varactor would be incorporated in one of the oscillators controlling the receiver frequency and its bias would be varied in response to the error signal. Suitable designs were not available at press time. A detailed discussion of the modulation techniques used on the packet radio links is contained in Chapter 15 and information on operating procedures can be found in Chapter 10.

The brief overview of equipment for satellite packet operation provided here will get you started and help explain many unusual features of the system. For example, if you find that you can upload packets reliably on distant passes, but are having difficulty on overhead passes, it probably means your transmitter is overdeviating. The problem is occurring because of Doppler shift: On overhead passes, where Doppler shift is large, your signal chops out of the satellite receiver's passband; on distant passes, where Doppler shift is small, your signal remains in the satellite receiver's passband.

OVERALL STATION DESIGN

The elements of a Mode JD packet-radio ground station were illustrated in Fig 9-20. Typical configurations will be shown for several additional stations. In Fig 9-21 we see a deluxe station for operating on Modes A, B, J, JD and L (only the RF equipment is indicated). Fig 9-22 shows

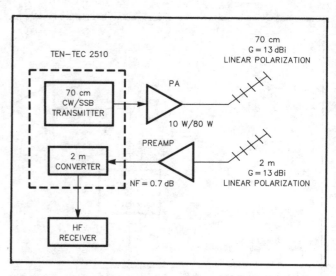

Fig 9-22—Modest Mode B ground station.

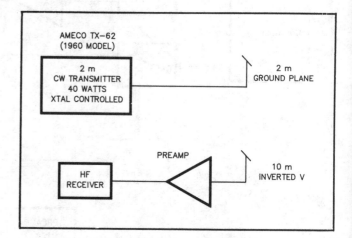

Fig 9-23—Simple Mode A ground station.

a relatively modest station capable of excellent performance on Mode B. A simple beginner's station for Mode A operation is shown in Fig 9-23. It happens to be the station I used back in the early '70s on OSCAR 6. The gear still works fine and it gets dragged out aperiodically for demonstrations.

Chapter 10

Operating Notes

USING A LINEAR TRANSPONDER

Operating through a linear transponder differs in several significant ways from terrestrial HF operation. We begin by looking at the problem of locating your downlink frequency.

Locating Downlink Frequency

When you're searching for your downlink, or trying to determine the frequency to set your transmitter to in order to respond to a CQ, it helps to have an understanding of how uplink and downlink frequencies are related. The relationship depends on whether the transponder is inverting or noninverting (see Chapter 5). Each type has a simple formula (called the translation equation) which predicts the approximate downlink frequency (f[down]) corresponding to a particular uplink frequency (f[up]).

For an inverting transponder (Modes B, J, L):

$$f[down] = f^* - f[up] \qquad \text{(Eq 10.1)}$$

For a noninverting transponder (Mode A):

$$f[down] = f^* + f[up] \qquad \text{(Eq 10.2)}$$

The constant, f*, associated with each transponder is listed in Appendix B. Many amateurs use Eq 10.1 (and/or Eq 10.2, as appropriate) to construct translation charts (like the one shown in Fig 10-1) for each transponder they're interested in. Most people find it more convenient to use such a chart rather than an equation to estimate downlink frequency.

The downlink frequencies predicted by Eqs 10.1 and 10.2 were referred to as approximate because they don't take Doppler shift into account. (See Chapter 13 for details on Doppler shift.) The maximum Doppler shifts observed on transponders aboard amateur satellites are about ±4 kHz on Mode A, ±7 kHz on Modes B and J, and ±22 kHz on Mode L. For spacecraft in circular low earth orbits, the maximums only occur on orbits where the spacecraft passes nearly overhead. For satellites in highly elliptical orbits (Phase 3), the maximums occur near perigee.

Most users working with Phase 3 satellites don't operate near perigee, so Doppler shift is a minor concern. If you use Eq 10.1 (or the equivalent chart) to estimate downlink frequency, you'll only have to search a couple of kHz while sending a few dits to locate the downlink.

Doppler corrections are significant when working with Phase 2 spacecraft. For example, if you use a Mode J transponder on a low-altitude satellite, you'll find that on

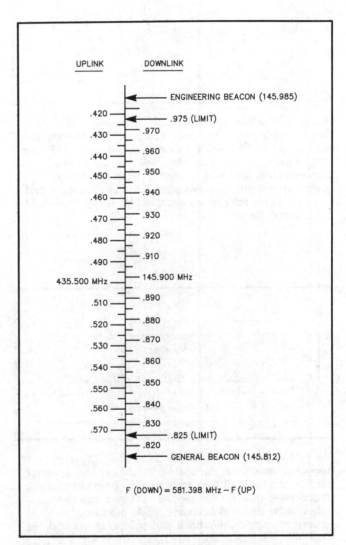

Fig 10-1—OSCAR 13 Mode B Frequency Translation Chart.

nearby passes the actual downlink frequency may be 7 kHz higher than predicted by Eq 10.1 at the start of a pass and 7 kHz lower than predicted at the end of the pass. For distant passes (where the satellite elevation angle never goes above 30 degrees), the observed Doppler shift will be much smaller; the downlink may start about 2 kHz high and end up about 2 kHz low. Keeping these facts in mind, it's possible to use Eqs 10.1 and 10.2 in conjunction with a rough knowledge of expected Doppler shift to predict downlink

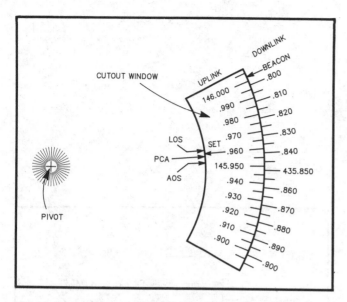

Fig 10-2—Frequency Translation Operating Aid: Fuji-OSCAR 12 Mode J. For overhead pass, set overlay to AOS at start of pass, PCA near midpoint and LOS near end of pass. For grazing pass, set overlay midway between AOS and PCA at start of pass, PCA near midpoint and midway between LOS and PCA near end of pass. Chart may be constructed from two sheets of polar graph paper.

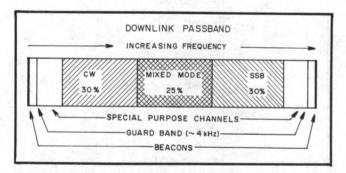

Fig 10-3—Linear Transponder Band Plan: General downlink band plan for OSCAR satellites. The special-service channels (SSCs) are meant to be used in a coordinated manner for special activities such as bulletins by national societies, code practice, emergency communications and computer networking. Channels are designated H1, H2,...(high end), L1, L2,...(low end); with H1 and L1 being closest to the beacons.

frequency. As a result, it will only be necessary to hunt through a limited band segment when searching for your downlink.

One can construct a variation of the frequency translation chart which takes Doppler shift into account. The modified chart is shown in Fig 10-2. When the inner circle, which can rotate, is set at the appropriate point—start of pass, middle of pass, or end of pass—the chart will provide a very good estimate of the downlink frequency corresponding to a particular uplink.

Transponder Band Plan

Amateurs have voluntarily adopted certain guidelines for use of different types of modulation on amateur satellites. SSB and CW are preferred because of their efficient use of transponder power and bandwidth. FM and slow-scan TV are discouraged, except for certain experimental applications, because they use a relatively large share of the transponder's power and bandwidth. The current band plan is shown in Fig 10-3. Note that only the downlink is considered; the band plan is independent of whether the transponder is inverting or noninverting.

On The Air

A satellite is a shared resource. The operating guidelines that follow all evolved from the need to share spacecraft power and bandwidth as efficiently and fairly as possible.

Whenever you transmit, check your downlink signal strength. Keep your signal below the level of the satellite beacon. Periodically check your signal strength and the beacon level to assure that changing conditions haven't caused you to exceed the recommended power level. If you hear stations whose downlink is louder than the beacon, mention that fact to them politely. A responsible operator will appreciate the information.

If all downlinks, including your own, sound weak, you're going to be tempted to turn up the power. It's therefore extremely important when setting up your ground station to concentrate your efforts on developing first-class receiving capabilities.

Never transmit on a transponder uplink frequency unless you're able to monitor the downlink. Be sure to check your downlink frequency before calling CQ.

When calling CQ or when responding to another station, don't send "dits" or croon "heelloo" up and down the band searching for your downlink. If you use a chart like the one shown in Fig 10-1 or 10-2 to estimate your downlink frequency, a few dits will be sufficient. If for some reason you have to search for your downlink, keep your receive frequency constant and vary your transmit frequency. This approach is suggested because it generates a minimal amount of QRM to a large number of stations rather than completely disrupting communications on a single downlink frequency. Remember, good dial calibration, a little experience, and a frequency translation chart will minimize the all-too-familiar clatter of background dits.

Because of Doppler shift, QSOs tend to creep across a transponder's passband. Since different ground stations observe different Doppler shifts, this leads to QRM as one QSO collides with another. The effect is most serious when working with low-altitude satellites, so the guidelines developed for minimizing this problem focus on Phase 2 spacecraft. However, the suggestions also work for Phase 3 satellites. The ideal way to minimize the QRM problem would be for all stations to maintain their downlink frequency at a fixed distance from the beacon. At present, many stations find this operationally difficult to accomplish. Therefore, the following interim guideline, which minimizes creep but does not eliminate it, is suggested.

Consider a QSO involving two or more stations. Since a station relatively distant from the ground track will see a smaller Doppler shift than one close to the ground track, select a distant station to be the frequency reference. The reference station will leave his or her transmitter or receiver frequency fixed during the course of the contact. The choice of receiver or transmitter is not arbitrary. Select the one that operates at a lower frequency. For example, on Mode J the reference station will keep the transmitter frequency fixed, while on Mode B the reference station will keep the receive frequency fixed. The procedure sounds a little awkward, but it's easy to implement and it can significantly reduce QRM due to creep.

Duplex operation (simultaneous transmit and receive) is not an option—it's a *necessity*. If your ground station isn't set up for duplex operation, you'll never know just what your downlink frequency is, or whether your signal is so strong that it's overloading the satellite, or so weak that it can't be heard. Since you'll be operating duplex it's important to prevent your receiver output from feeding back into your transmitter. Significant feedback leads to horrendous howls. A small amount of feedback may not be immediately noticeable, but it can result in severely garbled audio. Attempting to use a speaker on your receiver will leave you continuously lunging for the audio gain control and searching for the nonexistent level setting that will provide feedback-free listening. The sooner you find a comfortable set of headphones, the happier you'll be.

For general communications, adhere to the band plan (Fig 10-3) and stick to CW and SSB, since they use the transponder's power and bandwidth efficiently. An inverting transponder flips sidebands, so transmit on LSB if you want an USB downlink when using Modes B, J and L. Avoid any type of speech processing. If all users were to adopt speech processing, the average power per user would remain the same while the intelligibility of each signal would decrease. Even if speech processing temporarily increases your share of the available power, the results might not be what you expect. Experiments on 432-MHz EME have convinced most operators that any type of speech processing reduces signal intelligibility when working with weak-signal modes; a clean SSB signal works best.

Because DX windows for low-altitude satellites are very brief, you'll note a lot of contest-style short exchanges in the passband. Nevertheless, rag chewing is also common on Phase 2 satellites—the choice is yours. On Phase 3 satellites, you'll find that breaking into a QSO to collect a new DX preface is considered rude. Follow the lead of the DX station. If they want to chat, allow them to do so. Most will periodically run a string of contest-like QSOs to accommodate those who wish to log the contact.

Finally, remember that guidelines are only suggestions and that the important factor is the spirit of cooperation and effective, efficient use of the spacecraft. When a large number of users converge for a Science Education Net, the use of slow-scan TV is clearly justified. If a crystal-controlled CW Field Day station is stuck at the lower end of the passband, an SSB response can easily be rationalized.

USING A DIGITAL TRANSPONDER

There are a number of different types of digital transponders aboard amateur spacecraft. These include the Pacsat design, which was optimized for mailbox or bulletin board operation from a low-altitude spacecraft, the RUDAK I system, which was optimized for digipeating via a high-altitude spacecraft, the flexible PCE (Packet Communications Experiment) on UoSAT and the simple ROBOT on the RS satellites. Technical descriptions of the transponders and the modems required to access them are contained in Chapter 15. In this section we'll look at how the Pacsat and ROBOT transponders are used.

Pacsat

There are two aspects to setting up a ground station for operating through a Pacsat: hardware and operations. We'll consider each in turn.

From an operational viewpoint, a Pacsat operates very much like a terrestrial bulletin board. Before you attempt satellite packet communications, you should gather a little experience working with a terrestrial bulletin board on 2 meters. This will give you a chance to become familiar with the various details associated with operating a TNC, such as connecting the TNC to your rig, setting the audio signal levels, initializing the TNC and using basic TNC commands. It will also provide an opportunity to confirm that your TNC and 2-meter rig are working correctly. The instruction manual that comes with your TNC will probably provide all the information you need to get started on 2-meter packet, but a little help from a local amateur will certainly speed up the process. The following directions for setting up and operating a Pacsat ground station assume that you've set up and tested your 2-meter terrestrial packet station and verified that everything is working correctly.

The hardware requirements for a Pacsat ground station are shown in Fig 9-20. In addition to your 2-meter packet station, you need a 70-cm SSB receiver and a special PSK modem (Pacsat modem).

In early 1990, Pacsat modems were available in both kit and preassembled form (see Table 9-8). By the time you read this, it is likely that additional units, perhaps including a TNC, may be on the market. Extensive tests with Fuji-OSCAR 12 demonstrated that the G3RUH and TAPR kits provide excellent performance. The great majority of the more than 300 users who passed packets through FO-12 used one of these two units. The preassembled PacComm modem was released in early 1990. It also provides good performance. If you plan to purchase a preassembled modem, check whether it is designed to work with the TNC you're using and that instructions for connecting it to your TNC are included. Although the G3RUH and TAPR modems take very different approaches to decoding satellite signals, they are similar to set up and operate. The following directions apply to both units; where there are differences, they'll be described.

The first step is to build and check out the kit according to the instruction manual. This generally requires access to an oscilloscope and audio function generator. Next, disconnect the internal modem in the TNC and connect the special Pacsat modem. To do this, you'll have to consult the manuals accompanying the TNC and the new modem. Most modems (TNC-1, TNC-2, and clones) have a modem-disconnect jack, so this procedure should be relatively easy. However, in some cases this may require cutting PC-board traces. The audio output from the TNC should be shifted from the 2-meter FM receiver used for terrestrial packet to the 70-cm SSB receiver. The Pacsat ground station has one extra control line linking the digital gear (modem, TNC and computer or terminal) and the RF gear (transmitter and receiver). This is the digital AFC (automatic frequency control) line, which goes to the receiver. One potentially confusing part of setting up the ground station is establishing the correct connections between the AFC line and the receiver. Modem designers have devoted a great deal of attention to producing a flexible AFC system that will work with every known microprocessor-controlled rig produced by ICOM, Kenwood and Yaesu. Different rigs by the same manufacturer often require different AFC signals, so there are a large number of ways in which the modem kit can be configured. It's up to the user (that's you) to set various jumpers correctly so the modem will operate with your receiver. This will involve ferreting information out of the receiver manual and the modem manual.

Once the equipment is interconnected, you'll have to set some switches and signal levels. Carefully follow the directions accompanying your modem. Decrease the transmitter deviation to ± 3 kHz. The spacecraft will *not* accept packets from a transmitter that is overdeviating. In most cases, the receiver can be set to either upper or lower sideband. If the sideband filter is symmetrical, there shouldn't be any difference in performance. Once everything is operating, you might want to experiment with sideband selection since an asymmetrical filter may yield better performance on one setting. (The signals produced by the AFC circuit may depend on whether USB or LSB is selected, so you may have to flip a switch or reset a jumper when switching sidebands.)

Some of the TNC settings used for terrestrial packet should be modified. The TNC should be set for AX.25 Version 2. Because the mailbox has a single downlink and several users may be simultaneously connected, you should increase *FRACK* (the waiting time for the acknowledgment signal from the spacecraft) to 7 or higher. Reduce *MAXFRAME* (the number of packets that your station can transmit at one time) to three or less. *PACLEN* (the number of bytes of data in a single packet) should be less than 200. For comparison, Fuji-OSCAR 12 transmitted with *PACLEN* = 128 and MAXFRAME = 1.

We'll use the G3RUH modem (front panel shown in Fig 10-4) to illustrate how the controls function. The LOOP-BANDWIDTH switch on the G3RUH unit is set to W (wide)

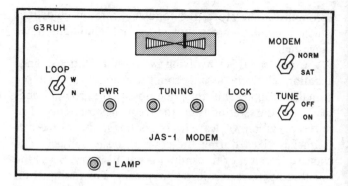

Fig 10-4—Front panel controls on G3RUH modem

for receivers that tune in 100-Hz steps and to N (narrow) for rigs that tune in 10- or 20-Hz steps. (The TAPR circuit uses a single setting for both step sizes, so it does not contain a loop-bandwidth switch.) The MODEM switch toggles between the regular modem contained in the TNC and the special Pacsat modem. It should be set to the SAT (satellite) position if you're going to operate through a Pacsat.

Suppose that a Pacsat pass is about to start and you want to listen in. Set the TUNE (AFC on/off) control to the OFF position. When you hear the satellite come into range, manually tune the receiver slowly until the lock indicator lights. This shows that the modem has locked onto the incoming signal. The tuning lamps and zero-center meter make manual tuning easy. When the lock indicator lights, flip the TUNE switch to "on" and the receiver will start to automatically track the satellite frequency. If your receiver tunes in 10-Hz steps, it may be easier to use the following alternate procedure to achieve lock. Start by placing the loop switch in the wide position and the tune switch off. Manually tune the receiver and when the lock light comes on, toggle the loop switch to narrow. The lock lamp will probably go out. Adjust the receiver tuning slightly until the lock indicator lights again. Now flip the tune switch to ON. If everything is working properly and your TNC is in the monitor mode, you'll periodically see some telemetry frames or downlink messages being sent.

To connect to the satellite, set your transmitter to one of the four uplink frequencies. Make sure your TNC is in the command mode then type

CONNECT 8J1JBS

and press the enter key on your computer. 8J1JBS is the call sign of Fuji-OSCAR 20. Use the proper call sign for the satellite you wish to connect to. If OSCAR 20 receives your request, it will announce that you've connected and provide a few lines of information followed by the prompt

JAS>

This is an invitation to you to continue. The actual prompt and the information preceding it will depend on the spacecraft and the version of the mailbox software operating. At this point you can request help, ask to see

Table 10-1

Commands the Satellite Mailbox Software Will Recognize Once a Connection Has Been Established

Satellite Mailbox Commands

These commands will only be accepted when you see the spacecraft prompt on the screen. For OSCARs 12 and 20 the prompt was JAS>. Each satellite will have its own prompt. Do not confuse the brief list of satellite commands with the long list of commands accepted by the TNC. The ground station must end all lines sent to the spacecraft by pressing ENTER. This key is sometimes referred to as the RETURN or CARRIAGE RETURN or <CR> key. Please be sure to disconnect before the spacecraft dips over your radio horizon.

Command Syntax

The general format for a command is a single letter followed by at least one space, followed by a qualifier. Only the single letter is required; the rest is optional. Following standard notation, qualifiers have been enclosed in < > in the list below. These < > should be omitted when the command is sent.

Available Commands

B		: List file headers addressed to ALL
F		: List latest 15 file headers
F	*	: List latest 50 file headers
F	<d>	: List file headers posted on day <d>
H		: Show help message (list of available commands). (See Note 3.)
K	<n>	: Kill a file numbered <n>. (Only files addressed to, or posted by, your station can be deleted.)
M		: List file headers addressed to current user
R	<n>	: Read a file numbered <n>
U		: List call signs of those currently connected
W		: Write a file

Notes

[1]The file number, <n>, may be obtained by using the F command.
[2]The list of commands and options may change slightly as new versions of the mailbox software are loaded on the spacecraft. You can always obtain an updated list by using the H (help) command.
[3]Future versions of the satellite software may contain an expanded help command. When this feature is implemented, you'll be able to follow help with a blank space and a single letter signifying a command of interest. The satellite will then provide details on the command and its various qualifiers.

---

a list of messages that have been deposited in the mailbox, request that a specific message be read out, and so on. Table 10-1 contains a list of commands that the satellite will accept. A sample script is shown in Fig 10-5.

Under certain conditions it may be possible to connect to the spacecraft without using AFC on your receiver. You have to select a pass where Doppler shift will be small. This means a pass near your horizon, where access time will be short (5 to 10 minutes). Place the TUNE switch in the OFF position and the LOOP switch in the wide position (G3RUH modem). Place the AFC switch in the OFF position (TAPR modem). Using the tuning lamps and the zero center meter, manually adjust the receiver to keep the lock lamp lit. Because of the awkward, unreliable nature of this approach, it should only be regarded as a temporary measure.

Radio amateurs who desire to optimize the performance of their ground stations may find that using a computer to adjust transmit frequency during a pass (to keep the uplink centered in the satellite receiver passband) improves their success rate at uplinking packets. This isn't too difficult to accomplish if you have a modern 2-meter transmitter whose frequency can be externally controlled in 10- or 20-Hz steps and appropriate software. One possible approach is to, prior to the pass, have your software prepare a frequency vs time file based on orbit-prediction data. During the pass the software would periodically interrupt communications activities and adjust the transmitted frequency. After OSCAR 16 is in orbit for a year or two, AMSAT plans to switch certain uplink channels aboard the spacecraft from 1200 bauds to 4800 bauds. Ground stations using the higher-rate uplink will have to track their transmit frequency as outlined.

You've probably noticed that we've never discussed the details of the design of the TAPR and G3RUH modems. You don't have to know the difference between PSK and FSK, the positive and negative features of a Costas loop demodulator, what the term Manchester coding means, or anything about link data rates to set up and operate a Pacsat ground station. If you are interested in this information, you'll find a brief introduction in Chapter 15. Details may be found in the references listed in Table 9-8.

Pacsat communications are just beginning. Keep in mind that the information just presented relates to two specific modem kits that were available in early 1990 and is based on limited operational experience. Major changes in ground-station equipment availability are likely to occur now that several pacsats are in orbit. Be sure to check recent periodicals for the latest information.

ROBOT

Several RS satellites have carried simple autotransponders, known as ROBOTs, which enable radio amateurs to "contact" the satellite. If you call the spacecraft using the correct protocol, an onboard computer will (1) acknowledge your call, (2) assign you a serial contact number and (3) store your call letters and contact number for later downlinking to a command station. QSLs for these contacts are available from Box 88, Moscow.

The ROBOT receive window is only 2- or 3-kHz wide, centered on the announced frequency (see Appendix B). Be sure to take Doppler shift into account by transmitting a few kHz low when the spacecraft is approaching you and a few kHz high when it's receding.

When the ROBOT is active (when it calls CQ), send a few dits on the uplink frequency (*only* a few!). If you hear your dits regenerated on the downlink, you're in the capture window. Call the satellite (clean CW at 10 to 30 WPM) as follows:

RS10 DE KA1GD A̅R̅

If you're successful, Radio 10 will respond:

```
001    cmd: CONNECT 8J1JBS
002
003    *** CONNECTED to 8J1JBS
004
005    FO-20/JAS1b Mailbox ver. 2.01
006    commands [B/F/H/M/R/U/W]
007    Use H command for Help
008
009    JAS> B
010
011    NO.    DATE    UTC     FROM     TO      SUBJECT
012    117    10/12   09:32   F8ZS     ALL     ARSENE update
013    121    10/12   09:37   DL3AH    ALL     Abgleichanleitung
014    123    10/12   09:42   WA2LQQ   ALL     New RS s/c due
015
016    JAS> W
017
018    TO? K2UYH
019
020    SUBJECT? Trenton EME Conference
021
022    Enter text, <CR>.<CR> to end.
023    Have dates been firmed up?
024    Do you need speaker on Pacsat procedures?
025    73, Marty, K2UBC
026
027
028    END
029    <Control>C
030    cmd: DISC
```

Explanation

Note: The ground station must end each line sent to the satellite by pressing the ENTER key. This key is frequently referred to as the RETURN or CARRIAGE RETURN or <CR> key.

line 001:   Session begins with TNC in command mode. Ground station enters CONNECT command.
line 003:   Satellite responds that connection has been made.
line 005-7: Log-on banner from satellite. Contains list of commands that will be accepted. Ground station may obtain an explanation of the commands by typing H for help.
line 009:   Satellite sends prompt JAS>. Ground-station response requests list of files addressed to ALL.
line 011-14: Satellite responds to request.
line 016:   Satellite sends prompt JAS>. Ground-station response indicates desire to write a file.
line 018:   Satellite asks destination of message. Ground station replies K2UYH. If ground station leaves this field blank, message will be addressed to "all."
line 020:   Satellite asks for subject. Ground station replies Trenton EME conference. (Subject must be 32 characters or less).
line 022:   Satellite requests body of message and reminds ground station to terminate message by sending carriage return, period, carriage return. This produces a line with only a period followed by a blank line.
line 023-27: Ground station sends message followed by carriage return, period, carriage return.
line 028:   Satellite signals receipt of message by sending END.
line 029:   Ground station returns to TNC command mode by entering <Control>C (simultaneously pressing control and C keys).
line 030:   TNC replies by printing cmd:. Ground station then enters the DISCONNECT command to indicate the session is over. Please be sure to disconnect before the spacecraft leaves your access circle.

Fig 10-5—Simulated script illustrating communication with OSCAR 20 Mailbox. Line numbers are not part of session—they were added so explanations could be appended.

KA1GD DE RS10 QSO NR 123 OP ROBOT TU FR QSC 73 $\overline{SK}$

The 3-digit QSO number is incremented after each contact.

When it is switched on, the ROBOT will call CQ about once per minute. Please do *not* hold your key down on the ROBOT input frequency, as this will simply cause the downlink to generate a continuous tone. If only a partial message is received by the ROBOT, you may hear a response of QRZ, QRM or RPT. In this case, just try again. If the ROBOT wants you to send faster or slower, it will respond QRQ or QRS. Clean, high-speed CW usually works best, probably because interference is less likely to be a problem. A memory dump of RS7 recorded the first 10 autotransponder QSOs for posterity:

00 UK3ACM	03 UA3XBU	07 G3IOR
01 UV3FL	04 UI8F	08 UK1BI
02 RS3A	05 G3IOR	09 KA1GD
	06 G3IOR	

CONTEST POLICY

Many amateurs enjoy competing in contests and working toward awards—activities that encourage one to improve operating skills and station performance. Certain features of satellite communication, however, make it very important for us to consider carefully what types of contests and awards are appropriate. Satellites are a shared resource and are most effectively used when everyone cooperates, especially in using the minimum necessary power levels. When communicating via the ionosphere on HF, a low-power operator can use skill and patience to successfully compete against high-power stations; no one can overload the ionosphere and make it useless for others. Satellites, however, can be overloaded by a few inconsiderate stations, and then no amount of skill or patience will let anyone maintain communications. Contest formats that tempt participants to use spacecraft resources unfairly are therefore frowned upon.

Does this mean satellites should not be used for contests or awards? Certainly not! It just suggests that we consider the effects of various activities carefully before endorsing them. Several types of contests can contribute to the advancement and general enjoyment of satellite communication for all. For example, we might wish to encourage emergency preparedness by continuing to support Field Day satellite use. QRP-only DXCC and similar awards might encourage operators to improve their ground station's performance and their operating skills. Contests that stimulate the occupancy of under-utilized transponders might also prove worthwhile. In any event, remember that AMSAT does not have control; anyone can sponsor a contest or award. Therefore, in a very real sense, the future is up to the user community. Your interest, support and tolerance will determine the future of satellite contests and awards.

Techno-sport

WA2LQQ coined the term *techno-sport* to describe contest-like activities that encourage the development of operational skills, the perfection of station performance and the enhancement of the operator's knowledge of satellite communications systems. AMSAT will actively support any group that wishes to sponsor such an activity as long as the proposed event does not undermine the cooperative nature of satellite communication.

One techno-sport activity that has proved immensely popular is the ZRO receiving tests. Named in memory of Kaz Deskur, K2ZRO, a long-time AMSAT member who made many major contributions in the area of user tracking aids and operations, the tests are designed to provide users with a quantitative measure of receive-system performance. This gives operators a chance to compare the performance of their systems to those of others and to evaluate improvements over long time periods. During the test a transmitting station carefully adjusts uplink power to provide a downlink equal in strength to the beacon. A series of 3-dB attenuators are then inserted in the transmitter line. After each attenuator is inserted, a series of code groups are sent.

The object is to receive the code groups correctly at the lowest possible level. Participants with super receive systems have been able to successfully copy transmissions sent at 24 dB below the beacon level (the satellite transmitting power is below 10 mW at this point). In order to achieve uniformity from test to test, dates and times must be selected to satisfy the following criteria: the spacecraft must be near apogee, the subsatellite point must be close to the region of the earth where participants are concentrated and the satellite orientation must produce low values of squint angle to prospective participants. In addition, times should not conflict with normal working hours. Although these criteria are restrictive, it's usually possible to identify several times each year that are appropriate. Dates and times are generally announced about two months in advance.

Another proposed techno-sport activity involves the simulation of the SARSAT/COSPAS search-and-rescue satellite system used to locate the position of downed civil aviation aircraft. One possible scenario involves the "downed flyer" activating a beacon signal (satellite uplink) on a pre-announced weekend. Participants monitoring the downlink would use every method at their disposal to determine the location of the transmitting station. Sponsoring such an activity is a big undertaking. Rules must be carefully designed so that they will interest amateurs and stimulate the types of activities desired, information brochures must be produced and distributed to prospective participants, the downed flyer station must be activated, results must be processed and disseminated, and so on. AMSAT would be pleased to cosponsor such an event if a group willing to accept primary responsibility for overseeing the activity comes forth.

INFORMATION RESOURCES

If you operate with amateur satellites you're going to need current information on orbital elements, operating schedules, new satellites, techno-sport activity dates, and so on. Information sources include the hard-copy world of magazines, newsletters, conference proceedings and technical journals; traditional radio activities such as HF and satellite nets; and computer communications, including 2-meter terrestrial packet, satellite packet and dial-up bulletin boards.

Regular and Special Nets

There are a large number of regularly scheduled nets devoted to the amateur satellite program. Some have been meeting for nearly 20 years. A few of the major ones are listed in Table 10-2.

Every time a new AMSAT-coordinated spacecraft is launched, special nets are organized to disseminate launch information around the world. These nets are part of ALINS, the AMSAT-sponsored Amateur Launch Information Net Service. With respect to input and output, ALINS is extensive. ALINS nets generally begin meeting several days before launch to provide up-to-the-minute launch-schedule information. The highlight, of course, is the actual launch and deployment of the amateur

Table 10-2

AMSAT Terrestrial Nets: Frequencies and Times

Service Area	Day	Time	Frequency
International	Sunday	1800 UTC	21.280 MHz
International	Sunday	1800 UTC	14.282 MHz
South Pacific	Saturday	2200 UTC	14.282 MHz
US East Coast	Tuesday	2100 EST[1]	3.840 MHz
US Central	Tuesday	2100 CST[1]	3.840 MHz
US West Coast	Tuesday	2100 PST[1]	3.840 MHz

[1]Net remains at 2100 when clocks are switched to Daylight time.

Table 10-3

ALINS Frequencies

AMSAT Launch Information Network Service (ALINS)
Participating Relay Stations (partial list)

AMSAT

      3.840 MHz
     14.282 MHz
     21.280 MHz

Goddard Space Flight Center ARC, Greenbelt, Maryland
  WA3NAN 3.860 MHz
          7.185 MHz
        14.295 MHz
        21.395 MHz

Jet Propulsion Laboratory ARC, Pasadena, California
  W6VIO   3.850 MHz
        14.282 MHz
        21.280 MHz

Johnson Space Center ARC, Houston, Texas
  W5RRR 3.840 MHz
        14.280 MHz

American Radio Relay League, Newington, Connecticut
  W1AW   See *QST* for bulletin frequencies (all modes)

spacecraft. Some examples of events covered by ALINS nets follow.

Shortly after FO-12 went up, a radio amateur in Argentina captured telemetry as the spacecraft was released from the launch vehicle. Minutes later, this telemetry was relayed back to Japan via a station in the US. Amateurs listening to the nets heard the traffic being relayed. During the launches of OSCARs 10 and 13, the net control operator had access to the European Space Agency network, the NASA network and strategically placed ground stations around the world. Those listening to the net during the launch of Phase 3A will probably never forget hearing the ESA control station in the background as it first reported "non-nominal" conditions and shortly thereafter "splashdown." On a happier note, we also heard each of the successful stagings leading to OSCAR 13 being deposited in its transfer orbit. Shortly after OSCAR 8 was launched, ALINS carried traffic from the command stations as they sent critically important signals to the spacecraft directing it to deploy its 10-meter antenna. Many amateurs in the eastern half of the US were able to monitor the spacecraft telemetry as the commands were successfully received and implemented.

Launch net information is carried by AMSAT and the ARRL on their regular broadcasting frequencies. Several additional groups have volunteered their facilities to help disseminate information during recent launches. These include the radio clubs of the Johnson Space Center, Houston, TX (W5RRR); the Goddard Space Flight Center, Greenbelt, MD (WA3NAN); and the Jet Propulsion Laboratory, Pasadena, CA (W6VIO). For a list of HF frequencies normally employed, see Table 10-3. In addition to the above, a great many 2-meter repeaters rebroadcast the net in real-time.

Other operating activities that have been quite popular are the Science Educational Nets and AMSAT Operation Nets currently on OSCAR 13 Mode B. These nets are scheduled aperiodically when the spacecraft is in a favorable location with respect to North America at a convenient weekend time. A special topic is selected for each net and individuals having in-depth knowledge of the subject are lined up to participate. Times and topics are generally announced about two months in advance on regularly scheduled nets.

Publications

While nets and electronic mail are excellent for rapid exchange of time-critical information, ink and paper remain a cost-effective medium for disseminating detailed technical data. As a result, periodicals will continue to be a very important source for information on amateur satellites. Since it was founded in 1969, AMSAT-NA has published either a newsletter or journal which is sent to members. It also aperiodically produces the *AMSAT Technical Journal*, which contains professional-level papers accessible to advanced amateurs. In addition, it has released beginner's guides, booklets focusing on specific satellites and a manual focusing on tracking programs.

The ARRL produces several publications of interest to satellite users. The annual *ARRL Handbook* is an excellent source for construction articles on VHF and UHF equipment. *The ARRL Antenna Book* contains detailed instructions for VHF and UHF antennas. *The Weather Satellite Handbook,* now in its 4th edition, is a key resource for those interested in receiving weather satellite images. All League members receive *QST*, which frequently carries an Amateur Satellite Communications column and often includes feature articles on satellite topics and VHF and UHF gear. The ARRL publishes the monthly magazine *QEX*, which focuses on advanced technology. Articles included are frequently of interest to satellite users, especially those involved in packet operations. Examples include the construction of a modem for packet operations through FO-12, VHF and UHF preamp and antenna design, and so on. The ARRL also publishes conference proceedings for the annual AMSAT, Computer Networking, and Central States VHF conferences.

Other major US Amateur Radio publications also frequently contain information on the amateur satellite program. *Ham Radio* published Jim Miller's (G3RUH) designs for FO-12 and Phase 3 satellite telemetry modems. *73* carries a monthly satellite column. *CQ* has had a number of major articles on weather-satellite receiving equipment. The biweekly newsletter *OSCAR Satellite Report* (PO Box 175, Litchfield, CT 06759) provides up-to-date information and orbital elements.

Several national AMSAT organizations publish newsletters with excellent technical and operating information. The most accessible of these to North American amateurs (because it's written in English, and because they accept MasterCard and VISA) is *OSCAR News*, published by AMSAT-UK. *OSCAR News*, which comes out about five times each year, is loaded with interesting articles and is a prime source for information on the UoSAT series of satellites. AMSAT-UK also publishes several booklets, manuals and computer tracking and telemetry-capture programs. Addresses for publications may be found in Table 10-4.

If you read German, the quarterly AMSAT-DL Journal is frequently a source of information that's available nowhere else, especially concerning Phase 3 spacecraft systems. English translations of Japanese language publications focusing on the Japanese amateur satellite program are generally available from Project OSCAR. The most authoritative information on the RS satellite program can be found in the Russian-language journal *RADIO*. In the US, subscriptions are available from the Victor Kamkin Bookstore.

As you can see, the number of publications available is large. If you need some specialized technical information, there's a good chance that it may be in print. If you do some innovative technical work that others may be interested in, there's an audience waiting to read about it.

Bulletin Boards

News items and other information of a time-critical nature is most effectively distributed by electronic bulletin boards. The number of bulletin boards accessible by phone, by terrestrial packet radio and by satellite packet radio is huge. Many of these bulletin boards carry information of interest to satellite users. If you really want to catch the latest-breaking news, you have two choices: (1) set up for satellite packet radio or (2) move to a location where a nearby Pacsat user is feeding news directly from the satellite to a local bulletin board. Table 10-5 contains a brief list of telephone-access bulletin boards which contain information of interest to satellite users. Since maintaining an up-to-date bulletin board is a very time-consuming task, the phone numbers tend to change every few years as responsibility for operations is passed on. As a result, Table 10-5 may be outdated by the time you see it. However, it's fairly easy to track down current sources. Most bulletin boards contain lists of other active boards that cover related topics. So, if you can tie into one board that focuses on Amateur Radio, it should be fairly easy to locate those covering satellite communications.

435-MHz TRANSMITTING RESTRICTIONS

For many years the FCC has restricted amateurs who use the 420-450 MHz band to 50 watts input power in certain parts of the United States. In response to growing

---

Table 10-4

Addresses for Selected Amateur Satellite Publications

AMSAT-DL
Holderstrauch 10
D-3550 Marburg 1
Germany

AMSAT-UK
R. J. C. Broadbent
94 Herongate Road
Wanstead Park
London E12 5EQ
England

(Japanese publications available in English)
Project OSCAR, Inc
PO Box 1136
Los Altos, CA 94022
USA

(Russian publication *RADIO*)
Victor Kamkin Bookstore, Inc
12224 Parklawn Dr
Rockville, MD 20852
USA

Table 10-5

Selected Bulletin Boards of Interest to Amateur Satellite Users

Telephone numbers are current as of early 1990.

Note: All are accessible at 300 or 1200 baud. 2400 baud may be available. If your modem/software requires presetting parameters, try: N/8/1 (no parity, 8-bit word, 1 stop bit) unless specified otherwise.

Celestial RCP/M Bulletin Board (excellent source for up-to-date orbital elements on large number of spacecraft)
513-427-0674

CompuServe's HamNet
via local CompuServe access number

DRIG/AMSAT
(DRIG = Dallas Remote Imaging Group)
Datalink Bulletin Board
214-394-7438

NOAA Direct Readout Users Bulletin Board
(mainly weather satellite information)
Provided by NOAA/NESDIS (see Chapter 14)
Access via PAC*IT Plus, an EDS (electronic data system) network. Contact EDS at 800-544-8953 for local number and access code.
Can also be accessed on SCIENCEnet via WXSAT gateway
In Washington, DC area, dial 834-9700
(7 data bits, 1 stop bit, even parity)

NOAA Space Environment Laboratory
(up-to-the-minute information on solar-terrestrial environment for propagation studies)
303-497-5000
See D. Rosenthal, "NOAA's Space Environment Laboratory Public Computer Bulletin board Service," *QST*, Aug 1989, pp 15-18.

satellite activity and a concurrent increase in requests for Special Temporary Authorizations (often referred to as STAs) to use higher power, the FCC has acted to "ease" the restrictions.

As a result of FCC actions, beginning in April 1981 (1) additional restricted areas were introduced and (2) the power limitations were divided into two categories, one for terrestrial operations and another for satellite communications. In August 1982, the restricted regions were increased in number and size.

The restricted areas now include:

1) Those portions of Texas and New Mexico bounded by latitudes 33°24′N and 31°53′N, and longitudes 105°40′W and 106°40′W.

2) The entire state of Florida, including the Key West area and the areas enclosed within circles of 320-kilometer (200-mile) radius of Patrick Air Force Base (28°21′N, 80°43′W) and Eglin Air Force Base (30°30′N, 86°30′W).

3) The entire state of Arizona.

4) Those portions of California and Nevada south of latitude 37°10′N, and the area within a 320-km (200-mile) radius of the US Naval Missile Center (34°09′N, 119°11′W).

5) In the state of Massachusetts within a 160-kilometer (100-mile) radius of Otis Air Force Base (41°45′N, 70°32′W).

6) In the state of California within a 240-kilometer (150-mile) radius of Beale Air Force Base (39°08′N, 121°26′W).

7) In the state of Alaska within a 160-kilometer (100-mile) radius of Elmendorf Air Force Base (64°17′N, 149°10′W).

8) In the state of North Dakota within a 160-kilometer (100-mile) radius of Grand Forks Air Force Base (48°43′N, 97°54′W).

The 50-watt input-power limit continues to apply to stations that are engaged in terrestrial communication in the restricted areas. Amateurs engaged in satellite communication on frequencies between 435 and 438 MHz in those areas, however, will be permitted to use 1000 watts EIRP, provided their antennas' elevations are adjusted so that the half-power points of the radiated pattern remain at least 10 degrees above the horizon. See Sections 97.3, 97.303 and 97.313 of the Amateur Rules and Regulations.

Chapter 11

Satellite Orbits

Using the step-by-step techniques of Chapter 6, radio amateurs can track OSCAR spacecraft without needing to know the basic physics of satellite motion or how a satellite moves in space. This chapter is for those amateurs interested in "why" as well as "how." Here we'll examine satellite motion from a more detailed physical/mathematical point of view.

Several of the topics we look at in this chapter are usually found in texts designed for graduate-level scientists and engineers. These texts, rigorous and generalized, are often incomprehensible to readers who don't have an advanced mathematical background. Yet most of the ideas and results can be expressed in terms that someone with a solid background in algebra, plane trigonometry and analytic geometry can understand. We'll keep the mathematics in this chapter as simple as possible, but—face it—mathematics is a key element in understanding satellite motion. Study the solved *Sample Problems* scattered throughout this chapter to see how key formulas are applied. As they also form the basis for later work, the problems may be the most valuable part of the chapter.

At several points we had to raise the mathematical level slightly higher than desired to avoid obscuring potentially useful information. Much of the material in this chapter is not serial in nature, however, so you can skip big chunks and still follow later sections. By now, you should realize that this chapter is not for the faint-hearted. If you elect to plow through, reviewing the tracking material of Chapter 6 before beginning will make the path a little easier. Also note that Table 11-1 summarizes repeatedly used symbols. Good luck!

The objectives of this chapter are:

1) to introduce the satellite-orbit problem (from a scientific point of view);

2) to provide the reader with an overview of satellite motion (including both an understanding of important parameters and an ability to visualize the motion in space and with respect to earth);

3) to summarize the important equations needed to compute orbital parameters so that these equations will be easily accessible when needed.

BACKGROUND

The satellite-orbit problem (determining the position of a satellite as a function of time and finding its path in space) is essentially the same whether we are studying the

Table 11-1
Symbols Used in This Chapter

Note: Abbreviations used only in computer programs are marked (*)

a	— primary: semimajor axis of ellipse (secondary: side of spherical triangle)
A	— angle in spherical triangle
b	— primary: semiminor axis of ellipse (secondary: side of spherical triangle)
B	— angle in spherical triangle
c	— primary: distance between center of ellipse and focal point (secondary: side of spherical triangle)
C	— angle in spherical triangle
e	— eccentricity of ellipse
E	— eccentric anomaly (angle)
EA	— eccentric anomaly (*)
ETY	— eccentricity (*)
G	— gravitational constant
h	— satellite height above surface of earth
h_a	— satellite height above surface of earth at apogee
h_p	— satellite height above surface of earth at perigee
i	— orbital inclination
$\bar{I}$	— longitude increment (rough estimate)
I	— estimated longitude increment
m	— primary: mass of satellite (secondary: abbreviation for meter)
M	— mass of earth
MM	— mean motion
m/s	— meters per second
r	— satellite-geocenter distance
r_a	— satellite-geocenter distance at apogee
r_p	— satellite-geocenter distance at perigee
R	— mean radius of earth = 6371 km satellite-geocenter distance (*)
R_{eq}	— mean equatorial radius of earth = 6378 km
s	— seconds
SMA	— semimajor axis (of ellipse) (*)
t	— elapsed time since last ascending node (circular orbits) or last perigee (elliptical orbits)
T	— period of satellite
v	— magnitude of satellite velocity with respect to static earth
w	— argument of perigee
$\dot{w}$	— rate of change of argument of perigee
θ	— polar angle in orbital plane
ϕ	— latitude
λ	— longitude
Ω	— right ascension of ascending node
$\dot{\Omega}$	— rate of orbital-plane precession about earth's N-S axis

motion of the planets around the sun, the moon around the earth, or artificial satellites revolving around either. The similarity arises from the nature of the forces affecting an orbiting body that doesn't have a propulsion system. In the early 17th century Kepler discovered some remarkable properties of planetary motion; they have come to be called *Kepler's Laws*.

I) Each planet moves around the sun in an ellipse, with the sun at one focus (motion lies in a plane);

II) The line from the sun to planet (radius vector, r) sweeps out across equal areas in equal intervals of time;

III) The ratio of the square of the period (T) to the cube of the semimajor axis (a) is the same for all planets in our solar system. (T^2/a^3) is constant.

These three properties summarize observations; they say nothing about the forces governing planetary motion. It remained for Newton to deduce the characteristics of the force that would yield Kepler's Laws. The force is the same one that keeps us glued to the surface of the earth— good old gravity.

Newton showed that Kepler's Second Law would result if the planets were being acted on by an attractive force always directed at a fixed central point: the sun (central force). To satisfy the First Law, this force would have to vary as the inverse square of the distance between planet and sun $(1/r^2)$. Finally, if Kepler's Third Law were to hold, the force would have to be proportional to the mass of the planet. Actually, Newton went a lot further. He assumed that not only does the sun attract the planets in this manner, but that every mass (m_1) attracts every other mass (m_2) with a force directed along the line joining the two masses and having a magnitude (F) given by

$$F = \frac{Gm_1m_2}{r^2} \text{ (Universal Law of Gravitation) (Eq 11.1)}$$

where G is the Universal Gravitational Constant.

THE GEOMETRY OF THE ELLIPSE

As Kepler noted in his First Law, ellipses take center stage in satellite motion. A brief look at the geometry of the ellipse is therefore in order (see Fig 11-1). The lengths a, b and c shown in Fig 11-1 are not independent. They're related by the formula

$$c^2 = a^2 - b^2 \text{ or } c = \sqrt{(a^2 - b^2)} \qquad \text{(Eq 11.2)}$$

Using Eq 11.2, any one of the parameters a, b or c can be computed if the other two are known. In essence, it takes two parameters to completely describe the shape of an ellipse. One could, for example, give the semimajor and semiminor axes (a and b), the semimajor axis and the distance from the origin to one focus (a and c), or the semiminor axis and the distance from the origin to one focus (b and c).

There's another convenient parameter, called eccentricity (e), for describing an ellipse. Eccentricity may be thought of as a number describing how closely an ellipse resembles a circle. When the eccentricity is 0, we've got a circle. The larger the eccentricity, the more elongated the

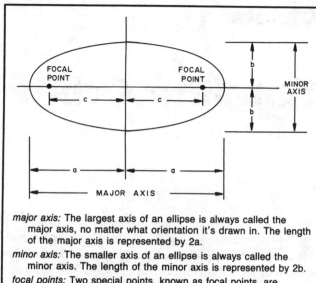

major axis: The largest axis of an ellipse is always called the major axis, no matter what orientation it's drawn in. The length of the major axis is represented by 2a.

minor axis: The smaller axis of an ellipse is always called the minor axis. The length of the minor axis is represented by 2b.

focal points: Two special points, known as focal points, are located on the major axis equidistant from the origin. The distance between the origin and each focal point is represented by c.

Fig 11-1—Geometry of the ellipse.

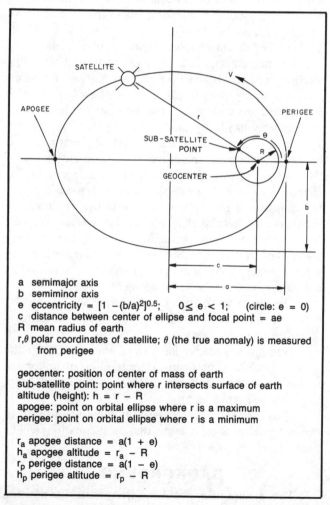

a   semimajor axis
b   semiminor axis
e   eccentricity = $[1 - (b/a)^2]^{0.5}$; $0 \leq e < 1$;   (circle: e = 0)
c   distance between center of ellipse and focal point = ae
R   mean radius of earth
r,θ polar coordinates of satellite; θ (the true anomaly) is measured from perigee

geocenter: position of center of mass of earth
sub-satellite point: point where r intersects surface of earth
altitude (height): h = r − R
apogee: point on orbital ellipse where r is a maximum
perigee: point on orbital ellipse where r is a minimum

r_a apogee distance = a(1 + e)
h_a apogee altitude = r_a − R
r_p perigee distance = a(1 − e)
h_p perigee altitude = r_p − R

Fig 11-2—Geometry of the orbital ellipse for an earth satellite.

ellipse becomes. To be more precise, eccentricity is given by

$$e^2 = 1 - (b/a)^2 \quad \text{or} \quad e = \sqrt{1 - (b/a)^2}$$
$$0 \leq e < 1 \qquad \qquad \text{(Eq 11.3)}$$

Because of its mathematical definition, e must be a dimensionless number between 0 and +1. Using Eqs 11.2 and 11.3, we can derive another useful relationship:

$$c = ae$$

As stated earlier, it always takes two parameters to describe the shape of an ellipse. Any two of the four parameters, a, b, c or e, will suffice.

Fig 11-2 shows the elliptical path of a typical earth satellite. Since the earth is located at a focal point of the ellipse (Kepler's Law 1), it is convenient to introduce two additional parameters that relate to our earth-bound vantage point: the distances between the center of the earth and the "high" and "low" points on the orbit. Fig 11-2 summarizes several useful relations and definitions. Note especially

apogee distance: $r_a = a(1 + e)$ (Eq 11.5a)

perigee distance: $r_p = a(1 - e)$ (Eq 11.5b)

We now have six parameters, a, b, c, e, r_a and r_p, any two of which can be used to describe an ellipse. With the information we've learned so far, many practical satellite problems can be solved. (See Sample Problem 11.1)

Sample Problem 11.1

AMSAT-OSCAR 10 has an apogee distance (r_a) of 6.57R and a perigee distance (r_p) of 1.62R. Specify the orbit in terms of the semimajor axis (a) and eccentricity (e). (**Note:** In studying earth satellites, distances are sometimes expressed in terms of R, the mean radius of the earth).

Solution

(Given r_a and r_b, solve for a and e)
Subtracting Eq 11.5b from Eq 11.5a we obtain:

$$r_a - r_p = 2ae \quad \text{or} \quad e = \frac{r_a - r_p}{2a}$$

Adding Eq 11.5a to Eq 11.5b gives

$$r_a + r_p = 2a; \quad a = 4.10 \text{ R}$$

so, $e = \dfrac{r_a - r_p}{r_a + r_p} = \dfrac{6.75R - 1.62R}{6.75R + 1.62R} = 0.604$

When the major and minor axes of an ellipse are equal, the ellipse becomes a circle. From Eq 11.2 we see that setting a = b gives c = 0. This means that in a circle, both focal points coalesce at the center. Setting a = b in Eq 11.3 yields e = 0, as we stated earlier.

Since the circular orbit is just a special case of the elliptical orbit, the most general approach to the satellite-orbit problem would be to begin by studying elliptical orbits. Circular orbits, however, are often simpler to work

with, so we'll look at them separately whenever it makes our work easier.

Our approach to the satellite-orbit problem involves first determining the path of the satellite in space and then looking at the path the sub-satellite point traces on the surface of the earth. Each of these steps is, in turn, broken down into several smaller steps.

SATELLITE PATH IN SPACE

The motion of an object results from the forces acting on it. To determine the path of a satellite in space, we will (1) make a number of simplifying assumptions about the forces on the satellite and other aspects of the problem, taking care to keep the most important determinants of the motion intact; (2) then solve the simplified model; and then (3) add corrections to our solution, accounting for the initial simplifications.

Simplifying Assumptions

We begin by listing the assumptions usually employed to simplify the problem of determining satellite motion in the orbital plane.

1) The earth is considered stationary and a coordinate system is chosen with its origin at the earth's center of mass (geocenter).

2) The earth and satellite are assumed to be spherically symmetric. This enables us to represent each one by a point mass concentrated at its center (M for the earth, m for the satellite).

3) The satellite is subject to only one force, an attractive one directed at the geocenter; the magnitude of the force varies as the inverse of the square of the distance separating satellite and geocenter ($1/r^2$).

The model just outlined is known as the two-body problem, a detailed solution for which is given in most introductory physics texts.[1,2] Some of the important results follow.

Solution to the Two-Body Problem

Initial Conditions. Certain initial conditions (the velocity and position of the satellite at burnout, the instant the propulsion system is turned off) produce elliptical orbits ($0 \leq e < 1$). Other initial conditions produce hyperbolic ($e > 1$) or parabolic ($e = 1$) orbits, which we will not discuss.

The Circle. For a certain subset of the set of initial conditions resulting in elliptical orbits, the ellipse degenerates (simplifies) into a circle ($e = 0$).

Satellite Plane. The orbit of a satellite lies in a plane that always contains the geocenter. The orientation of this plane remains fixed in space (with respect to the fixed stars) after being determined by the initial conditions.

Period and Semimajor Axis. The period (T) of a satellite and the semimajor axis (a) of its orbit are related by the equation

$$T^2 = \frac{4\pi^2}{GM} a^3 \qquad \qquad \text{(Eq 11.6a)}$$

where M is the mass of the earth and G is the Universal Gravitational Constant. For computations involving a satellite in earth orbit, the following equations may be used (T in minutes, a in kilometers).

$$T = 165.87 \times 10^{-6} \times a^{3/2} \qquad \text{(Eq 11.6b)}$$

$$a = 331.25 \times T^{2/3} \qquad \text{(Eq 11.6c)}$$

Note that the period of an artificial satellite that is orbiting the earth depends only on the semimajor axis of its orbit. For a circular orbit, a is equal to r, the constant satellite-geocenter distance. Sample Problems 11.2 and 11.3 show how Eq 11.6 is used.

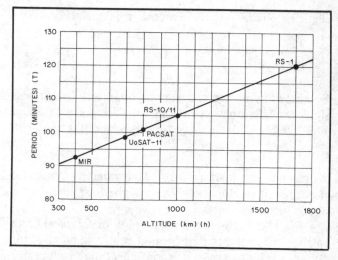

Fig 11-3—Period vs altitude for satellites in low-altitude circular orbits.

Sample Problem 11.2

Given that RS-10/11 is in a circular orbit at a height of 1003 km, find its period.

Solution

Eq 11.6b provides the period when the semimajor axis is known. To obtain the orbital radius (geocenter-satellite distance) we have to add the radius of the earth to RS-10/11's altitude: r = 6371 + 1003 = 7374 km. Plugging this value into Eq 11.6b yields a period of 105.0 minutes:

$$T = 165.87 \times 10^{-6} \times (7374)^{3/2} = 105.0 \text{ minutes.}$$

Sample Problem 11.3

OSCAR 13's orbit is characterized by an apogee height (h_a) of 36,265 km and a perigee height (h_p) of 2,545 km. What is its period?

Solution

R = radius of earth = 6371 km

r_a = 36,265 + 6371 = 42,636 km

r_p = 2545 + 6371 = 8916 km

$2a = r_a + r_p$ (see Sample Problem 11.1)

2a = 51,552 km; a = 25,776 km

Applying Eq 11.6b we obtain

$$T = 165.87 \times 10^{-6} \times (25,776)^{3/2} = 686.4 \text{ minutes}$$
$$= 11 \text{ hours } 26.4 \text{ minutes.}$$

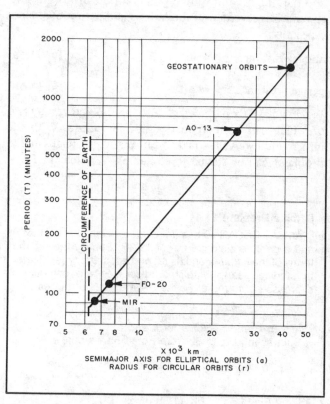

Fig 11-4—Period vs semimajor axis.

A graph of period vs height for low-altitude spacecraft in circular orbits is shown in Fig 11-3. In Fig 11-4 we plot period vs semimajor axis. Both of these plots were obtained from Eq 11.6b. The *mean motion,* MM, was defined in chapter 6 as the number of revolutions (perigee to perigee) completed by a satellite in a solar day (1440 minutes). A satellite's mean motion is related to its period by Eq 11.7. (MM in revolutions per solar day, T in minutes.)

$$MM = 1440/T \qquad \text{(Eq 11.7)}$$

Since many sources of orbital elements provide the mean motion, it is often necessary to compute period and semimajor axis from it. A short BASIC program that does this is shown in Table 11-2.

Table 11-2

Program to Calculate Period and Semimajor Axis from Mean Motion

Language: Microsoft BASIC

```
100      'Program to calculate period and semimajor
110      'axis, SMA, from mean motion, MM
120 INPUT "mean motion (rev/day) = ? ", MM
130 PERIOD = 1440/MM              'See Eq. 11.7
140 PRINT "Period (minutes) ="; PERIOD
150 SMA = 331.25 * PERIOD^(2/3)   'See Eq. 11.6c
160 PRINT "semimajor axis (km) ="; SMA
170 END
```

Sample Problem 11.4

OSCAR 7 (no longer operating) will remain in its 1460-km circular orbit for centuries. What is its velocity?

Solution

In a circular orbit we can use the simplified form of Eq 11.8.

$$v^2 = \frac{GM}{r} = (3.986 \times 10^{14})(1/r)$$

$$r = 1460 \text{ km} + 6371 \text{ km} = 7.831 \times 10^6 \text{ m}$$

$$v^2 = \frac{3.986 \times 10^{14}}{7.831 \times 10^6} = 0.5090 \times 10^8 \text{ (m/s)}^2$$

$$v = 0.7134 \times 10^4 = 7134 \text{ m/s}$$

Sample Problem 11.5

Shortly after launch, OSCAR 13 had an apogee height (h_a) of 36,265 km and a perigee height (h_p) of 2545 km. What was its velocity at apogee? At perigee? Compare the perigee velocity of OSCAR 13 (h_p = 2545 km) to that of OSCAR 7 (h = 1460 km).

Solution

(See Sample Problem 11.3)

$$a = 25,776 \text{ km}$$
$$r_a = 42,636 \text{ km}$$
$$r_p = 8916 \text{ km}$$

Use Eq 11.8: $v^2 = 3.986 \times 10^{14} \left(\frac{2}{r} - \frac{1}{a} \right)$

At Apogee

$$v^2 = 3.986 \times 10^{14} \left(\frac{2}{42,636,000} - \frac{1}{25,776,000} \right)$$

$$= 3.2338 \times 10^6 \text{ (m/s)}^2$$

$$v = 1798 \text{ m/s}$$

At Perigee

$$v^2 = 3.986 \times 10^{14} \left(\frac{2}{8,916,000} - \frac{1}{25,776,000} \right)$$

$$= 73.948 \times 10^6 \text{ (m/s)}^2$$

$$v = 8599 \text{ m/s}$$

At perigee, OSCAR 13 is moving about 20% faster than OSCAR 7. Doppler shift (see Chapter 13) on OSCAR 13 Mode B near perigee is therefore about 20% greater than it was through the Mode B transponder on OSCAR 7.

Velocity. The magnitude of a satellite's total velocity (v) generally varies along the orbit. It's given by

$$v^2 = GM \left(\frac{2}{r} - \frac{1}{a} \right) = 3.986 \times 10^{14} \left(\frac{2}{r} - \frac{1}{a} \right)$$
(Eq 11.8)

where r is the satellite-geocenter distance, r and a are in meters, and v is in meters/sec (see Fig 11-2). Note that for

a given orbit, G, M and a are constants, so that v depends only on r. Eq 11.8 can therefore be used to compute the velocity at any point along the orbit if r is known. The range of velocities is bounded: The maximum velocity occurs at perigee and the minimum velocity occurs at apogee. The direction of motion is always tangent to the orbital ellipse. For a circular orbit r = a and Eq 11.8 simplifies to (r in meters, v in m/s)

$$v^2 = \frac{GM}{r} = (3.986 \times 10^{14}) \left(\frac{1}{r} \right)$$
(circular earth orbit only)

Note that for circular orbits, v is constant. Sample Problems 11.4 and 11.5 illustrate the use of Eq 11.8.

Position. Fig 11-2 shows how the satellite position is specified by the polar coordinates r and θ. Note that θ is measured counterclockwise from perigee. Often, it's necessary to know r and θ as a function of the elapsed time, t, since the satellite passed perigee (or some other reference point when a circle is being considered).

For a satellite in a circular orbit moving at constant speed:

$$\theta \text{ [in degrees]} = \frac{t}{T} (360°) \text{ or}$$
$$\theta \text{ [in radians]} = 2\pi \frac{t}{T}$$
(Eq 11.9)

and the radius is fixed.

The elliptical-orbit problem is considerably more involved. We know (Eq 11.8) that the satellite moves much more rapidly near perigee. The relation between t and θ can be derived from Kepler's Law II. For details of the derivation, see references 3 and 4.

In an elliptical orbit, time from perigee, t, is given by

$$t = \frac{T}{2\pi} [E - e \sin E]$$
(Eq 11.10)

where the angle E, known as the eccentric anomaly, is defined by the associated equation

$$E = 2 \arctan \left[\left(\frac{1 - e}{1 + e} \right)^{0.5} \tan \frac{\theta}{2} \right] + 360°n \quad \text{(Eq 11.11)}$$

$$n = \begin{cases} 0 \text{ when } -180° \leq \theta \leq 180° \\ 1 \text{ when } 180° < \theta \leq 540° \end{cases}$$

Eq 11.11 may also appear in several alternate forms:

$$E = \arcsin \left[\frac{(1 - e^2)^{0.5} \sin \theta}{1 + e \cos \theta} \right] \text{ or}$$

$$E = \arccos \left[\frac{e + \cos \theta}{1 + e \cos \theta} \right]$$

Note that here, "anomaly" just means angle. Eqs 11.10 and 11.11, taken together, are commonly referred to as Kepler's Equation. Fig 11-5 shows the position of a satellite in an elliptical orbit (similar to those used for Phase 3 spacecraft) as a function of time. It should give Kepler's Equation some physical meaning.

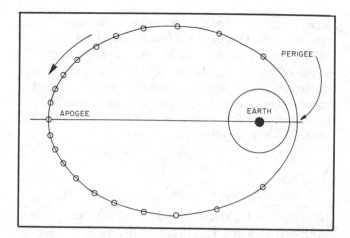

Fig 11-5—This orbital plane diagram shows the position of a satellite in a 12-hour elliptical orbit at half-hour intervals. Note that near apogee the satellite moves relatively slowly.

Table 11-3
Program to Calculate True Anomaly and Satellite-geocenter Distance when Time from Perigee is Given

Illustrates Newton-Raphson method of solving Kepler's Equation.

Language: Microsoft BASIC

```
100     'Program to calculate true anomoly (THETA) and satellite-
110     'geocenter distance (R) for OSCAR 13 when time from perigee
120     'is given.  For other satellites change lines 150 & 160.
130     'EA=Eccentric Anomaly; SMA=semimajor axis; ETY=eccentricity
140     PI=3.14159
150     ETY=.6541
160     PERIOD=686.4  'minutes
170     SMA=331.25*(PERIOD)^(2/3)        'see Eq. 11.6c
180     INPUT "minutes after perigee = ? ", TIME
190     EAINIT = 2*PI*TIME/PERIOD        'initial estimate for EA
200     EA = EAINIT
210     FOR I = 1 TO 20                  'loop to improve estimate for EA
220       CORRECTION = (EA-EAINIT-ETY*SIN(EA))/(1-ETY*COS(EA))
230       EA = EA - CORRECTION
240       IF ABS(CORRECTION) < .0001 THEN GOTO 270
250     NEXT I
260     PRINT "Loop did not converge." : STOP
270     PRINT "Iterations = ";I
280     IF ABS(EA-PI) < .0001 THEN THETA =PI : GOTO 310
290     THETA = 2*ATN(SQR((1+ETY)/(1-ETY))*TAN(EA/2))    'Eq. 11.11 inverted
300     IF THETA < 0 THEN THETA = THETA + 2*PI
310     R = SMA*(1-ETY*ETY)/(1+ETY*COS(THETA))           'Eq. 11.12
320     PRINT "theta (degrees) ="; THETA*180/PI
330     PRINT "Satellite-geocenter distance (km) ="; R
340     END
```

There are two common mistakes that people frequently make the first time they try to solve Eqs 11.10 and 11.11. Eq 11.10 contains the first pitfall. Since the expression e sin E is a unitless number, the E term standing by itself inside the brackets *must* be given in *radians*. The second pitfall is encountered when working with the various forms of Eq 11.11. Although all inverse trigonometric functions are multi-valued, computers and hand calculators are programmed to give only principal values. For example, if sin θ = 0.99, then θ may equal 82° or 98° (or either of these two values ± any integer multiple of 360°), but a calculator only lights up 82°. If the physical situation requires a value outside the principal range, appropriate adjustments must be made. Eq 11.11 already includes the adjustments needed so that it can be used for values of θ in the range −180° to +540°. If the alternate forms of Eq 11.11 are used, it's up to you to select the appropriate range. A few hints may help: (1) E/2 and θ/2 must always be in the same quadrant; (2) as θ increases, E must increase; (3) adjustments to the alternate forms of Eq 11.11 occur when the term in brackets passes through ±1.

We now have a procedure for finding t when θ is known: Plug θ into Eq 11.11 to compute E, then plug E into Eq 11.10 to obtain t. The reverse procedure, finding θ when t is known, is more complex. The key step is solving Eq 11.10 for E when t is known. Unfortunately, there isn't any way to neatly express E in terms of t. We can, however, find the value of E corresponding to any value of t by drawing a graph of t vs θ, then reading it "backwards," or by using an iterative approach. An iterative technique is just a systematic way of guessing an answer for E, computing the resulting t to determine how close it is to the desired value, then using the information to make a better guess for E. Although this procedure may sound involved, it's actually simple. The iterative technique usually employed to solve Kepler's Equation is known as the Newton-Raphson method. A BASIC language subroutine that calculates θ when t is known is shown in

Table 11-3. Most modern introductory calculus texts explain how the Newton-Raphson method is used.

We now turn to r, the satellite-to-geocenter distance. Rather than attempt to express r as a function of t, it's simpler and often more useful just to note the relation between r and θ.

$$r = \frac{a(1 - e^2)}{1 + e \cos \theta} \qquad \text{(Eq 11.12)}$$

Now try Sample Problem 11.6.

Corrections to the Simplified Model

Now that we've looked at the solutions to the two-body problem (the simplified satellite-orbit model), let's examine how a more detailed analysis would modify our results.

1) In the two-body problem, the stationary point is the center of mass of the system, not the geocenter. The mass of the earth is so much greater than the mass of an artificial satellite that this correction is negligible.

2) Treating the earth as a point mass implicitly assumes that the shape and the distribution of mass in the earth are spherically symmetrical. Taking into account the actual asymmetry of the earth (most notably the bulge at the equator) produces additional central force terms acting on the satellite. These forces vary as higher orders of $1/r$ (for example, $1/r^3$, $1/r^4$, and so on). They cause (i) the major axis of the orbital ellipse to rotate slowly in the plane of the satellite and (ii) the plane of the satellite to rotate about the earth's N-S axis. Both of these effects are observed readily, and we'll return to them shortly.

3) The satellite is affected by a number of other forces in addition to gravitational attraction by the earth. For example, such forces as gravitational attraction by the sun, moon and other planets; friction from the atmosphere (atmospheric drag); radiation pressure from the sun, and

so on enter into the system. We turn now to the effects of some of these forces.

Atmospheric Drag. At low altitudes the most prominent perturbing force acting on a satellite is drag caused by collisions with atoms and ions in the earth's atmosphere. Let us consider the effect of drag in two cases: (i) elliptical orbits with high apogee and low perigee and (ii) low-altitude circular orbits. In the elliptical-orbit case, drag acts mainly near perigee, reducing the satellite velocity and causing the altitude at the following apogee to be lowered (perigee altitude initially tends to remain constant). Atmospheric drag therefore tends to reduce the eccentricity of elliptical orbits having a low perigee (makes them more circular) by lowering the apogee. In the low-altitude circular orbit case, drag is of consequence during the entire orbit. It causes the satellite to spiral in toward the earth at an ever-*increasing* velocity. This is not a misprint. Contrary to intuition, drag causes the velocity of a satellite to *increase*. As the satellite loses energy through collisions it falls to a lower orbit; Eq 11.8 shows that velocity increases as the height decreases.

A satellite's lifetime in space (before burning up on re-entry) depends on the initial orbit, the geometry and mass of the spacecraft, and the composition of the earth's ionosphere (which varies a great deal from day to day and year to year). Fig 11-6 provides a very rough estimate of the lifetime in orbit of a satellite similar in geometry and mass to AMSAT-OSCAR 7 or 8 as a function of orbital altitude.[5] As the altitudes of AMSAT-OSCAR communication spacecraft are greater than 800 km, their lifetimes in orbit should not be a serious concern.

Solar activity has a very big effect on the composition of the earth's atmosphere at altitudes between 300 and 600 km. High solar activity results in increased atmospheric density and greater drag on spacecraft. The effect was clearly visible on OSCAR 9. See Fig 11-7. Early predictions, for a satellite lifetime of three to five years, had to be revised because of the very low level of solar activity recorded during the 1984-87 time period. Re-entry occurred in October 1989 with spacecraft electronics subsystems fully functional until the final orbits.

Gravitational Effects. The effects on a satellite's orbit

Sample Problem 11.6

Consider the OSCAR 13 orbit. (See Sample Problem 11.3)

r_a = 42,636 km
r_p = 8916 km
a = 25,776 km
T = 686.4 minutes

(a) Compute the satellite altitude (h) when $\theta = 108°$
(b) How long after perigee does this occur?

Solution

(a) *Step 1*: Solve for the eccentricity
(see Sample Problem 11.1).

$$e = \frac{r_a - r_p}{2a} = 0.6541$$

Step 2: Solve for r using Eq 11.12

$$r = \frac{a(1 - e^2)}{1 + e \cos \theta} = 18,484 \text{ km}$$

h = r − R = 18,484 km − 6371 km = 12,113 km

(b) *Step 3*: Compute the eccentric anomaly using Eq 11.11

$$E = 2 \arctan \left[\left(\frac{1 - e}{1 + e} \right)^{0.5} \tan \frac{\theta}{2} \right]$$

$$= 2 \arctan \left[\left(\frac{1 - 0.6541}{1 + 0.6541} \right)^{0.5} \tan 54° \right]$$

E = 64.37° = 1.123 radians

Step 4: Compute t from Eq 11.10

$$t = \frac{T}{2\pi} [E - e \sin E]$$

$$= \frac{686.4}{2\pi} [1.123 - (0.6541) \sin (64.37°)]$$

= 58.3 minutes

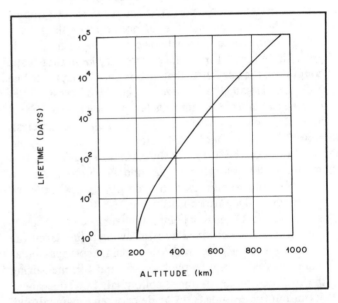

Fig 11-6—Satellite lifetime for circular orbit with the satellite geometry and mass similar to AMSAT-OSCARs 7 and 8.

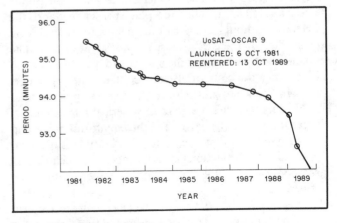

Fig 11-7—UoSAT-OSCAR 9.

from gravitational attraction by the sun and moon are most prominent when the apogee distance is large. The sun and moon have a significant effect on the orbit of AMSAT Phase-3 satellites. The casual user need not worry about this, but AMSAT scientists must investigate the long-term effects of these forces in detail to ensure that the chosen orbit is stable. Instabilities because of resonant (cumulative) perturbations can cause the loss of a satellite within months. Table 11-4 shows the relative strengths of selected perturbing forces. Now that the motion of the satellite in space has been described, we turn to the problem of relating this motion to an observer on the surface of the earth.

SATELLITE MOTION VIEWED FROM EARTH

Terrestrial Reference Frame

To describe a satellite's movement as seen by an observer on the earth, we have to establish a terrestrial reference frame. Once again we simplify the situation by treating the earth as a sphere. The rotational axis of the earth (N-S axis) provides a unique line through the geocenter that intersects the surface of the earth at two points that are designated the *north* (N) and *south* (S) geographic *poles*. The intersection of the surface of the earth and any plane containing the geocenter is called a *great circle*. The great circle formed from the *equatorial plane*, that plane containing the geocenter that also is perpendicular to the N-S axis, is called the *equator*. The set of great circles formed by planes containing the N-S axis are also of special interest. Each is divided into two *meridians* (half circles), connecting north and south poles.

Points on the surface of the earth are specified by two angular coordinates, *latitude* and *longitude*. As an example, the angles used to specify the position of Washington, DC are shown in Fig 11-8.

Latitude. Given any point on the surface of the earth, the latitude is determined by (i) drawing a line from the given point to the geocenter, (ii) dropping a perpendicular from the given point to the N-S axis and (iii) measuring the included angle. A more colloquial, but equivalent, definition for latitude is the angle between the line drawn from the given point to the geocenter and the equatorial plane. To prevent ambiguity, an N or S is appended to the latitude to indicate whether the given point lies in the northern or southern hemisphere. The set of all points having a given latitude lies on a plane perpendicular to the N-S axis. Although these latitude curves form circles on the surface of the earth, most are *not* great circles. The equator (latitude = 0°) is the only one to qualify as a great circle, since the equatorial plane contains the geocenter. The significance of great circles will become apparent later in this chapter when we look at spherical trigonometry. Better models of the earth take the equatorial bulge and other asymmetries into account when latitude is defined. This leads to a distinction between geodetic, geocentric and astronomical latitude. We won't bother with such refinements.

Longitude. All points on a given meridian are assigned the same longitude. To specify longitude one chooses a reference or "prime" meridian (the original site of the Royal Greenwich Observatory in England is used). The longitude of a given point is then obtained by measuring the angle between the lines joining the geocenter to (i) the point where the equator and prime meridian intersect and (ii) the point where the equator and the meridian containing the given point intersect. For convenience, longitude is given a suffix, E or W, to designate whether one is measuring the angle east or west of the prime meridian.

The Inclination

As the earth rotates about its N-S axis and moves around the sun, the orientation of both the plane containing the equator (*equatorial plane*) and, to a first approximation, the plane containing the satellite (*orbital plane*) remain fixed in space relative to the fixed stars. Fig 11-9(A) shows how the orbital plane and equatorial plane are related. The line of intersection of the two planes is called the *line of nodes*, since it joins the ascending and descending nodes. The relative orientation of these two planes is very important to satellite users. It is partially specified by

---

Table 11-4

The Approximate Magnitudes of Various Forces Acting on Two Satellites

Source of perturbing force	Relative force on satellite at specified height	
	Satellite I (h = 370 km)	Satellite II (h = 37,000 km)
Sun	7×10^{-4}	3×10^{-2}
Moon	4×10^{-6}	1×10^{-4}
Earth's oblateness	1×10^{-3}	4×10^{-6}

$$\text{Relative force} = \frac{\text{Average force exerted by perturbation}}{\text{Force exerted by symmetrical earth}}$$

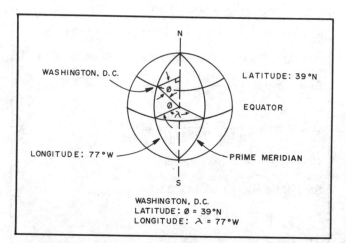

Fig 11-8—The location of Washington, DC on the earth can be described by giving its latitude and longitude coordinates.

giving the inclination. The *inclination*, i, is the angle between the line joining the geocenter and north pole and the line through the geocenter perpendicular to the orbital plane (to avoid ambiguity, the half-line in the direction of advance of a right-hand screw following satellite motion is used). An equivalent definition of the inclination, the angle between the equator and the sub-satellite path on a static (non-rotating) earth as the satellite enters the northern hemisphere, is shown in Fig 11-9(B).

The inclination can vary from 0° to 180°. To first order, none of the perturbations to the simplified model we discussed earlier cause the inclination to change, but higher-order effects result in small oscillations about a mean value. Diagrams of orbits having inclinations of 0°, 90° and 135° are shown in Fig 11-10. A quick analysis of these three cases yields the following information. When the inclination is 0°, the satellite will always be directly above the equator. When the inclination is non-zero the satellite passes over the equator twice each orbit, once heading north and once heading south. When the inclination is 90°, the satellite passes over the north pole and over the south pole during each orbit.

Orbits are sometimes classified as being polar (near polar) when their inclination is 90° (near 90°), or equatorial (near equatorial) when their inclination is 0° (near 0° or 180°). The maximum latitude (ϕ_{max}), north or south, that the sub-satellite point will reach equals (i) the inclination when the inclination is between 0° and 90° or (ii) 180° less the inclination when the inclination is between 90° and 180°. This can be seen from Fig 11-11.

Argument of Perigee

The angle between the line of nodes (the segment joining the geocenter to the ascending node) and the major axis of the ellipse (the segment joining the geocenter and perigee) is known as the *argument of perigee*. Fig 11-9(C) shows how the argument of perigee serves to locate the perigee in the orbital plane. In the simplified two-body model of satellite motion, the argument of perigee is constant. In reality however, it does vary with time, mainly as a result of the earth's equatorial bulge. The rate of precession (variation) is given by

$$\dot{w} = 4.97\left(\frac{R_{eq}}{a}\right)^{3.5} \frac{(5\cos^2 i - 1)}{(1 - e^2)^2} \qquad \text{(Eq 11.13a)}$$

where

$\dot{w}$ = rate of change of argument of perigee in degrees per day

R_{eq} = mean equatorial radius of earth in same units as a

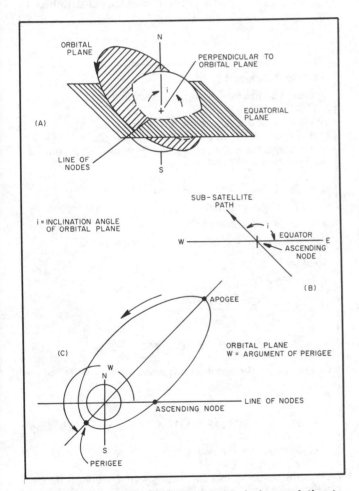

Fig 11-9—The orientation of the orbital plane relative to the equatorial plane is given by i, the inclination angle. The position of the perigee in the orbital plane is given by w, the argument of perigee.

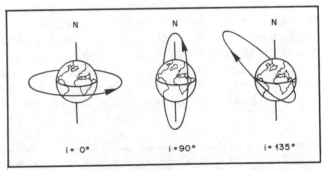

Fig 11-10—Satellite orbits with inclination angles of 0°, 90° and 135°. Orbits with 0 < i < 90° are called prograde or direct. Orbits with 90° < i ≤ 180° are called retrograde.

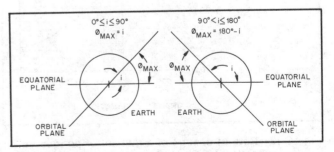

Fig 11-11—The maximum latitude reached by the subsatellite point depends only on the inclination angle of the orbital plane. The cross sections shown in the diagram are edge views of the orbital and equatorial planes.

a = semimajor axis
i = inclination
e = eccentricity

Focusing on the ($5 \cos^2 i - 1$) term, we see that no matter what the values of a and e, when i = 63.4° the argument of perigee will be constant. The position of the perigee rotates in the same direction as the satellite when i < 63.4° or i > 116.6°, and in the opposite direction when 63.4° < i < 116.6°.

Let w_o represent the value of w at a specific time. Future values of w can be obtained from

$$w(t) = w_o + \dot{w}t \qquad \text{(Eq 11.13b)}$$

where t is the elapsed time in days.

In Sample Problem 11.7 we calculate the rate of change of argument of perigee for OSCAR 13. The daily rate of change of the argument of perigee as a function of inclination angle for a typical Phase 3 orbit (h_a = 35,800 km, h_p = 1500 km) is shown in Fig 11-12.

Nodal vs Anomalistic Period

Once we've seen how the earth's equatorial bulge affects the argument of perigee, we have to introduce a new term, *nodal period*, to refer to the elapsed time as a satellite travels from one ascending node to the next. The period that we've been referring to up to this point in this chapter is the *anomalistic period* (elapsed time from perigee to perigee). The adjectives "nodal" and "anomalistic" are often omitted in technical literature and conversation when the meaning is clear from the context. For example, when we discussed the various graphic tracking devices back in Chapter 6, the term period referred to nodal period since we considered an orbit to begin at one ascending node and end at the next. In the equations in this chapter, we'll be explicit when we refer to nodal period. The term period

by itself will refer to anomalistic period.

The numerical differences between anomalistic period and nodal period are generally quite small. However, if one is making long-term predictions using the wrong period, the error is cumulative. After a few weeks, the predictions will be useless. As a result, it's often necessary to calculate nodal period from the information distributed with classical orbital elements which refers to anomalistic period. An example showing how this is done can be found in Sample Problem 11.8.

Solar and Sidereal Time

Living on earth we quite naturally keep time by the sun. So when we say the earth undergoes one complete rotation about its N-S axis each day, we're actually referring to a mean *solar day*, which is arbitrarily divided into exactly 24 hours (1440 minutes). Fig 11-13 illustrates how a solar day can be measured. The time interval known as the solar day begins at A, when the sun passes our meridian, and ends at C, when the sun next passes our meridian. Note that, because of its motion about the sun, the earth rotates slightly more than 360° during the solar day. The number is calculated in Sample Problem 11.9. The time for the earth to rotate exactly 360° is known as the *sidereal day*. When the word "day" is used by itself, solar day is meant. For example, orbital elements distributed by

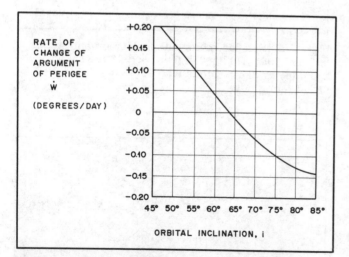

Fig 11-12—Rate of change of argument of perigee vs orbital inclination for Phase 3-type elliptical orbit (h_a = 35,800 km, h_p = 1500 km). See Eq 11.13. When $\dot{w}$ is positive, the argument of perigee rotates in the same direction as the satellite. When $\dot{w}$ is negative, the argument of perigee rotates in the opposite direction.

Sample Problem 11.7

a) Calculate the rate of change of argument of perigee for OSCAR 13.
b) Given that the argument of perigee was 198.6° on day 40 of 1989, when will the apogee occur at the northernmost point on the orbit? (When will argument of perigee = 270°?)

a = 25,776 km (semimajor axis)
i = 57.37° (inclination)
e = 0.654 (eccentricity)
R_{eq} = 6378 km (equatorial radius of earth)

Solution

a) Use Eq 11.13a

$$\dot{w} = 4.97 \left(\frac{R_{eq}}{a}\right)^{3.5} \frac{(5 \cos^2 i - 1)}{(1 - e^2)^2}$$

$$= 0.0519°/\text{day} = 18.953°/\text{year}$$

b) The required change in argument of perigee is

270° − 198.6° = 71.4°

This will take 71.4/0.0519 = 1376 days = 3 years + 280 days

Northernmost apogee will occur 1992 day 320.

Since the numbers this calculation were based on change slightly with time, this result should only be considered an approximation. It is much more realistic to conclude that northernmost apogee will occur in late 1992 or early 1993.

Sample Problem 11.8

Calculate the nodal period for RS-10/11 given the following classical orbital elements and constants.

MM = 13.71883140 (mean motion)

i = 82.9265° (inclination)

e = 0.0010301 (eccentricity)

R_{eq} = 6378 km (equatorial radius of earth)

Solution

First calculate the anomalistic period using Eq 11.7

Anomalistic period = 1440/MM = 104.9652086 minutes

Next calculate the semimajor axis using Eq 11.6c

SMA = 331.25 × (anomalistic period)$^{2/3}$ = 7370.88 km

Now calculate the rate of change of argument of perigee using Eq 11.13a

$$\dot{w} = 4.97 \left(\frac{R_{eq}}{a}\right)^{3.5} \frac{(5 \cos^2 i - 1)}{(1 - e^2)^2}$$

$$\dot{w} = -2.76816°/day$$

Divide this by the mean motion to obtain the total precession during one orbit = −0.201778°

The negative sign means the rotation is opposite to that of the satellite. During one anomalistic period the satellite will therefore rotate 360 − 0.201778 = 359.798°.

The following proportion yields the nodal period

$$\frac{359.798°}{104.965 \text{ min}} = \frac{360°}{\text{nodal period}}$$

Nodal period = 105.024 minutes

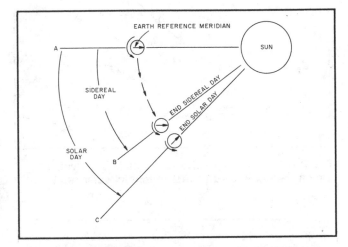

Fig 11-13—The figure shows the relation between the solar day and the sidereal day from the vantage point of an observer on the North Star. The measured day begins at A as the reference meridian aligns with the sun. The sidereal day ends at B when the reference meridian rotates 360°. The solar day ends at C when the reference meridian again aligns with the sun. (Not to scale.)

Sample Problem 11.9

(a) How many degrees does the earth rotate in one solar day?

(b) How many minutes are there in a sidereal day?

Solution

The difference between the solar day and sidereal day occurs because of the earth's rotation about the sun. To an observer off in space viewing a scene like that in Fig 11-13, the yearly circuit about the sun is equivalent to adding 360° of extra axial rotation to the earth each year. Since there are approximately 365.25 days per year, the earth's movement about the sun adds

$$\frac{360°}{365.25 \text{ days}} = 0.98563°/day$$

to the earth's axial rotation. The earth therefore rotates 360.98563°/day on the average. To find the number of minutes in a sidereal day, we set up a proportion

$$\frac{\text{number of minutes in sidereal day}}{360°} = \frac{\text{number of minutes in solar day}}{360.98563°}$$

Solving, we obtain approximately 1436.07 minutes for the sidereal day.

NASA give the units for mean motion as revolutions/day. The day referred to is the solar day of 1440 minutes.

Precession: Circular Orbits

Fig 11-14 shows a satellite whose orbital plane is fixed in space as the earth moves about the sun. In the illustration, the satellite closely follows the terminator (day-night line) in summer. As a result, passes accessible to a ground station will be centered near 6 AM and 6 PM each day. Three months later, the satellite passes over the center of the day and night regions. Accessible passes now occur near 3 AM and 3 PM each day.

Although the two-body model predicts that the orbital plane will remain stationary, we've already noted that when the earth's equatorial bulge is taken into account, the plane precesses about the earth's N-S axis. Fig 11-15 shows an example of such precession. For circular orbits the precession is given by

$$\dot{\Omega} = -9.95 \left(\frac{R_{eq}}{r}\right)^{3.5} \cos i \qquad \text{(Eq 11.14)}$$

(circular earth orbits only)

where

$\dot{\Omega}$ = orbital plane precession rate in °/day. A positive precession is shown in Fig 11-15 (counterclockwise as seen from the North Star).

R_{eq} = mean equatorial radius of earth = 6378 km

r = satellite-geocenter distance in same units as R_{eq}

i = orbital inclination

Sun-Synchronous Orbits

By choosing the altitude and inclination of a satellite, we can vary $\dot{\Omega}$ over a considerable range of values.

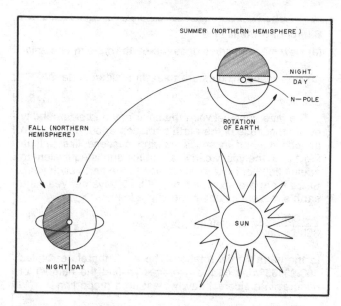

Fig 11-14—The illustration shows a satellite whose orbital plane is fixed in space. The view is that of an observer looking down from the North Star.

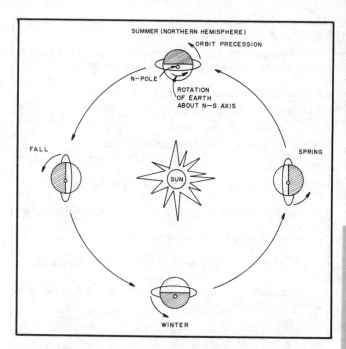

Fig 11-15—Sun-synchronous orbit like the one chosen for OSCARs 6-9, 11 and 14-19. The view of the sun-earth satellite system is from the North Star. Note how the orbital precession can keep the satellite near the twilight line year-round when total precession for a year is 360°.

Looking at the example of Fig 11-15, you may have noted that the orbital plane precessed exactly 360° in one year. As a result, the satellite followed the terminator the entire time. An orbit that precesses very nearly 360° per year is called sun-synchronous. Such orbits pass over the same part of the earth at roughly the same time each day, making communication and various forms of data collection convenient. They can also provide nearly continuous sunlight for solar cells and good sun angles for weather satellite photos when the injection orbit is similar to that shown in Fig 11-15. Because of all these desirable features, orbits are often carefully chosen to be sun-synchronous.

To obtain an orbital precession of 360° per year, we need a precession rate of 0.986°/day (360°/365.25 days). Substituting this value in Eq 11.14 and solving for i we obtain

$$i^* = \arccos\left[-(0.09910)\left(\frac{r}{6378}\right)^{3.5}\right] \qquad \text{(Eq 11.15)}$$

where i* is the inclination needed to produce a sun-synchronous circular orbit. In this form, we can plug in values of r and calculate the inclination which will produce a sun-synchronous orbit. Graphing Eq 11.15 (Fig 11-16), we see that for low-altitude satellites sun-synchronous orbits will be near polar. You may have noted that the 0.986°/day precession rate needed to produce a sun-synchronous orbit exactly corresponds to the amount in excess of 360° that the earth rotates each solar day (Sample Problem 11.9). This is no accident; the precession rate was chosen precisely for this purpose.

Precession: Elliptical Orbits

The precession of the orbital plane about the earth's N-S axis for elliptical orbits is given by

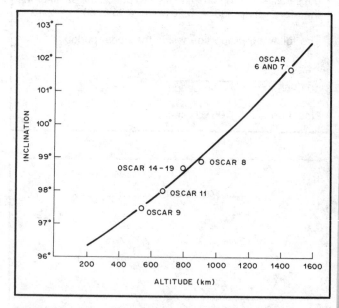

Fig 11-16—This graph shows the inclination value which results in a sun-synchronous circular orbit.

$$\dot{\Omega} = -9.95\left(\frac{R_{eq}}{a}\right)^{3.5}\frac{\cos(i)}{(1-e^2)^2} \qquad \text{(Eq 11.16)}$$

If a = r and e = 0 (that is, when the ellipse becomes a circle), Eq 11.16 simplifies to Eq 11.14.

Longitude Increment

We now know how the satellite moves in the orbital plane and how the orientation of the orbital plane changes with time. Our next objective is to relate this information to the longitude increment. The *longitude increment* (I), or simply *increment*, is defined as the change in longitude between two successive ascending nodes. In mathematical terms

$$I = \lambda_{n+1} - \lambda_n \qquad \text{(Eq 11.17)}$$

where λ_{n+1} is the longitude at any ascending node in degrees east of Greenwich [°E], λ_n is the longitude at the preceding ascending node in °E, and I is in degrees east per revolution [°E/rev].

In Chapter 6 we saw that the increment is an important parameter to those working with the OSCARLOCATOR and similar tracking aids. There are two ways to obtain the increment: experimentally by averaging observations over a long period of time, or theoretically by calculating it from a model. Though the best numbers are obtained experimentally, the calculation approach is needed; we, after all, want a value for I before a spacecraft is launched, and in the early weeks or months of its stay in orbit when observations haven't accumulated over a long time period.

The computations that follow will use the sign convention specified with Eq 11.16. However, when convenient, the final result may be expressed in degrees west by simply changing the sign.

If we neglect precession of the orbital plane, the increment can be estimated by computing how much the earth rotates during the time it takes for the satellite to complete one revolution from ascending node to ascending node. (In this section, period refers to nodal period.)

$$\bar{I} = (T/1440)(-360.98563\,°E) \qquad \text{(Eq 11.18a)}$$

$$\bar{I} = -(0.250684\,°E)\,T \qquad \text{(Eq 11.18b)}$$

The period, T, must be in minutes; the negative sign means that each succeeding node is further west; and 360.98563° is the angular rotation of the earth about its axis during a solar day (1440 minutes). From Eq 11.18b we see that it's easy to get a quick estimate of $\bar{I}$ by computing T/4 and expressing the result in degrees west per revolution.

The value for the orbital increment provided by Eq 11.18 can be improved by taking into account the fact that the precession of the orbital plane (Eq 11.14 or Eq 11.16) will affect the apparent rotation of the earth during one solar day. The result is given in Eq 11.19.

$$\bar{I} = (T/1440)(-360.98563 + \dot{\Omega}) \qquad \text{(Eq 11.19)}$$

Sample Problem 11.10 illustrates how calculations involving the longitude increment proceed.

Once the increment is known we can compute the longitude of any ascending node, λ_m, given the longitude of any other ascending node, λ_n. The orbit reference integers, m and n, may either be the standard ones beginning with the first orbit after launch, or any other convenient serial set.

$$\lambda_m = \lambda_n + (m - n)\,I \qquad \text{(Eq 11.20)}$$

This formula works either forward or backward in time. When future orbits are being predicted, m > n. The right side of Eq 11.20 must be brought into the range of 0-360° by successive subtractions or additions of 360° if necessary (see Sample Problem 11.11).

GROUND TRACK

To study the ground track, we have to (1) look at the geometry involved when the orbital plane intersects the surface of the earth, (2) consider the motion of the satellite about its orbit and (3) take into account the rotation of the earth. The best way to handle complex problems of this type is to use mathematical objects known as vectors. All advanced texts in orbital mechanics proceed in this manner. However, many simple problems can be treated using spherical trigonometry. If you have a reasonable background in plane trigonometry, a brief introduction to spherical trigonometry will provide you with all the information needed to understand how the ground-track equations are derived. Since many more readers have experience with trigonometry than with vectors, we'll use the spherical trigonometry approach.

A bare-bones introduction to spherical trigonometry follows. The results for circular orbits are then generalized and summarized. (Readers who just need access to the ground-track equations for programming a computer can skip the spherical trigonometry and derivation sections and jump right to the summary.) We then go on to derive and summarize the ground-track equations for elliptical orbits.

Sample Problem 11.10

About four weeks after the launch of the Soviet RS-1, observations had yielded the nodal period to six significant digits, T = 120.389 minutes, and had confirmed the TASS and NASA reports that gave i as 82.6° (i is not very critical in Ω computations). Using this data, compute the longitudinal increment.

Solution

Step 1. Compute r from the period using Eq 11.6c.

$$r = 331.25 \times T^{2/3} = 8076 \text{ km}$$

Step 2. Calculate $\dot{\Omega}$ using Eq 11.14.

$$\dot{\Omega} = -9.95 \left(\frac{6378}{8076}\right)^{3.5} \cos(82.6) = -0.5610\,°E/day$$

Step 3. Calculate I from Eq 11.19

$$I = \frac{120.389}{1440}(-360.9856 - 0.5610)$$

$$= -30.227\,°E/rev$$

$$= 30.227\,°W/rev$$

Note: Observations of RS-1 over several months yielded the same value for I.

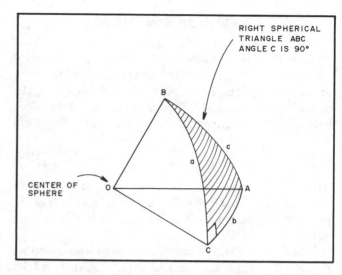

Fig 11-17—Right spherical triangle ABC lies on the surface of a sphere. The three sides are formed from segments of great circles.

This section on spherical trigonometry will also be referred to in the next chapter when we discuss terrestrial distance, bearing and "spiderwebs."

Spherical Trigonometry Basics

A triangle drawn on the surface of a sphere is called a spherical triangle *only* if all three sides are arcs of great circles. A great circle is *only* formed when a plane containing the center of a sphere intersects the surface. The earth's equator is a great circle; other latitude lines are not. The intersection of a satellite's orbital plane and the surface of a static (non-rotating) earth is a great circle. Range circles drawn around a ground station are not great circles.

Spherical trigonometry is the study of the relations between sides and angles in spherical triangles. The notation of spherical trigonometry closely follows that of plane trigonometry. Surface angles and vertices in a triangle are labeled with capital letters A, B and C, and the side opposite each angle is labeled with the corresponding lower-case letter, as shown in Fig 11-17. Note that the arc length of each side is proportional to the central angle formed by joining its end points to the center of the sphere. For example, side b is proportional to angle AOC. The proportionality constant is the radius of the sphere, but because it cancels out in the computations we'll be interested in, the length of a side will often be referred to by its angular measure.

The rules governing the relationships between sides and angles in spherical triangles differ from those in plane triangles. In spherical trigonometry, the internal angles in a triangle do *not* usually add up to 180° and the square of the hypotenuse does *not* generally equal the sum of the squares of the other two sides in a right triangle.

A spherical triangle that has at least one 90° angle is called a right spherical triangle (see Fig 11-17).

Recall how in plane trigonometry the rules for right triangles were simpler than those for oblique triangles. In spherical trigonometry the situation is similar: The rules for right spherical triangles are simpler than those for general spherical triangles. Fortunately, since the spherical triangles we'll be working with have at least one right angle, we need only consider the laws for right spherical triangles. A convenient method for summarizing these rules, developed by Napier, is shown in Fig 11-18. Sample Problem 11.12 illustrates how Napier's Rules can be applied.

Two major pitfalls await newcomers attempting to apply spherical trigonometry for the first time. The first pitfall, the degree-radian trap, comes from overlooking the fact that angles must be expressed in units appropriate to a given equation *and* a given computing machine. For example, focus on the angle $\theta = 30° = \pi/6$ radians. Consider the machine-dependent aspect first. To evaluate $\sin(\theta)$ on most simple scientific calculators, you must input "30" since the calculator expects θ to be in degrees unless you're instructed otherwise. To evaluate $\sin(\theta)$ in BASIC on a microcomputer, you must input $\pi/6$ (or 0.52360), because the BASIC language expects θ to be in radians. In some situations, especially in cases where θ is *not* the argument of a trigonometric function, the form of the equation determines whether θ must be in degrees or radians. Consider a radio station at 30° N latitude trying to use the equation $S = R\theta$ to find the surface distance (S) along a meridian (earth radius = R) to the equator. The equation only holds for θ in radians, so the input must be $\pi/6$.

The second trap awaiting spherical trigonometry

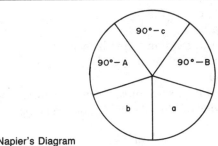

Napier's Diagram

Napier's Rules for right spherical triangle ABC (see Fig 11-17)

Rule 1: The sine of any of the five angles in Napier's diagram is equal to the product of the tangents of the two angles adjoining it.

Rule II: The sine of any of the five angles in Napier's diagram is equal to the product of the cosines of the two angles opposing it.

Three identities that are useful in conjunction with Napier's Rules are:

$\cos(90° - x) = \sin(x)$
$\sin(90° - x) = \cos(x)$
$\tan(90° - x) = \cot(x)$

Fig 11-18—Napier's Rules and this diagram provide an easy way to remember and apply the rules for right spherical triangles.

novices is using a latitude line as one side of a spherical triangle. The only latitude line that will serve in this manner is the equator. All other latitude lines do not work, since they are not arcs of great circles.

Circular Orbits: Derivation

The most important step in deriving the ground-track equations for circular orbits is drawing a clear picture. In Fig 11-19 we've chosen to show i between 90° and 180° and a satellite headed north in the Northern Hemisphere. Our object is to compute the latitude and longitude of the subsatellite point (SSP)—ϕ_S and λ_S—when the spacecraft reaches S, t minutes after the most recent ascending node. We assume that the period T, orbit inclination i, and the longitude of the ascending node, λ_o, are known.

Since arc AS, along the actual ground track, is *not* a section of a great circle, we first consider the situation for a static earth (one not rotating about its N-S axis). On such an earth, the SSP would be at point B at t minutes after the ascending node. Triangle ABC is a right spherical triangle. Angle A is given by 180° − i. Arc AB (side c of the spherical triangle) is a section of the circular orbit with

$$c = 2\pi\frac{t}{T} \quad \text{(see Eq 11.9)}$$

By definition, the latitude of point B is equal to a.

The problem of finding the latitude of point B, ϕ_B, in terms of i, t and T is identical to the problem of finding a in terms of A and c. This was solved in Sample Problem 11.12 where we found that

$$a = \arcsin [\sin (c) \sin (A)]$$

Substituting the variables ϕ_B, T, i and t we obtain

Sample Problem 11.12

Given a right spherical triangle like the one in Fig 11-16, assume that A and c are known. (a) First solve for a in terms of A and c. (b) Then solve for b in terms of a and c.

Solution

(a) To find a, apply Napier Rule II to the indicated segment of the Napier diagram

$$\sin(a) = \cos (90° - c) \cos (90° - A)$$

Using the identity for $\cos (90° - x)$

$$\sin (a) = \sin (c) \sin (A)$$
$$a = \arcsin [\sin (c) \sin (A)]$$

(b) To find b apply Napier Rule II again, this time to the segment shown

$$\sin (90° - c) = \cos (a) \cos (b)$$

Using the identify for $\sin (90° - c)$

$$\cos (c) = \cos (a) \cos (b)$$

$$\cos (b) = \frac{\cos (c)}{\cos (a)}$$

$$b = \arccos \left[\frac{\cos (c)}{\cos (a)}\right]$$

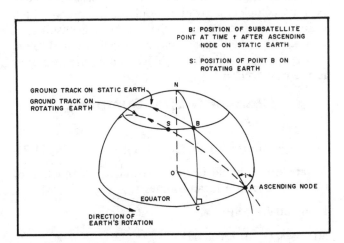

Fig 11-19—Illustration for applying the principles of spherical trigonometry to the circular-orbit ground-track problem.

$$\phi_B = \arcsin \left[\sin \left(2\pi\frac{t}{T}\right) \sin (180° - i)\right]$$

Using the symmetry of the sine function, this simplifies to

$$\phi_B = \arcsin \left[\sin \left(2\pi\frac{t}{T}\right) \sin (i)\right] \text{ (for non-rotating earth).}$$

If for computations we wish to specify c in degrees, we would replace

$$2\pi\frac{t}{T} \quad \text{by} \quad 360°\frac{t}{T}$$

To solve for the longitude at B, λ_B, we note that $b = \lambda_o - \lambda_B$. So, our problem of solving for λ_B in terms of ϕ_B, t and T is equivalent to solving for b in terms of a and c. This was also done in Sample Problem 11.12 where we found that

$$b = \arccos\left[\frac{\cos(c)}{\cos(a)}\right]$$

Making the appropriate substitutions, this yields

$$\lambda_o - \lambda_B = \arccos\left[\frac{\cos(2\pi t/T)}{\cos(\phi_B)}\right] \text{ (for non-rotating earth)}$$

The effect of the earth's rotation is to move the SSP from B to S. The latitude remains constant ($\phi_s = \phi_B$); only the longitude changes. The longitude change is simply the angular rotation of the earth during the time t. To a first approximation, the rotation rate of the earth is 0.25°/minute, so if we measure t in minutes, $\lambda_S = \lambda_B - t/4$. For long-term predictions a more accurate figure for the earth's rotation should be used (as we saw when the longitude increment was discussed). This value is 0.25068°/minute. (See Eq 11.18b.)

This completes the derivation for the case illustrated. A more complete derivation would consider several additional cases: satellites in the southern hemisphere, i between 0° and 90°, spacecraft headed south, and so on. As the approach is similar, we'll just summarize the results in the next section.

Circular Orbits: Summary

Latitude of SSP:

$$\phi(t) = \arcsin[\sin(i)\sin(360° t/T)] \qquad \text{(Eq 11.21)}$$

Note: "$\phi(t)$" should be read "latitude as a function of time"; it does *not* mean ϕ times t.

Longitude of SSP:

$$\lambda(t) = \lambda_o - (0.250684)T$$
$$+ (S1)(S2)\arccos\left[\frac{\cos(360° t/T)}{\cos(\phi(t))}\right] \quad \text{(Eq 11.22)}$$

$$S1 = \begin{cases} +1 \text{ when } 0° \leq i \leq 90° \\ -1 \text{ when } 90° < i \leq 180° \end{cases}$$

$$S2 = \begin{cases} +1 \text{ when } \phi(t) \geq 0° \text{ (Northern Hemisphere)} \\ -1 \text{ when } \phi(t) < 0° \text{ (Southern Hemisphere)} \end{cases}$$

Sign Conventions

   Latitude
      North: positive
      South: negative

   Longitude
      East: positive
      West: negative

All angles are in degrees and time is in minutes
   i = inclination of orbit
   T = period
   t = elapsed time since most recent ascending node
   λ_o = longitude of SSP at most recent ascending node

COMMENTS

1) Please note the sign conventions for east and west longitudes. Most maps used by radio amateurs in the US are labeled in degrees west of Greenwich. This is equivalent to calling west longitudes positive. Because there are important computational advantages to using a right-hand coordinate system, however, almost all physics and mathematics books refer to east as positive, a custom that we follow for computations. When calculations are completed it's a simple matter to re-label longitudes in degrees west. This has been done for all user-oriented data in this book.

2) Eq 11.22 should only be applied to a single orbit. At the end of each orbit, the best available longitude increment should be used to compute a new longitude of ascending node. Eq 11.22 can then be re-applied.

3) Eq 11.21 and Eq 11.22 can be solved at any time, t, if i, λ_o and T are known. In other words, it takes four parameters to specify the location of the SSP for a circular orbit. The four we've used are known as the "classical orbital elements." They were chosen because each has a clear physical meaning. There are several other sets of orbital elements that may also be employed.[6]

4) If you have a microcomputer or programmable hand calculator, you can use Eqs 11.21 and 11.22 to run your own predictions, either to follow a particular satellite pass or to produce data for an OSCARLOCATOR ground-track overlay. The flow chart of Fig 11-20 outlines one simple approach. All sorts of refinements can be added, but it's best to get the basic program running first. You might, for example, input the time increment instead of using a fixed value of two minutes. Or you might add a time-delay to the loop to produce a real-time display. The tables in Appendix C, for circular orbit spacecraft, were produced using an algorithm based on the flow chart of Fig 11-12 by inputting $\lambda_o = 0°$. It's a good idea to use these tables to check any program you write. A simple program following the logic of Fig 11-20 has been included in Appendix D to illustrate how Eqs 11.21 and 11.22 are applied.

Elliptical Orbits: Derivation

Now that we've seen how the ground-track equations for a circular orbit are derived, we go on to look at the additional parameters and steps required for elliptical orbits.

Once again, a clear diagram is essential. In Fig 11-21 we've chosen an inclination between 90° and 180°, a satellite perigee in the northern hemisphere, and the spacecraft headed north in the northern hemisphere. A diagram of

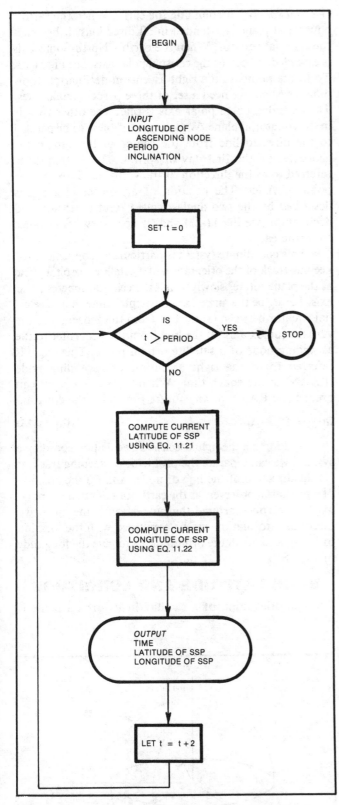

Fig 11-20—Flow chart for circular-orbit ground-track program. See Appendix D for program in BASIC.

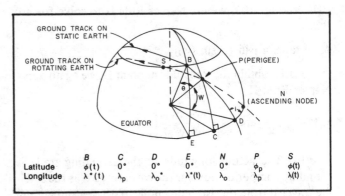

	B	C	D	E	N	P	S
Latitude	$\phi(t)$	0°	0°	0°	0°	ϕ_p	$\phi(t)$
Longitude	$\lambda^*(t)$	λ_p	λ_o^*	$\lambda^*(t)$	λ_o	λ_p	$\lambda(t)$

Fig 11-21—Illustration for applying the principles of spherical trigonometry to the elliptical-orbit ground-track problem.

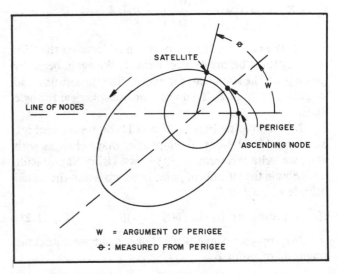

W = ARGUMENT OF PERIGEE

θ : MEASURED FROM PERIGEE

Fig 11-22—Satellite position in orbital plane.

the orbital plane (Fig 11-22) is also very helpful.

We assume that the following parameters are known: T (period in minutes), i (inclination in degrees), λ_p (SSP longitude at perigee), w (argument of perigee) and e (eccentricity). Our object is to solve for the latitude and longitude of the SSP—$\phi(t)$ and $\lambda(t)$—at any time t. We will measure t from perigee.

The actual ground track is not a great circle, so our strategy will again be to focus first on a static-earth model where the principles of spherical trigonometry can be applied. The results will then be adjusted to take into account the rotation of the earth. In Fig 11-19, which was drawn for circular orbits, we elected to let the static-earth ground track coincide with the true ground track at the ascending node. In Fig 11-21 we've chosen to let the two ground tracks coincide at perigee.

Step 1. Our object here is to relate our perigee-based parameters to the ascending node. More specifically, we wish to calculate (a) elapsed time as the satellite moves from D to P, (b) the latitude at perigee and (c) the longitude at the ascending node.

1a) Consider the static-earth model and focus on spherical triangle CPD. From Fig 11-22 we see that arc PD is, by definition, equal to the argument of perigee, w. Using Kepler's Equation (Eqs 11.10 and 11.11), we can plug the value of w in for θ and calculate the elapsed time between perigee and the ascending node, which we call t_p.

1b) The latitude at perigee is, by definition, the length of arc PC. Angle PDC is equal to 180° − i. Knowing angle

PDC and arc PD, we use Napier Rule II to solve for arc PC.

$$\phi_p = \arcsin [\sin (i) \sin (w)]$$

1c) To obtain the longitude at point D, we again apply Napier Rule II.

$$\lambda_o^* = \lambda_p + \arccos \left[\frac{\cos (w)}{\cos (\phi_p)} \right]$$

1d) The actual longitude at the ascending node is found by computing how far the earth rotated as the satellite traveled from the ascending node to perigee and adding this to the preceding static-earth result. To simplify the following equations we approximate the rotation of the earth by 0.25°/min.

$$\lambda_o = \lambda_p + \arccos \left[\frac{\cos (w)}{\cos (\phi_p)} \right] + |t_p|/4$$

Step 2. We now turn to the problem of locating the SSP at S, any time, before or after perigee. We again begin by focusing on the static-earth model to find the latitude and longitude of point B. To do this we use spherical triangle BDE.

2a) Comparing Figs 11-21 and 11-22, we see that arc BD is equal to $(\theta + w)$. To emphasize that θ changes with time, we write this term as $(\theta(t) + w)$. Using Napier Rule II we obtain the latitude of point B, which is also the actual latitude of SSP at S.

$$\phi(t) = \arcsin [\sin (i) \sin (\phi(t) + w)] \qquad \text{(Eq 11.23)}$$

2b) Applying Napier Rule II once again we obtain the longitude of point B.

$$\lambda^*(t) = \lambda^*_o - \arccos \left[\frac{\cos (\theta(t) + w)}{\cos (\phi(t))} \right]$$

2c) Finally, correcting for the rotation of the earth we obtain the actual longitude of the SSP at S.

$$\lambda(t) = \lambda_o - \arccos \left[\frac{\cos (\theta(t) + w)}{\cos (\phi(t))} \right] - t/4 - t_p/4 \qquad \text{(Eq 11.24)}$$

Eq 11.24 only gives the correct results when the spacecraft is in the northern hemisphere, $\theta > w$, and both are less than 90°, as shown in Fig 11-21. A computer program which takes into account all configurations of θ and w is included in Appendix D.

RIGHT ASCENSION-DECLINATION COORDINATE SYSTEM

We've now discussed all classical orbital elements except for the term *right ascension*. Calculations by astronomers and those working with satellites are best carried out in an inertial coordinate system (one that has fixed directions with respect to the distant stars). The right ascension-declination coordinate system is often used for this purpose. The position of the center of the system isn't important. For convenience we take it to be the geocenter. We now imagine a sphere of infinite radius, called the

celestial sphere, surrounding the earth. When the earth's equatorial plane is extended in all directions, it becomes the celestial equator. When the earth's North-South axis is extended, it become the celestial polar axis. See Fig 11-23. To locate points in the right ascension-declination coordinate system, we need a set of three perpendicular axes. The extended north polar axis is one. The other two lie in the equatorial plane. We take one of these to be parallel to the directed line from the center of the sun to the geocenter on the first day of spring. This is frequently referred to as the direction of the vernal equinox or first point of Aries. The position of an object in space is described by the two angles, called right ascension and declination (see Fig 11-23) and, if necessary, its distance from the earth.

This coordinate system is particularly convenient for keeping track of the orientation of a satellite's orbital plane as the plane rotates slowly about the celestial spheres polar axis. Let Ω_o be the angle (at a particular time) in the celestial equator between (1) the line from the geocenter to the vernal equinox and (2) the line from the geocenter to the ascending node of a satellite orbital plane. This angle is referred to as the right ascension of ascending node (RAAN) at the epoch time. When Ω_o is known, we can predict the RAAN at any future time with the equation

$$\Omega(t) = \Omega_o + \dot{\Omega}t \qquad \text{(Eq 11.25)}$$

By adopting the right ascension-declination coordinate system, we can separate the problems of keeping track of (1) the direction of the line of nodes and (2) the position of a terrestrial observer, as the earth rotates about its polar axis and revolves around the sun. At any time these two pieces of information can be combined with the location of the spacecraft in the orbital plane to give the longitude of the SSP.

BAHN LATITUDE AND LONGITUDE

The orientation of a satellite in its orbital plane is

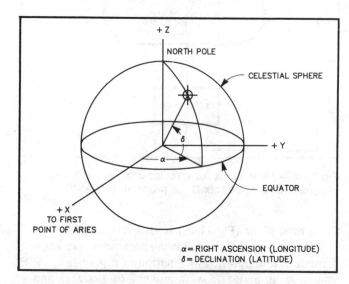

Fig 11-23—Celestial sphere and Right Ascension-Declination coordinate system.

specified by two angles, *Bahn latitude* and *Bahn longitude*. Bahn latitude is the angle between the spacecraft +Z axis and the orbital plane. Suppose we rotate the spacecraft through this angle so that the Z axis is now in the orbital plane (Fig 11-24) and then rotate the spacecraft a second time so that its +Z axis is aligned with the directed line running from the geocenter to the spacecraft at apogee. This second rotation angle is the Bahn longitude.

To understand why the terms "latitude" and "longitude" are used, refer to Fig 11-23 and note the connection to the ascension and declination coordinates. Declination, rotation of a line into a plane, is essentially a latitude. Ascension, rotation in a plane, is essentially a longitude.

ORBITAL ELEMENTS

The parameters used to describe the position and motion of a satellite, rocket, planet or other heavenly body are called orbital elements. There's a lot of flexibility in selecting particular parameters to serve as orbital elements. The choice depends on the characteristics of the problem being examined, the information available and the coordinate system being used. If we focus our attention on satellites that do not contain propulsion systems and are not affected by atmospheric drag, we find that it requires a set of six parameters specified at a particular time (called the epoch time) to specify the current spacecraft location and to accurately predict its future positions (the basic satellite tracking problem).

From a computational viewpoint, a desirable set of six elements consists of three position coordinates and three velocity components expressed in an inertial Cartesian coordinate system. One shortcoming of this set of elements is that it doesn't provide any immediate clues as to what the orbit looks like. In contrast, classical orbital elements, which involve parameters like eccentricity, inclination, argument of perigee, mean motion (or semi-major axis), RAAN and time since last perigee are extremely helpful for visualizing an orbit.

The orbital elements used by amateurs in tracking problems (see Chapter 6) are almost always a variation of the classical set. In many instances, more than six parameters are provided by the source (or requested by the program) at a particular epoch time. One reason that many programs require more than six elements is that they do more than solve the basic problem. For example, they may take drag into account or keep track of the spacecraft orientation in the orbital plane so that squint angle can be determined. Some element distributors provide redundant data to allow for the fact that one tracking program may, for example, request mean motion as an input while another might request semimajor axis. In any event, a set of orbital elements must include at least six parameters specified at a particular time.

THE OBLATE EARTH

As a first approximation, it is reasonable to treat the earth as a sphere having a mean radius of 6371 km. A model of satellite motion based on a spherical earth predicts that the orientation (right ascension and inclination) of a satellite's orbital plane will remain fixed in space and that the position of a satellite's perigee in the orbital plane (argument of perigee) will not change. Such a model might be acceptable for a single orbit, but it is not adequate for long-term predictions.

In order to obtain equations for important factors such as rate of change of perigee (Eq 11.13) and precession of the line of nodes (Eq 11.16), one has to adopt a more complex model. The next step in complexity is to use an ellipsoidal earth (an ellipse rotated about the polar axis). Higher-order models are possible, but it turns out that an oblate earth model based on the ellipse gives very good results. The semimajor axis of the ellipse is the equatorial radius of the earth: 6378 km; the semiminor axis is the polar radius: 6357 km; and the earth's eccentricity is 0.08182. (For more precise numbers see Appendix E.)

An oblate earth model enters our calculations two ways. First, it affects the motion of the satellite in space. We have taken these effects into account by our use of Eqs 11.13 and 11.16. Second, it affects calculations that involve the position of a ground station on the surface of the earth, such as antenna aiming parameters and range, which we'll be considering in the next chapter. If one compares antenna pointing predictions (azimuth and elevation) using a spherical earth model to those using an oblate earth model, the differences turn out to be small fractions of a degree, an amount that will never be visible with any amateur antenna. Using an oblate earth model for these calculations does lead to complications, such as a need to

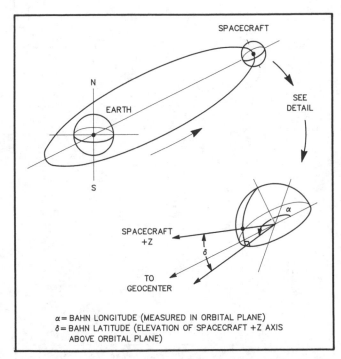

α = BAHN LONGITUDE (MEASURED IN ORBITAL PLANE)
δ = BAHN LATITUDE (ELEVATION OF SPACECRAFT +Z AXIS ABOVE ORBITAL PLANE)

Fig 11-24—Geometry used to define Bahn Latitude and Longitude, the angles which specify the orientation of a Phase 3 spacecraft. If Bahn Longitude >180°, then OSCAR 13 is earth pointing *after* apogee. If Bahn Longitude <180°, then OSCAR 13 is earth pointing *before* apogee.

distinguish between geocentric latitude, geodetic latitude and astronomical latitude.

A reasonable approach to designing tracking programs for amateurs would therefore be to incorporate the oblate earth by using Eqs 11.13 and 11.16, but to treat the earth as a sphere for calculations involving range and antenna aiming. This would suffice for 99.9% of amateur needs. However, for the critical orbit determination process leading to the rocket burns used to boost Phase 3 satellites from a transfer orbit to an operating orbit, it's desirable to use the most accurate model possible. Since key early tracking programs were specifically developed for the orbit transfer process, they incorporated an oblate earth model at all steps. This refinement is included in most programs being distributed today. For a discussion of the mathematics involved, see references 3 and 4.

SPECIAL ORBITS

With at least four parameters to vary—eccentricity, inclination, semimajor axis and argument of perigee—there are many possible ways of classifying orbits. We've already paid considerable attention to low earth circular orbits with inclinations near 90°. We've also discussed the special characteristics of sun-synchronous orbits. In this section, we briefly look at two additional types of orbits of special interest to radio amateurs.

The Geostationary Orbit

A satellite launched into an orbit with an inclination of zero degrees will always remain directly above the equator. If such a satellite is in a circular orbit (constant velocity), traveling west to east, at a carefully selected height (35,800 km), its angular velocity will equal that of the earth about its axis (period = 24 hours). As a result, to an observer on the surface of the earth the spacecraft will appear to be hanging motionless in the sky. Satellites in such orbits are called *geostationary* (or stationary for short).

The geostationary orbit has a number of features that make it nearly ideal for a communications satellite. Of prime importance, Doppler shift on the radio links is nonexistent, and ground stations can forget about orbit calendars and tracking. These features have not gone unnoticed—so many commercial spacecraft are spaced along the geostationary arc above the equator that a severe "parking" problem exists. From an Amateur Radio point of view, a geostationary satellite is not without problems. The biggest shortcoming is that a single spacecraft can only serve slightly less than half the earth. It's sometimes stated that a geostationary satellite provides poor east-west communications coverage to radio amateurs at medium to high latitudes. This may be true when Molniya-type orbits (see next section) are the standard of comparison, but take a good look at the map shown in Fig 11-25 before adopting an opinion.

Amateurs are currently engaged in the Phase 4 project (see Chapter 4), which involves the construction of geostationary spacecraft. The goal is to have the first satellite in orbit by the late 1990s.

If the orbital inclination of a satellite is not zero, the spacecraft cannot appear stationary; stationary satellites can only be located above the equator. A 24-hour-period circular orbit of non-zero inclination will have a ground track like a symmetrical figure eight (see Fig 11-26). Note

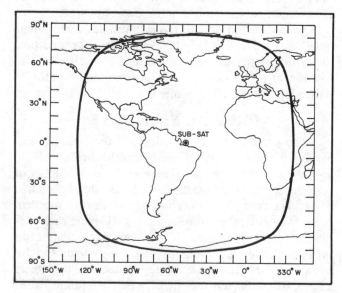

Fig 11-25—Coverage provided by a geostationary satellite at 0°N, 47°W. The access region corresponds to 0° elevation angle at ground station.

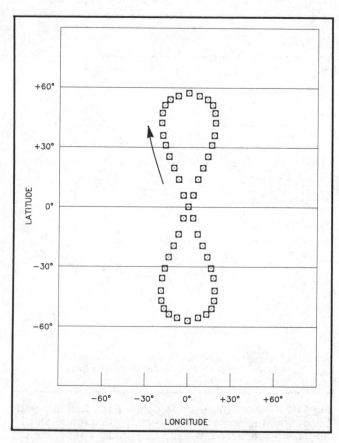

Fig 11-26—Ground track for satellite in circular orbit with a period of 24 hours and an inclination of 57°. Position shown at 0.5-hour intervals.

that the ascending and descending nodes of such an orbit coincide and the longitude of ascending node is constant (the increment is nearly zero). The 24-hour circular path is known as a *synchronous orbit*. The geostationary orbit is a special type of synchronous orbit, one with a zero-degree inclination.

Note that some authors apply the term synchronous (or geosynchronous) to other types of orbits, ones that are circular or elliptical with periods that are an exact divisor of 24 hours, such as 8 hours or 12 hours. Because this might lead to unnecessary confusion, we'll avoid this use of the term synchronous.

Molniya-Type Orbits

Looking at elliptical orbits earlier in this chapter, we noted that the position of the perigee in the orbital plane (the argument of perigee) changes from day-to-day at a rate given by Eq 11.13. An interesting feature of this equation is that when $i = 63.4°$, the argument of perigee remains constant regardless of the values of the period and eccentricity. As a consequence, the argument of perigee, period and eccentricity can be chosen independently to satisfy other mission requirements.

Orbits with $i = 63.4°$, eccentricities in the 0.6 to 0.7 range and periods of 8 to 12 hours have a number of features that make them attractive for communications satellites. Spacecraft in the Russian Molniya series were designed to take advantage of this type of orbit.

Let's take a brief look at a Molniya II series communications satellite of the type used for the Moscow-Washington Hotline. (The Hotline uses redundant Molniya and Intelsat links.) The spacecraft is maintained in an orbit with an inclination of 63.4°, an argument of perigee constant at 270° and a period of 12 hours. Because of the 12-hour period, the ground track tends to repeat on a daily basis (there is a slow drift in longitude of ascending node). Apogee, where the satellite moves slowly, always occurs over 63.4° N latitude. At apogee nearly half the earth, most in the northern hemisphere, is in view. A typical Molniya-orbit ground track is shown in Fig 11-27. The primary Washington-Moscow mutual visibility window lasts 8 to 9 hours. Fig 11-28 represents the orbit geometry for two consecutive orbits, with apogees occurring 180° apart in longitude. In reality, the orbit remains fixed as the earth rotates, so don't take the illustration too literally. Notice how on the Eurasian circuit there's a second good four-to five-hour Washington-Moscow window. A single spacecraft is accessible to the Washington station about 16 hours per day, to the Moscow station about 18 hours per day, and simultaneously to both stations about 12 hours each day. Thus, a three-satellite Molniya system is used to provide a reliable Washington-Moscow link 24 hours a day.

Interestingly, the Russians also use Molniya spacecraft for relaying domestic TV signals in the 4-GHz range, and these satellites are active on the North American apogee as well as on the Eurasian apogee. Since relatively wide-beamwidth antennas are used and the format (except for color) is compatible with the US system, anyone with a

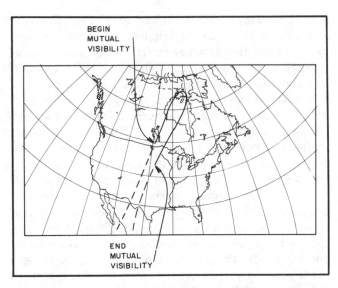

Fig 11-27—Typical Molniya II ground track with apogee over North America. The mutual visibility window is for the Washington, DC-to-Moscow path.

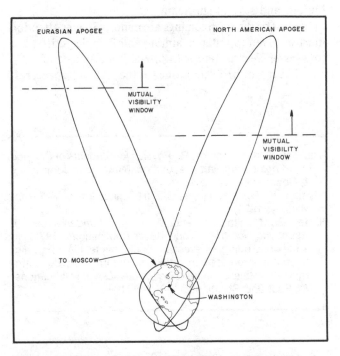

Fig 11-28—The relative positions of two successive Molniya II apogees. In actuality, the orbit plane remains fixed in space while the earth rotates. The mutual visibility windows shown are for the Washington, DC-to-Moscow path.

TVRO (4-GHz satellite TV receive only terminal) can monitor these transmissions by pointing his antenna northward (see Chapter 14 for additional information).

From an Amateur Radio point of view, the Molniya-type orbit has a number of attractive features. The orbits selected for Phase 3 spacecraft have been variations on the Molniya theme.

There has been a great deal of discussion as to the specific orbit desired. Some amateurs preferred an orbit with $i = 63.4°$. With this inclination, the overlays on graphic tracking devices never have to be changed and the apogee can be set to continuously favor the northern hemisphere. Other amateurs preferred a different value of inclination, since a changing argument of perigee eventually gives one access to a considerably larger portion of the world. Thinking in terms of a long-term Phase 3 system, such a satellite would begin to favor the southern hemisphere six or seven years after launch as the apogee drifted south of the equator. A new spacecraft would then be launched to take over in the northern hemisphere.

Meanwhile, the spacecraft design engineers investigated the tradeoffs involving the thrust required to reach a high inclination orbit and the sun angles the spacecraft would encounter during the year. The sun-angle numbers have a big impact on the transmitter power available on the spacecraft. The final decisions for OSCAR 13 were made nearly a year before the spacecraft was launched. Transfer maneuvers after launch placed the spacecraft precisely in the targeted orbit.

When compared to a geostationary orbit, the Molniya-type orbit has several advantages, which we've been focusing on, and several shortcomings.

Most of the shortcomings are minor. Greater attention must be given to antenna aiming and Doppler shifts, but to a lesser degree than with low-altitude spacecraft.

The major problem is one for the AMSAT spacecraft engineers. A satellite in a Molniya orbit traverses the Van Allen radiation belts twice each orbit, subjecting many of the onboard electronic subsystems, especially those associated with the central computer, to damage from the high-energy particles that may be encountered. Extensive shielding of the computer chip is necessary, but this shielding increases the weight, restricting access to desirable orbits.

The tradeoff involved here is so important that AMSAT undertook a special research program to look into the effects of radiation on the RCA CMOS integrated circuits used on OSCAR 10. Chips of the type which were to be flown were exposed to radiation under conditions that simulated the anticipated space environment. These failure-rate studies, performed at Argonne and Brookhaven National Laboratories, used various amounts and types of shielding. The results provided the data used to design OSCAR 10. Using every option possible, the engineers were able to project a three- to five-year lifespan for the spacecraft computer. The projections were remarkably accurate. When OSCAR 13 was built, integrated circuits having a much higher resistance to radiation damage were available. With respect to radiation damage, OSCAR 13 has a projected lifetime of nearly a century. Other spacecraft systems will, of course, considerably reduce this value.

Notes

[1]Halliday, D. and Resnick, R., *Physics for Students of Science and Engineering Part I* second ed. (New York: John Wiley & Sons, 1962) Chap 16.

[2]Symon, K. R., *Mechanics*, 3/E (Reading, Mass: Addison-Wesley, 1971).

[3]Bate, R., Mueller, D. and White, J., *Fundamentals of Astrodynamics* (New York: Dover Publications, 1971). In addition to being an excellent book, this text is a bargain. If you're interested in additional information on astrodynamics, this is the first book to buy. Dover Publications, 31 East 2nd St, Mineola, NY 11501.

[4]Escobal, P., *Methods of Orbit Determination*. New York: John Wiley & Sons, 1976. This text is also an excellent introduction to astrodynamics. The price is typical for technical books at this level, about three times that of Reference 3.

[5]Kork, J., "Satellite lifetimes in elliptic orbits," *J Aerospace Science*, Vol 29, 1962, pp 1273-1290.

[6]Corliss, W. R., *Scientific Satellites* (NASA SP-133), National Aeronautics and Space Administration, Washington, DC, 1967, p 104.

Chapter 12

Tracking Topics

To communicate through an Amateur Radio satellite, you must know where it is, whether (or when) it's accessible from your location and, if you're using directional antennas, what arc it will trace in the sky above. For a geostationary satellite, you'll need to know where to point and fix your antennas; for other types of spacecraft, you'll want to know where to aim your antennas at various times. This task, tracking, is met and solved by every satellite user.

This chapter is divided into two major sections focusing on tracking. The first section builds on the information related to graphic tracking methods provided in Chapter 6. Our main objective in Chapter 6 was to describe a single general-purpose graphic method that worked. Our aim here is to look at several different graphic approaches to tracking, including variations on the OSCARLOCATOR and modifications that will permit its use with satellites in elliptical orbits. With this information, we will be able to determine which approach is most appropriate to a given situation. The methods which will be discussed, most of them developed by radio amateurs, constitute an important contribution to the technology of low-cost satellite tracking. Readers involved in developing graphic output screens for computer tracking programs will discover that these "old fashioned" methods provide many stimulating ideas.

The second part of this chapter focuses on several computational topics essential to tracking that have not yet been covered. These include determining satellite coverage, predicting azimuth and elevation angles from your ground station to a satellite at any time, and obtaining data to produce a range-circle "spiderweb" around any point on the earth.

INTRODUCTION

Every tracking method presents selected pieces of information in a format that's designed to be convenient to a particular group of users. One reason that radio amateurs have developed and tested a great many approaches to satellite tracking is that there are several groups of amateurs, each with different interests and access to different resources (maps, computers, and so on). The amateur satellite community can be partitioned in several ways: amateurs interested in communications vs those

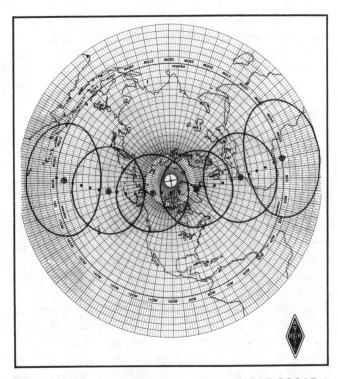

Fig 12-1—Access circles drawn about AMSAT-OSCAR 8 every 10 minutes beginning at ascending node at 0° W longitude.

interested in spacecraft system design and management, amateurs with microcomputers and those without, amateurs at high latitudes vs those near the equator, amateurs interested in low earth orbit spacecraft and those interested in Phase-3 satellites, and so on. Each group has a need for tracking methods with different features.

To illustrate the contrasting needs of different groups, consider tracking approaches for a low-altitude, near-polar, circular-orbit satellite such as OSCAR 8. Both the system design/management and general-user viewpoints will be taken. The general user, following the suggestions in Chapter 6, might choose an OSCARLOCATOR and draw a spiderweb around his location. The spacecraft perspective, however, is much more useful to the design/management group. They might, for example, modify the OSCARLOCATOR (see Fig 12-1) so that they can see what

A satellite tracking method should enable one to predict:

1) when the satellite will be in range (times for AOS and LOS);
2) proper antenna azimuth and elevation at any time; and
3) the region of the earth that has access to the spacecraft at any instant.

The tracking aid should, in addition, be simple to construct or program, easy to use and inexpensive.

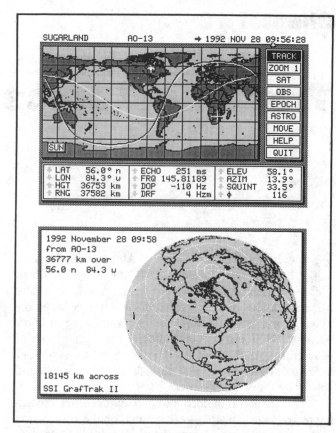

Fig 12-2—Two graphics screens available in the powerful tracking program GrafTrak II (Silicon Solutions, Inc, PO Box 742546, Houston, TX 77274) illustrate the coverage that OSCAR 13 will provide when apogee latitude nears its maximum value. The rectangular view includes the position of the sun (over South Africa) and the terminator (day/night line).

kind of coverage OSCAR 8 provides. Note that Fig 12-1 shows access circles drawn about the spacecraft at fixed time intervals. Alternatively, by drawing a set of access circles at selected latitudes on the OSCARLOCATOR, a system designer could use it to compute the average daily access time as a function of ground station latitude. The averages can be figured quite accurately by setting the ground-track overlay for ascending nodes at longitude increments of 10°, noting the access times and tabulating the results. Of course, the same information could be obtained using a computer, but the map-based approaches are often considerably quicker.

Since readers of this handbook are most likely in the user group, we'll focus on this viewpoint. User requirements for a tracking device were specified in Chapter 6, and are restated in Table 12-1.

Earlier, we found it convenient to categorize tracking approaches by focusing on whether they were computer-based or map-based. Although this appears to be a clear and distinct partition of approaches, it really isn't. Map-based methods have their roots in a set of calculations, performed one time, using the formulas in Chapter 11 (ground track) and Chapter 12 (azimuth, elevation, spiderweb). The results of these calculations are then used to construct ground-track overlays and spiderwebs that are keyed to a particular map. Most computer methods are based on the same set of formulas, but the calculations are repeated for each orbit and the results are presented in tabular or graphic form, or possibly even used to drive an antenna automatically. Depending on the power, flexibility and human engineering that has gone into programming, microcomputer methods may or may not be more convenient than map-based approaches.

Each tracking method attempts to present data in a format that will be most useful to a particular set of users. The problem of data presentation is not confined to tracking, Amateur Radio or even science in general. It's one that occurs wherever a large amount of numerical material is handled. As computers make number crunching easier and quicker, developing methods for presenting results in a meaningful format becomes more and more important. Recognizing that a good picture is often worth considerably more than a long list of numbers, amateurs have recently begun to develop excellent graphic displays for

tracking software. Consider, for example, Fig 12-2, which is taken from the GrafTrak program. Not only does it tell us where the spacecraft is, it also tells us where it was, where it's going, the position of the solar terminator, the current values of several key orbital and operational parameters, and the direction of change for these parameters.

Anyone interested in computer map displays for tracking should read the informative articles by W. Johnston: "Computer Generated Maps," Part I, *Byte,* May 1979, pp 10-12, 76, 78, 80, 82-84, 86, 88, 90, 92-94, 96, 98, 100-101; Part II, *Byte,* June 1979, pp 100, 102, 104, 106, 108, 110, 112, 114, 118-119, 122-123.

MAP-BASED METHODS

Map-based tracking methods such as the OSCAR-LOCATOR have passed the test of time. Their popularity endures because they do a good job of satisfying user requirements.

Every map-based approach shows (1) a ground track and (2) a spiderweb, usually drawn about a ground station, but sometimes shown in reference to the satellite position. The popularity of a particular approach depends on how easy it is to construct, to reposition the ground track(s) and spiderweb(s) when necessary, and to use. The type of map employed turns out to be a very important parameter, so we need some knowledge of maps. All two-dimensional maps distort the globe. Most map projections are designed to minimize particular distortions such as area, distance or bearing on at least certain portions of the map. We'll discuss the important characteristics of each map as we look at it.

On certain types of maps (polar and rectangular), the shape of the ground track for circular orbits does not change. With these projections, it is possible to draw a permanent ground track on a transparent overlay that can then be repositioned for each pass. On a polar map, repositioning means rotating the overlay about the pole; on a rectangular-coordinate map, it involves shifting the overlay horizontally along the equator.

Polar Projection Maps

Polar maps have proved more popular than rectangular maps among amateurs working with graphic tracking devices for several reasons: Their ground-track overlays are easier to reposition; mid-latitude ground stations can approximate spiderwebs with circles and incur only a minor penalty in accuracy; over-the-pole paths are clearly visible; and there is a quick and simple way to sketch ground tracks, which we will outline shortly.

Polar projection maps, centered about either the north or south poles, are readily available (see Table 6-7). On these maps, latitude curves are represented by a set of concentric circles centered on the pole, and longitude curves (meridians) by lines that radiate straight out from the pole. The various projections differ primarily in the spacing between latitude circles.

The three most common polar map projections are the equidistant, the stereographic and the orthographic. The equidistant is designed to show true distances from the pole, the stereographic is designed so that all circles on the surface of the globe will be shown as circles, and the orthographic shows what the earth would actually look like from a particular height above the pole. The first two projections are excellent for constructing OSCAR-LOCATORs. We'll look at them in detail shortly. Because the orthographic projection severely compresses geographic features near the equator and does not show any of the opposite hemisphere, it is generally regarded as poorly suited to tracking. However, if you were to plan an Arctic expedition, you might find the characteristics of the orthographic polar projection useful.

As alluded to earlier, rough ground tracks can be sketched on polar maps using a shortcut that bypasses the calculations of Chapter 11. The shortcut, which is suitable only for low-altitude circular orbits, has proven useful in many instances where it was necessary to get a quick fix on a new spacecraft. Assume that a northern hemisphere polar projection map is being used and that you have a rough estimate of the satellite's period, T (in minutes), and orbit inclination, i (in degrees). If an *ascending node* occurs at latitude $\phi_{an} = 0°N$, longitude $\lambda_{an} = 0°W$, then a *descending node* will occur T/2 minutes later at $\phi_{dn} =$

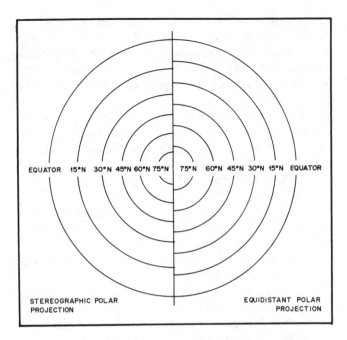

Fig 12-3—A comparison of latitude-circle spacing on stereographic and equidistant polar-projection maps.

$0°N$, $\lambda_{dn} = 180° + (T/8)°W$. Midway between these two points, the spacecraft will be at its *northernmost point*: $\phi_{np} = i°N$, $\lambda_{np} = 270°W + (T/16)°W$ when i is between $0°$ and $90°$, or $\phi_{np} = (180-i)°N$, $\lambda_{np} = 90°W + (T/16)°W$ when i is between $90°$ and $180°$. A roughly sketched curve joining just these three points gives a surprisingly good picture of the ground track of this particular orbit. Once this curve is transferred to a rotatable overlay, you're all set to track the satellite when it's over the northern hemisphere.

EQUIDISTANT POLAR PROJECTION

An equidistant polar projection map is characterized by equal spacing between the concentric latitude circles. Fig 12-3 illustrates the difference between equidistant and stereographic projections. The difficulty in using this type of map is that accurate spiderwebs are tedious to draw, since range circles about a specific location are distorted. Despite this drawback, equidistant polar maps have been very popular. We've standardized on this projection for the master ground-track overlays and spiderwebs in Appendix C. Geographic coverage is out to $30°S$ latitude.

STEREOGRAPHIC POLAR PROJECTION

Stereographic polar projection maps are characterized by increased spacing between latitude lines as one gets farther from the pole (Fig 12-3). The formula for latitude line position is

$$S = S_o\tan[(90° - \phi)/2]$$

where

S = distance between pole and latitude line

ϕ = latitude

S_o = distance between pole and equator (arbitrarily chosen to adjust overall size of map).

The stereographic projection has the interesting characteristic of preserving circles: A circle on the globe is also a circle on the map. Therefore, drawing range circles, acquisition circles or elevation circles about a particular location, or locating a mutual window for two ground stations, is relatively easy.

To draw a range circle about a particular ground station (latitude ϕ_g, longitude λ_g), note that the center of the circle (ϕ_o, λ_o) does *not* coincide with the ground station; both do lie along the same meridian ($\lambda_o = \lambda_g$), however.

To find the latitude of the center of the circle (ϕ_o) and the radius of the circle:

1) Transform the range you're interested in into degrees of arc along the surface of the earth using $1.000°$ of arc = 111.2 km. Call the result $\Delta\phi$.

2) Compute $\phi_g + \Delta\phi$ and $\phi_g - \Delta\phi$ and plot both these points along meridian λ_g.

3) Bisect the line joining the two points found in Step 2. This gives the center of the circle.

4) With the center of the circle (Step 3) and two points on the circumference (Step 2), sketch the circle using a drawing compass.

The main drawback of the stereographic projection is that the mid-northern latitudes become significantly compressed if one draws a map extending out to latitude $30°$ S.

W2GFF Plotter

One variation of the OSCARLOCATOR, developed by R. Peacock, has gained widespread acceptance. It provides users of circular-orbit spacecraft the convenience of a real-time read-out. The key elements of the W2GFF Plotter© are shown schematically in Fig 12-4. An equidistant polar map is attached to a mounting board at the pole so that it's free to rotate. A ground-track overlay, drawn on transparent stock, is permanently affixed to the mounting board above the map. An adjustable real-time

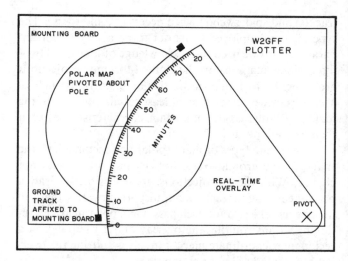

Fig 12-4—Key elements of W2GFF Plotter.

scale, also drawn on transparent stock, is placed on top of the other components and pivoted so it can be set at the ascending node. To preview a particular orbit, one uses an orbit calendar to set first the polar map for the correct longitude of ascending node and then the real-time overlay for time at ascending node. The position of the spacecraft can then be followed in real time without the burden of mental arithmetic. Though it isn't shown in Fig 12-4, the polar map of the W2GFF Plotter includes a spiderweb to provide azimuth, elevation, AOS and LOS data. An equidistant projection must be used so that the equally spaced time ticks on the real-time overlay will provide accurate SSP position data.

Rectangular Coordinate Maps

Rectangular coordinate maps (Mercator, Miller Cylindrical, and so on) have also been used for tracking by radio amateurs. A ground-track overlay for AMSAT-OSCAR 7

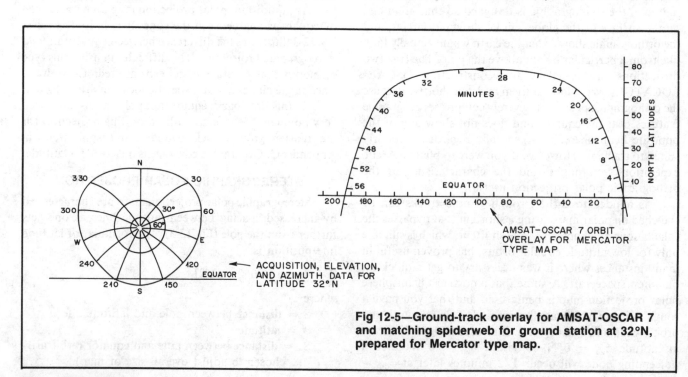

Fig 12-5—Ground-track overlay for AMSAT-OSCAR 7 and matching spiderweb for ground station at 32°N, prepared for Mercator type map.

and a matching spiderweb as they would appear on a Mercator map are shown in Fig 12-5. Note the severe distortion of the spiderweb. Users who have tried trackers of this type and the polar-map models for low-altitude, near-polar, circular-type orbits almost universally prefer the polar-map trackers. With high-altitude, high-inclination satellites, the preference for polar maps will likely be even more emphatic, since over-the-pole communications paths are certain to be of special interest, and rectangular coordinate maps are poorly suited for analyzing what happens beyond the pole.

All this negative publicity for Mercator-type maps may tempt you to dump your cache into the nearest wastepaper basket. Don't do it. Rectangular coordinate maps may be very useful to mid-latitude ground stations, for satellites in low-inclination orbits, and to ground stations near the equator, for a variety of orbits.

Equidistant Projection
(Ground-Station Centered)

Earlier, we discussed equidistant projections that are centered on the pole. Maps of this type can also be drawn using any point on the earth as the center. On such a map, azimuth curves radiating from the center will be straight lines and range curves about the center will be true circles. Azimuthal equidistant projection maps are available centered on many large cities and a computer-generated map made to order for your particular location can be obtained at modest cost (see Table 12-2).

The following approach is suitable only for low-altitude circular orbits. Although it hasn't received the publicity of the OSCARLOCATOR, a good percentage of the people who have tried both prefer the method based on the ground-station-centered equidistant projection. One of the most tedious parts of building a map-based tracker is plotting the spiderweb. When using an equidistant projection centered on your location, spiderweb construction is trivial: Azimuth lines are already in place and range circles corresponding to particular distances or elevation angles are easily added using a drawing compass.

The ground track situation is more involved. Unlike all the other map-based methods so far considered, the shape of the ground track depends on the longitude of the ascending node. As a result, we *cannot* draw a single ground-track overlay to reposition for each pass. Instead, we draw representative ground tracks for ascending nodes that enter our acquisition circle (see Fig 12-6). Latitude lines may be labeled with the time to nearest ascending or descending node. As the last step, the map is covered with a sheet of clear plastic. To preview any particular orbit, one uses "eyeball interpolation" with respect to the representative orbits previously drawn to locate the ground track of the orbit in question. Sketching in the orbit of interest with a felt-tipped marker, and erasing it with a tissue when no longer needed, works well. For additional information, see K. Nose, "Making Your Own Satellite Tracking Nomogram," *QST*, Mar 1974, pp 40-41, 78.

Table 12-2

Sources for Equidistant Projection Maps

Source: Defense Mapping Agency Hydrographic Center, Washington, DC 20390.

Central city and identification information:

Fairbanks, Alaska	WOXZP5180
Seattle, Washington	WOXZP5181
Honolulu, Oahu, Hawaii	WOXZP5182
San Francisco, California	WOXZP5184
Washington, DC	WOXZP5185
San Diego, California	WOXZP5190
Balboa, Panama	WOXZP5192
Yosami, Japan	WOXZP5193

Source: National Ocean Survey, NOAA, Chart Distribution Division-C44, Rockville, MD 20852. This agency also handles USAF Aeronautical Charts and Publications.

Central city and identification information:
New York City NOS 3042 (This map also contains two small, 6″ diameter maps centered on London and Tokyo)

Source: Rand McNally & Co, PO Box 7600, Chicago, IL 60680.

Central City:
Wichita, Kansas

Source: William D. Johnston, N5KR, PO Box 370, Las Cruces, NM 88002.

Central city:
Anyplace: Johnston can provide a custom computer-generated map centered about any coordinates you desire.

1990 prices for most are under $10, except for the custom computer-generated map, which is under $15.

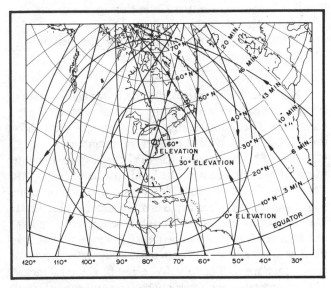

Fig 12-6—Azimuthal equidistant projection map centered on Washington, DC used as orbit calculator for AMSAT OSCAR 7.

P3 Tracker

As mentioned earlier, the OSCARLOCATOR can be modified to handle satellites in elliptical orbits. The resulting device, which we'll call a P3 Tracker, is most useful for orbits where the argument of perigee is changing slowly. This makes it very functional for OSCAR 13. The P3 Tracker is designed to take into account (1) the constantly changing height of a satellite in an elliptical orbit and (2) slow changes in the shape of the ground-track overlay. It consists of:

1) Map board. (The same map board can be used for the P3 Tracker and the OSCARLOCATOR.)

2) Ground-track overlay. (This overlay, usually drawn on transparent material, is mounted on the map board so that it can be rotated about the pole. It must be periodically replaced.)

3) Elevation angle table.

The P3 Tracker is used in conjunction with an orbit calendar that provides information needed to set the ground-track overlay. Because the height of a satellite in an elliptical orbit constantly changes, its acquisition distance (access range) is not constant. Drawing acquisition and elevation circles about a particular ground station, as was done with the OSCARLOCATOR, is therefore impossible. In their place, the P3 Tracker uses a set of *range circles* at fixed distances about a given location. The range circles are color coded. The format given in Table 12-3 is suggested. The reason for the color coding will be clear shortly. The ground-track overlay for the P3 Tracker is also color coded as per Table 12-3, but this time the colors signify the minimal access range during each segment of the orbit. During the green section of the orbit, for example, the access range will be *at least* 7000 km. This means that any ground station located within 7000 km of the SSP will have access to the spacecraft. The color code is the *key* to understanding the operation of the P3 Tracker:

The green range circle about your ground station is a rough acquisition circle during the green segment of the orbit, the yellow range circle is a rough acquisition circle during the yellow segment of the orbit, and so on.

Note that we're interested only in matching colors: It's not necessary to memorize the distances associated with each color. The following step-by-step example shows how the P3 Tracker is used. Consider an imaginary satellite, OSCAR P3*, in an elliptical orbit similar to the one that had been planned for the ill-fated AMSAT-OSCAR Phase 3A mission in 1980. Focus your attention on orbit 1607, which is described in the orbit calendar (Table 12-4). The orbit begins at 02:37 UTC (one half period before apogee) and ends at 13:33 (one half period after apogee). Note that the time and longitude entries in the calendar refer to *apogee* rather than ascending node. As a result, the P3 Tracker is most accurate near apogee, when the majority of users are active. Orbit reference numbers treat each orbit as beginning at perigee and ending at the next perigee.

Table 12-3
Color Coding Format for P3 Tracker Range Circles and Orbit Overlay

Color	Map board (range circle radius)	Orbit overlay (minimum communication range)
blue	9000 km	9000 km
green	8000 km	8000 km
yellow	7000 km	7000 km
orange	6000 km	6000 km
red	5000 km	5000 km
brown	4500 km	4500 km
black	4000 km	4000 km
not coded	3000 km	—
not coded	2000 km	—
not coded	1000 km	—

Table 12-4
Orbit Calendar Entry for Imaginary Satellite OSCAR P3*

OSCAR P3*   Period = 656 minutes

increment = 164° West per orbit

inclination = 57°

Orbit reference number	Apogee time (UTC)	Apogee longitude (°West)
1 July 1982 (182) Thursday		
1607	08:05	16°
1608	19:01	180°

THE GROUND-TRACK OVERLAY

If a satellite is in a circular orbit, a single ground-track overlay will work indefinitely. When a satellite is in an elliptical orbit, its ground track changes daily as a result of variations in argument of perigee. The target orbits for Phase 3 spacecraft were selected so that argument of perigee changes are very slow. As a result, a single overlay will serve for several months. With OSCAR 13, changing the overlay once every six months is sufficient. (OSCAR 10 did not reach its desired orbit, so overlays had to be changed about once per month.) A ground-track overlay for P3* is shown in Fig 12-7. Overlays for OSCAR 13 through 1997 are contained in Appendix C.

To view orbit 1607, rotate the ground-track overlay on the map board so the "set apogee" arrow points to the longitude value specified in the calendar: 16°W (see Fig 12-8).

During orbit 1607, AOS for Washington occurs roughly at point A, where the green segment on the overlay crosses the green range circle. The time marks on the ground-track overlay show that this happens about 1 hour and 30 minutes before apogee (at 06:35 UTC). The bearing of point A is read directly from the map board (approxi-

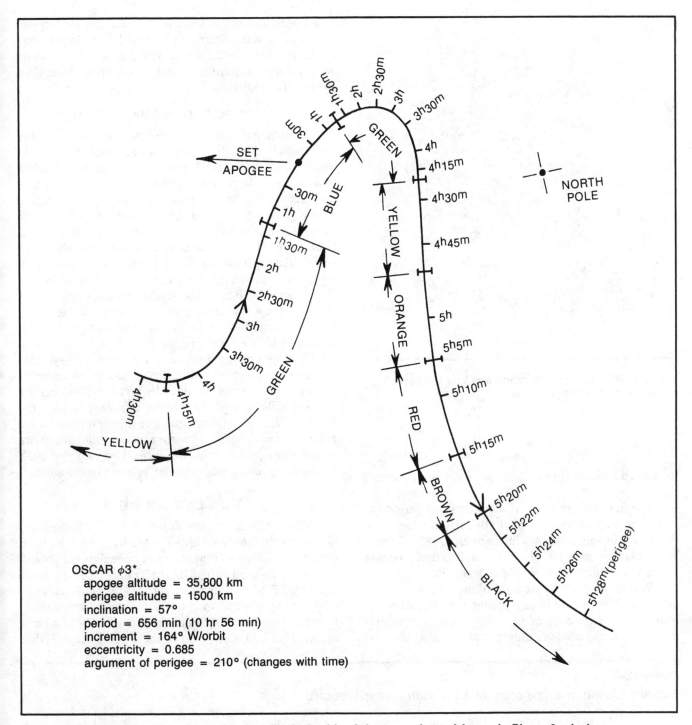

Fig 12-7—Typical ground-track overlay for elliptical orbit of the type planned for early Phase 3 missions.

mately 87°). At AOS, the ground-station antenna should, of course, be just above the horizon.

At 9:05 UTC (apogee plus 1 hour), the satellite will be at position B. The ground-track color is blue and the SSP is far inside the blue range circle, so the satellite is well within range. The bearing of the spacecraft is again read directly from the map board (about 83°). The elevation of the satellite cannot be obtained directly from the map; you must refer to Table 12-5. As the color of the overlay at point B is blue, locate the blue row of Table 12-5. The closest range circle on the map to point B is 5000 km,

so we locate the 5000-km column of Table 12-5. The box where the blue row meets the 5000-km column provides the elevation angle: 38°.

LOS for orbit 1607 will occur at position C, about 4 hours and 45 minutes past apogee (at 12:50 UTC), when the yellow orbit overlay segment crosses the yellow range circle. On this typical orbit, the Washington station has an opening that lasts over six hours. Near apogee, the satellite is simultaneously available to stations in North America (except for Alaska and the West Coast), Central America, Europe, Africa, the Middle East, a large part of Asiatic

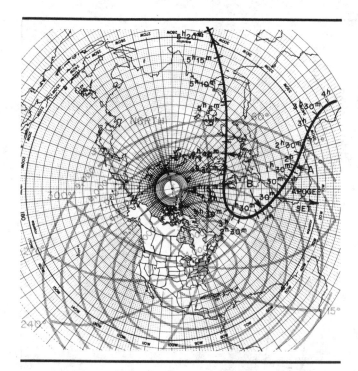

Fig 12-8—P3 Tracker with range and bearing curves drawn about Washington, DC.

acquisition distance at any time will actually lie between the range circle of matching color and the next larger range circle. In other words, a user with a good radio horizon will probably acquire the spacecraft a little earlier than previously predicted.

SIMPLIFIED P3 TRACKER

Most amateurs contemplating the use of a P3 Tracker for OSCAR 13 will find the simplified model discussed in this section completely adequate for their needs. A satellite in a Phase 3 orbit spends about 80% of its life in the blue and green segments of the ground-track overlay. Because of the geometry involved, ground stations will find the satellite in this region more than 90% of the time that it is in range. The great majority of users spend almost all of their time operating during an 8.5-hour window centered on apogee, because problems associated with spin modulation, Doppler shift and rapid satellite motion become more troublesome as the spacecraft approaches perigee.

The Simplified P3 Tracker is designed to provide information for the 8.5-hour apogee window. It's essentially a P3 Tracker stripped down to two range circles—the blue and green ones—with only the blue and green segments of the orbit overlay color coded. This bare-bones P3 Tracker will tell us if the satellite is in range during the 8.5-hour apogee window and provide information on antenna bearing at any time. An amateur using this device would most probably ignore Table 12-5 and just peak received signals by scanning the antenna in elevation.

Fuji-OSCAR 20

The orbit chosen for Fuji-OSCAR 20 is unusual. While most recent low-altitude AMSAT spacecraft have been placed in orbits that are nearly circular, FO-20 was inserted in an orbit with distinct elliptical characteristics. Although FO-20's orbital eccentricity is small (close to 0.05), at low altitudes even this amount produces a significant difference in height between apogee (1745 km) and perigee (912 km). One result of the varying height is that

Russia, and South America (except for the southernmost tip).

The next apogee (orbit 1608) occurs one period (10 hours and 56 minutes) later at 10:01 UTC, at a longitude that is one increment (164°) further west (179°). Given a P3 Tracker and an orbit calendar, you should now be able to track any elliptical-orbit Phase 3 satellite.

We can improve the accuracy of the P3 Tracker if we take a little extra care in interpreting the information it provides. Since the color of an overlay segment represents minimum access distance during that segment, the true

Table 12-5

Satellite Elevation Angle (may be used with any spacecraft)

GROUND TRACK OVERLAY	ELEVATION ANGLE (DEGREES)																	
BLUE	85	79	74	69	64	58	53	48	43	38	33	28	23	19	14	9	5	0
GREEN	84	78	72	67	61	55	50	44	39	34	29	24	19	14	9	5	0	
YELLOW	83	75	68	61	55	48	42	36	30	25	19	14	9	5	0			
ORANGE	81	71	63	54	46	39	32	26	20	15	9	5	0					
RED	77	65	54	44	36	28	21	15	10	5	0							
BROWN	73	58	46	35	26	19	13	7	2			OUT OF RANGE						
BLACK	70	52	39	28	20	13	7	2										
	0.5	1	1.5	2	2.5	3	3.5	4	4.5	5	5.5	6	6.5	7	7.5	8	8.5	9

DISTANCE BETWEEN GROUND STATION AND SUBSATELLITE POINT (x 1,000 km)

the acquisition circle associated with FO-20 changes noticeably between apogee and perigee.

Most amateurs interested in tracking FO-20 will use a computer tracking program that automatically takes the elliptical nature of the orbit into account. However, if circumstances dictate the use of graphic tracking techniques, it is possible to obtain reasonable data from an OSCARLOCATOR type device. The OSCARLOCATOR overlay should be constructed for a circular orbit with a period and inclination equal to that of FO-20 (see Appendix C). FO-20's actual ground track will be very similar to the one shown except for the fact that FO-20 may arrive at a specific point up to 3 minutes before or after the time indicated. Acquisition circles should be drawn for both the perigee and apogee heights. At any time, the actual acquisition circle will lie somewhere between these two extremes.

To use the OSCARLOCATOR, you'll need some reference orbit data. You can obtain this data either by (1) asking someone with a computer tracking program to generate it for you, or (2) estimating it from observations over a few orbits.

Since FO-20's argument of perigee is changing by approximately 2.3° per day, it takes about two and a half months for the argument of perigee to change by 180°. As a result, over periods of several weeks FO-20 will be at approximately the same height as it passes nearby. If you access FO-20 on a regular basis, you'll find it easy to estimate the position of the acquisition circle and, if you're using computer-generated ascending-node data, the small time correction required. Using this approach, the OSCARLOCATOR can provide estimates of AOS and LOS that are accurate to within a minute or two.

BEARING AND SURFACE DISTANCE

A ground station that is using directional antennas needs to know where to point them. This information is usually presented as two angles: azimuth (angle in a plane tangent to the earth at the ground station, measured *clockwise* with respect to true north) and elevation (angle above the tangent plane). Bearing (the azimuth angle) can be computed when four quantities are known—the positions in latitude and longitude of both the ground station and the SSP. If, in addition, one knows the instantaneous height of the satellite, then elevation can also be computed. We'll look at each of these problems separately.

Given two points on the surface of the earth, only one great circle goes through both. If we were to look at the surface distances along various paths joining the two points, the minimum distance would be along the great-circle path. When we use the term distance, we mean great-circle distance.

Finding the distance between two points on the surface of the earth and the bearing from one to the other is the basic problem of navigation. Although the problem can be solved using the information on right spherical triangles presented in Chapter 11, there are more direct methods. Since the derivation is readily available in books on navigation and in texts covering vectors in three dimensions, we'll just present the results.

Using a spherical-earth model, the formula for distance between two points along a great circle arc is

$$s = R\beta \qquad \text{(Eq 12.1)}$$

where

s = surface distance in the same units as R

R = radius of earth in kilometers, statute miles or nautical miles

β = central angle at geocenter in *radians* (angle between the line segments joining the geocenter to the two points of interest).

With this equation in hand, we can discuss surface distance in terms of either s or the associated central angle, β. The formula relating surface distance to the coordinates (latitude and longitude) of the two points is

surface distance:
$$\cos \beta = \sin\phi_1\sin\phi_2 + \cos\phi_1\cos\phi_2\cos(\lambda_1 - \lambda_2)$$
$$\text{(Eq 12.2)}$$

where

ϕ_1, λ_1 = latitude and longitude of point 1 (ground station)

ϕ_2, λ_2 = latitude and longitude of point 2 (SSP or second ground station)

β = central angle representing the short path (β between 0° and 180°) distance. Note: 1.000° of arc = 111.2 km = 60.00 nautical miles = 69.05 statute miles.

Sign conventions

Latitude: north (positive), south (negative).
Longitude: east (positive), west (negative).

The azimuth of point 2 as seen from point 1 is given by

$$\text{Azimuth: } \cos A = \frac{\sin\phi_2 - \sin\phi_1 \cos\beta}{\cos\phi_1 \sin\beta} \qquad \text{(Eq 12.3)}$$

where A is the azimuth parameter. To obtain the true azimuth of point 2 as seen from point 1, measured clockwise from north, we must account for the fact that the arccos function only returns a value between 0° and +180°. Check to see if $\lambda_1 - \lambda_2$ is between -180° and +180°. If not, add or subtract 360° to bring it into this range. If the resulting value of $\lambda_1 - \lambda_2$ is

1) negative or zero, then true azimuth is given by A;
2) positive, then true azimuth is given by 360° - A.

COVERAGE

Because the radio frequencies used in conjunction with most satellites normally propagate over line-of-sight paths only, we will consider a communications satellite to be within range whenever the elevation angle at the ground station is greater than zero degrees. Depending on one's radio horizon and propagation conditions, communications might not be possible until the satellite is well above an elevation angle of 0°. Commercial satellite users typically use an elevation angle of +5° as their cutoff point for formally specifying when a satellite is in range.

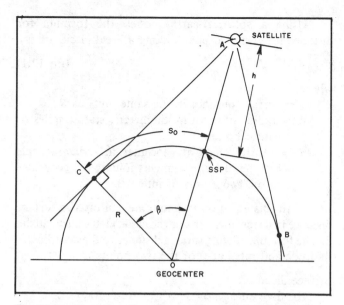

Fig 12-9—Cross section of satellite-coverage cone.

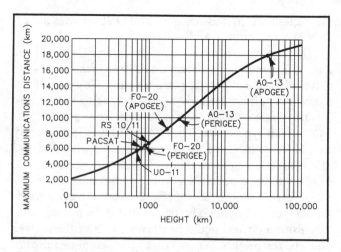

Fig 12-10—Maximum communications distance vs instantaneous satellite altitude.

The locus of all lines through a satellite and tangent to the earth at a specific instant of time forms a cone (see Fig 12-9). The intersection of this cone with the surface of the earth is a circle whose center lies on the line through the satellite, SSP and geocenter. Any ground station inside the circle has access to the satellite. Any two suitably equipped ground stations inside the circle can communicate via the satellite. The maximum terrestrial distance (between ground station and SSP) at which one can hear signals from the satellite is s_o. The maximum surface distance over which communication is possible is $2s_o$; see, for example, stations B and C in Fig 12-9.

Solving for s_o as a function of satellite height requires only plane trigonometry. Since line AC is tangent to the earth, triangle AOC is a right triangle, $\cos\beta = R/(R+h)$, and s_o is given by $R\beta$ (Eq 12-1). Therefore,

Maximum communications distance:

$$2s_o = 2\,R\,\arccos\,[R/(R+h)] \qquad \text{(Eq 12.4)}$$

where s_o = maximum access distance. See Sample Problem 12.1. A graph relating maximum communications distance to height is provided in Fig 12-10.

ELEVATION AND SLANT RANGE

We now consider satellite elevation angle and slant range. The elevation angle (ϵ) of a satellite can be obtained if (1) the height, h, of the satellite above the surface of the earth and (2) the surface distance, s, between the SSP and one's ground station are known. We just saw (Eq 12.2) that s can be found if the latitude and longitude of the ground station and SSP are known. Our object is to express ϵ in terms of s (or β) and h. In the course of determining the elevation angle, the slant range (line-of-sight distance between satellite and ground station) will also be found.

Once again, this problem can be solved using only plane trigonometry. The parameters involved are shown in Fig 12-11. Note the difference between this figure and Fig 12-9. The solution is obtained as follows.

Sample Problem 12.1

Find the maximum communications distance for Fuji-OSCAR 20 at apogee, RS-10/11 and AMSAT-OSCAR 13 at apogee and at perigee given the following information

Fuji-OSCAR 20, h_a = 1745 km,
RS-10/11, h = 1003 km,
AMSAT-OSCAR 13, h_a = 36,265 km,
h_p = 2545 km.

Solution

Plugging the given values for h into Eq 12.4 and using R = 6371 (spherical earth model), we obtain for

Fuji-OSCAR 20: $2s_o$ = 8514 km;
RS-10/11: $2s_o$ = 6724 km;
AMSAT-OSCAR 13:
   at apogee $2s_o$ = 18,104 km,
   at perigee $2s_o$ = 9873 km.

If your answers don't agree, you probably forgot to convert the $\arccos[R/(R+h)]$ term into radians (see Eq 12-1) before multiplying by 2R.

Elevation Circles

Focus attention on triangle AOC formed by the satellite, the geocenter and the ground station. Since the angles in a plane triangle must add up to 180°, the included angle at the satellite is:

$$A = 180° - \beta - (\epsilon + 90°) = 90° - \beta - \epsilon$$

Applying the Law of Sines to sides R and R + h we get

$$\frac{R + h}{\sin\,(\epsilon + 90°)} = \frac{R}{\sin\,(90° - \beta - \epsilon)}$$

Using the trigonometric identity

$\sin(90° \pm x) = \cos(x)$, we reduce this to

$$\frac{R + h}{\cos\,(\epsilon)} = \frac{R}{\cos\,(\epsilon + \beta)}$$

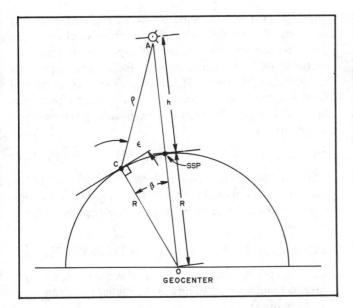

Fig 12-11—Diagram for determining satellite elevation angle and slant range as a function of height and distance to subsatellite point.

Next, the addition formula for the cosine function,

$$\cos(x + y) = \cos(x)\cos(y) - \sin(x)\sin(y)$$

is applied. This gives

$$\frac{R + h}{\cos(\epsilon)} = \frac{R}{\cos(\epsilon)\cos(\beta) - \sin(\epsilon)\sin(\beta)}$$

Finally, isolating all terms containing ϵ on the left-hand side, we arrive at the desired formula

elevation angle:

$$\tan(\epsilon) = \frac{[(R + h)\cos(\beta)] - R}{(R + h)\sin(\beta)} \quad \text{(Eq 12.5a)}$$

Using Eq 12.1, we can rewrite the elevation-angle formula in terms of surface distance instead of central angle:

elevation angle:

$$\tan(\epsilon) = \frac{(R + h)\cos(s/R) - R}{(R + h)\sin(s/R)} \quad \text{(Eq 12.5b)}$$

Note that the arguments of the angles in Eq 12.5b are given in radians.

Slant Range

An expression for slant range can be obtained by applying the Law of Cosines to the satellite-ground-station-geocenter triangle of Fig 12-11.

slant range:

$$\rho = [(R+h)^2 + R^2 - 2R(R+h)\cos(s/R)]^{1/2} \quad \text{(Eq 12.6)}$$

Elevation angle and slant range depend only on (1) the

height of the satellite and (2) the surface distance between SSP and ground station.

Eqs 12.5 and 12.6, therefore, are valid for elliptical as well as circular orbits. See Sample Problems 12.2, 12.3 and 12.4.

Elevation Circle Area

In Chapter 8 we discussed a method for estimating the amount of time a satellite in a low-altitude circular orbit

---

Sample Problem 12.2

A ground station using a satellite in a circular orbit may find it convenient to have a graph of elevation angle vs surface distance between ground station and SSP, since the latter is easily estimated from an OSCARLOCATOR. On a single set of axes, prepare graphs for OSCAR 16 (Pacsat), RS-10/11 and UoSAT-OSCAR 11. Plot similar curves for AMSAT-OSCAR 13 and Fuji-OSCAR 20 using apogee and perigee heights. Since the height of OSCAR 13 changes very slowly near apogee, the apogee curve will serve for several hours during each orbit.

Solution

Use Eq 12.5b in conjunction with the following values.

R = 6371 km
Pacsat: h = 800 km
Fuji-OSCAR 20: h_a = 1745 km (apogee)
Fuji-OSCAR 20: h_p = 912 km (perigee)
RS-10/11: h = 1003 km
UoSAT-OSCAR 11: h = 690 km
AMSAT-OSCAR 13: h_a = 36,265 km (apogee)
AMSAT-OSCAR 13: h_p = 2545 km (perigee)

The results are shown in Fig 12-12. Note that the curve for OSCAR 13 is nearly a straight line. As a result, the following approximate expression

$$\epsilon = 90° - 0.01s \quad (\epsilon \text{ in degrees, } s \text{ in km})$$

can be used for determining satellite elevation during the six hours centered on apogee.

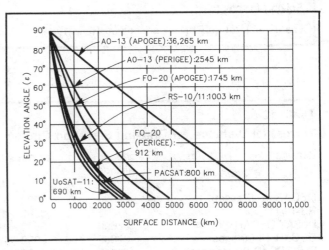

Fig 12-12—Elevation angle as a function of surface distance for several satellites (see Eq 12.5b).

Sample Problem 12.3

The concentric circles that form a spiderweb on an OSCARLOCATOR are iso-elevation curves. Find an expression that gives the surface distance, s, corresponding to a given elevation angle, ϵ, for a satellite at height h.

Solution

To solve this problem, we just rearrange Eq 12.5b to obtain s as a function of ϵ and h. This is easier said than done. The result is

$$s = R \left[arccos \left(\left[\frac{R}{R+h} \right] \cos \epsilon \right) - \epsilon \right] \qquad \text{(Eq 12.7)}$$

Once again, it's very important to pay attention to units; the term inside the brackets and the isolated ϵ must be expressed in radians. Data for several satellites of interest are included in Table 12-6. Note how the 0° elevation curve gives the same results as Sample Problem 12.1 and that Eq 12.7 simplifies to Eq 12.4 when $\epsilon = 0°$.

Sample Problem 12.4

The ground-track overlay on a P3 Tracker is divided into segments that are color coded to show the minimum access range. One step in designing a P3 Tracker is to find the polar angle in the orbital plane, θ, that corresponds to various access ranges, s_o. Given an orbit with eccentricity e and semi-major axis a, find θ as a function of s_o.

Solution

Take a look at Fig 11-2 before starting. The satellite orbital radius is given by Eq 11.2

$$r = \frac{a(1 - e^2)}{1 + e \cos(\theta)}$$

Also note Eq 12.4 for maximum access distance

$$s_o = R \ arccos[R/(R+h)]$$

Combining these equations to eliminate r we obtain

$$\cos(s_o/R) = R/(R+h) = R/r = \frac{R(1 + e \cos(\theta))}{a(1 - e^2)}$$

Solving for $\cos(\theta)$,

$$\cos(\theta) = (1/e)[(a/R)(1 - e^2)\cos(s_o/R) - 1]$$

would appear in a selected elevation angle interval, say between 30° and 60°. This information is helpful when planning the antenna system for a ground station. The method involved calculating areas for circular regions on the surface of the earth. The relevant formulas are derived in most introductory calculus texts. Results are presented here for reference. The surface area inside a given elevation angle circle is

$$area \ [circle] = 2\pi R^2 [1 - \cos(s/R)] \qquad \text{(Eq 12.8)}$$

where s is the surface distance between the ground station and the elevation circle. The surface area inside the ring (washer) shaped region between two elevation circles is given by

$$area \ [ring] = 2\pi R^2 [\cos(s_2/R) - \cos(s_1/R)] \qquad \text{(Eq 12.9)}$$

where s_1 corresponds to the lower elevation angle (larger distance) and s_2 corresponds to the higher elevation angle (smaller distance).

SPIDERWEB COMPUTATION

A set of range circles about a specific location on the earth, and a set of azimuth curves that radiate outward from it, are commonly referred to as a *spiderweb*. Most map-based tracking techniques use spiderwebs as an effective way to present information on satellite azimuth and elevation. To draw a spiderweb about a specific location on a map, we need coordinates for a large set of points on each range circle and azimuth line. The data in Appendix C will enable you to plot spiderwebs on any convenient map. If you want to know where this information comes from, or if you need access to the generating equations for a specialized tracking program you're putting together, the necessary information follows.

The desired information can be obtained from Eqs 12.2 and 12.3. Let ϕ_1, λ_1 be the coordinates of the central point and let ϕ_2, λ_2 be the coordinates of a representative point on the spiderweb at azimuth A_z. For convenience, temporarily set λ_1 equal to zero. This is equivalent to calculating our spiderweb for a point on the Greenwich Meridian. After we finish, we can just add the longitude of the central point to the longitude of all points on the spiderweb.

Table 12-6

Terrestrial Distance for Drawing Elevation Circles

Satellite	Mean Altitude (km)	Surface Distance (km) at Elevation Angle					
		0°	15°	30°	45°	60°	75°
OSCAR 16 (Pacsat)	800	3038	1767	1078	676	403	190
RS-10/11	1003	3362	2050	1286	817	490	231
UoSAT-OSCAR 11	690	2840	1597	958	596	354	166
OSCAR 13 (apogee)	36265	9052	7417	5845	4329	2859	1421
Fuji-OSCAR 20 (apogee)	1745	4257	2857	1909	1255	766	364
Fuji-OSCAR 20 (perigee)	912	3223	1927	1195	755	452	213

Let's assume that we've used Eq 12.7 to calculate appropriate distances for the spiderweb. When we set out to calculate the coordinates of a specific point on the spiderweb, we know A_z, β, ϕ_1 and λ_1. We wish to find ϕ_2 and λ_2. Eq 12.3 may be solved for the sine of the latitude of point 2

$$\sin\phi_2 = (\sin\phi_1)(\cos\beta) + (\cos A_z)(\sin\beta)(\cos\phi_1) \quad \text{(Eq 12.10)}$$

Note that all quantities on the right-hand side of Eq 12.10 are known. We therefore know the latitude of point 2.

To find the longitude of point 2, we solve Eq 12.2 for the cosine of the longitude of point 2

$$\cos \lambda_2 = \frac{\cos\beta - (\sin\phi_1)(\sin\phi_2)}{(\cos\phi_1)(\cos\phi_2)} \quad \text{(Eq 12.11)}$$

For additional details concerning spiderweb calculations, see D. Zachariadis, "Spiderweb—The Range Circle Calculation," *QST*, Feb 1986, pp 36-38. A computer program for drawing spiderwebs is included in Appendix D.

Chapter 13

Satellite Radio Links

This chapter focuses on the radio signals linking satellites and ground stations. The topics we'll cover include basic physical phenomena such as Doppler shift and Faraday rotation, unusual forms of propagation that may be encountered and a discussion of the process of selecting transponder frequencies.

THE DOPPLER EFFECT

Have you ever noticed how the pitch of a whistle on a passing train appears to decrease? A passenger on the train, listening to the same whistle at the same time, wouldn't notice a change in frequency. Whose perceptions are correct? Both of yours. The frequency of the sound you hear depends on the relative motion between the source (the whistle) and you (the observer). Since the train passenger moves along with the source, while the distance between you and the source is continually changing, each of you observes a different audio frequency. This phenomenon is known as the Doppler effect (after Johann Doppler, 1803-1853).

Though radio waves are very different from the sound waves we've just been discussing, they do exhibit a similar effect: An observer that is at rest with respect to a transmitter will measure a frequency f_o, while an observer who is moving with respect to the transmitter will measure a different frequency, f^*. The relation is given by

$$f^* = f_o - \frac{v_r}{c} f_o \qquad \text{(Eq 13.1)}$$

where

- f_o = frequency as measured by an observer at rest with respect to the source (source frequency)
- f^* = frequency as measured by an observer who is moving with respect to the source (apparent frequency)
- v_r = relative velocity of observer with respect to source
- c = speed of light = 3.00×10^8 m/s

Eq 13.1 is often written

Doppler shift: $\Delta f = f^* - f_o = -\dfrac{v_r}{c} f_o$ $\quad$ (Eq 13.2)

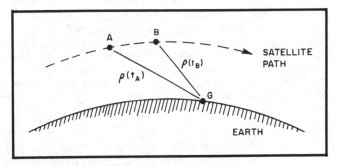

Fig 13-1—Doppler shift is observed when the distance between the satellite and ground station is changing.

Note that v_r is negative when a spacecraft is approaching. The apparent frequency will therefore be higher than the source frequency. When a spacecraft is receding, v_r is positive and the apparent frequency is lower than the source frequency. We'll settle for an intuitive understanding of Eq 13.1 and leave the details of the derivation to a physics text. The situation is easier to grasp when expressed in terms of period; since period = 1/frequency, we don't lose anything by doing so. Note that the period we use here is *not* the orbital period of the satellite, but the time for one complete cycle of the transmitted radio wave to pass. Refer to the moving satellite and fixed observer located at G in Fig 13-1. For convenience, consider that the source, aboard the satellite, is transmitting a linearly polarized wave and that the period is the time interval between two successive crests (occurring at A and B) in the transmitted electric field (E-field). From the diagrams, it's clear that the slant range $\overline{AG}$ is longer than the slant range $\overline{BG}$. Therefore, it takes less time for a signal that is sent from B to reach G than it does for a signal that is sent from A. Our observer at G, recording the time interval between the two successive E-field crests, will therefore record the time as being shorter than that measured by someone who remains equidistant from the satellite or who is sitting on it. So, for an approaching satellite, our observer records a shorter period than that measured at the spacecraft. Consequently, the observed frequency is higher.

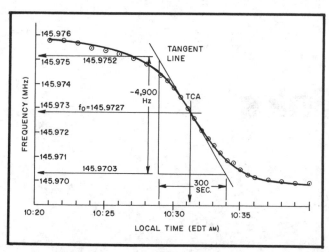

Fig 13-2—AMSAT-OSCAR 7 Doppler curve, orbit 7603, 14 July 1976, as observed from Baltimore, MD. Using the triangle shown, we evaluate the slope at TCA: m* = (−4900 Hz)/(300 sec) = −16.3 Hz/s. The satellite velocity was determined in Sample Problem 11.4, v = 7.13 km/s. Applying Eq 13.4, we obtain the slant range at closest approach, ρ_o = 1520 km.

Two primary questions related to Doppler shift are of interest to satellite users: (1) What will the actual Doppler shift be on a given link at a given time? and (2) What is the maximum Doppler shift that can be expected on a given link? We'll look at each of these questions in turn.

Doppler Shift

To calculate the instantaneous Doppler shift (the shift at any instant), we apply Eq 13.1 to Fig 13-1. If we assume that the source frequency is known, then the only unknown quantity in Eq 13.1 is v_r, the relative velocity. Relative velocity during a short time interval can be approximated by dividing the change in slant range by the change in time, in other words:

$$\bar{v}_r = \frac{\rho(t_B) - \rho(t_A)}{t_B - t_A} \qquad (Eq\ 13.3)$$

where

    $\bar{v}_r$ = approximate relative velocity
    t_B = time satellite passes point B
    t_A = time satellite passes point A
    $\rho(t_A)$ = slant range at time t_A
    $\rho(t_B)$ = slant range at time t_B

To calculate the slant range at two times, we can apply Eq 12.6, which gives us slant range as a function of satellite height and the ground-station-to-SSP distance. In sum, if you have a computer or calculator program that is written to predict basic tracking information (latitude and longitude of SSP and satellite height), adding Eq 12.6 for slant range and Eqs 13.2 and 13.3 for Doppler shift is a simple matter.

Doppler Shift at Closest Approach

A graph of apparent frequency against time for a specific satellite pass and ground station is called a Doppler curve. A typical Doppler curve, plotted from observations made during an AMSAT-OSCAR 7 pass, is shown in Fig 13-2. For circular orbits, the steepest part of the graph occurs at the point of closest approach (position where slant range is a minimum), and the observed frequency at this point is equal to the actual source frequency. Referring to Fig 13-2, we can determine that closest approach occurred at time 10:31:20 and the source frequency is 145.9727 MHz ± the accuracy of our frequency measurement. From the steepness (slope) of the curve at closest approach, we can compute the minimum slant range using the formula.

$$\rho_o = -\frac{f_o\, v^2}{c\, m^*} \qquad (Eq\ 13.4)$$

where

    ρ_o = slant range at closest approach (minimum slant range)
    f_o = transmitter frequency
    v = magnitude of satellite velocity (note: this is *not* the relative velocity discussed above)
    c = speed of light = 3.00×10^8 m/s
    m^* = slope of tangent line at TCA

To obtain m^* from an experimental graph like the one in Fig 13-2, align a transparent ruler over the central part of the curve until you get the steepest match and draw that line. Using any two convenient times, complete the right triangle as shown. Slope m^* is given by the ratio of the vertical side to the horizontal side of the triangle.

While situations occur in which one may want to predict the actual Doppler shift, radio amateurs using a satellite transponder are often satisfied to monitor the downlink and just twiddle their transmitter frequency control while sending a few dits until the downlink is at the desired spot. One particularly useful piece of information for operators, however, is the value of the maximum Doppler shift that any ground station might see on a given link.

Doppler Shift Limits

At any given time, there's a maximum and minimum Doppler shift that can be seen by any ground station. Most stations will observe a shift somewhere between these two extremes. For a circular orbit, the two limits remain constant; for an elliptical path, the limits vary over the course of the orbit.

We'll consider both cases, but first let's look at the two factors that contribute to the relative velocity term in Eq 13.1: (1) satellite motion in the orbital plane and (2) rotation of the earth about the N-S axis. In any given situation, these two factors can be combined (velocities add as vectors) to produce a relative velocity having a magnitude that can range from the arithmetic difference to the arithmetic sum of the two components. Since our objective is to determine the worst-case limits for a practical situation, we need only calculate each contribution separately and then form the sum and difference. First we look at the angular rotational velocity of the earth.

Rotation of Earth: The earth rotates about its N-S axis at an angular velocity of approximately

$$\dot{w}_E \sim 360°/day = 15°/hour = 0.25°/min$$
$$= 0.000073 \text{ radian/sec}$$

The tangential velocity of a point on the surface of the earth at latitude ϕ is

$$v_E = \dot{w}_E R \cos\phi$$

where

    R = radius of earth
    ϕ = latitude
    $\dot{w}_E$ = angular velocity of earth (expressed in radians)

The maximum value of v_E will occur at the equator, v_E (max) = 465 m/s. To get a handle on the size of the Doppler shift that arises exclusively from the rotation of the earth, assume a link frequency of 146 MHz and a ground station on the equator that sees a satellite due east on the horizon. (In this position, the tangential velocity of the earth and the relative ground-station-to-satellite velocity are equal.) Using Eq 13.2,

$$\Delta f = \frac{465}{3.00 \times 10^8} \times 146 \times 10^6 = 226 \text{ Hz}$$

So, at 2 meters, the worst-case contribution to Doppler shift produced by the rotation of the earth is less than a quarter kilohertz. As shown in Fig 13-2, observed Doppler shifts are often much larger. The contribution of the satellite orbital motion to Doppler shift must therefore be very important. Let's look at this contribution, first for a circular orbit, then for the case of elliptical motion.

Satellite Motion: Circular Orbits. Fig 13-3 shows the geometry of this problem in the orbital plane. The ground station that sees the largest relative velocity lies in the orbital plane and sees the spacecraft at 0° elevation. The velocity (v in meters per second) of a satellite in a circular orbit is given by

$$v^2 = \frac{GM}{r} = 3.986 \times 10^{14} (1/r) \qquad \text{(see Eq 11.8)}$$

From Fig 13-3, the relative velocity seen by the ground station is

$$v_r = v \cos\theta = v \cos(180° - \beta) = -v \cos\beta = -v\frac{R}{r}$$

If the direction of the satellite in Fig 13-3 were reversed, the satellite would be receding from the ground station. The Doppler shift would be equal in magnitude, but would represent a decrease in frequency. A short sample problem shows how this information can be applied. See Sample Problem 13.1.

Combining the results of Sample Problem 13.1 (3.22 kHz) and the earlier calculation of the maximum shift

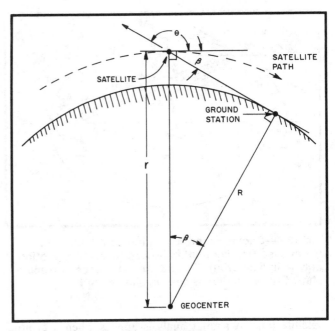

Fig 13-3—Geometry for computing contribution to worst-case Doppler shift from satellite motion only (circular orbit).

Sample Problem 13.1

Consider OSCAR 16 (Pacsat). Find the contribution to the maximum Doppler shift from orbital motion on the 146-MHz uplink.

Solution:

h = 800 km
R = 6371 km
r = R + h = 7171 km

$$v^2 = 3.986 \times 10^{14} \left(\frac{1}{7.171 \times 10^6}\right)$$

$$= 0.5558 \times 10^8 \text{ (m/s)}^2$$
v = 7456 m/s
v_r = -v(R/r) = -7456(6371/7171) = -6624 m/s
Δf = -(vr/c)f_o = (6624 × 146 × 10^6)/(3.00 × 10^8)
= 3.22 kHz

contributed by the rotation of the earth (226 Hz), we see that the Doppler shift on the OSCAR 16 (Pacsat) 146-MHz link will never exceed 3.45 kHz. For our needs, it's more appropriate to express the maximum as less than ±3.5 kHz. Note that this is the maximum shift from the actual transmitted frequency. An "imaginary observer" aboard OSCAR 16 listening to an uplink as the spacecraft approached a ground station would record a frequency between the transmitted frequency and a value 3.5 kHz higher. As the satellite recedes, the observer would measure the frequency as being between the transmitted frequency and a value 3.5 kHz lower. The range of observed frequencies would be less than 7.0 kHz on this 146-MHz

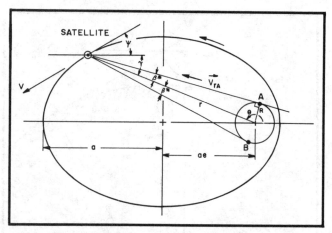

Fig 13-4—Geometry for computing contribution to worst-case Doppler shift produced by satellite motion only (elliptical orbit). Ground station A sees maximum shift. Ground station B sees minimum shift.

satellite link. A similar analysis applies when a ground station is monitoring a downlink.

Satellite Motion: Elliptical Orbits. We now consider the contribution to Doppler shift provided by satellite motion when the orbit is an ellipse. The geometry, shown in Fig 13-4, is somewhat more involved. The ground station observing the largest velocity (station A) lies in the orbital plane and sees the spacecraft at 0° elevation. The ground station observing the smallest relative velocity (station B) also lies in the orbital plane and sees the spacecraft at 0° elevation. The following series of steps enables one to compute the Doppler for stations located at the special points A and B. Note that the locations of A and B change; we're not considering two fixed stations. We assume that the semimajor axis (a) and eccentricity (e) are known.

Step 1. Use Eqs 11.10 and 11.11 to solve for θ.

Step 2. Use Eq 8.12 to solve for r.

Step 3. Solve for the angle ψ shown in Fig 13-4.

$$\psi = \arctan\left[(e^2 - 1)(r\cos\theta + ae)/(r\sin\theta)\right]$$

(Note: This equation was derived using the techniques of elementary calculus to solve for the slope of a line that is tangent to an ellipse.)

Step 4. Solve for β^* using $\sin\beta^* = \dfrac{R}{r}$

Step 5. Solve for $\gamma = 180° - \theta$

Step 6. Solve Eq 11.8 for the satellite velocity, v.

Step 7. Solve for the relative velocity seen by ground station at A: $v_{rA} = v\cos(\psi + \gamma - \beta^*)$ and ground station at B: $v_{rB} = v\cos(\psi + \gamma + \beta^*)$

Step 8. Solve for the Doppler shift using Eq 13.2.

Fig 13-5—Doppler-shift limits for Phase-3 injection orbit as seen by ground station anywhere on earth.

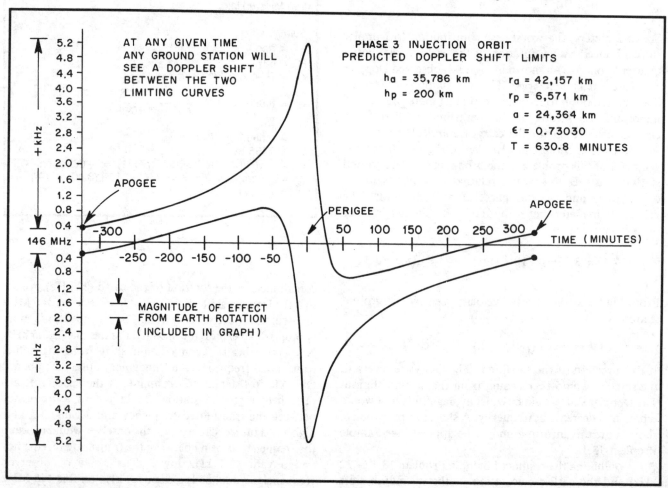

In preparation for the AMSAT Phase 3A launch, this procedure was applied to the transfer orbit, since Doppler-shift limits were needed to design the command-station network. The resulting graph that includes the effects of satellite motion and rotation of the earth is shown in Fig 13-5. Keep in mind that this is *not* the Doppler shift seen by a particular ground station. The graph represents the maximum possible shifts that could be seen from somewhere on the earth at any instant during the orbit. Note that for elliptical orbits, Doppler-shift limits are generally inversely related to satellite height: Greatest near perigee, least near apogee.

Doppler Shift and Transponders

So far, all our Doppler-shift calculations have focused on a single link. When communicating via a transponder, there are two links involved—the uplink and the downlink. Let f_d represent the downlink frequency corresponding to an uplink frequency of f_u when the relative velocity between satellite and ground is zero.

We begin by looking at the situation where you are monitoring your own downlink. If the spacecraft is approaching, both links will be shifted up in frequency. To calculate the total shift for a non-inverting transponder, just add f_u and f_d and plug the result in Eq 13.2 for f_o. In most situations, just plugging in the sum of the center frequencies of the uplink and downlink passbands is sufficiently accurate. For example, to calculate the total shift on Mode A, use 175 MHz (146 + 29) for f_o. The resulting shift is simply the sum of the two shifts calculated separately.

Now consider the case where you're monitoring your own downlink through an inverting transponder. We'll use Mode B (435 MHz up, 146 MHz down) aboard a low-altitude satellite as an example. Assume the spacecraft is approaching you on a nearby pass. The spacecraft will see your uplink frequency shifted up about 6 kHz (f_u + 6); after inversion in the transponder, the frequency is f_d − 6. On the downlink, the signal is again shifted up, this time about 2 kHz, so the signal you receive will appear at f_d − 4.

Table 13-1

Maximum Doppler Shifts on Various Satellite Radio Links

Maximum Doppler Shift (kHz)
Satellite/Height

	Mir 370 km	OSCAR 11 690 km	Pacsat 800 km	RS-10/11 1003 km	OSCAR 12 1495 km	OSCAR 13 (perigee) 2545 km	OSCAR 13 (apogee) 36265 km
Beacon freq[1]							
29.5 MHz (0.045)	0.76	0.72	0.70	0.67	0.62	0.52	0.09
146. MHz (0.226)	3.76	3.53	3.45	3.32	3.04	2.56	0.45
435. MHz (0.674)	11.2	10.5	10.3	9.88	9.03	7.60	1.33
1.27 GHz (1.967)	32.7	30.7	30.1	28.9	26.4	22.2	3.9
2.40 GHz (3.722)	61.8	57.9	56.7	54.5	49.8	41.9	7.4
10.5 GHz (15.500)	269.5	252.8	247.4	237.8	217.3	182.8	31.5
Transponder							
Mode A[2] (0.271)	4.50	4.23	4.13	3.97	3.63	3.06	0.54
Mode B/J (0.448)	7.44	6.98	6.83	6.57	6.00	5.05	0.89
Mode L (1.293)	21.5	20.1	19.7	19.0	17.3	14.6	2.56
Mode S (1.755)	29.2	27.4	26.8	25.8	23.6	19.8	3.48
Mode S†[2,3] (3.047)	74.0	68.5	66.9	64.4	58.9	49.6	8.72

[1]Numbers in ( ) are maximum Doppler shift, in kHz, due to rotation of earth. Doppler-shift values in table include contributions from satellite motion in the orbital plane and rotation of the earth.
[2]Noninverting transponders. All other modes are for inverting transponders.
[3]Mode S† refers to AO-13 only (435 MHz/2.4 GHz).
Not all links shown are in existence—values quoted are for comparison purposes.

In other words, your downlink signal is about 4 kHz lower than it would be if Doppler were neglected. This might be somewhat surprising, since we've gotten used to thinking that an approaching spacecraft means the frequency is shifted up. If we reworked this example using a Mode-J transponder, we would find that the downlink signal was 4 kHz higher than predicted if Doppler were neglected. The direction of the resulting shift clearly depends on how the spacecraft link frequencies are assigned. For an inverting transponder, we get the expected result—signal shifted up in frequency when the spacecraft is approaching—only when the downlink is the higher frequency.

An easy way to remember the expected shift direction is to note that the higher link frequency always dominates. If the higher frequency is on the downlink (after transponder frequency inversion), we see the normal shift direction (up shift as spacecraft approaches). If the higher frequency is on the uplink (before transponder frequency inversion), the normal shift direction is reversed.

No matter how the link frequencies are assigned, replacing f_o with $\pm (f_d - f_u)$ in Eq 13.2 provides the shift for an inverting transponder. Many tracking programs provide a readout of Doppler shift for the spacecraft beacon. If you replace the spacecraft beacon frequency by $\pm (f_d - f_u)$ (passband center frequencies work fine), you can use this utility to compute the Doppler shift on your own signal. Select the sign that produces results consistent with the shift direction specified in the previous paragraph.

As we noted in Chapter 10, when two or more stations are in contact, Doppler shift and frequency retuning cause downlink signals to creep across the band. Since different geometries lead to different shifts, QRM often results as QSOs collide. Based on the information presented in this chapter, an operating protocol that minimizes creep was suggested in Chapter 10. Briefly, when using a low-altitude satellite, the station farthest from the SSP at closest approach should keep his/her highest link frequency constant (that is, on Mode J keep the 435-MHz receive frequency constant; on Mode B, keep the 435-MHz transmit frequency constant). For reference, the maximum expected Doppler shifts for several beacons and transponders of interest are presented in Table 13-1.

Anomalous Doppler

In 1972, an experimenter who was collecting Doppler data from the 435-MHz beacon aboard AMSAT-OSCAR 6 noticed a strange effect on a northbound pass. For the first few minutes after AOS, the frequency of the observed signal increased, instead of decreasing as would normally be expected. Departures of up to 700 Hz from predicted values were observed. After thoroughly checking a number of factors that could have accounted for the observations (for example, drift in ground-station frequency-measuring equipment, change in satellite temperature that would affect the beacon oscillator frequency, and so on), it was concluded that an interesting physical anomaly was being seen. The effect was later observed on navigational satellites that were operating near 400 MHz.

An exhaustive experimental investigation was undertaken to delineate the spatial and temporal (time of day, season of year, and so on) extent of the anomaly and to determine the frequency range over which it occurred. It was hoped that this data would make it possible to correlate the effect with physical changes in the ionosphere that were suspected of being related. Although a great deal of data was collected, no firm conclusions have ever been reached as to the cause of anomalous Doppler. For additional information, see W. Smith, "Doppler Anomaly on OSCAR 6 435-MHz Beacon," *QST*, May 1973, pp 105-106; J. Fox and R. Dunbar, "Inverted Doppler Effect," *Proceedings of the ARRL Technical Symposium on Space Communications*, Reston, VA, Sep 1973, Newington, CT: ARRL, pp 1-30.

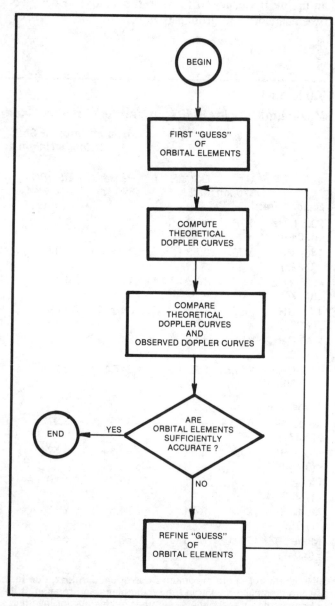

Fig 13-6—Flowchart illustrating how Doppler observations are used to compute the orbital elements of a satellite.

Doppler: Orbit Determination and Navigation

We saw earlier how a Doppler curve enables us to determine time of closest approach (TCA), slant range at closest approach and the actual transmission frequency. A single Doppler curve actually provides us a unique signature for a satellite's orbit. Applying a sophisticated model, one can use Doppler data collected over one or more orbits to determine the six parameters needed to characterize an elliptical orbit, or the four parameters needed to characterize a circular orbit. A flowchart illustrating how this is done is shown in Fig 13-6. The task of determining the orbital parameters (elements) for a satellite is usually simplified if a combination of Doppler and ranging measurements are used.

A closely related problem is that of using Doppler data from a satellite whose orbital elements are known very accurately to determine the latitude and longitude of a ground station. This technique is used with navigation satellites such as the Transit series, and by search-and-rescue satellites in the cooperative SARSAT (US, Canada, France) and COSPAS (USSR) series. Radio amateurs are justified in being proud that the satellite-aided search-and-rescue concept was first tested using the Mode-A transponders aboard AMSAT-OSCARs 6 and 7. For additional information, see R. Bate, D. Mueller and J. White, *Fundamentals of Astrodynamics*, New York: Dover, 1971; P. Escobal, *Methods of Orbit Determination*, New York: Krieger, 1975; P. Karn, KA9Q, B. McGwier, N4HY, "Spread Spectrum Ranging for Phase IV and Nonlinear Filtering for Orbit Determination in Phase III-C and Phase 4," *Proceedings of the AMSAT-NA Fifth Space Symposium*, Southfield, MI, Nov 1987, ARRL, Newington, CT; S. Eckart, DL2MDL, "Determination of Orbital Parameters Through Distance Measurements," *AMSAT-DL Journal*, No. 6/16, Dec 88/Feb 89, English translation in *Oscar News*, Apr 1989, pp 22-25.

FARADAY ROTATION

When a linearly polarized radio wave travels through the ionosphere, its plane of polarization rotates about the line of travel. The effect, known as Faraday rotation, depends on the frequency of the radio wave, the strength and orientation of the earth's magnetic field over the path, and the number of electrons encountered. As a satellite moves along its orbit, the downlink path changes. As a result, the amount of rotation of the polarization plane changes also. This can lead to severe signal fading when the antennas at both ends of a satellite downlink are linearly polarized.

Other factors being equal, the number of revolutions of the plane of polarization for an uplink or downlink signal is roughly proportional to λ^m (λ = wavelength), with the exponent m having a value between 2 and 3. The predictions of a simple model of Faraday rotation are summarized in Table 13-2. The absolute number of revolutions that the plane of polarization undergoes is not of much practical interest. The change in this number is, however, because we expect to experience two deep fades

Table 13-2

Faraday Effect: Revolutions of Plane of Signal Polarization*

Let N = Total number of revolutions of plane of polarization as signal passes through ionosphere at specified elevation angle.

ΔN = Change in N as satellite moves from horizon to overhead

Frequency	N at 0° elevation	N at 90° elevation	ΔN
29 MHz	89	28	61
146 MHz	4.4	1.4	3
435 MHz	0.8	0.25	0.5

*As predicted by the simple model outlined in G. N. Krassner and J. V. Michaels, *Introduction to Space Communications Systems*, New York: McGraw-Hill 1964.

Table 13-3

Variables Affecting Strength of the Downlink Received Signal

1) Satellite (antenna) orientation with respect to ground station.
2) Satellite spin producing a time-dependent antenna pattern.
3) Changing slant range (inverse power law).
4) Signal absorption in the ionosphere.
5) Ground-station antenna pattern.
6) Faraday rotation.

in signal strength for each unit change in the total number of revolutions. While the model is very simple, comparing the predictions with observations is interesting.

Let's consider the 29-MHz beacon on AMSAT-OSCAR 8. The model (Table 13-2) predicts 60 deep fades in the 9 minutes it takes AO-8 to go from horizon to overhead. This amounts to one deep fade every 5 seconds on average. The model actually predicts a shorter time interval between fades when the elevation angle is near 90° and a longer interval at low elevation angles. Observations of the beacon generally showed a time interval between fades of 20 to 100 seconds. This is not necessarily in contradiction to the model, since the maximum spacecraft elevation angle on most passes was relatively low and the results do depend, to a significant extent, on the actual satellite path and the location of the ground station.

From an operational viewpoint, Faraday rotation is important at 29 MHz, of minor concern at 146 MHz, and of little effect at higher frequencies. Variations in downlink signal strength may be caused by many factors, including those listed in Table 13-3. From a communications standpoint, we're interested mainly in minimizing fading. Using circular polarization at the ground station or the spacecraft does reduce fading from several of the factors listed.

Experimenters who are interested in observing Faraday rotation directly will look at Table 13-3 in a

From an operational viewpoint, Faraday rotation is important at 29 MHz, of minor concern at 146 MHz, and of little effect at higher frequencies. Variations in downlink signal strength may be caused by many factors, including those listed in Table 13-3. From a communications standpoint, we're interested mainly in minimizing fading. Using circular polarization at the ground station or the spacecraft does reduce fading from several of the factors listed.

Experimenters who are interested in observing Faraday rotation directly will look at Table 13-3 in a different light. How can the effects of Faraday rotation be separated from the other factors? One strategy would be to concentrate on those links where Faraday rotation is very prominent. A 29-MHz beacon on a low-altitude satellite that uses a linearly polarized antenna is clearly the link of choice. If the ground station uses two linearly polarized antennas that are mounted at right angles to one another and perpendicular to the incoming wave, it's possible to switch back and forth between them to monitor the signal strength alternately at each polarization. With this information, we can separate out most of the factors listed in Table 13-3. Faraday rotation will appear as fading on the two antennas, one reaching a peak as the other reaches a minimum, with the period changing slowly in a regular manner. Studies of the Faraday effect are often used to deduce electron concentrations in specific regions of the ionosphere. For additional information on the Faraday effect, see G. N. Krassner and J. V. Michaels, *Introduction to Space Communication Systems*, New York: McGraw-Hill, 1964 (this text discusses the model upon which Table 13-2 is based); J. D. Kraus, *Radio Astronomy*, New York: McGraw-Hill, 1964, Chapter 5, section 5; and W. A. S. Murray and J. K. Hargreaves, "Lunar Radio Echoes and the Faraday Effect in the Ionosphere," *Nature*, Vol 173, no. 4411, May 15, 1954, pp 944-945.

SPIN MODULATION

Satellites are often stabilized by being spun about a particular axis. When the spacecraft spin axis is not pointing directly at your ground station (non-zero squint angle), you're likely to see changes in amplitude, and possibly polarization, resulting from the spacecraft rotation. These changes, which affect the up- and downlinks, occur at an integer multiple of the satellite spin frequency. The magnitude of the effect will generally become greater as the squint angle increases.

Polarization changes that result from a linear component of polarization at the spacecraft can, to a large extent, be handled at the ground station by using a circularly polarized antenna. Gain variations are considerably more difficult to compensate for. Using a very sensitive receiving system and a very short AGC time constant will always reduce the problem, but not necessarily to a manageable level.

With Phase-2 satellites having magnetic-bar stabilization systems, spin rates were on the order of 0.01 Hz. This resulted in relatively short duration deep fades every few minutes. Phase-3 satellites are spin stabilized at approximately 20 RPM. Due to their antenna configuration, spin-modulation variations tend to cycle at a 1.0 Hz rate. The effect is most noticeable when squint angles are greater than 20°.

UNUSUAL PROPAGATION

While ionospheric effects on 29-MHz satellite links were both expected and observed, most discussions of VHF and UHF links treat the ionosphere as if it ceases to have any impact, other than Faraday rotation, above 40 MHz. Contrary to traditional thought, amateur measurements show that satellite links are clearly affected by the ionosphere at both 146 and 435 MHz. Signal attenuations of 12 dB or greater that were attributable to the state of the ionosphere were frequently observed on the downlinks of Phase-2 spacecraft. S. Eckart, in his orbit-determination article (previously mentioned), noted that refraction at 146 MHz often resulted in ranging errors amounting to 100 km.

When the F2 layer is efficiently reflecting terrestrial 10-meter signals back to the earth, it's also reflecting 10-meter signals arriving from space back to whence they came. As a result, an open 10-meter band often coincides with an absence of observable 29-MHz Mode-A downlink signals.

Turning to higher frequencies, John Branegan (GM8OXQ/GM4IHJ) collected detailed quantitative information on 70-cm downlink signal strength over a large number of orbits involving OSCARs 7 and 8 and other spacecraft. Statistical procedures were then used to separate the temporal and spatial extent of the attenuation region(s) from antenna orientation and other effects. For details, see J. Branegan, "Reception of 70-cm Signals from Satellites, Summary of Results March to Oct. 1978," *AMSAT Newsletter*, Vol X, no. 4, Dec 1978, pp 10-14.

Sporadic E. In a later study, Branegan reported that high attenuation levels on VHF/UHF satellite downlinks were correlated positively with enhanced terrestrial propagation attributable to sporadic E. Sporadic E refers to relatively dense clouds or patches of ionized particles that often form at heights approximately the same as the E layer. To monitor terrestrial sporadic E, he selected VHF TV and FM stations located so that the same general region of the ionosphere was shared by both satellite and terrestrial links. For additional information, see J. Branegan, "Sporadic-E Impact on Satellite Signals," *Orbit,* Vol 1, no. 4, Nov/Dec 1980, pp 8-10.

FAI. OSCAR operation was directly responsible for the discovery of a new mode of VHF propagation, called FAI (magnetic-field-aligned irregularities), after the mechanism thought to be responsible. The first observations of signals via this medium were reported by stations in equatorial zones who listened for direct signals from amateurs uplinking to OSCAR spacecraft at 146 MHz. The positive results led to direct terrestrial experiments at 2 meters and 70 cm, which helped determine the properties of the FAI mechanism. For details, see J. Reisert and G. Pfeffer, "A Newly Discovered Mode of VHF Propagation," *QST*, Oct 1978, pp 11-14; and T. F. Kneisel, "Ionospheric Scatter by Field-Aligned Irregularities at 144 MHz," *QST*, Jan 1982, pp 30-32.

Antipodal Reception. Soon after Sputnik I was launched, observers noticed that the 20-MHz signal from the satellite was often heard for a short period of time when the satellite was located nearly antipodal to the observer. (If you were to dig a hole right through the center of the earth, the spot where you'd emerge is called the antipodal point). The phenomenon was quickly dubbed the "antipodal effect," and a number of articles appeared in IEEE journals during the late 1950s discussing its causes. Antipodal effects were later observed on the 29.5-MHz beacon of OSCAR 5. See R. Soifer, "Australis-OSCAR 5 Ionospheric Propagation Results," *QST*, Oct 1970, pp 54-57.

The likelihood of antipodal reception is correlated positively with solar activity. During sunspot maxima, it is relatively common on 29.5-MHz satellite beacon signals. Although most occurrences are thought to result from normal multihop propagation under the influence of a favorable maximum usable frequency, signal strength is at times exceptionally high. This suggests that a ducting mechanism may sometimes be responsible.

Auroral Effects. Radio signals that pass through zones of aurora activity acquire a characteristic distorted sound, described as raspy, rough, hissy, fluttery, growling and so on. Low-altitude satellites in near-polar orbits are excellent tools for studying auroral effects. One can, for example, use the various beacons on UoSAT to map the extent of and note changes in the auroral zone experimentally at various frequencies. The changes at particular frequencies or locations may turn out to be excellent predictors of HF or VHF openings that are caused by various modes. See K. Doyle, "10 Meter Anomalous Propagation with Australis OSCAR [AMSAT-OSCAR 5]," *CQ*, May 1970, pp 60-64, 89.

General. A great deal is still to be learned about ionospheric propagation of VHF and UHF signals, and amateurs are in a unique position to collect important data. OSCAR 9, with its array of beacons ranging from HF to microwave frequencies, was especially well suited to propagation studies. Beginning in about 1985, UoSAT spacecraft began collecting and downlinking whole-orbit telemetry data on a regular basis from many of the onboard experiments. As a result, amateurs interested in studying propagation now have access to measurements made outside their normal radio horizon. See R. J. Diersing, N5AHD, "Processing UoSAT Whole-Orbit Telemetry Data," *Proceedings of the 4th Annual AMSAT Space Symposium,* Dallas, TX, Nov 1986, ARRL, Newington, CT.

This discussion of unusual propagation has touched upon only a few of the curious phenomena that occur. Other topics of possible interest include RF noise (atmospheric, manmade, cosmic, terrestrial, oxygen and water vapor, solar), attenuation (electron, condensed water vapor, oxygen and water vapor), refraction (ionospheric, tropospheric) and scintillation. For a general overview of all these propagation effects, see G. N. Krassner and J.

V. Michaels, *Introduction to Space Communications Systems,* New York: McGraw-Hill, 1964.

FREQUENCY SELECTION

The selection of frequencies for an Amateur Radio spacecraft transponder is a complicated process. Consideration must be given to

1) legal constraints (national and international laws governing the use of the RF spectrum)

2) technical factors (including propagation, type of orbit, ability of amateur community to produce required flight hardware, and so on)

3) user-community needs and preferences

4) frequency management (cooperative agreements concerning frequency use among the worldwide amateur community).

We'll look at how each of these factors affects the selection process.

Table 13-4

International Telecommunication Union Amateur-Satellite Service Frequency Allocations

1971 WARC	1979 WARC
7.000-7.100 MHz	7.000-7.100 MHz
14.000-14.250 MHz	14.000-14.250 MHz
	18.068-18.168 MHz
21.000-21.450 MHz	21.000-21.450 MHz
	24.890-24.990 MHz
28.000-29.700 MHz	28.000-29.700 MHz
144.000-146.000 MHz	144.000-146.000 MHz
435.000-438.000 MHz*	435.000-438.000 MHz* (3644/320A)
	1.26-1.27 GHz* [uplink only] (3644/320A)
	2.40-2.45 GHz* (3644/320A)
	3.40-3.41 GHz [* in Region 2 and 3 only] (3644/320A)
	5.65-5.67 GHz* [uplink only] (3644/320A)
	5.83-5.85 GHz [downlink only] (3761C)
	10.45-10.50 (3780A)
24.00-24.05 GHz	24.00-24.05 GHz
	47.0-47.2 GHz [Amateur Exclusive]
	75.5-76.0 GHz [Amateur Exclusive]
	76-81 GHz
	142-144 GHz [Amateur Exclusive]
	144-149 GHz
	241-248 GHz
	248-250 GHz [Amateur Exclusive]

*The amateur-satellite service may use these frequencies subject to *not* causing harmful interference to other services operating in accordance with provisions of Allocation Table. This applies to space stations and ground stations.

For additional information, see "Extracts From the International Radio Regulations for the Amateur and Amateur-Satellite Services," *QST*, Feb 1980, pp 62-71. Paragraph numbers ( ) refer to relevant sections of the Regulations.

Legal Constraints

The amateur-satellite service was formally recognized by the ITU (International Telecommunication Union) at the 1971 WARC (World Administrative Radio Conference) for Space Telecommunications. The US amateur-satellite service was established by the FCC in 1973. Prior to this date, amateur satellites were licensed under the rules and regulations governing the amateur service.

At the 1971 WARC, the ITU allocated frequencies for the amateur satellite service. These allocations are shown in column 1 of Table 13-4. The tremendous gap between 438 MHz and 24.0 GHz placed serious limitations on the future development of amateur space communications.

At WARC-79, the frequencies allocated to the amateur satellite service were opened for discussion. A concerted effort by IARU, ARRL, AMSAT and other interested amateur groups succeeded in securing several additional frequency bands. (See Chapters 3 and 4 for a review of the 1971 and 1979 WARCs.) The complete list of allocations adopted at this meeting is given in column 2 of Table 13-4.

US amateur access to WARC allocations is, of course, not immediate. New allocations have to be ratified by the Senate and then implemented by the FCC. For specific details, see Appendix F: FCC Rules and Regulations Governing the Amateur Satellite Service. As of 1990, the frequencies listed in column 2 of Table 13-4 are available for US Amateur Radio space projects.

Technical Factors

A great many technical factors are involved in the selection of frequencies for satellite links. We'll look at several of these factors, beginning with some general considerations and then moving on to specifics.

Many of the desirable features of a satellite communication link depend on operating the transponder and ground station in a duplex mode (simultaneous transmission and reception). A cross-band transponder is needed to allow duplex operation without inordinately complex equipment at each ground station. Therefore, our objective here will be to pick (from the list in Table 13-4) the two optimal bands, based purely on technical considerations.

We'll also consider how other bands compare to our optimal choices in case we encounter obstacles that prevent their use.

From a system viewpoint, the downlink is the "weak link" in the communication chain. If necessary, ground-station transmitter power levels can exceed the power allocated to a single user at the spacecraft by considerably more than 20 dB. And, even with a sophisticated attitude-stabilization system, satellite antenna gain must be limited in order to provide a sufficiently broad pattern (footprint) for full earth coverage during most of the orbit. Consequently, the "best" band should be used for the downlink unless there are compelling reasons to do otherwise. Frequencies will therefore be evaluated as downlinks. If a band provides good downlink performance, it almost certainly will be excellent as an uplink.

We now focus our attention on high-altitude spin-stabilized spacecraft of the Phase 3 series and compare downlink performance at several frequencies. For the comparison, we'll also assume that transponder power and bandwidth are constant and that ground stations provide good, but not necessarily state-of-the-art, performance. Most of the following observations apply equally well to Phase 4 spacecraft.

Free-space path loss increases with frequency. Column 2 of Table 13-5 presents information on relative path loss at the frequencies under consideration. Since we're interested in relative performance, we choose a convenient reference level—2 meters in this case. Though beam antennas can be used on spin-stabilized high-altitude satellites, for reasonable earth coverage satellite antenna gain must be limited to approximately 12 dBi. (We will discuss the possibility of higher-gain antennas later). This gain can be achieved at UHF and higher frequencies, but at 146 MHz, antenna size becomes a problem. At 29 MHz, a gain antenna presents monstrous mechanical problems that so far have made it impossible to place such a device on an AMSAT satellite. Reasonable estimates for achievable antenna gain on a spin-stabilized Phase 3 satellite have been included in column 3 of Table 13-5.

The ground-station antenna-gain entries in Table 13-5 are based on a constant boom length of 8 feet at 10 meters,

Table 13-5

High-Altitude Satellite Link Performance at Several Frequencies†

Band[1]	Relative free-space path loss	Spacecraft antenna gain	Ground-station antenna gain	Relative performance
29 MHz	−14 dB	0 dB	+5 dB	+19 dB
146 MHz	0 dB	+7 dB	+12 dB	+19 dB
435 MHz	10 dB	+10 dB	+16 dB	+16 dB
1.26 GHz	19 dB	+12 dB	+21 dB	+14 dB
2.4 GHz	24 dB	+12 dB	+26 dB	+14 dB

Notes

†These figures take into account path loss, practical spacecraft antennas and ground-station antennas of similar physical size.
[1]50 MHz and 220 MHz are not included because these frequencies are not authorized for amateur satellite service use.

2 meters and 70 cm. At 23 cm, a 4-ft-diameter dish produces about the same gain as an 8-ft-boom loop Yagi; therefore we've selected a 4-ft dish as a comparable ground station antenna at 23 cm and 13 cm. Finally, the last column in Table 13-5 summarizes relative link performance taking into account the factors listed in columns 2-4: path loss, satellite antenna gain and ground-station antenna gain.

Based on the information presented so far, 2 meters would be our choice for a downlink and 10 meters for an uplink. Recall, however, the information on sky noise arriving at an antenna, which was presented in Table 9-1. As a result of the steep increase in noise below 1 GHz, a good receiver is capable of recovering much weaker signals at 2 meters than it is at 10 meters. The relative advantage of 2 meters over 10 meters can reach 15 dB. Also, atmospheric absorption at 10 meters may amount to as much as 20 dB, especially during peaks of the sunspot cycle. Taking these facts into account, we'd conclude that 146 and 435 MHz are the best link frequencies available. Based entirely on Table 13-5, we'd once again choose 2 meters as a downlink. Taking sky noise into account, the selection of a downlink, 2 meters or 70 cm, is really a toss-up.

Our analysis so far has considered only relative performance. The conclusion that 2 meters and 70 cm will provide similar performance does not mean that either will provide an acceptable downlink signal, so we have to make contact with the absolute levels of the real world.

Absolute signal levels can be predicted if factors such as receiver noise figure and bandwidth, and cosmic noise, are taken into account. We will show how to do this in the example that concludes this section. Although calculations of absolute link performance are very useful, there's no substitute for experience. Amateurs have always attempted to collect data on beacon performance before a frequency was used on a transponder. Data of this type is an extremely valuable aid in predicting transponder performance. When calculations of relative performance are coupled with measurements made on beacons, excellent projections of transponder performance are possible.

As a result of calculations of this type, it was possible to predict in the mid-1970s that a Phase 3 200-kHz-wide transponder with a 50-W PEP downlink on 146 MHz would provide users a signal-to-noise ratio of 18-23 dB. Such a link would provide excellent performance. Experience with OSCARs 10 and 13 has confirmed these figures. Other possible downlink frequencies will be compared to this link.

As stated earlier, a transponder using 2 meters and 70 cm is therefore a prime choice, with the selection of downlink a toss-up. Three important reasons led to the selection of 2 meters as the downlink (Mode B) back in the mid-1970s. First, the number of amateurs having receiving equipment for 2 meters was considerably greater than that for 70 cm; second, low-noise preamps that could take advantage of the sky noise advantage at 70 cm were not widely available; third, receiver desensing was a serious obstacle to Mode-J ground stations.

The reason why Mode B was employed on OSCAR 7 and on the Phase 3A spacecraft should be clear. Over a two-decade time span, the factors entering into frequency selection are apt to change. Today, we have to take into account the fact that equipment for 70 cm is as easily and widely available as that for 2 meters, and that low-noise preamps for 70 cm are relatively inexpensive. Also, in some portions of the world 145.8-146.0 MHz is so congested that a 2-meter downlink is unusable.

Finally, Mode J is a more efficient stepping stone to Mode L than Mode B. In sum, there are excellent arguments for switching to Mode J. However, there is also a new important factor favoring Mode B—the need to provide continuity of service to those who have supported the amateur satellite program over the years. It would have been unfair to ask amateurs who set up for Mode B on OSCAR 10 to switch to Mode J for OSCAR 13. For these reasons, we're likely to see both modes supported in coming years.

In Chapter 5, we discussed many of the advantages and disadvantages of various modes. One key factor was total available bandwidth. Whether we opt for Mode B or Mode J, we're restricted to 200 kHz total spectrum, and this is already clearly inadequate for supporting CW and SSB communications. Using a 70-cm and 23-cm combination for a transponder provides 3 MHz of bandwidth. There's no downlink choice here, since the allocation table states that 23 cm can only be used as an uplink. Mode L is clearly the preferred mode for direct amateur-to-amateur CW, SSB and digital communications over the coming decade.

If amateurs were to restrict themselves to communicating via CW, SSB and today's digital modes, Mode L would suffice for a long time into the future. However, if we wish to use satellites for real-time video, digital backbone systems and terrestrial repeater linking, the bandwidth provided by Mode L is not sufficient. (These are just a few of the ideas being pursued for Phase 4 spacecraft—see Chapter 4.) Mode S (23 cm up, 13 cm down) can provide 10 MHz per spacecraft, and several spacecraft can operate on a non-interfering basis. Even if this is not a high immediate priority, Mode S must be utilized as a satellite link if we wish to retain it for the future. OSCAR 13 includes a transponder with a 13-cm downlink that will be scheduled on a regular basis. This transponder will provide valuable data for future missions. Phase 4 spacecraft will include a major commitment to Mode S.

As we go to higher frequencies, the link performance figures from Table 13-5 look a little disturbing. However, we can use spot beam antennas and/or shaped beam antennas to improve the situation. With spot beam antennas on the spacecraft, operation at even higher frequencies is possible and will be looked into. Shaped beam antennas are being seriously considered for the Phase 3D spacecraft. With such antennas, it will be possible to increase spacecraft antenna gains by about 3 dB at apogee on Mode L (see Table 13-5), and then switch to a fan-shaped pattern when the spacecraft moves away from apogee. Additional information on the shaped beam antennas for Phase 3D is contained in Chapter 15.

The considerations leading to frequency selection for low-altitude satellites are considerably different. First, there are actually at least two distinct classes of low-altitude spacecraft. One type carries linear transponders with at least one HF link. This class continues to have widespread appeal and it serves an important function by attracting new blood to the amateur satellite service, so it deserves continuing support. The second class is the Pacsat, which requires the reliable link performance provided by VHF and UHF links. Frequency selection for this service is not clear. The Mode-J combination is the current choice, but increased bandwidths associated with rising data rates and a proliferation of spacecraft may result in a need to switch to Mode L.

PREDICTING SIGNAL LEVELS: AN EXAMPLE

The performance of a 435-MHz satellite downlink from geostationary altitude (Phase 3 or 4) may be calculated in the following manner.

Spacecraft Characteristics

Transponder
Total power = 35 watts average
Bandwidth = 300 + kHz

Antenna
Gain = 10 dBi
Apogee height = 35,800 km

Ground Station Characteristics

Antenna
Gain = 13 dBi
Sky temperature as seen by antenna (T_s) = 150 K
(often much better—see Fig 9-1)

Receiver
Total noise figure (F_T) = 2.2 dB
Bandwidth (B) = 3 kHz (for SSB)

Ground station-Satellite distance

(slant range) = 42,000 km (ground station at edge of coverage cone at apogee)

Our objective is to calculate the expected signal-to-noise ratio (SNR) of a typical downlink SSB signal.

$$SNR = \frac{Received\ Signal\ Power}{Received\ Noise\ Power} = \frac{W_s}{W_n}$$

$$SNR\ [in\ dB] = 10 \log W_s - 10 \log W_n = P_s - P_n$$

Our approach will be to (1) compute P_s, (2) compute P_n and (3) evaluate SNR.

Step 1: Computation of Received Signal Power

Assume that the transponder is handling 70 equal-power SSB contacts. The average power allocated to each user is therefore 0.5 watt. For unprocessed SSB, this represents about 3 watts PEP (34.8 dBm). (Note: dBm =

dB above 1 milliwatt.) Free-space path-loss may be calculated from

$$L = 10 \log \left(\frac{4\pi\rho}{\lambda}\right)^2$$

where
L = free-space path-loss in decibels
ρ = slant range in meters
λ = wavelength in meters

For calculation, it's easier to use the equivalent formula

$$L = 32.4 + 20 \log f + 20 \log \rho \qquad (Eq\ 13.5)$$

where
f = frequency in MHz
ρ = slant range in km

When applying Eq 13.5, be sure to express the variables in the units indicated. In our example

$$L = 32.4 + 20 \log (435) + 20 \log (42,000) = 177.6\ dB.$$

We now evaluate P_s

P_s = transmitted signal power [in dBm]
+ satellite transmit antenna gain [in dBi]
+ ground station receive antenna gain [in dB]
− free-space path-loss [in dB].

P_s = 34.8 dBm + 10 dBi + 13 dBi − 177.6 dB
= −119.8 dBm.

Step 2: Computation of Received Noise Power

Received noise power is given by

$$W_n\ [in\ milliwatts] = k\ T_e\ B$$

where
k = Boltzmann's constant = 1.38×10^{-20}
[in (milliwatts)/(hertz)(kelvin)]
T_e = effective system temperature (discussed below)
B = receiver bandwidth [in hertz]

Note: All temperatures are in kelvins. Temperatures in the Kelvin scale are referenced to absolute zero (Celsius temperature + 273°). Room temperature is defined as 17°C = 290 K. "290 K" is read as "290 kelvins"—there's no degree sign and the expression "degrees kelvin" is *not* used.

The effective system temperature (T_e) takes into account (1) noise picked up by the receive antenna (cosmic noise plus radiation from the earth at ~290 K that enters the main or side lobes) and (2) noise generated in the receiver. The temperature of the receive system (T_R) can be computed when the system noise figure (F_T in dB) is known. (See Fig 9-3 for a discussion of F_T.)

$$T_R\ [in\ K] = 290\ (10^{F_T/10} - 1) \qquad (Eq\ 13.6)$$

When using Eq 13.6, any feed-line losses between the antenna and first active receiver stage must be included in the receiver noise-figure computation, as illustrated in Fig 9-3. Applying Eq 13.6 to our example, we obtain

$$T_R = 290\ (10^{0.22} - 1) = 191\ K$$

The sky temperature was given as $T_s = 150$ K, so we now have everything needed to evaluate T_e

$$T_e = 191 + 150 = 341 \text{ K}$$

The total received noise power can now be calculated

$$\begin{aligned}
P_n \text{ [in dBm]} &= 10 \log W_n \text{[in milliwatts]} = 10 \log (K\, T_e\, B) \\
&= 10 \log k + 10 \log T_e + 10 \log B \\
&= 10 \log (10^{-20}) + 10 \log (1.38) + 10 \log (341) \\
&\quad + 10 \log (3000) \\
&= -200 + 1.4 + 25.3 + 34.8 = -138.5 \text{ dBm}
\end{aligned}$$

Step 3: Calculation of SNR
$$\begin{aligned}
\text{SNR [in dB]} &= P_s \text{ [in dBm]} - P_n \text{ [in dBm]} \\
&= -119.8 - (-138.5) \\
&= 18.7 \text{ dB}
\end{aligned}$$

An 18.7-dB SNR indicates a very good quality signal. Note that the calculations were based on a good (but in no way exotic) 70-cm receive setup situated as far as possible from the spacecraft. For paths where the slant range is shorter, the sky noise temperature behind the satellite is lower or the preamp in use is better, the link SNR will exceed the value calculated. For situations where a user doesn't have sufficient uplink EIRP to drive the transponder to the indicated output power, or where the number of stations simultaneously using the transponder is greater than the 70 assumed, the link SNR will be less than the calculated value.

For additional information on calculating link performance, see the following references:

1990 ARRL Handbook (Newington: ARRL, 1989), Chapter 23: Space Communications, EME Path Loss, pp 23-30.

B. Atkins, "Estimating Microwave System Performance," *QST*, Dec 1980, p 74. Contains a brief but very clear example of calculating link performance at 10 GHz.

J. D. Kraus, *Radio Astronomy* (New York: McGraw-Hill, 1966). Chapters 3 and 7 (by M. E. Tiuri) contain advanced-level information on calculating ultimate receiver sensitivity. Also see references in Table 9-3.

Frequency Management

Now that the legal constraints and the technical trade-offs have been considered, we get down to the difficult problem: All frequencies allocated to the amateur-satellite service are shared with the Amateur service. Therefore, it's extremely important for satellite users and the general amateur community to establish guidelines for frequency use. Satellite buffs can pursue two paths: (1) Use bands that are sparsely populated, or (2) educate the general amateur population as to the goals and constraints of the amateur-satellite service so that, even if not personally interested in space communications, they'll understand how space activities help justify the existence of the amateur service.

The significance and difficulty of the educational task should not be underestimated. When a local radio club has pioneered in the development of fast-scan ATV repeaters over many years, it will take considerable tact to convince them to invest time and cash in switching to a different segment of the 70-cm band. Likewise, when a country has 100 times as many amateurs equipped for 2-meter FM as for satellite operation, a dedicated effort will certainly be needed to explain why certain segments of the crowded 2-meter band should be considered off-limits.

The amateur satellite service is taking a balanced approach. For several years, worldwide support for establishing exclusive satellite segments at 29.300-29.500 MHz, 145.800-146.000 MHz, 435.000-438.000 MHz, 1.26-1.27 GHz and 2.40-2.45 GHz has been growing. Two decades ago, when the 10-meter and 2-meter proposals were made, the band segments were almost empty. Today, nearby crowding often leads amateurs who are unaware of the "gentlemen's agreements" on frequency management to move into the "open space," not realizing that they're disrupting satellite links. A continuing, tactful educational program is a necessity.

Returning to our primary concern, choosing transponder frequencies, it's clear that the 2-meter segment utilized by Modes B, J and A has already become severely overcrowded. Mode L will clearly be the mode of choice for direct amateur-to-amateur communication via high-altitude satellites over at least the next decade. Keep in mind that although this mode was not our first choice, this does not imply inferior performance once we take into account terrestrial 2-meter congestion and steerable spacecraft antennas.

Summary

An understanding of the frequency selection process should make it clear that AMSAT designers have not rushed up the frequency ladder with callous disregard for users' needs. The complex constraints and trade-offs involved in transponder frequency selection have always been considered carefully, and some difficult decisions have had to be made.

For detailed information on the frequency selection process, see the papers by R. Soifer "Frequency Planning for AMSAT Satellites" (pp 101-127) and K. Meinzer, "Spacecraft Considerations for Future OSCAR Satellites" (pp 137-143) in the *Proceedings of the ARRL Technical Symposium on Space Communications*, Reston, VA, Sep 1973, Newington, CT: ARRL. Despite advances in low-noise microwave transistors, the new emphasis on high-altitude Phase 3 and geostationary satellites, the availability of new bands, and the introduction of Pacsats, the conclusions of both of these studies remain valid.

Chapter 14

Weather, TV and Other Satellites

O ur focus so far in this text has been on OSCAR satellites. However, a great deal of general information has been presented that applies to all spacecraft systems. Two programs which radio amateurs have expressed considerable interest in are weather and TV-broadcast satellites. This chapter will provide an overview of these two systems. In the space allotted we can't begin to provide the details needed to set up a complete receiving and processing station, but we can explain what signals are available and direct you to sources of detailed construction information. In addition to the two satellite systems mentioned, we'll also discuss a related activity often referred to as satellite sleuthing: determining the origin and nature of unidentified radio signals from space.

WEATHER SATELLITES

When people refer to "weather satellites," they're actually referring to two separate classes of spacecraft, each carrying several instruments for earth monitoring and broadcasting or relaying downlink data. Direct broadcast services (DBS) are currently provided by low-altitude polar-orbiting US and USSR satellites, and by geostationary satellites operated by the US, Japan and the European Space Agency. The broadcasts of image data provided by these spacecraft are our primary interest.

The polar orbiting satellites include the US TIROS-N series (currently NOAA-9, 10 and 11) and the Soviet Meteor satellites. Both satellite series transmit images in the Automatic Picture Transmission (APT) format and use frequencies near 137 MHz. Setting up a ground station for APT reception from polar-orbiting spacecraft gives you access to both series. The NOAA spacecraft also transmit images in a digital High Resolution Picture Transmission (HRPT) format at 1691 MHz. Data for APT and HRPT transmissions comes from an imaging device known as the AVHRR (Advanced Very High Resolution Radiometer). We'll discuss the NOAA spacecraft shortly. Table 14-1 summarizes the myriad of acronyms employed by those operating and using environmental spacecraft.

The geostationary program includes the US GOES (Geostationary Operational Environmental Satellite) series. These satellites downlink two services near 1.7 GHz—WEFAX and VISSR/VAS. WEFAX (Weather Facsimile) pictorial data is prepared on the ground from various sources, placed in the APT format, and relayed through a transponder on the spacecraft. The VISSR/VAS (Visible-Infrared Spin Scan Radiometer) service is used to downlink high-resolution digital data from an imager aboard the spacecraft.

Table 14-1

Weather Satellite Acronyms and Abbreviations

US Agencies

NESDIS (formally NESS): The National Environmental Satellite Data and Information Service. The agency responsible for operating and disseminating information related to US Weather Satellites. NESDIS is part of NOAA.

NOAA: The National Oceanic and Atmospheric Administration, a branch of the US Department of Commerce.

NTIS: The National Technical Information Service acts as a central clearinghouse for publications by NOAA and other government agencies.

Spacecraft Systems

GMS: Geostationary Meteorological Satellite (Japan).

GOES: Geostationary Operational Environmental Satellite (US).

GOMS: Geostationary Operational Meteorological Satellite (USSR).

Meteor: Low-altitude, polar-orbiting meteorological satellites (USSR).

METEOSAT: European Geostationary Meteorological Satellite (European Space Agency).

NOAA-n: TIROS-N series, low-altitude, polar-orbiting (US).

TIROS: Television Infrared Observation Satellite. Early US low-altitude, polar-orbiting satellite series. Generic name for all US low-altitude, polar-orbiting satellites.

Miscellaneous

AVHRR: Advanced Very High Resolution Radiometer. The imaging instrument on NOAA-n satellites

APT: Automatic Picture Transmission. A data-encoding scheme used to downlink pictorial information in an analog format. Designed to minimize complexity of receiving and decoding equipment.

DBS: Direct Broadcast Services provided by NESDIS

DRS: Direct Readout Services provided by NESDIS

HRPT: High Resolution Picture Transmission. A high-resolution digital data-encoding scheme used for downlinking information from NOAA-n spacecraft.

SVISSR (stretched VISSR): VISSR data that has been processed at an earth station and relayed through a GOES satellite.

VISSR: Visible and Infrared Spin Scan Radiometer. The imaging instrument on GOES spacecraft.

VISSR/VAS: See VAS.

VAS: Short for VISSR Atmospheric Sounder. A new version of VISSR with additional capabilities.

WEFAX: Weather Facsimile. This term is used to refer to (1) a communications system aboard GOES spacecraft and (2) a service provided by NESDIS through GOES spacecraft.

Background

In the early 1960s, when satellites designed to provide cloud-cover pictures were first being developed, a group of far-sighted individuals realized that such spacecraft held tremendous potential for saving lives and for improving the quality of life of people all over our planet. The question naturally arose as to whether the downlink information should be encoded and marketed commercially to partially offset the costs of the project or made freely available. Many believed that it was not proper to withhold such information, since in many cases lives could be directly affected.

As a result, decisions were made to provide US weather-satellite information freely to anyone who wished to use it and to adopt the APT image-encoding system, which required relatively simple receiving and processing equipment at the ground station. Potential beneficiaries around the world were encouraged to use the system without charge; information on the satellite system and ground-station construction was widely distributed; and a commitment was made to support the APT format for an extended period of time.

The first APT system was put into service in late 1963 aboard the low-altitude, polar-orbiting spacecraft TIROS VIII (Television Infrared Observation Satellite). Current US polar-orbiting spacecraft are still frequently referred to as TIROS satellites. Even though satellite-borne imaging equipment has changed drastically over the years, the APT format for encoding downlink pictorial information has undergone only minor revision. Indeed, receiving equipment built in the mid-1960s can still be used with only slight changes, and plans are to continue supporting the APT format indefinitely.

Pictures from space have proven valuable for several applications other than weather forecasting. Images have been used for natural-resource assessment, crop management, efficient air/sea navigation, forest-fire spotting, and so on. A detailed list of applications may be found in "The Economic Benefits of Operational Environmental Satellites" by J. Hussey (March 1983), Chief, Satellite Services Division, NESDIS (National Environmental Satellite Data and Information Service).

Early on in the weather-satellite program, it was found that different surface features register most clearly in different parts of the electromagnetic spectrum. A great deal of effort has gone into identifying spectral segments of greatest utility for discerning important features and into developing instruments that respond to these segments. By using instruments that are sensitive to infrared (IR) and near-IR radiation in addition to visual light spectra, we can choose to focus on cloud cover, land/water boundaries, ice/water transitions, forests, snow cover, forest fires, and so on. The downlink images we see are more likely to have been recorded in the IR or near-IR portions of the spectrum than in the visible part.

APT Polar Orbiting Spacecraft

The US low-altitude APT system is operated by

Table 14-2

APT Spacecraft that can be heard in the US (mid-1990)

Spacecraft	Location	Status
Low-altitude polar orbits		
NOAA-11	———	APT: 137.62 MHz; HRPT: 1707.0 MHz
NOAA-10	———	APT: 137.50 MHz; HRPT: 1698.0 MHz
NOAA-9	———	APT: 137.62 MHz; HRPT: 1707.0 MHz
Meteor 2-19	———	APT: 137.40 MHz
Meteor 2-18	———	APT: 137.30 MHz
Meteor 2-17	———	APT: 137.40 MHz; APT: 137.30 MHz
Meteor 2-16	———	APT: 137.40 MHz
Meteor 3-3	———	APT: 137.85 MHz
Meteor 3-2	———	APT: 137.85 MHz
OKEAN 1 (USSR)	———	APT: 137.40 MHz
Geostationary Orbits		
GOES-7	108.6°W	WEFAX: 1691.0 MHz; (only operating GOES imager); stretched VISSR/VAS: 1687.1 MHz
GOES-6	135.7°W	WEFAX: 1691.0 MHz

Notes

(1) NOAA and GOES spacecraft operate 24 hours per day.
(2) Other spacecraft operate on limited and changing schedules.
(3) Most spacecraft have alternate transmit frequencies available.

NESDIS. Satellites are known as NOAA-n, where "n," which is a letter prior to launch, is changed to a number after launch. Current NOAA (National Oceanic and Atmospheric Administration) spacecraft are part of the third generation of TIROS spacecraft, known as the TIROS-N series. Satellites transmitting APT data as this is written are listed in Table 14-2.

NOAA spacecraft operate 24 hours per day. The imaging instruments cover a ribbon-like area centered on the ground track. This area is scanned from side-to-side (perpendicular to the ground track) twice per second. Each scan line covers a strip approximately 2 km by 3000 km. The AVHRR records data from five spectral channels during each scan. Data from all five channels (plus additional information) is immediately transmitted down via the digital HRPT system. APT data consists of samples from two channels.

During the day, one of the channels is usually in the visible spectrum, but at night both are generally IR or near-IR. The downlink information is multiplexed so a line from one sensor channel is concatenated with a line from the second sensor channel to produce a single downlink line. As a result, when the pictures are printed at the same rate they're recorded at (two lines per second or 120 lines per minute), you see a pair of side-by-side images covering the same geographical region at different spectral frequencies.

Whenever the satellite is in range, these transmissions can be received, so most ground stations are able to obtain coverage of a region about 6000-km long and 3000-km wide, centered on their geographic location, on a near-overhead pass at least twice a day. The downlink signal contains sync pulses indicating the beginning of each side-to-side strip and separating the two images, but the ribbon is continuous—there are no sync signals to mark the top or bottom of a picture.

The HRPT system downlinks data at the full resolution of the AVHRR: about 1 km at the subsatellite point, decreasing to about 4 km at the edges of the coverage area. The ground resolution of the APT system is about 4 km. APT signals are processed before downlinking to remove panoramic distortion.

Soviet Meteor satellites also transmit in the APT format. Several spacecraft in the Meteor-2 series, which has been in operation since 1976, are currently transmitting (see Table 14-2). The operating parameters of the Meteor spacecraft were first determined and publicized by amateurs equipped for TIROS reception. Grant Zehr, WA9TFB, has been reporting on the Soviet weather-satellite program since the early 1980s. For the past several years, his activity summaries have been appearing in the *Journal of the Environmental Satellite Amateur Users' Group*.

The APT format employed by the Soviet polar orbiters is nearly identical to that of the NOAA spacecraft, but there are a few differences that it is important to be aware of. At times, the Meteor satellites may use 240 or 20 lines-per-minute transmission rates. Meteor spacecraft generally operate only when in sunlight. The downlink signal of the Meteor spacecraft has a slightly narrower bandwidth than that of the NOAA spacecraft. This will be discussed when we focus on ground-station RF equipment. Finally, some Meteor spacecraft use an audio subcarrier in the vicinity of 2500 Hz in place of the exact 2400-Hz signal used by NOAA spacecraft. Most approaches to image reconstruction are not sensitive to this parameter, but those that are may have to be modified.

Tracking of NOAA and Meteor spacecraft is easily accomplished using either the OSCARLOCATOR or computer techniques described for OSCAR Phase-2 spacecraft. Since the NOAA, Meteor and OSCAR satellites are in similar orbits, we again see the familiar pattern of two or three ascending passes, followed half a day later by two or three descending passes. Most sources (published and computer bulletin board) that provide orbital elements for the OSCAR spacecraft also provide data on weather satellites.

APT Geostationary Satellites

GOES satellites contain two subsystems of primary interest: the VISSR imaging system and a transponder for downlinking WEFAX images in the APT format. The VISSR system collects data on the full earth disk (the segment of the earth in view). The ground resolution is 1 km in the visible and 7 km in the IR. This data is downlinked to a special processing station in Wallops Island, VA, where it is processed (reformatted, calibrated and gridded) and immediately sent back up to transponders on one or more GEOS spacecraft. The retransmitted data,

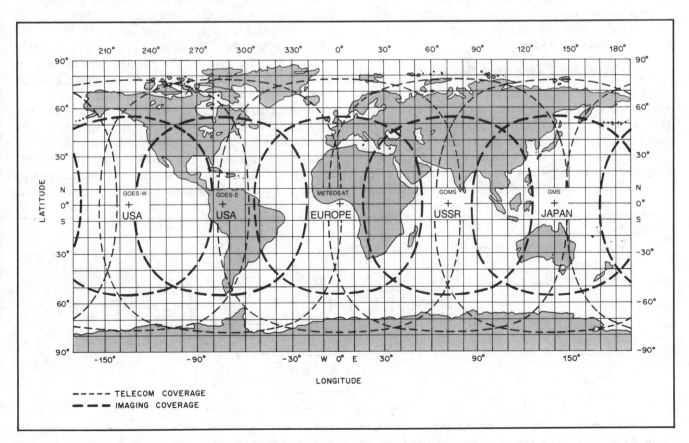

Figure 14-1—Five geostationary WEFAX satellites provide weather-satellite pictures worldwide.

called "stretched VISSR" (SVISSR), is available to any user with a ground terminal capable of receiving and demodulating the digital 1.7-GHz signal.

The WEFAX transponder simply relays whatever NESDIS sends up. This information is similar to the APT format except for the fact that pictures have a discrete beginning and end (top and bottom) and are sent at 4 lines per second (240 lines per minute). GOES WEFAX currently devotes about 50 percent of its time to relaying processed VISSR images, 30 percent to relaying images from the NOAA spacecraft and 20 percent to National Meteorological Center charts and text information related to scheduling of services, TIROS orbit information, and operational plans.

The term WEFAX identifies both a communications subsystem and a service provided by, and through, a cooperative system of worldwide geostationary spacecraft. Fig 14-1 shows the communications and picture coverage provided by geostationary weather spacecraft located at the five internationally agreed-upon positions listed in Table 14-3. Satellites are permitted to deviate slightly from these positions as long as such deviations do not interfere with scheduled international relaying of information. The US currently maintains an additional satellite at a central position, near longitude 250 °E or 260 °E. (The only GOES spacecraft with an operational imaging system [4/90] is switched between the two positions in response to priorities for hurricane and winter storm watch). Current status of the GOES system is summarized in Table 14-2.

When a GOES transponder is relaying both WEFAX data and SVISSR data, the available power is shared between the two services. If you have a marginal receiving setup, it is therefore preferable to choose a downlink that is only transmitting WEFAX.

The APT Ground Station

A ground station set up to receive APT direct-readout services from low-altitude satellites or WEFAX services from geostationary satellites will have two major, distinct subsystems: (1) the RF receive system and (2) the image-reconstruction system. Fig 14-2 shows the RF portion of the system. Fig 14-3 shows two popular approaches to image reconstruction. We'll look at these systems in detail.

Table 14-3

International WEFAX (Weather Facsimile) Geostationary Satellite System

Location Designation	Operated by	Nominal Subpoint Longitude
GOES-East	USA	285°E (75°W)
GOES-Central	USA	255°E (105°W)
GOES-West	USA	225°E (135°W)
METEOSAT	ESA	0°E
GMS	Japan	140°E (220°W)
GOMS	USSR	70°E (290°W)

Notes

(1) The designations refer to locations. Different spacecraft may occupy a given location at different times.
2) GOES-Central is operated in addition to the formal international agreement outlined in Fig 14-1.

The RF System

A conventional FM receiving system is used. The usual procedure is to set up a receiver for the 137-MHz NOAA downlink and then, if desired, add a converter to receive the 1691-MHz GOES WEFAX signal. The RF characteristics of the two links are listed in Table 14-4.

We begin by focusing our attention on the 137-MHz link. The 5-W satellite transmitter produces a strong downlink signal. A signal-to-noise ratio of about 10 dB is required for noise-free images. Many users report that an omnidirectional antenna with an antenna-mounted preamp having a 1.5-dB noise figure provides good results. Because of cosmic noise, there is no advantage to using a preamp with a lower noise figure. Most of the 146-MHz designs mentioned in Chapter 9 can be peaked at 137 MHz without any changes. Suitable antennas are the simple ground plane or Lindenblad. If you feel that a beam is necessary, stick to a low-gain, broad-beamwidth model. Crossed Yagis, three or four elements in each plane, will work fine. So will a 3-turn helix. If the antenna is set at a fixed elevation of 25°, only azimuth rotation will be required. The spacecraft is transmitting an RHCP signal, so an RHCP receive antenna is preferred, but many users report satisfactory results with a linearly polarized antenna.

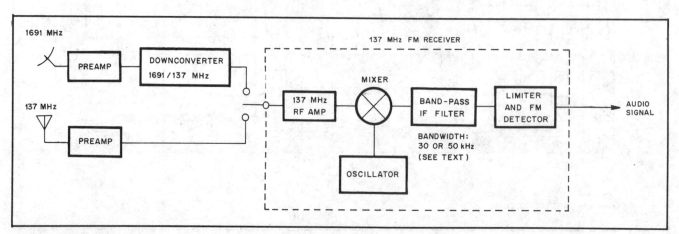

Figure 14-2—RF section of an APT ground station for 137 MHz and 1691 MHz.

Table 14-4
Link Characteristics: TIROS and GOES

	TIROS (low-altitude)	GOES (high-altitude)
Satellites of prime interest[1]	NOAA-9, 10, 11	GOES-6, 7
Name of APT Service	Direct readout	WEFAX
Frequencies	137.500 MHz	1691.00 MHz
	137.620 MHz	1694.50 MHz[2]
Frequency stability	±3 kHz	———
Maximum Doppler	±4 kHz	0 kHz
Transmitted bandwidth	±17 kHz	±9 kHz
Transmitter power	5 W	———
Satellite EIRP (nominal)	37 dBm	56 dBm (GOES)[3]
		52 dBm (METEOSAT)
Satellite antenna polarization	RHCP	Linear
Free-space path loss	141.3 dB[4]	189.5 dB
Ground-station receiver IF bandwidth	50 kHz	30 kHz[5]

Notes

[1]Operating 4/90
[2]METEOSAT only
[3]Shared with SVISSR
[4]At 2000 km
[5]300 kHz for GMS

Table 14-5
Weather-Satellite Information Resources: US Agencies

NESDIS is a key resource for operational information on NOAA and GOES spacecraft. The following offices may be of assistance.

APT/HRPT Coordinator
NOAA/NESDIS
World Weather Bldg, Room 806
Washington, DC 20233
(Ms Mona F. Smith E/SP21, 301-763-8447)

WEFAX Coordinator
NOAA/NESDIS
World Weather Bldg, Room 806
Washington, DC 20233
(Mr Jim Green E/SP21, 301-763-8131)

Educational Support Services
NESDIS Constituent Affairs Officer
NOAA Office of Legislative Affairs
Federal Office Bldg 4, Room 0158
Washington, DC 20233
(Mr Doug Brown LA3(E), 301-763-4690)

Technical Reports and Memorandums
Training and Information Services
NOAA/NESDIS
World Weather Bldg, Room 703
Washington, DC 20233
(Ms Nancy Everson E/RA22, 301-763-8204)

A periodical called *APT Information Notes* is issued quarterly by NOAA/NESDIS. Subscriptions are free. Request an application from the APT/HRPT Coordinator.

NOAA publishes a series of Technical Memorandums (TMs) and Technical Reports (TRs) on weather satellites. For lists of NOAA publications, contact Training and Information Services. NOAA generally has a limited supply of publications available for free distribution. Once this supply is exhausted, copies may be obtained from NTIS.

NTIS will supply copies in microfiche or hard-copy format. Since microfiche copies generally cost less than 10 percent of the hard-copy price, you might wish to check your local library to see if they have a microfiche reader available. Ordering is usually a two-step process. First request the price of the documents (specify as completely as possible), then send your check. The address is: NTIS, Dept of Commerce, 5285 Port Royal Rd, Springfield, VA 22151.

No matter what antenna is used, be sure to use a preamp mounted at the antenna.

You should strive for the lowest possible receiver noise figure. The FM receiver system front end (through the mixer) is identical to that of a CW/SSB receiver and its performance can be quantified in a similar manner (see Chapter 9). Aim for the same performance you would for OSCAR 13 Mode B—about a 2-dB overall noise figure.

The band-pass characteristics of the FM receiver IF section are extremely important to overall receiver performance. If the IF is too wide, you lose overall signal-to-noise ratio. This is acceptable as long as the signal remains above a threshold level. If the IF is too narrow, you improve signal-to-noise ratio, but lose resolution. Imperfections in the shape of the band-pass curve and the response of the discriminator (FM detector) will degrade the picture. Many users with operational systems attribute the poor quality of their images to the limitations of the APT system, when the real problem is IF shape. The APT system is capable of producing excellent images.

The preferred IF system for NOAA reception will have a 50-kHz bandwidth with relatively steep skirts. This value allows for Doppler shift and some frequency drift in the satellite and at the ground station. With a 50-kHz bandwidth, you do not have to retune to compensate for Doppler. Unfortunately, 50 kHz is not a standard bandwidth. Most FM communications receivers designed for this part of the spectrum use 15- or 30-kHz-wide ceramic filters operating at 10.7 MHz or 455 kHz to establish the IF band-pass. Quality commercial receivers for weather-satellite reception will therefore generally use conventional tuned circuits (LC) to establish the correct IF response. A 30-kHz band-pass will work fine for GOES and for the

Soviet Meteor APT spacecraft, but will cause some loss of detail with NOAA spacecraft. The design and construction of a 50-kHz-wide IF strip and matching discriminator is not overly difficult, but cookbook information is not readily available. If the topic interests you, check old (1960s) *Radio Amateur's Handbooks* or *RadioTron Design Handbooks* of similar vintage for the necessary information.

As stated earlier, details of ground-station construction will not be covered here. Rather, an overview of approaches that have been successful and references to construction information will be presented. The references in the following paragraphs are contained in Tables 14-5 through 14-8.

The most popular approach to APT receiver procurement is to purchase an inexpensive pre-built or kit receiver. Hamtronics produces a popular kit and Vanguard manufactures a ready-to-plug-in unit. Both are widely used. There are a number of companies in the market, so it's wise to check current prices. The DBS office at NESDIS maintains a list of current manufacturers of receiving equipment. The lower-cost receivers are usually crystal controlled. They generally solve the IF-shape problem by using a single 30-kHz-wide ceramic filter with wide skirts. As a result, their resolution and interference rejection are not up to the better commercial units. If you go this route, it may be helpful to use two preamps, one at the antenna, and one at the receiver input preceded by a sharp bandpass filter centered at 137.5 MHz.

Amateurs have also converted scanner radios for APT reception. Used crystal-controlled units covering the appropriate frequencies are often available very inexpensively. If you go this route, you will of course have to purchase crystals for the frequencies of interest and make other modifications. A circuit diagram is really critical here. Some receivers use single conversion, with a single 15-kHz ceramic filter in a 10.7-MHz IF strip. Others use dual conversion, with most filtering taking place at 455 kHz. Modification depends on the basic design.

In the *Weather Satellite Handbook*, Taggart details several approaches. In one case involving a single-conversion receiver, it was only necessary to replace a 15-kHz ceramic filter with a 30-kHz model. In another case, this time involving a dual-conversion model where the 10.7-MHz IF had a 30-kHz filter and the 455-kHz IF had a 15-kHz filter, it was only necessary to replace the filter in the second IF with a pair of capacitors. In a carefully designed receiver, it would also be necessary to modify the discriminator to match the IF response, but this does not seem necessary in these conversions. It appears that most manufacturers are using general-purpose designs that will operate at 15-kHz or 30-kHz bandwidths with the appropriate filter. Front-end circuits in these units have usually been stagger tuned. They should be peaked at 137.5 MHz. The previous comments about a second preamp and additional front-end selectivity also apply here. Scanner receivers vary tremendously in quality. The better ones will work well for APT, while the poorer ones are useless.

In the late 1960s, several commercial FM services (such as taxi and police communications) switched from wideband-FM (30 kHz) to narrowband-FM (15 kHz) operation, and a lot of the early equipment reached the surplus market. Although less available than they once were, receivers from this service are generally excellent for APT reception. Conversion is usually relatively simple (peaking tuned circuits in the front end and crystal-multiplier chain) for someone with RF experience and the schematic(s) for the equipment being operated on. The resulting receiver will be a little narrow for NOAA APT and this may result in some lost definition, but overall

Table 14-6

Weather-Satellite Information Resources: Computer Bulletin Boards, Books and Periodicals

Computer Bulletin Boards
(Also see Table 10-5)
NOAA Direct Readout Users' Bulletin Board (NOAA.DRUSER) via PAC*IT Plus network or SCIENCEnet via WXSAT gateway
Celestial RCP/M: 513-427-0674
Dallas Remote Imaging Group: 214-394-7438

Key Books and Periodicals on Weather Satellites
Summers, R. J., *Educator's Guide For Building and Operating Environmental Satellite Receiving Stations,* NOAA Technical Report NESDIS 44, 1989. Focuses on APT reception from NOAA and GOES spacecraft.

Taggart, R. E., *Weather Satellite Handbook*. This is the best source of practical information on constructing APT ground stations for reception from GOES, NOAA and Meteor Spacecraft. The current fourth edition (1990) is published by ARRL.

C. H. Vermillion, *Weather Satellite Picture Receiving Stations: Inexpensive Construction of Automatic Picture Transmission Ground Equipment*, NASA SP-5080, 1969. NTIS Accession no. N69- 31985. The material in this classic on TIROS reception is somewhat dated, but sections are still helpful.

JESAUG (Journal of the Environmental Satellite Amateur Users' Group) has been published quarterly since late 1973. It contains a great deal of practical operational and construction information. For subscription information, contact the editor, Dr J. Wallach, PO Box 117088, Carrollton, TX 75011-7088.

NASA APT Information Notes, prepared quarterly by NOAA/NESDIS. Contact APT/HRT Coordinator for subscription application.

J. Barnes and M. Smallwood, *TIROS-N Series Direct Readout Services Users' Guide*, March 1982. Prepared under contract for NOAA/NESDIS. For information, contact Direct Readout Services, NESDIS.

J. Clark, *The GOES User's Guide*, NOAA/NESDIS, June 1983.

picture quality often turns out better than the other inexpensive approaches because of superior interference rejection and limiting and discriminator characteristics. Note that these comments apply only to the older wideband equipment; modern gear using 15-kHz-wide IF strips cannot be used for APT reception. Modifications are generally very difficult due to specialized circuitry, use of nonstandard IF center frequencies and the need to match the discriminator to the IF passband.

We now turn our attention to the 1691-MHz GOES WEFAX downlink. The usual approach to receiving GOES is to convert the 1691-MHz downlink to 137 MHz and feed the signal into a NOAA APT receiver. If the receiver is a commercial unit with a 50-kHz-wide IF, we might lose a few dBs of signal-to-noise ratio, but this can usually be tolerated, especially now that excellent low-noise preamps for 1691 MHz are available. Since many 137-MHz receivers

Table 14-7

Weather-Satellite Information Resources: Selected Construction Articles

Christieson, M., "A METEOSAT Earth Station," *Wireless World*, Jun 1979, Jul 1979.

Christieson, M., "High-Resolution Weather Satellite Pictures," *Wireless World*, Nov and Dec 1981, Jan 1982.

Dahl, J., "A Weather-Facsimile Package for the IBM PC," *QST*, Part 1, Apr 1990, pp 15-24; Part 2, May 1990, pp 17-21.

DuBois, J., "World's First 'GOES VISSR' Amateur Station Built by Massachusetts Amateur," *JESAUG*, 1985, no. 4.

DuBois, J., "A Low Cost GOES VAS Imaging System for PC/XT/AT host Systems," *JESAUG*, Part I, 1988 no. 1; Part II, 1988, no. 2; Part III, 1988, no. 3.

DuBois, J., "Amateur HRPT is Alive and Well in the United States," *JESAUG*, 1988, no. 3.

Ehrler, G. and J. DuBois, "A Very High Performance LNA for 1500-1750 MHz," *JESAUG*, 1989, no. 1. Also see: DuBois, J., "Recent Improvements to the Ehrler LNA," *JESAUG*, 1989, no. 1.

Emilani, G., and M. Righini, "An S-band Receiving System for Weather Satellites," *QST*, Aug 1980, pp 28-33.

Emilani, G., and M. Righini, "Printing Pictures from 'Your' Weather Geostationary Satellite," *QST*, Apr 1981, pp 20-25.

Johnson, N., "Soviet Satellites and APT," *JESAUG*, 1988, no. 1.

Schwittek, E., "WEFAX Pictures on Your IBM PC," *QST*, Jun 1985, pp 14-18.

Schwittek, E., "HF WEFAX on the IBM PC," *QST*, Dec 1986, pp 46-47.

Shuch, H. P., "A Weather-Facsimile Display Board for the IBM PC," *QEX*, Sep 1988, pp 3-7, 15.

Shuch, H. P., "A Cost-Effective Modular Downconverter for S-Band WEFAX Reception," *IEEE Transactions on Microwave Theory and Techniques*, Dec 1977, p 1127.

Taggart, R., "New Weather Eye in the Sky," *73*, Nov 1980, pp 176-181. A primer on TIROS-N series spacecraft.

Taggart, R., "Direct Printing FAX," *73*, Part I, Nov 1980, pp 90-98; Part II, Dec 1980, pp 52-56; Part III, Jan 1981, pp 54-57. Contains complete construction information for a FAX printer (using electrosensitive paper) and all associated electronics.

Wallach, J., "A Simple Modification To The Vanguard FMR-260 Receiver To Improve Audio Bandwidth," *JESAUG*, 1988, no. 2.

G. Zehr, "The VIP Image Processor," *QST*, Aug 1985, pp 25-31.

G. Zehr, "ASAT: An Apple-based Satellite Imaging System," *QEX*, Mar 1988, pp 3-9.

G. Zehr, "A Weather Satellite Display System Using A Dedicated CRT," *1990 ARRL Handbook*, pp 28-13 to 28-22.

Table 14-8

Weather-Satellite Information Resources: Equipment and Software Suppliers Mentioned in Text

RF Equipment
Vanguard Labs
196-23 Jamaica Ave
Hollis, NY 11423
(Inexpensive VHF weather-satellite receiver)

Hamtronics
65 Moule Rd
Hilton, NY 14468
(Inexpensive VHF weather-satellite receiver kit, preamps)

Down East Microwave
Box 2310, RR 1
Troy, ME 04987
(1691/137-MHz converters; loop-Yagi antennas and preamps for 1691 MHz)

Spectrum International
PO Box 1084
Concord, MA 01742
(Microwave Modules line of 1691/137-MHz converters and preamps)

Also see "Dallas Remote Imaging Group" and "Satellite Data Systems" below.

A list of suppliers of commercial weather-satellite equipment is available from the DBS coordinator at NESDIS.

Computer Interface and Display Software
David E. Schwittek, NW2T
1659 Waterford Rd
Walworth, NY 14568
(MultiFax software for IBM PC)

A&A Engineering
2521 W La Palma, Unit K
Anaheim, CA 92801
(Interface unit for IBM [Multifax software] and Amiga [Dallas Remote Imaging Group Software])

MacLean and Atkinson
2112 Parsons Ave
Melbourne, FL 32901
(Interface and Software for IBM PC)

Dallas Remote Imaging Group
PO Box 118053
Carrollton, TX 75011
(Amiga Software and turnkey system)

Satellite Data Systems, Inc
800 Broadway St
PO Box 219
Cleveland, MN 56017
(Turnkey display systems for Apple and IBM PCs. Distributor for Vanguard and Microwave Modules RF equipment)

are somewhat under the suggested bandwidth, they're perfect for use in a GOES receiving system.

Because of the greater distance to geostationary orbit and the added path losses at 1691 MHz, the signals on this link are considerably weaker than the 137-MHz NOAA downlink. However, this is, to a large extent, compensated for by the fact that a fixed high-gain antenna can be used at the ground station. Very significant advances in the design of low-noise preamps for these frequencies have occurred in the late 1980s. Designs for high-gain, low-noise, multi-stage preamps that use readily available components and are easy to reproduce have been published by Dubois and Shuch. With one of these preamps and at least a 3-ft dish or 6-ft long loop Yagi, GOES reception is straightforward. Converters may be purchased or built. Several manufacturers producing 23-cm converters for the radio amateur market also produce units for GOES reception; Microwave Modules and Microcomm are two such manufacturers. Though not many converter construction articles have been published specifically for 1691 MHz, several designs have appeared for units which convert the amateur 1.27- and 2.4 GHz bands down to 146 MHz, and these can generally be easily modified for GOES applications.

Image Reconstruction System

The APT receiver output is an audio signal with a 2400-Hz subcarrier (near 2500 Hz in the case of Meteor). The subcarrier is amplitude modulated with video information. Modulation percentage varies from 5 percent (black level) to 80 percent (white level). In addition to the video modulation, trains of square-wave pulses are used to indicate the beginning of each line, separation between two images in the NOAA two-channel signal, and the start and end of the GOES picture. The beginning-of-line sync signal is critical for keeping the printing device synchronized with the transmitted video. Video- and sync-processing equipment depends, to some extent, on the image-reconstruction system being used.

Users have experimented with a great many approaches to image reconstruction over the years. These include the cathode-ray tube (CRT) and camera combination, homemade drum recorders, commercial electrostatic recorders and microcomputer systems.

Camera/CRT System

In the camera/CRT system, the picture is painted, line-by-line as it was sent, on the face of a standard oscilloscope. Since each picture takes several minutes to complete, and the glow of most CRT phosphors dies out after a fraction of a second, this approach isn't appropriate for direct viewing. To see the entire picture, you have to take a time-exposure photograph of the CRT screen. Good points: Detailed designs are available; excellent picture quality is possible; different line rates are easily accommodated. Bad points: Polaroid™ film is expensive; other film formats require darkroom facilities. Reference: G. Zehr, *1990 ARRL Handbook*.

Homemade Drum Recorders

Recorders of this type use a rolling-pin-like drum wrapped with special sensitized paper. The APT signal consists of several hundred lines of information sent in sequence. The drum rotates once during each incoming line; during the rotation, video modulation is transferred to the sensitized paper by a special stylus. For the next line, the stylus is moved slightly and the whole process is repeated. Three types of sensitized paper are in common use: photo-sensitized, electrosensitive and electrolytic. In each case, the incoming video information controls the exposure. With photo-sensitized paper, the stylus is a carefully focused beam of light that varies in intensity. Standard photographic techniques must be used to print the picture. With electrosensitive and electrolytic paper, the stylus is a pin-like device that directs a variable-intensity arc. Electrolytic paper must be treated chemically to view the image; electrosensitive paper has the image burnt directly into it, so no processing is necessary. Good points: Excellent picture quality is possible; construction information is available; electrosensitive paper requires no processing (cost is modest compared to Polaroid film, but can mount up). Bad points: Mechanical construction is an obstacle; difficult to handle different line rates; photosensitized and electrolytic paper require messy processing facilities. References: [electrosensitive paper] R. Taggart (Nov and Dec 1980 and Jan 1981 *73*); [photo-sensitive paper] G. Emiliani and M. Righini (Apr 1981 *QST*).

Commercial Electrostatic Recorders

Commercial electrostatic recorders using electrosensitive paper are similar to the drum-type devices described above. The cost of new units is generally prohibitive, but it has been possible to obtain used units on the surplus market. See comments regarding drum printers. For additional information, see R. Summers, 1989.

Microcomputers

Microcomputers have radically altered image-reconstruction options over the past few years, and changes over the next few years are likely to be more rapid. The computer is more than an image-reconstruction medium; it is a tool that, for the first time, gives users the ability to process and enhance images. Once the raw data is saved, the computer can be used to define the relationship between video-signal intensity and screen brightness in any way we wish. We can experiment with linear, logarithmic, or any other relationships imaginable. Different relationships often make certain cloud or surface features very apparent. Using computers, we can also experiment with averaging procedures on a single image or develop methods for having the computer compare images taken months apart.

Over the next few years, I expect that computer display systems will begin to dominate other approaches to image reconstruction at amateur receiving stations. Computer-storage and video-display technology are evolving extremely rapidly. The first 10-megabyte hard drives costing under $1000 appeared in about 1984. In 1989, a

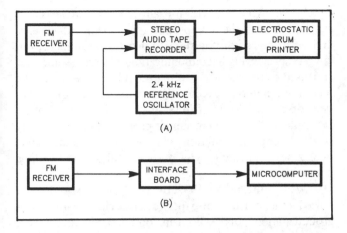

Figure 14-3—Two common methods of image decoding and reconstruction. (A) Electrostatic drum printer and (B) microcomputer.

60-megabyte hard drive was about $400. A 650-megabyte optical drive costs about $4000 as this is written; extra optical disks (it is a simple matter to change the disks, unlike conventional hard drives) for such drives are about $100. As the price of optical drives continues to drop, they will provide a practical way to store gigabytes of weather-satellite pictorial data in an easily accessible format. Computer video displays of the mid-1980s had rather crude resolution. By the late 1980s, major improvements had appeared, but current video standards still force us to deal with images that are clearly inferior to photographs and to ignore 75 percent of the available data when viewing full-coverage APT images. We will no doubt continue to see major improvements in weather-satellite screen displays.

The cost of a microcomputer imaging system often greatly exceeds the cost of other imaging approaches. It is not reasonable, however, to compare the price of a dedicated picture-reconstruction device to a computer, which is probably used for a variety of applications unrelated to weather satellites, and which also provides capabilities for image processing and archiving. The real cost of the computer approach is the incremental investment in a superior video system or optical-storage unit that might not otherwise be needed. To this, one must add the cost of the electronics needed to interface the receiver to the computer and the appropriate software. When considered this way, the computer and other approaches are in the same cost ballpark.

Software for the IBM PC has been described by E. Schwittek, K2LAF, in *QST* (Jun and Dec 1985). A revised and enhanced version of this software is being sold under the name Multifax®. The price is approximately $50, and updates, when issued, are available at very modest cost. The Multifax manual describes a simple interface unit which is available from A&A Engineering for about $70. P. Shuch has described an interface board with additional features that plugs directly into an expansion slot on an IBM PC (*QEX*, Sep 1988). The *FaxBoard* contains the subcarrier demodulator, analog-to-digital converter and

support circuitry needed. Complete details for construction and operation with the Multifax software are included in the article.

Let's take a brief look at some of the major functions provided by Multifax software. Each incoming video line is sampled 1280 times. All data is stored in memory while a sample is routed to the screen. Sections of the screen can then be expanded to the full resolution of the monitor and memory. Program menus provide access to the various options available, which enable you to record a picture, expand sections of a picture, and select line rates suitable for any operating NOAA, GEOS or Meteor spacecraft and for US Navy HF WEFAX broadcasts. Computer displays have drastically reduced the complexity of setting up a weather-satellite receiving system. Today, anyone with an IBM PC or clone can purchase everything needed to assemble a TIROS APT receiving station at modest cost and start collecting images without even touching a soldering iron. For example, a simple system might include a Vanguard receiver, Hamtronics preamp, A&A Engineering interface, Multifax software and a homemade ground-plane antenna.

Other Weather Imaging Services

The first report of amateur HRPT reception from a NOAA satellite was by M. Christieson in 1981. His series of articles in *Wireless World* provides an excellent tutorial on HRPT reception. With today's digital ICs and the powerful microcomputers available, duplicating his feat is now considerably easier, but it was a remarkable accomplishment at the time.

The first, and to date the only, report of amateur SVISSR reception was by John DuBois, W1HDX, in September 1985. Details of his system were published in the *Journal of the Amateur Satellite Users' Group* (1985, no. 4; 1988, nos. 1, 2 and 3).

SATELLITE TV

There are two distinct groups of geostationary satellites designed to downlink TV program material to US ground stations. One group, transmitting in the 3.7- to 4.2-GHz band allocated to the Common Carrier Service, is currently providing more than 40 channels of first-run movies, sports and other special programming. This was not meant to be a broadcast service; it was meant to be a private distribution system. The other group consists of direct-broadcast satellites (DBS) operating in the 11.7- to 12.2-GHz band. These spacecraft were designed to provide direct-to-home services in conjunction with relatively simple, low-cost receiving equipment. But, as we'll see, results don't always coincide with intentions. The two systems are of interest to radio amateurs because of the technical progress they've stimulated in microwave RF techniques and because the story of the evolution of the 4-GHz service in the decade between 1978 and 1988 is intriguing.

4-GHz Band

More than a dozen geostationary satellites, parked

between 70°W and 135°W, are capable of relaying TV programming in the 4-GHz band (3.7-4.2 GHz) down to North American ground stations. The heart of each spacecraft is an FM repeater designed for crossband (uplink: 5.9- to 6.4-GHz) operation. A typical satellite has 24 channels, each of which is 36 MHz (plus a 4-MHz guard band) wide. Each channel, or transponder as it is often called, has its own 5-W amplifier. If you're following closely, you may wonder how we fit all these channels (40 MHz × 24 = 960 MHz) in a band that's only 500 MHz wide. The trick is to use linear polarization, with alternate transponders, spaced 20 MHz center-to-center, polarized at right angles. In effect, we double the available frequency spectrum by using each segment twice, once with each polarization. Interference between neighboring channels is negligible.

Anyone familiar with US NTSC color-TV standards, based on a 4-MHz-wide amplitude-modulated vestigial-sideband video signal, will realize that the 36-MHz-wide FM downlink just described is quite different. As a result, you *cannot* just build a downconverter to shift one of the 4-GHz satellite channels to an unoccupied TV channel. A TVRO (*TV Receive Only*) terminal must capture some downlink RF, filter out the 36-MHz-wide channel of interest, demodulate the video and audio information and reconstruct a standard NTSC TV signal to feed into a regular TV set. If this sounds like a big job, it was. Back in the mid-1970s, commercial stations cost $50,000 and up.

A 24-transponder satellite costs more than 50 million dollars by the time it reaches geostationary orbit; the expense splits about 50-50 between the spacecraft hardware and the launch. With all this money being invested and with the technical expertise available, you may wonder why the system wasn't designed to be easier to receive. The reason is that there was no economic incentive to do so. The system was meant for limited private distribution in accordance with the objectives of the Common Carrier allocations. The satellites are owned by large corporations such as RCA (SATCOM), Western Union (WESTAR) and AT&T/GTE (COMSTAR), with a total investment of over half a billion dollars in orbiting hardware. These organizations aimed to recoup construction and launch costs and generate profits by renting transponders. With the total ground-station market envisioned at well under 10,000 over the next several years, no one thought about mass construction, and saving money on ground stations didn't add up to all that much. Renting transponder time proved to be a viable business because TV networks found them to be cost-effective for distributing programming to affiliates and to the rapidly growing cable industry. As a result, lots of first-run movies and other types of programming began being beamed down.

History

Imagine that it's 1975 and that you've just been involved in the installation of a $100,000 earth station for your employer. Seeing what's involved, you decide to play around to see if you can put together a ground station for a few thousand dollars. The first step is to obtain a license,

which the FCC requires for all common-carrier earth stations. After a pile of paperwork, an experimental license arrives. It gives you the right to develop equipment, but lawyers specializing in communications law seem somewhat confused as to whether you're even allowed to watch the incoming material except as necessary for equipment development. Word filters in about several other experimenters involved in similar projects, so you naturally start trading information via newsletters and, later, magazines. Eventually, you're successful—your TVRO is operational.

Meanwhile, in late 1979, the FCC, finding itself being buried under a deluge of paperwork associated with 4-GHz ground stations, makes licensing optional. However, there are laws governing the common-carrier service that must be followed. In an effort to comply, you write program originators to ask how you can sign up to become a "specific intended recipient." For various reasons, most of the big outfits totally ignore you, but you find one station willing to grant you permission for a modest yearly fee, so your station has some legal standing.

Meanwhile, equipment costs are rapidly decreasing as many of the early experimenters set up for "garage manufacturing." The market for ground stations rapidly expands. Congress starts getting tens of thousands of letters from citizens complaining about the absurd legal situation. In late 1984, legislation is passed legalizing home TVRO terminals. The number of home terminals is on an exponential growth curve, poised to go through the roof. Manufacturers in the Far East begin producing ground-station equipment with specifications and at prices that were undreamed of a few years earlier. A whole new industry is born.

At this point, the corporations building and operating satellites, and the big cable companies, can't ignore the situation any longer. In about 1986, the "premium" channels begin to scramble their transmissions, and the bottom falls out of the home-terminal market. After a few years, a new, more modest equilibrium is reached, and home TVRO becomes just another facet of the consumer home electronics market. (Except for the emergence of a descrambling underground—but that's another story.)

4-GHz TVRO Terminal

One possible TVRO terminal configuration is shown in Fig 14-4. A TV picture with a 48-dB video signal-to-noise ratio (SNR) is defined as being of excellent quality. To obtain a 48-dB SNR at the detector output, we need an 11-dB link margin on the downlink signal; this margin must be maintained up to the detector. It's critically important that the antenna gain and LNA (Low Noise Amplifier = preamp) noise temperature be chosen to provide the needed link margin. LNAs are almost always specified in noise temperature and gain, with 50-dB gain being typical.

A sample link calculation was shown at the end of Chapter 13. One required input was the satellite EIRP in the direction of the ground station. For the 4-GHz satellites of interest, this information is provided in the form of

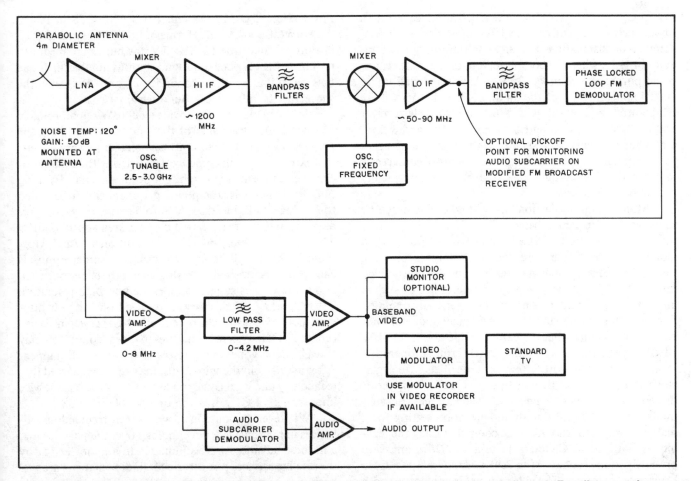

Figure 14-4—Block diagram of typical double-conversion TVRO with nominal values indicated. Excellent results can be obtained using either single- or double-conversion approaches if good engineering practices are followed.

contour charts (in dBW EIRP) like the one shown in Fig 14-5 for SATCOM F1. The central region (where the strongest signal is received) is referred to as the *boresight*. The entire pattern is called the *footprint*. From Fig 14-5, we see that signal levels across the continental US vary by more than 4 dB, making it considerably easier for stations in Chicago to put together a ground station than those in Miami.

As an example, let's consider a station in Chicago that is working with a 35-dBW EIRP signal. A link calculation assuming a 30-MHz receiver IF bandwidth would show that this station could obtain the desired 11-dB link margin using a 4-m diameter parabolic dish antenna and a 200-K LNA. Note that there's a direct trade-off between antenna size and LNA temperature: The Chicago station could obtain the same 11-dB link margin with a 5-m dish and a 300-K LNA, or with a 3-m dish and a 120-K LNA. A station in Miami would need at least a 5-m dish and a 120-K LNA to produce the desired 11-dB margin.

On most Amateur Radio communications circuits (excluding EME), we don't worry about a 1- or 2-dB change in signal level. On a TVRO downlink, however, a difference of this magnitude can be extremely significant. Since (1) the signal level on a particular downlink path rarely varies by more than ±0.5 dB from propagation, (2) the picture does not perceptibly improve once we exceed

the "magic" 11-dB link margin threshold and (3) each decibel of extra link margin may cost several hundred dollars, there's a strong incentive to design the system with

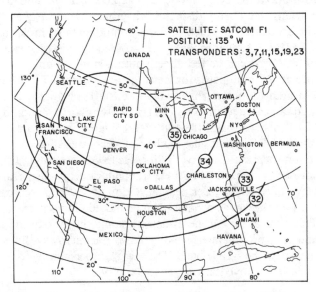

Figure 14-5—This figure shows the footprint (EIRP contours in dBW) for one set of transponders on SATCOM F1. The boresight is near Rapid City, South Dakota.

just enough sensitivity. While system performance degrades rapidly below the 11-dB margin level, most viewers of noncommercial installations find an 8- or 9-dB margin acceptable. In 1986, when the bottom fell out of the market, one could purchase a complete TVRO station for well under 1000 dollars, including $100 for an LNA with a 55-K noise temperature and 50-dB gain, $300 for a tunable receiver and the rest for a dish antenna between 2 m and 3.5 m. Costs have actually risen slightly as the market stabilized, and an antenna positioner and a descrambler can greatly increase the figures quoted. Good buys are frequently available at hamfests.

Although our discussion has focused on US domestic satellites that are parked between 70°W and 135°W, other satellites, including INTELSAT and Soviet birds parked along the geostationary arc and in Molniya orbits (see Fig 11-26 and accompanying text), also distribute video information. The Soviet Molniya satellites are of special interest to experimenters, since they are generally operated during both Eurasian and North American apogees. Full-earth disk antennas that typically provide about 30-dBW EIRP are generally employed. TVRO-terminal operators in most of the US, with a few extra decibels in margin, can catch these transmissions by searching for spacecraft using nominal values for height (35,800 km) and subsatellite latitude (62°N). Azimuth and elevation setting for a search from your location can be computed using the techniques outlined in Chapters 11 and 12. Downlinks are primarily in the range 3.75 to 3.95 GHz, with 3.895 GHz being most common. The video format is compatible with US systems, so you won't have any trouble obtaining a picture. Decoding the color information is complex, however, so most experimenters settle for black-and-white viewing.

Let's look at how the downlink signal from a domestic 4-GHz TV satellite is decoded. The output of the TVRO terminal detector (Fig 14-4) consists of (1) a video waveform, (2) a frequency-modulated audio subcarrier and (3) a triangular energy-dispersal waveform. The *baseband* video waveform, which contains components from dc to 4.2 MHz, is similar to the signal provided by video cameras or VCRs. It can be fed directly into a studio monitor or into an RF modulator for viewing on a standard TV. The FM audio subcarrier is usually at 6.8 MHz or 6.2 MHz. With a peak deviation of 75 kHz, it's similar to standard FM broadcast-band (88-108 MHz) signals. Some homemade TVRO systems use modified FM broadcast receivers to tune across the receiver's 70-MHz low IF searching for subcarrier signals. The 30-Hz triangular energy-dispersal waveform, which has a peak deviation of 750 kHz, serves to move the carrier around when no modulation is present. This reduces the RF energy density at any single frequency and helps prevent interference to terrestrial microwave links that share the 4-GHz band. The energy-dispersal waveform must be removed at the ground station.

12-GHz Direct Broadcast Satellites

Geostationary Direct Broadcast Satellites are assigned the 12-GHz band (11.7 to 12.5 GHz). Experiments with this service began in the late 1970s with CTS in North America (also known as Hermes), BSE in Japan and OTS in Europe. Since the 12-GHz TV downlinks are meant to be a broadcast service to ground stations numbering in the tens of millions, the economics dictate designing the spacecraft to minimize ground-terminal cost.

Initial plans were to provide a 60-dBW signal, roughly 25 dBW higher than that of the 4-GHz service. Since the 4-GHz and 12-GHz spacecraft have similar power budgets, this extra power will come from restricting the number of channels and using high-gain spot-beam antennas. By using only four channels, the power per channel can be raised to 30 W, a 7.8-dB increase. A 50-fold reduction in coverage area (from the entire US to a circular area approximately 300 km in diameter) provides an additional 17 dB. How these channels will be used is not clear. Some proposals call for a service similar to that currently being operated at 4 GHz. Others suggest adopting a new high-resolution TV system. While discussions dragged on for year after year, potential customers opted for a 4-GHz system. Work on the 12-GHz systems continues to creep along. You may as well make your own guesses as to what will happen. After all, back in the mid-1970s, the experts all agreed that technology and legal issues made a 4-GHz broadcast service impossible; 12 GHz was the only viable option.

Whatever the outcome, we've seen tremendous advances in microwave RF technology over the past decade, and these advances were stimulated to a major extent by the satellite services just described. Radio amateurs are just beginning to benefit from the spin-off. LNA performance at frequencies between 0.1 and 10 GHz has increased dramatically over the past decade; the higher the frequency, the larger the advance. Low-cost GaAsFET transistors now provide amateurs with everyday capabilities that professional radio astronomers couldn't obtain at any price 15 years ago. Chip capacitors and strip-line circuitry, coupled with computer design aids, now make the construction of microwave circuits nearly as simple as of those for audio frequencies. With monolithic microwave integrated circuits, schematics and block diagrams are nearly indistinguishable. The latest introduction is the GaAs HEMT (High Electron Mobility Transistor). A thousand dollars for a 1.8-dB noise figure at 20 GHz may not meet your current needs, but if past experience is any guide, you'll be able to obtain better performance for the price of a McDonald's hamburger by the turn of the century. Meanwhile, those bargain-basement 4-GHz 55-K LNAs are certainly helping to populate the amateur 3.4-GHz band.

SATELLITE SLEUTHING

Ever since the early days of the space program, radio amateurs have been captivated by the challenge of identifying "unknown" transmissions from space, an activity sometimes referred to as satellite sleuthing. In Chapter 2 we mentioned a group of US amateurs who were asked in the late '50s and early '60s to apply their HF-radio expertise to help locate unannounced low-altitude satellites. They were to do so by noting propagation anomalies produced by the ionized trails such spacecraft left behind.

A school group in Kettering, England, led by Geoff Perry, has become famous for often providing details of Soviet space launches long before official announcements. By studying orbital and launch data carefully, and by correlating this information with known astronomical facts and available details of the Soviet space program, the Kettering group has been able to predict mission objectives with uncanny accuracy.

Another well-known satellite detective is Greg Roberts, ZS1BI, a professional astronomer who first became involved in Amateur Radio and the OSCAR program as a result of his satellite-monitoring activities. Roberts regularly shares his quarter century of experience through articles in various amateur magazines. References are listed in Table 14-9.

Although eavesdropping is usually approached as a fascinating game in which the objectives are to (1) locate an unidentified transmission, (2) determine the orbital elements and (3) identify the transmitting spacecraft, its real value is as an educational tool. Those who pursue this activity gain a practical understanding and knowledge of space communications that few professionals ever acquire. The Kettering group has on occasion, for example, left the professionals dumfounded by decoding downlink telemetry without prior knowledge of the content or format. (C. Wood and G. Perry, "The Russian Satellite Navigation

---

Table 14-9

Information Resources: Satellite Sleuthing

Orbital Data: 1957 to Date

Satellite Situation Report. Published by NASA. Contains orbital data on more than 2000 space objects. A copy may be obtained from NASA, Project Operations Branch, Code 513, Goddard Space Flight Center, Greenbelt, MD 20771.

TRW Space Log. Extensive unofficial compilation of orbital data dating back to 1957. Published annually by the Public Relations staff of TRW Systems group. Copies are available to professional personnel in the aerospace industry, military and other government agencies. Requests must be on organization letterhead. Write: Editor, Space Log, Public Relations Dept, TRW Systems Group, One Space Park, Redondo Beach, CA 90278.

Orbital Data: Recent Launches and Re-entries

Spacewarn Bulletin. Published by IUWDS (World Warning Agency for Satellites); World Data Center A for Rockets and Satellites, Code 630.2; Goddard Space Flight Center; Greenbelt, MD USA 20771. Distribution is limited. Contains data on recent launches and re-entries and on satellites nearing re-entry. Also includes a section titled "Spacecraft with essentially continuous radio beacons on frequencies less than 150 MHz, or higher frequencies if especially suited for ionospheric or geodetic studies."

Aviation Week and Space Technology. Published weekly by McGraw-Hill. Contains orbital data for all announced launches. Often carries good descriptions of new satellite systems, launch vehicles and facilities. This magazine is very widely distributed; check your local library.

General Information

Aviation Week and Space Technology. (See previous description)

L. Van Horn, *Communications Satellites* (latest edition), Grove Press, Brasstown, NC. This book is an excellent resource for those monitoring satellite signals throughout the RF spectrum. It contains a great deal of background information on specific satellite systems, a complete list of frequency assignments, lists of additional resources, and so on. Van Horn has been updating the information on a yearly basis. Very highly recommended.

Spaceflight. Published monthly by the British Interplanetary Society. This journal contains excellent in-depth descriptions of US, USSR and other satellite systems. Articles are frequently more informative than official system documentation. For subscription information, contact: British Interplanetary Society, 27/29 South Lambeth Rd, London SW8 1SZ, England.

JESAUG (Journal of Environmental Satellite Amateur Users' Group). Issued quarterly. Contains information on APT systems. Includes METEOR System notes compiled by G. Zehr. See Table 14-6 for address.

NASA APT Notes. Contains information on NOAA and GOES satellites. Issued quarterly. See Table 14-6 for address.

Interferometer Design

Swenson, G. W., "An Amateur Radio Telescope," *Sky & Telescope*, Part I: May 1978, pp 385-390; Part II: Jun 1978, pp 475-479; Part III: Jul 1978, pp 28-33; Part IV: Aug 1978, pp 114-120; Part V: Sep 1978, pp 201-205; Part VI: Oct 1978, pp 290-293. This series provides the best practical introduction to amateur radio-astronomy instrumentation available. Parts I and II contain information on interferometers that is valuable to serious experimenters interested in passive orbit-determination techniques. These articles were reprinted in booklet form by Pachart Publishing House, Tucson, Arizona (1980). Unfortunately, the reprint booklet omits the valuable photographs accompanying the original articles.

Swenson, G. W., "Antennas for Amateur Radio-Interferometers," *Sky & Telescope*, Apr 1979, pp 338-341.

Table 14-10
Satellite Radio Transmissions: 136- to 138-MHz Band

Frequency (MHz)	International designation	Name	Period (minutes)	Inclination (degrees)	Apogee/perigee (km)
136.110	77014A	KIKU-2	1436.3	8.0	35851/35731
136.112	86061C	MABES	116.0	50.0	1500/1470
136.112	90013A	MOS-1B	103.3	99.1	940/913
136.138	77080A	SIRIO	1437.2	5.7	35849/35766
136.140	62BU1	RELAY-1	185.1	47.5	7439/1320
136.230	62AA1	TIROS-5	99.8	58.1	916/577
136.260	67040D	ERS-20	2840*1	32.9	111529/8619
136.370	67111A	ATS-3	1436.1	12.6	35849/35724
136.380	69046B	OV5-6	3115*1	32.9	113084/15460
136.380	75100A	GOES-1	1436.7	8.7	35814/35781
136.380	77048A	GOES-2	1436.2	7.0	35812/35762
136.380	78062A	GOES-3	1436.1	5.8	35811/35761
136.410	69009A	ISIS-1	127.9	88.4	3484/576
136.410	71024A	ISIS-2	113.5	88.2	1423/1354
136.440	66077B	EGRS-7	167.5	89.7	3697/3674
136.500	70025A	NIMBUS-4	107.1	99.8	1097/1086
136.620	62BU1	RELAY	185.1	47.5	7439/1320
136.625	70009A	SERT-2	106.1	99.2	1045/1039
136.650	64083D	TRANSIT	106.2	89.8	1081/1017
136.694	71080A	SHINSEI	113.1	32.0	1867/873
136.770	79057A	NOAA-6	100.9	98.5	812/794
136.770	84123A	NOAA-9	101.9	99.1	861/838
136.770	88089A	NOAA-11	102.0	98.9	863/843
136.800	66077C	ERS-15	167.6	89.7	3698/3682
136.800	69037B	EGRS-13	107.2	99.9	1127/1067
136.860	78012A	IUE	1435.9	31.6	42428/29137
136.890	90013B	DEBUT	110.5	99.1	1614/903
137.050	77108A	METEOSAT-1	1435.9	7.8	35888/35675
137.080	71110C	DOD	104.8*1	70.0	993/978
137.080	81057A	METEOSAT-2	1436.2	2.3	35808/35767
137.170	81122A	MARECS A	1436.1	3.0	35803/35770
137.170	84114B	MARECS-B2	1436.1	2.7	35799/35778
137.190	78062A	GOES-3	1436.1	5.8	35811/35761
137.230	81115A	BHASKARA2	94.4	50.7	497/480
137.300 APT	89018A	METEOR 2-18	104.0	82.5	958/936
137.330 APT	84105A	COSMOS 1602	97.6	82.5	658/624
137.350	66110A	ATS-1	1250.8	12.5	38075/26076
137.350	67111A	ATS-3	1436.1	12.6	35849/35724
137.380	69082B	TIMATION	103.2	70.0	926/895
137.400	83033A	ROHINI 3	95.0	46.6	668/368
137.400 APT	87068A	METEOR 2-16	104.0	82.6	959/938
137.400 APT	88005A	METEOR 2-17	104.0	82.5	959/933
137.400 APT	88056A	OKEAN-1	97.6	82.5	661/626
137.400 APT	90057A	METEOR 2-19	104.1	82.3	974/951
137.410	69082E	S69-4	103.3	70.0	930/899
137.440	75033A	ARIABAT	94.8	50.7	516/496
137.500 APT	79057A	NOAA-6	100.9	98.5	812/794
137.500 APT	86073A	NOAA-10	101.1	98.6	823/802
137.530	75049B	SRET-2	737.8	62.8	40825/513
137.560	71093A	PROSPERO	105.2	82.1	1469/536
137.560	79047A	ARIEL-6	95.0	55.0	532/500
137.620 APT	84123A	NOAA-9	101.9	99.1	861/838
137.620 APT	88089A	NOAA-11	102.0	98.9	863/843
137.770	84123A	NOAA-9	101.9	99.1	861/838
137.770	88089A	NOAA-11	102.0	98.9	863/843
137.795 APT*2	88080A	FENGYUN-1	102.7	99.2	897/877
137.850 APT	85100A	METEOR 3-1	109.3	82.5	1209/1178
137.850 APT	88064A	METEOR 3-2	109.3	82.5	1209/1178
137.850 APT	89086A	METEOR 3-3	109.5	82.6	1228/1191

International designations: First two digits give year. Next three digits label launches in order. Letter indicates object when multiple spacecraft are launched on a single vehicle.

Data for this chart was obtained from: G. Roberts (ZS1BI), G. Zehr (WA9TFB), *Spacewarn Bulletin*, JESAUG, *NOAA/NESDIS APT Information Notes* and other sources listed in Table 14-9.

Orbit data obtained from: *Satellite Situation Report*, NASA, March 31, 1989.

*1Outdated orbital data; current elements not available.   *2Believed to be no longer operating.

System," *Phil Trans Royal Society*, Vol A 294 [1980], pp 307-315.) Hearing Perry and Leo Labutin, UA3CR, swapping stories after lunch at the 1988 AMSAT Annual Space Symposium was an experience this author will never forget.

Monitoring interest always perks up when astronauts or cosmonauts are involved in space missions. Chris van den Berg has carefully monitored radio transmissions from Soviet cosmonauts in recent years and provided a running update in *OSCAR News*. As mentioned in Chapter 3, a few radio amateurs successfully monitored lunar missions in the late 1960s and early 1970s. Undoubtedly, there will be eavesdroppers when the first visitors from Earth set foot on Mars.

136- to 138-MHz Satellite Band

The 136- to 138-MHz band is probably the best place to begin searching for unidentified transmissions from space. It is suggested as a starting point because (1) it's relatively easy to obtain monitoring equipment, (2) there are lots of signals, (3) signals are often quite strong and (4) resource information is available to get you started.

For monitoring, I use an old vacuum-tube 2-meter converter picked up at a hamfest nearly 20 years ago. The unit was originally configured for 144 to 146 MHz in, 28 to 30 MHz out. The front-end circuits were peaked at 137 MHz without any changes, the oscillator chain wasn't touched, and a single capacitor added to the output coupling network enabled me to move it down to 21 MHz. The Drake R4-C being used as a tunable IF had crystals for 20.0 to 20.5 MHz and 21.0 to 21.5 MHz, so I could immediately check half the band. Using a 2-meter ground-plane antenna, several spacecraft were heard over the next few days. Two extra crystals for the R4-C gave full coverage, and a JFET preamp produced a big improvement in sensitivity. If you have an 2-meter/10-meter converter lying around, just plug it in line and tune your HF receiver to the 21-MHz range. If the converter is an inexpensive model with poor selectivity, chances are you'll hear some 137-MHz activity without any modifications.

Some of the satellite signals you may hear are listed in Table 14-10. All entries are known to have been active during or after 1986. Although NASA plans to phase out VHF downlinks, this segment of the spectrum should continue to be well populated with space signals until at least 2000. Some additional frequencies of interest to APT users and to those interested in following the Soviet space program are listed in Table 14-11.

Becoming a proficient monitor involves several steps, many of which can be pursued in parallel. If you monitor known signals over a period of time, you'll learn to recognize many of them by the sound of their modulation. The APT signal, for example, is very distinctive. Deducing orbital information is also a key. If the signal appears for relatively short periods of time and exhibits Doppler shift, you're dealing with a low-altitude spacecraft. If the signal is continuous and no Doppler is observed, it may be a birdie in your receiver, a terrestrial-based signal, or a synchronous-orbit spacecraft. A directional antenna will help sort out some of these parameters.

Table 14-11
Additional Monitoring information

Non-US APT Notes

APT transmissions have occurred on frequencies not currently in use and from spacecraft other than NOAA (US) and METEOR (USSR):

137.080	80051A	METEOR 1-30
137.080	84105A	COSMOS 1602 (USSR)
137.130	80051A	METEOR 1-30
137.130	81065A	METEOR 1-31
137.330	84105A	COSMOS 1602 (USSR)
137.150	80051A	METEOR 1-30
137.400	83099A	COSMOS 1500[*1] (USSR)
137.400	84105A	COSMOS 1602 (USSR)
137.400	88056A	OKEAN-1 (USSR)
137.400	86055A	COSMOS 1766 (USSR)
137.780	88080A	FENGYUN-1 (China)
137.795	88080A	FENGYUN-1 (China)

[*1]See *Aviation Week and Space Technology*: 15 Oct 1984, p 23; 12 Nov 1984, p 212; 18 Mar 1985, p 121; 10 Jun 1985, p 25.

MIR VHF voice and telemetry frequencies

121.750 MHz
142.400 MHz
142.417 MHz
142.600 MHz
143.144 MHz
143.625 MHz * FM voice channel
143.825 MHz
166.000 MHz
192.040 MHz

Soviet HF Frequencies

19.954 MHz was used on the SALYUT 7 (82033A) as recently as 1986-1989. In 1985, 20.008 MHz was used. Check the frequencies near 20.000 MHz for new signals whenever significant Soviet space activities occur in near-earth orbit.

ATS Transponders

ATS-1 and ATS-3 have 90-kHz-wide transponders centered on 149.195/135.600 MHz. These are usually assigned as five designated channels to users employing ±5-kHz frequency modulation. Although these spacecraft no longer have fuel to remain in position, NASA may still accept proposals for transponder use, especially those that demonstrate the utility of the spacecraft in its marginally operative condition. For information, request "ATS VHF Experiments Guide," ATS Experiments Manager, Office of Applications, Code ECS, NASA, Washington, DC 20546.

The following method illustrates one approach to the problem. When Cliff Ranft of the Kettering Group noted unidentified transmissions with characteristics suggesting a low-altitude spacecraft, he prepared the log shown in Table 14-12. Note that no assumptions are made concerning the number of spacecraft being observed or whether transmissions are on a continuous basis. Reviewing the data, Ranft noted consecutive intervals at less than 90

Table 14-12

Observations by Cliff Ranft of the Kettering Group on 153.6 MHz

Date	Time of Closest Approach
5 May 1986	1925.0 Z
	2057.0 Z
6 May 1986	0307.0 Z
	0936.5 Z*
7 May 1986	0330.0 Z**
	0906.0 Z*
	1915.0 Z
	2000.0 Z
8 May 1986	0300.5 Z**
	0506.0 Z
	0920.0 Z

*See text.

From: G. E. Perry, "Kettering Methods and Soviet Space [Activities]," *OSCAR News*, Dec 1987, pp 4-9.

minutes. This implies that more than one spacecraft is involved. Next, he noticed the pairs marked by asterisks. If we assume the descending passes for one case and ascending passes for the other, we see that spacecraft closest approach occurs 30 minutes earlier each day. This implies an orbital period of $(1440 - 30)/n$ minutes, where n is an integer in the range from 12 to 16. The possibilities are 117.5, 108.46, 100.71, 94.0 or 88.13 minutes. Perry and Ranft noted that "Of these, 100.7 minutes immediately sprang to mind as characteristic of the class of Cosmos satellites believed to perform store-dump communications missions." This last deduction was based on experience. Checking published orbital information, Ranft saw that the orbital data logged correlated precisely with passes of Cosmos 1624, Cosmos 1680 and Cosmos 1741.

Clearly, a good information file is an important requirement for sleuthing. Several helpful sources are mentioned in Table 14-9. Orbital data is most useful if stored on a computerized data base. This makes it easy to, for example, pull out all satellites having periods between 118 and 121 minutes. The method used to organize a computerized data base will eventually determine how helpful it is. Being a good detective requires meticulous record keeping.

Identifying the source of a signal is not always possible. An expert like ZS1BI may be able to identify 95 percent of signals heard, but newcomers can take pride in every new identification. The ability to recognize modulation schemes develops with experience. Orbit determination using passive techniques involves applying ideas found in Chapters 6, 11 and 12. Valuable information can be collected by using highly directional antennas with accurate azimuth and elevation readouts, or by using an array of antennas in the form of an interferometer (see the reference by Swenson). Information on launch sites may also prove helpful for identifying new spacecraft. A list is included in Chapter 15. Good Hunting!

Sputnik I: A Detective Story

One day while preparing a lecture, I wanted to pinpoint the birth of the space age. The launch time of Sputnik I was the piece of information required. A trip to the library followed. Sitting in front of the microfilm reader, I started by searching through *The New York Times*. When this proved unsuccessful, I went on to *Science*, then *Scientific American*, then *Physics Today*, then *Sky & Telescope*, then *Science News*. I was surprised to find that this basic piece of information was not readily available. Then it dawned on me: The Russians didn't announce the time because it would disclose the position of the launch site, and Western governments didn't provide estimates because their figures would disclose how well they knew the Soviet launch site. However, there were two estimates: 21:05 GMT (attributed to "western scientists," *Science News*, Oct 19, 1957, p 243) and about 20:00 GMT (attributed to astronomers at Bonn University, *Sky & Telescope*, Nov 1957, p 11).

At this point, I started checking books and encyclopedias that were published some years later, after the launch site was well known. One reference (P. Smolders, *Soviets in Space*, NY, Taplinger, 1974 [original edition, Kluwer, Holland, 1971]) did quote a time. The information and photos included in this book implied that the author had direct access to Soviet sources. Unfortunately, the time quoted, "6 am Moscow time," which equates to 03:00 GMT, is clearly incorrect. A newspaper article on the 25th anniversary of the launch of Sputnik I by A. Pietila, Moscow correspondent to the *Baltimore Sun*, gives 10:28 PM Moscow time for the launch (Oct 4, 1987, p A1). This is 19:28 GMT. Pietila didn't respond to queries concerning the source of his information.

My main conclusion at this point was that I should have stopped after I found the first estimate if I wanted my lecture to sound definitive. I did have some information I considered reliable, including estimates of the spacecraft's orbital inclination, altitude, period and the first sighting. This data is summarized in Table 14-13.

With the information in Table 14-13, we can construct an OSCARLOCATOR and work out the launch time to within about 10 minutes. To begin, we estimate the location of the Tatsfield listening station at 51 °N, 0 °E. Since the satellite inclination is 65 °, the argument of perigee will be changing slowly; therefore the perigee occurred near 40 °N at the time of the first sighting. At 51 °N, the satellite height will be roughly 250 km so, using Eq 12-4, we draw an acquisition circle about Tatsfield at 1760 km. Even though the height changes substantially as the satellite moves along its path, the orbital eccentricity is relatively small, so the ground track will be very similar to that of a satellite in a circular orbit with a period of just over 96 minutes. A ground-track overlay for the OSCARLOCATOR can be constructed using the quick method discussed in Chapter 12. Setting the overlay to pass 1 ° or 2 ° west of Tyuratam to allow for the time it took Sputnik I to reach orbit, we note that the first orbit wasn't visible to Tatsfield. We reset the orbit overlay so the ascending node is 24 ° further west

Table 14-13

Sputnik I data

First Western Sighting
    Location: Tatsfield (just south of London)
    Time: 0015 GMT, Oct 5, 1957

Inclination: approximately 65°

Period: 96 minutes 10 seconds
    (I believe this to be the nodal period)

Longitude increment: 24°W per orbit

Perigee: 170 km (above 40°N latitude [Oct 10 ?])

Apogee: 990 km (above 40°S latitude [Oct 10 ?])

Launch Site: Tyuratam (Baikonur) [See chapter 15]
    45°55.3′N latitude, 63°20.5′E longitude

Source: G. C. Sponsler "Sputniks over Britain," *Physics Today*,
    Vol 11, no. 7, pp 16-21, 1958. [Does not include information
    on launch site.]

and check the next orbit. Still not visible from Tatsfield. Again set the overlay 24° further west. This time the spacecraft enters Tatsfield's access circle about 10 minutes after ascending node. When this occurs, the satellite has completed just under 2 complete orbits. A little arithmetic

yields approximately 21:10 GMT for the launch time. This is quite close to the 21:05 GMT time attributed to "western scientists" in the Oct 19, 1957 *Science News*.

There was quite a bit of speculation as to why the 65° inclination was chosen for this mission. Some scientists thought this was done to keep the perigee over mid-northern latitudes so optical tracking from the Soviet Union would be much easier. I opt for a simpler reason. The Soviets most likely preferred to launch due east to obtain maximum assist from the earth's rotation. If this were done, the orbit inclination would be 46°. Such a launch passes over China. This would, of course, make tracking more difficult and could lead to serious consequences in case of a launch failure. Checking the OSCARLOCATOR, we see that selecting a 65° inclination keeps the ground track over Soviet territory until the spacecraft passes out over the Pacific Ocean.

[Note added at press time.] Geoff Perry reports (personal communication dated 26 August 1990) that the Sputnik I launch time was announced by M. Tikhonravov at a conference in Baku in 1973 at 22 hr 28 min 04s Moscow time, or 19:28:04 GMT. The launch time we deduced therefore appears to be off by approximately one orbit. This is most likely due to the fact that the first reception report from Tatsfield involved subhorizon signals.

Chapter 15

Satellite Systems

Even a simple satellite is a complex collection of hardware (and possibly software). To manage the design and construction of communications and scientific satellites, it's convenient to think of a spacecraft as being built from a standard set of subsystems, each with a specific function.[1] See Table 15-1. This makes it possible to parcel out the tasks of analyzing and optimizing each subsystem. Several design objectives almost always apply: minimizing weight and cost, maximizing reliability and performance, and ensuring compatibility. These aims often result in conflicts. On OSCAR 10, for example, radiation shielding of the central computer was desired, but the total amount carried was limited by weight constraints. Even when designers focus on a single subsystem, they must keep in mind how it impacts other subsystems. For example, a small reduction in transponder power-

amplifier efficiency may have little effect on downlink signal strength, but it might completely upset spacecraft thermal design.

In this chapter we'll look at each of the systems listed in Table 15-1. Various methods of accomplishing system objectives will be discussed. Methods that were used on previous OSCAR satellites and those suitable for future missions will be emphasized.

COMMUNICATIONS, ENGINEERING AND MISSION SUBSYSTEMS

The communications subsystem provides ground stations a direct link to a satellite. It enables us to observe what's happening inside the spacecraft, often as it happens, and to modify the operation of the spacecraft. Three communications links are of interest: (1) downlink beacons

---

Table 15-1

Satellite Subsystems: Emphasis OSCAR

Subsystem	Function	OSCAR Series Equipment
Attitude control	To modify and stabilize satellite orientation	Phase 2: magnet, gravity boom, torquing coils Phase 3: torquing coils
Central computer	To coordinate and control other subsystems; provides memory, computation capability	Digital logic, microprocessor, D/A converter, command decoder
Communication	To receive uplink commands and transmit downlink telemetry	Command receiver, transmitters (beacons), antennas
Energy supply	To provide power for all onboard subsystems	Batteries, solar cells, conditioning electronics
Engineering telemetry	To measure operating status of onboard subsystems	Electronic sensors, telemetry encoders
Environment control	To regulate temperature levels, provide electromagnetic shielding, provide high-energy particle shielding	Mechanical design, thermal coatings
Guidance and control	To interface computer with sensors, attitude-control and propulsion subsystems	Hard-wired electronics, sun and earth sensors
Mission-unique equipment	To accomplish objectives	Transponders, scientific and educational instruments
Propulsion	To provide thrust for orbit changes	Phase 2: none Phase 3: solid-fuel kick motor, liquid-fuel kick motor
Structure	To provide support and packaging function, thermal control, protect modules from stress of launch, mate to launch vehicle.	Mechanical structure, aluminum sheet wherever possible to minimize machining

---

providing telemetry (TTY), (2) uplink telecommand and (3) communications links supported by transponders.

Beacons: Function

The beacons aboard the OSCAR satellites serve a number of functions. In the *telemetry mode*, they convey information about onboard satellite systems (solar cell panel currents, temperatures at various points, storage battery condition, and so on); in the *Codestore mode,* they can be used for store-and-forward broadcasts; in either mode, they can be used for tracking, for propagation measurements and as a reference signal of known characteristics for testing ground-station receiving equipment. Beacon functions are summarized in Table 15-2.

Beacon telemetry. Amateurs have used several telemetry encoding methods. From the user's point of view, each method can be characterized by the data transfer rate and the complexity of the decoding equipment required at the ground station. To a large extent, there's a trade-off between these two factors (see Table 15-3).

Table 15-2

Beacon Functions

1) Telemetry
   a) Morse code
   b) Radioteletype (RTTY)
   c) Advanced encoding techniques
   d) Digitized (digitally synthesized) speech

2) Communications
   a) Store-and-forward broadcasts

3) Miscellaneous
   a) Tracking
   b) Propagation measurements
   c) Reference signal

Each OSCAR includes some, but not necessarily all, of the telemetry options listed.

Table 15-3

Telemetry Encoding Methods: Ground Station Complexity vs Data Rate

Telemetry Encoding Method	Relative Ground Station Complexity	Telemetry Data Rate	First OSCAR Utilization
Beacon characteristic	Low	Very low	OSCAR I
Digitized speech	Very low	Very low	OSCAR 9
Morse code	Low	Low	OSCAR 6
Radioteletype	Moderate	Moderate	OSCAR 7
Advanced encoding techniques	High	High	OSCAR 9, OSCAR 10

Beacon characteristic. The method first used to encode telemetry on an amateur satellite was to vary the speed of OSCAR I's CW HI in response to a temperature sensor. Absolute temperature data, collected while the spacecraft was in sunlight and in shadow, provided information on reflective and emissive performance of thermal coatings. Rate of change of temperature as spacecraft passed from sunlight to shadow provided data on thermal conductivity of structural elements. Note that measurement of a single characteristic can provide a great deal of information, since we can record both absolute value and rate of change under various stable and changing conditions.

Morse code telemetry. The Morse code telemetry systems employed on OSCARs 6, 7 and 8 (and on all RS spacecraft to date) have made these satellites very valuable to educators and amateur scientists.[2] In the CW telemetry mode, data measurements at several key points on the spacecraft are downlinked in Morse code. Restricting the code to a numbers-only format, usually at either 25 or 50 numbers/minute (about 10 or 20 words/minute), enables individuals with no prior training to learn to decode the contents relatively quickly. As a result of the design of the telemetry processing equipment aboard the satellite, ground stations do not need any specialized decoding electronics. The information capacity of this mode is inherently limited, in that any attempt to speed up the Morse code transmission would interfere with the ability of untrained users to decode it without special equipment.

Radioteletype telemetry. During the mid 1970s, as OSCAR spacecraft grew more complex it became necessary to adopt higher-speed telemetry methods. Since many amateurs owned radioteletype equipment, this mode was adopted and flown on OSCAR 7. The information provided by the RTTY system was especially valuable to the advanced experimenter, to stations that were responsible for managing the satellite and to the engineers and scientists who would design and build future AMSAT spacecraft.

Advanced encoding techniques. A number of factors have acted to displace radioteletype as a primary means for downlinking satellite data. Recall that radioteletype was chosen for the convenience of users in the mid 1970s. A careful analysis shows that RTTY is relatively inefficient when considered in terms of data rate per unit of power. With Phase-3 satellites and the UoSAT series, it became apparent that a higher-speed, more power-efficient link was required. This need arose at the same time that microcomputers were becoming commonplace at ground stations. Since the new series of spacecraft would be controlled by onboard computers, it was natural to switch to encoding techniques suitable for computer-to-computer communications. Of course, once a ground station uses a microcomputer to capture telemetry, it's only a small step to have the computer process the raw telemetry, store it, automatically check for values that indicate developing problems, graph data over time, and so on. Both the UoSAT and Phase-3 series used ASCII encoding, but different modulation schemes were adopted. Designers of the low-altitude UoSAT spacecraft, which have powerful beacons, selected a 1200-bps system which permitted the use of standard Bell 202 Modems. Power efficiency was much more important with Phase-3 series spacecraft, so a special 400-bps optimized system was developed. (References to articles discussing modem construction are contained in Appendix B: Satellite Profiles.) Amateurs who desire access to Phase 3 and UoSAT telemetry, RUDAK and Pacsat communications, and terrestrial VHF/UHF packet operation, currently need at least four separate modems (modems used in TNCs designed for terrestrial VHF/UHF packet operation can sometimes be used for UoSAT telemetry reception).

A project currently underway, which involves the development of a device known as a digital signal processor (DSP), may alleviate this situation. When the DSP is inserted between an analog signal and a digital device, it will (if appropriate software is available) take the place of any presently used or envisioned modem. The DSP will also decode weather satellite APT pictures, dig weak EME signals out of the noise, copy slow-scan TV, and so on. It may sound like we're discussing the electronic equivalent of snake oil here, but rest assured, this one's for real. It's a very powerful microprocessor controlled analog-to-digital and digital-to-analog converter, optimized for signal processing and designed to run under the control of an external computer. Software commands will permit the operator to reconfigure the DSP to serve the various functions mentioned. The DSP project is being sponsored by AMSAT and TAPR. Its eventual impact will be felt far beyond the amateur satellite service.[3]

Digitized speech mode. In the digitally synthesized speech mode, telemetry is simply spoken. This produces the ultimate simplicity in ground station decoding requirements (assuming the telemetry is being spoken in a language you're familiar with). This mode is excellent for demonstrations to general audiences and for educational applications at lower grade levels, but the extremely low data rate makes it unsuitable for real communications needs. Digital speech telemetry systems have been used on OSCARs 9 and 11. A more sophisticated digital speech synthesizer has been placed on DOVE (OSCAR 17).

Beacon communication mode. Beacons can also be used for store-and-forward broadcasts. Messages are generally loaded via the command link. With Phase-3 spacecraft, these messages can be downlinked either as CW, RTTY or ASCII. With DOVE, the downlink is synthesized speech.

Beacon: Miscellaneous functions. In either the telemetry or store-and-forward broadcast modes, a beacon with stable intensity and frequency can serve a number of functions. For example, it can be used for Doppler-shift studies, propagation measurements and testing ground-based receiving equipment. In addition, stations communicating via a satellite transponder can properly adjust their uplink power by comparing the strength of their downlink signal to that of the beacon.

Beacons: Design

Beacon power levels are chosen to provide adequate signal-to-noise ratios at well-equipped ground stations.

Overkill (too much power) serves only to decrease the power available for other satellite subsystems, reduce reliability and cause potential compatibility problems with other spacecraft electronics systems. Typical power levels at 146 and 435 MHz are 40 to 100 mW on low-altitude spacecraft and 0.5 to 1.0 W on high-altitude spacecraft. As with all spacecraft subsystems, high power-efficiency is essential. Phase-3 spacecraft have tended to use two beacons, a relatively high power unit called the *engineering beacon,* which is mainly operated when the omni antenna is used at high altitudes (during the early orbit transfer stages or in case of emergency), and the *general beacon,* which is operated continuously. The frequencies of these two beacons sandwich the transponder passband. This position is convenient both to users (the same ground-station receiving system can be used for both transponder and beacon downlinks) and satellite designers (the same satellite antenna can serve both systems). Since the telemetry system is a key diagnostic tool for monitoring the health of a spacecraft, redundant beacons, often at different frequencies, are generally flown to enhance reliability. This approach has paid off in a number of instances. Beacon failures occurred on OSCARs 5 and 6, but neither mission was seriously affected since both spacecraft carried alternate units. Beacon power output is usually one of the readings sampled by the telemetry system.

Command Links

OSCAR spacecraft are designed so that authorized volunteer ground stations with the necessary equipment can command them. The first OSCAR satellite with a command link was OSCAR 5. OSCARs 5-8 responded to a relatively limited set of commands. More recent spacecraft, those controlled by onboard computers, can accept programs via the command link. As a result, the number of possible operating states is very great.

Satellite commandability is a necessity for several reasons. First, it's a legal requirement for spacecraft operating in the amateur satellite service. Regulations state that, should the situation arise, we must be able to turn off a malfunctioning transmitter that is causing harmful interference to other services. Second, it would be impossible to accomplish the orbital-change or attitude-adjustment procedures critical to the existence of the Phase-3 system without the ability to uplink software. The capabilities of today's sophisticated Phase-2 spacecraft are also critically dependent on our ability to transmit software to the satellite via a command channel. Even with relatively simple satellites, the existence of a command link can mean the difference between an operative mission and a failure. A constant stream of commands sent to OSCAR 6 turned a marginally usable spacecraft (one that was continually shutting itself down) into a reliable performer. Via the command system, one can turn off malfunctioning subsystems, adjust operating schedules to meet changing user needs, employ attitude-control systems that require periodic adjustment, and so on.

Command stations are built and manned by dedicated volunteers. Though command frequencies, access codes and formats are considered confidential, they are available to responsible stations for projects approved by AMSAT.

Transponders

On most amateur spacecraft, the primary mission subsystem is the transponder. OSCARs 3, 6, 7, 8, 10, 12, 13 and 20 and the RS spacecraft carried open-access linear transponders. OSCARs 12, 16 (Pacsat), 19 (Lusat) and 20 carry digital transponders.

Linear Transponders

A linear transponder receives signals in a narrow slice of the RF spectrum, shifts the frequency of the passband, amplifies all signals linearly, and then retransmits the entire slice. Total amplification is on the order of 10^{13} (130 dB). A linear transponder can be used with any type of signal when real-time communication is desired. From the standpoint of conserving valuable spacecraft resources such as power and bandwidth, the preferred user modes are SSB and CW. Transponders are specified by first giving the approximate input frequency, followed by the output frequency. For example, a 146/29-MHz transponder has an input passband centered near 146 MHz and an output passband centered near 29 MHz. The same transponder could be specified in wavelength as a 2/10-meter unit.

Transponder design is, in many respects, similar to receiver design. Input signals are typically on the order of 10^{-13} watts, and the output level is several watts. A major difference, of course, is that the transponder output is at

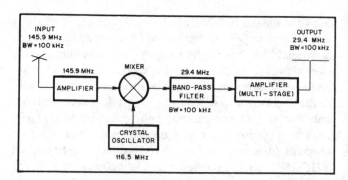

Fig 15-1—Block diagram of a simple 2-meter/10-meter linear transponder. Input passband 145.850-145.950 MHz, output passband 29.350-29.450 MHz.

radio frequency, while the receiver output is at audio frequency. A block diagram of a simple transponder is shown in Fig 15-1. For several reasons, flight-model transponders are more complex than the one shown. As with receiver design, such considerations as band-pass filter availability, image response, wide variations in input signal level and required overall gain often cause designers to use multiple-frequency conversions. A block diagram of the basic Mode-A transponder used on OSCARs 6, 7 and 8 is shown in Fig 15-2.

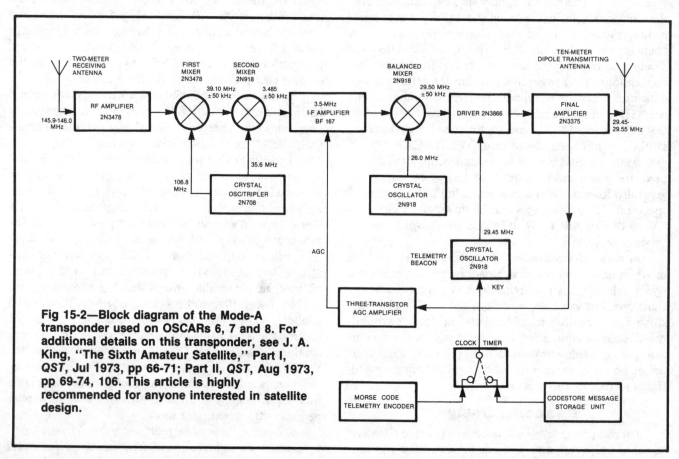

Fig 15-2—Block diagram of the Mode-A transponder used on OSCARs 6, 7 and 8. For additional details on this transponder, see J. A. King, "The Sixth Amateur Satellite," Part I, *QST*, Jul 1973, pp 66-71; Part II, *QST*, Aug 1973, pp 69-74, 106. This article is highly recommended for anyone interested in satellite design.

In spacecraft applications, a key characteristic of a linear amplifier is its overall efficiency (RF-output/dc-input). Once we reach power levels above a few watts, the use of class A, AB or B amplifiers cannot be tolerated. Dr Karl Meinzer (DJ4ZC) and his coworkers at the University of Marburg in the Federal Republic of Germany have developed a series of special high-efficiency linear transponders.

The first method, known as envelope elimination and restoration (EER), operates somewhat like a class D amplifier.[4] Although individual stages are not linear, the overall transponder is a linear device. This technique proved very successful on the Mode-B transponder on OSCAR 7, but it wasn't suitable for scaling to the high-power, wide-bandwidth system needed for Phase-3 spacecraft. For Phase 3A, a new system was developed that used EER and Doherty amplifiers.[5] This technique was later superseded by HELAPS (high efficiency linear amplification by parametric synthesis).[6]

Transponder operating characteristics. The power and bandwidth of a transponder must be compatible with each other and with the mission. That is, when the transponder is fully loaded with equal-strength signals, each signal should provide an adequate signal-to-noise ratio at the ground. Selecting appropriate values accurately on a theoretical basis using only link calculations is error-prone. Experience with a number of satellites, however, has provided AMSAT with a great deal of empirical data from which it's possible to extrapolate accurately to different orbits, bandwidths, power levels, frequencies and antenna characteristics.

In general, low-altitude (300 to 1600 km) satellites that use passive magnetic stabilization and omnidirectional antennas can provide reasonable downlink performance with from 1 to 10 watts PEP at frequencies between 29 and 435 MHz, using a 50- to 100-kHz-wide transponder. A high-altitude (35,000 km) spin-stabilized satellite that uses modest-gain (7 to 10 dBi) antennas should be able to provide acceptable performance with 35 watts PEP using a 300-kHz-wide transponder downlink at 146 or 435 MHz. Transponders are usually configured to be inverting in order to minimize Doppler shift.

Dynamic Range. The dynamic-range problem for transponders is quite different from that for HF receivers. At first glance, it may seem that satellite transponders pose a simpler problem. After all, an HF receiver must be designed to handle input signals differing in strength by as much as 100 dB, while a low-altitude satellite will encounter signals in its passband differing by perhaps 40 dB. Good HF receivers solve the problem by filtering out all but the desired signal before introducing significant gain. A satellite, however, has to accommodate all users simultaneously. The maximum overall gain can, therefore, be limited by the strongest signal in the passband. Considering the state-of-the-art in transponder design and available power budgets aboard the spacecraft, an effective dynamic range of about 25 dB is about the most that can be currently obtained. The satellite AGC will normally be set to accommodate the loudest user. As a result, stations 25 dB weaker will not be able to put a usable signal through the spacecraft, even though they may be capable of doing so when the AGC is not activated. In the ideal situation, users would adjust uplink power so that the spacecraft AGC is never activated.

The "power-hog" problem is a serious one. Karl Meinzer (DJ4ZC) noted that during the first few years of OSCAR 10 operation, the Mode-B transponder gain was almost always reduced by approximately 15 dB. He stated "...if all stations would reduce their power by a factor of 30, their strength would not change one whit. Weaker stations would then also have a chance to use the transponder." One approach to alleviating the high-power problem is educational. Many users may not realize they're overloading the spacecraft, or may not know the effects that such overloading produces (it doesn't necessarily increase the strength of their downlink; it often merely depresses the signal level of other stations). Education is very important, but technical approaches to reducing the overload problem are also needed.

One approach being considered is to divide the transponder IF into a number of discrete channels, each with its own AGC system. An overloading signal would then only depress signal levels in one channel. A second approach, which Meinzer plans to implement on Phase 3D, is known as LEILA, for *LEIstungs Limit Anzeige* (Power Limit Indicator). LEILA will operate as follows. The spacecraft onboard computer will continuously monitor the passband. When the computer encounters a signal whose level exceeds a predetermined level, it will insert a CW pulse over the signal to indicate to the station that the spacecraft is being overloaded. The station can then reduce power until the pulse signal disappears and know that the optimal power level is being used. If the signal exceeds an even higher preset level, LEILA will activate a notch filter tuned to the offending station's frequency. As planned, LEILA will be able to handle several high-power stations simultaneously.[7]

Redundancy. Since the transponder is the primary mission subsystem, reliability is extremely important. One way to improve system reliability is to include at least two transponders on each spacecraft; if one fails, the other would be available full time. There are significant advantages to *not* using identical units. Consider for example the Mode-A/Mode-B combination on OSCAR 7 and the Mode-B/Mode-L combination used on OSCARs 10 and 13. When OSCAR 7 was launched, not many amateurs had equipment to access Mode B, so Mode A was scheduled the great majority of the time. However, Mode B's superior performance was very apparent and the number of users increased over the years. By OSCAR 7's third year in orbit, Mode B was being scheduled nearly 70% of the time.

The situation with OSCAR 13 is somewhat different. Because the spacecraft is generally available many hours each day, transponder scheduling is based on performance. Mode L is scheduled when it provides the best link, Mode B at other times.

Digital Transponders

Digital transponders of the Pacsat or RUDAK type differ significantly from the linear transponders we've been discussing. A digital transponder demodulates the incoming signal. The data can be stored aboard the spacecraft (Pacsat Mailbox) or used to immediately regenerate a digital downlink signal (RUDAK digipeater). The mailbox service is best suited to low-altitude spacecraft; digipeating is most effective on high-altitude spacecraft. Like linear transponders, digital units are downlink-limited. A key step in the design procedure is to select modulation techniques and data rates to maximize the downlink capacity. Using assumptions about the type of traffic expected, one then selects appropriate uplink parameters. An analysis of both Pacsat and RUDAK suggests that, due to collisions, the uplink data capacity should be about four or five times that of the downlink.

PACSAT MAILBOX

For Pacsat mailbox operation, the designers decided to use similar data rates for the uplink and downlink and couple a single downlink with four uplinks. Fuji-OSCAR 12 ran both links at 1200 bps. Pacsat and Lusat began operation at 1200 bps, but will switch selected uplinks to 4800 bps after experience has been gained with the spacecraft. These Pacsats contain an FM receiver with a demodulator that accepts Manchester-encoded FSK on the uplink. To produce an appropriate uplink signal, the ground station uses an FM transmitter and a TNC-1 or TNC-2. However, minor modifications are necessary. The changes consist of extracting two signals from the host TNC—the TX data line and a 1200-Hz signal derived from the TXclock. These two signals are combined in an exclusive OR gate. The output of the OR gate is passed through a low-pass filter and then to the transmitter mike input. Most ground-station Pacsat demodulators contain the simple circuitry needed to accomplish this.

The Pacsat downlink uses binary phase-shift keying (BPSK) and can run either 1.5 or 4 watts. This modulation method was selected because, at a given power level and bit rate, it provides a significantly better bit error rate than other methods under consideration. One way of receiving the downlink is to use an SSB receiver and pass the audio output to a PSK demodulator. The SSB receiver is just serving as a linear downconverter in this situation. Other methods of capturing the downlink are possible, but the systems now operating use this approach.

In Chapter 10 we discussed the Pacsat modems currently available from TAPR and G3RUH. Both have been reproduced in significant numbers and are known to work well. The designs take different approaches to demodulating the signal. The G3RUH unit uses a limiter to immediately convert the received signal into a digital format so that all processing is digital. The more complex TAPR modem employs a Costas loop in the initial processing. Both designs provide good performance. Preliminary

Fig 15-3—Block diagram of a Pacsat digital transponder.

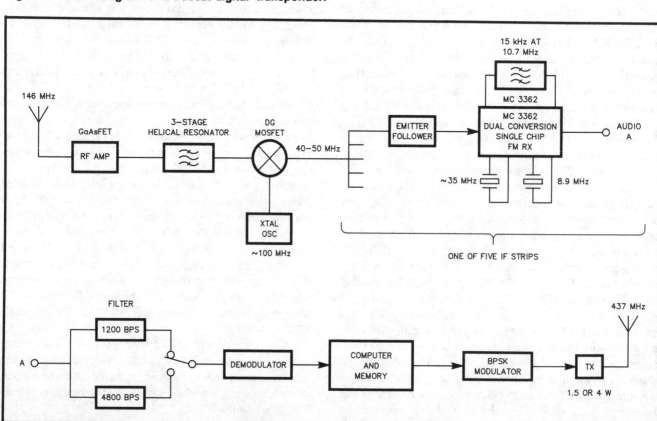

theoretical analyses suggest that the TAPR approach may be 3 or 4 dB more sensitive. However, this has not been demonstrated in actual tests, and it's very possible that the TAPR advantage is simply due to an incomplete theoretical analysis.

Until some definitive real-world comparisons are available, it's probably wisest to give other factors a high priority when selecting a modem. One should, for example, consider availability, alignment ease, and access to local amateurs familiar with a particular unit who can help should debugging be necessary. A block diagram of a pacsat transponder is shown in Fig 15-3.

RUDAK

RUDAK is an acronym for *Regenerativer Umsetzer fur Digitale Amateurfunk Kommunikation* (Regenerative Transponder for Digital Amateur Communications).[8,9] The RUDAK-I system takes a different approach to achieving the desired ratio of uplink to downlink capacity. It uses one uplink channel and one downlink channel, with the data rate on the uplink (2400 bps) roughly six times that on the downlink (400 bps). The 400-bps rate on the downlink was chosen because this is the standard that has been used for downlinking Phase-3 telemetry since the late 1970s. Users currently capturing Phase-3 telemetry are therefore already equipped to receive the RUDAK transmissions. The downlink modulation scheme is similar to that used on the Pacsats, but because of the speed difference, different modems are required for the two series. Although the RUDAK unit on OSCAR 13 failed during the launch, similar units are planned for the Phase-3D and RS-14 spacecraft.

Engineering/Telemetry Systems

The function of a spacecraft engineering system is to gather information about all onboard systems, encode the data in a format suitable for downlinking (engineering subsystem) and then transmit the encoded data on the spacecraft beacons (communications subsystem). In this section we look at engineering aspects of the telemetry subsystem. A block diagram of a telemetry encoder of the type used on OSCARs 6, 7 and 8 is shown in Fig 15-4.

Each parameter of interest aboard the spacecraft is monitored by a sensor. The sensor output must be converted to a voltage proportional to the measured value. This voltage then goes to a variable-gain amplifier and then to an analog-to-digital converter. During prelaunch calibration, the gain associated with each sensor is selected for optimum range and accuracy. The digital output is then converted to Morse code, RTTY or some other format for transmission via a beacon.

Phase-2 satellites used hard-wired logic to convert the output of the analog-to-digital converter to Morse code or RTTY. Phase-3 satellites are designed to perform the conversion with software in their onboard computers. Telemetry control logic (either hard-wired or in software) selects the proper input sensor, chooses the appropriate amplifier gain and conducts other bookkeeping chores.

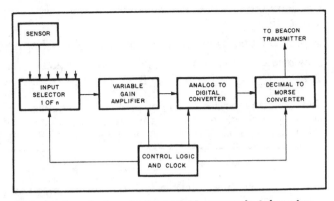

Fig 15-4—Typical early OSCAR Morse code telemetry encoder.

On early Phase-2 satellites, sensors were sampled sequentially (*serial mode*) and the measurements were transmitted as they were made. Under flexible computer control, the sampling strategy can be modified by instructions on the command link. If the situation warrants, we can dwell on a particular sensor (*dwell mode*) or sample it frequently so short-term changes can be studied. It's also possible to store readings in the onboard computer. This is especially valuable with a low-altitude spacecraft, where one can collect data over one or more orbits and then "dump" the information via the communications link while the satellite is in range. The *whole orbit data* (WOD) mode has been used extensively with OSCARs 9 and 11.

It's expected that future spacecraft will be controlled by onboard computers and that telemetry system operation will be controlled by software that's under ground control. As a result, AMSAT will have a great deal of flexibility in selecting channels, sampling strategies and encoding formats. For information on the telemetry capabilities of currently flying spacecraft, see Appendix B: Spacecraft Profiles.

Morse code format. The hard-wired Morse-code telemetry systems aboard OSCARs 6, 7 and 8 had several features in common. The parameters being measured were sampled in a fixed sequence. One complete series of measurements was called a *frame*. The beginning and end of each frame were marked by a distinctive signal; the Morse code letters HI were used on OSCARs 6, 7 and 8. Each transmitted value consisted of three integers, called a *channel*. To interpret a channel, we needed to identify which parameter was being monitored and obtain a raw data measurement that could be converted into a meaningful value. AMSAT used the first integer in a channel for parameter identification. When the number of channels is small (OSCAR 8 had six channels), a single digit can uniquely identify the parameter being measured. When the number of channels is large, the user must also note the order in which the channels are being sent in order to identify the particular parameter being sampled. Channel ID numbers are not needed, since one can always count from the previous frame marker, but the redundancy they provide is useful when QRM or QSB is present. However, including channel ID information does reduce the data rate.

Table 15-4

A Morse Code Telemetry Frame with Nine Channels

Raw data	HI	142	116	178	239	202	216	392	352	365	HI
Channel ID		1A	1B	1C	2A	2B	2C	3A	3B	3C	

The top row is the actual data as received. The bottom row assigns a unique label to each channel. Channel 1A is the first one received, channel 1B is the second, 1C is the third, 2A is the fourth, and so on. Such data is sometimes written in the form of a 3 × 3 matrix, in which case the ID integer is a line number and the ID letter is a column label.

As an example, a telemetry frame for an imaginary satellite using a nine-channel system is shown in Table 15-4. The first digit in the channel contains ID information, but it does not uniquely identify the channel. By noting the order in which the channels are received, we can assign unique labels consisting of the ID number and a letter (not transmitted). Our labeling system is shown in the second row of Table 15-4. The last two digits in each channel represent the encoded information. To decode a channel, one refers to published information about the specific satellite to determine which parameter is being measured and to obtain the simple algebraic equation needed to translate the raw data into a meaningful quantity. For example, a description of the imaginary satellite might tell us that channel 1A is total solar panel current, and that multiplying the raw data (42) by 30 will yield the value of total solar panel current in milliamperes (1260 mA) at the time the measurement was taken. The procedure outlined here applies to OSCARs 6, 7 and 8 and, with minor modifications, to all RS satellites to date. (The RS spacecraft used one or two letters to identify each channel.)

Advanced telemetry formats. With the telemetry system in recent satellites being controlled by the onboard computer, the traditional concept of a fixed telemetry frame is essentially obsolete. A frame may contain a fixed number of channels, but the channels are selected by software. A channel generally consists of at least five integers, two or three to provide unique identification and three or more to encode the data to a high accuracy. The telemetry system on the Microsats allows for 200 telemetry pick-off points. Ground-station software is usually designed to identify the channel being sent, apply the correct equation to the raw data and display the contents in a convenient format or save it for later analysis.

In the future, many of these chores may be handled by the spacecraft. For example, Pacsat telemetry will consist of plain-text packets similar to other mail packets. There are three encoding systems currently being used to downlink telemetry data from amateur spacecraft: the 1200-bps Bell-202 system on UoSATs, the 400-bps BPSK system on Phase-3 spacecraft, and the Pacsat BPSK system. Each requires a different modem. References to specific ground-station modem designs are contained in Appendix B. In the future, a single digital signal processor may replace the various modems.

Antennas

Satellite antenna selection is closely linked to mission objectives, type of orbit, attitude stabilization, frequencies being used, desired coverage, spacecraft structure, launch-vehicle constraints and other factors. With low-altitude satellites (OSCARs 5-8, 12, 20 and the Microsat series), spacecraft designers were able to choose relatively omnidirectional antennas and passive magnetic stabilization schemes, a combination that greatly simplified other aspects of satellite design. When mechanical considerations permitted, antennas that produced a circularly polarized wave, such as the canted turnstile (a turnstile with drooping elements), were used. The advantages of circular polarization are only partially realized however, since radiation from the canted turnstile is only fully circular along the main axis. At 10 meters, mechanical constraints made it impossible to provide circular polarization on the spacecraft. Dipole antennas have been used at this frequency.

Since the relatively large 10-meter antenna must be folded out of the way during launch, several schemes for antenna deployment have been tried (all successfully). With OSCARs 5 and 6, springy flexible elements made from metal similar to that used in a carpenter's rule were employed. The explosive bolts that released the satellite from the launch vehicle also released the folded antenna, allowing it to extend to its full, precut length.

On OSCAR 7, a different technique, producing a much stiffer antenna, was tried. Imagine a sheet of newspaper rolled into a two-inch diameter tube. Grab an inside corner and pull until the tube reaches about three times its original length. You've just modeled the OSCAR 7 10-meter antenna elements. The design and fabrication of antenna elements of this type, using springy metals that self-deploy when released by explosive bolts, is a very difficult and specialized procedure. Commercial products are prohibitively expensive for most OSCAR applications. The unit used on OSCAR 7 was a donation.

On OSCAR 8, yet another method was tried. The 10-meter antenna consisted of motor-deployed concentric tubes much like the car radio antennas that automatically extend when the radio is turned on. Producing motors that work reliably in the vacuum and temperature extremes of space is also a tricky and expensive business. A lot of nail biting occurred during the hours between OSCAR 8's launch and the successful commanding of the 10-meter antenna deployment mechanism. In sum, while all the approaches to 10-meter dipole design for OSCAR spacecraft have worked, none of the methods are completely satisfactory with respect to cost, operation and reliability.

At 146 and 435 MHz, quarter-wavelength monopoles (frequently called stubs) can be formed from springy, whip-like material that is bent back against the spacecraft during launch and freed when the spacecraft is released from the launch vehicle. At 146 MHz and higher frequencies, the spacecraft structure has a major effect on antenna pattern, so sophisticated theoretical models and empirical testing must be used to evaluate designs.

High-altitude spacecraft require beam antennas. Because of its size, the 2-meter antenna usually presents the biggest problem. The antenna selected for the 2-meter link on Phase-3 spacecraft is essentially an array of three Yagis. Each Yagi has a driven element and a director. The feed system introduces appropriate phase delays (120° and 240°) so that circular polarization is produced. The springy elements, which extend from each of the three arms on the satellite, are simply bent out of the way during launch. The 70-cm antenna on Phase-3 satellites can also be thought of as an array of three Yagis. Each Yagi consists of a driven element in front of a reflecting plate. Because of its position on the spacecraft, the 70-cm antenna has a cleaner pattern. The phasing system produces circular polarization. At 23 cm and 13 cm, helix antennas have worked well. (The quadrifilar helix isn't used because it doesn't have sufficient gain for use near apogee.) These spacecraft also include quarter-wave monopoles for use near perigee and shortly after launch before the spacecraft has been properly oriented.

SELECTING ANTENNAS FOR PHASE 3

When selecting an antenna for a Phase-3 spacecraft, one must choose a radiation pattern that strikes a good balance between coverage and signal level on the ground. Consider the geometry shown in Fig 15-5. Because of the large slant range at apogee, we want a beam antenna on the spacecraft. The narrow beamwidth of a high-gain antenna, however, can lead to poor results when the satellite is away from apogee, because ground stations will be far off to the side of the satellite antenna pattern (large squint angle). Let's look at one approach to modeling the situation.

J. Kraus has shown[10] that the radiation patterns of a great many common beam antennas can be approximated by the expression

$$2 (n + 1) \cos^n (\theta) \qquad \text{(Eq 15.1)}$$

In other words, the gain in a given direction can be calculated approximately using only θ and n, where θ is the angle between the antenna axis and the observer. The parameter n is chosen to produce the correct gain when θ is zero (on axis). See Table 15-5. With this formula and our knowledge of satellite orbits, we can calculate the signal power at the subsatellite point (SSP) as the satellite travels around its orbit (see Fig 15-5). The results for beams having gains of 6 dBi and 10 dBi, and for an isotropic antenna, are shown in Fig 15-6.[11]

With Phase-3 spacecraft, a beam is used during the apogee portion of the orbit and a quarter-wavelength whip is switched in at the point where it provides better signals.

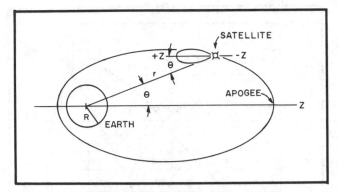

Fig 15-5—Orbit geometry for evaluating potential Phase-3 satellite antennas.

Table 15-5

Gain Pattern Simulation Equation Factors

n	Gain 2(n + 1)	Gain dBi	Half-power beamwidth
isotropic	—	0	—
0	2	3.0	180°
0.5	3	4.8	151°
1	4	6.0	120°
2	6	7.8	90°
3	8	9.0	74.9°
4	10	10.0	65.5°
6	14	11.5	54.0°
8	18	12.6	47.0°
12	26	14.1	38.6°
16	34	15.3	33.5°
20	42	16.2	30.0°
24	50	17.0	27.4°

The radiation patterns of many common beam antennas can be approximated by $2(n + 1) \cos^n (\theta)$, where n is related to the maximum gain as specified in the table. (See Table 8-1.)

Radiation from a whip along the $+Z$ axis spills over into the hemisphere centered about the $-Z$ axis. As a crude approximation, let's assume that the signal level from the whip is similar to that of the isotropic antenna. Referring to Fig 15-6, we see that the switch from beam to whip should be made when θ is approximately 56° for the 10-dBi beam and 76° for the 6-dBi beam. Also, the 6-dBi antenna begins to outperform the 10-dBi beam when θ increases past 43°. Each angle corresponds to a specific time from apogee. From Fig 15-6, we see that the 10-dBi beam will provide superior performance (by 0 to 4 dB) during a 7.5-hour segment of each orbit (3.75 hours on either side of apogee). The 6-dBi antenna will provide the best performance (by 0 to 3.5 dB) during 2.0 hours of each orbit (a one-hour segment centered at $\theta = +60°$ and a one-hour segment centered at $\theta = -60°$.

From the viewpoint of stations at the SSP, the 10-dBi antenna appears preferable. In fact, you might wonder why we don't consider higher-gain antennas. A more careful analysis would take into account (1) signal levels at ground stations located away from the SSP and (2) the fact that it may be necessary to align the spacecraft Z-axis in a

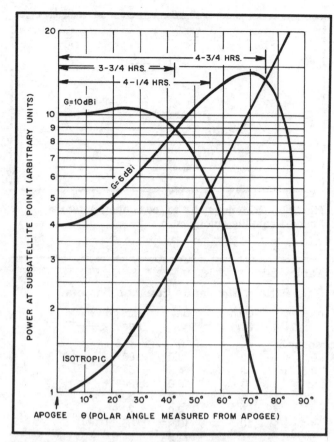

Fig 15-6—Relative power at subsatellite point as a function of Phase-3 satellite position in orbital plane (measured from apogee) for three possible antennas. Based on a period of 11.0 hours and an eccentricity of 0.688.

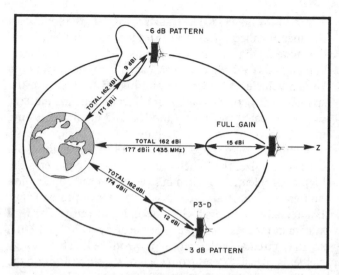

Fig 15-7—Phase-3D synthesized antenna patterns. (From *AMSAT-NA Technical Journal*, winter 1987-88)

slightly different orientation to account for poor sun angles on the solar panels. When this is done, it makes higher gains less appealing. There is no clear-cut "best" choice, but gains between 6 and 12 dBi appear to be a good compromise for Phase-3 spacecraft. Mechanical constraints limit the amount of gain that can actually be obtained aboard the spacecraft at 2 meters. Mode-L and Mode-S links will use higher gain, but, as Fig 15-6 demonstrates, these modes will therefore only provide acceptable performance near apogee.

Mode L will be the primary operating mode on Phase 3D. To improve the communications performance provided by this spacecraft, very-high-gain antennas will be used on both 70 cm and 23 cm. If the gain is greater than 18 dBi, the main lobe becomes so narrow that it no longer covers the entire earth at apogee. Maximum usable gain for full earth coverage (allowing for satellite attitude adjustments for proper sun angle) is therefore about 15 dBi. Such an antenna could only be used for a few hours near apogee. During the rest of the orbit, the spacecraft will shift to a lower-gain antenna. One possible approach is to place an array of medium-gain antennas on the spacecraft. By varying the phase of the power fed to each, it's possible to synthesize an optimal pattern for each section of the orbit. See Fig 15-7. This approach may provide a 5-dB im-

provement in Phase-3D link performance over that observed on OSCAR 13.[12]

PREDICTING RELATIVE LINK SIGNAL LEVELS

Eq 15.1 can be used to construct a model predicting relative uplink or downlink signal levels if we replace theta with the squint angle ψ. Link signal strength at time t will be proportional to

$$\frac{2(n+1)\cos^n(\psi)}{(\rho)^2}$$

where ρ is the range.

For our reference signal level, S_o, we'll assume (1) the satellite is directly overhead at apogee and (2) the squint angle is zero. The ratio of signal level at time t, $S(t)$, to reference signal level is

$$\frac{S(t)}{S_o} = \frac{(r_o - R)^2}{(\rho)^2} \cos^n (\psi) \qquad \text{(Eq 15.2)}$$

The predicted relative signal level (PRSL) in dB is then given by

$$\text{PRSL} = 20 \log(r_o - R) - 20 \log(\rho) + 10n \log[\cos(\psi)] \qquad \text{(Eq 15.3)}$$

As an example, we apply Eq 15.3 to the OSCAR-13 Mode-L uplink. Replace $r_o - R$ by the apogee height and use Table 15-5 and the published antenna gain (or half-power beamwidth) of OSCAR 13's 23-cm helix antenna (12.2 dBi) to select a value of 7.3 for n. For the OSCAR-13 Mode-B downlink (with ρ in km)

$$\text{PRSL [dB]} = 91.2 - 20 \log(\rho) + 73 \log[\cos(\psi)] \qquad \text{(Eq 15.4)}$$

The two variables appearing on the right hand side of Eq 15.4, squint angle and range, are available in most tracking programs. Incorporating a value for PRSL in these programs is therefore relatively simple.

STRUCTURAL, ENVIRONMENTAL-CONTROL AND ENERGY-SUPPLY SUBSYSTEMS

Structural Subsystem

The spacecraft structural subsystem, the frame that holds it all together, serves a number of functions, including physical support of antennas, solar cells and internal electronic modules; protection of onboard subsystems from the environment during launch and while in space; conduction of heat into and out of the satellite interior; mating to the launch vehicle, and so on. Structural design (size, shape and materials) is influenced by launch-vehicle constraints, by the spacecraft's mission and by the orbit and attitude-stabilization system employed. OSCARs 6, 7 and 8 have fallen in the 18- to 30-kg range. Phase-3 spacecraft, with their kick motors and fuel, had masses between 90 and 125 kg at launch. The first Phase-4 spacecraft will probably have a mass of about 400 kg. Spacecraft in the Microsat series have a mass of approximately 9 kg, contained in a cube roughly 23 cm on edge.

The prominent features one observes when looking at a satellite are the attach fitting used to mate the satellite to the launch vehicle, antennas for the various radio links, solar cells, the heat-radiative coating designed to achieve the desired spacecraft thermal equilibrium and, for Phase 3 and 4, the nozzle of the apogee kick motor. Insofar as possible, AMSAT satellite structures are fabricated from sheet aluminum to minimize the complexity of machining operations.

The Microsat structure is unique. It consists of five modules (trays) formed into a composite stack held together with stainless steel tie bolts. See Fig 15-8. Each frame is approximately 210 × 210 × 40 mm. Aluminum side panels about 5 mm thick holding the solar panels cover the sides. The entire structure is a rectangular solid 230 × 230 × 213 mm. The trays form an extremely sturdy structure; no additional spacecraft frame is needed.

Each tray contains an electronic sub-assembly. The basic modules, needed by every Microsat, are the (1) battery/BCR unit, (2) receiver unit, (3) CPU/RAM and (4) transmitter unit. The fifth module is the TSFR tray. TSFR stands for "this space for rent." The TSFR tray is essentially space set aside for experimental mission-specific subsystems. One significant aspect of the Microsat structure is that, by its very nature, it invites small amateur groups around the world to built a TSFR module for inclusion on future spacecraft. It also establishes a standard structure, which greatly simplifies the work of groups who wish to build their own Microsat, and reduces the cost of future Microsats.

Environmental Control

The function of a spacecraft environmental control subsystem is to regulate temperatures at various points, shield against high-energy particles and protect the onboard electronics from RF interference. We'll focus on thermal control.

The temperature of a satellite is determined by the inflow and outflow of energy. More specifically, the satellite temperature will adjust itself so heat inflow equals heat outflow. Although we talk about the "temperature" of the satellite, different parts of a spacecraft are at different temperatures, and these temperatures are constantly varying. The objective of the spacecraft designer is to create a model of the spacecraft and its environment that will accurately predict the average and extreme temperatures that each unit of the spacecraft will exhibit during all phases of satellite operation. This includes prelaunch, where the satellite may sit atop a rocket more than a week, baking under a hot tropical sun; the launch and orbit-insertion sequence of events; and the final orbit where, during certain seasons of the year, the spacecraft may go for months without any eclipse time (an eclipse occurs when the earth passes between the spacecraft and the sun),

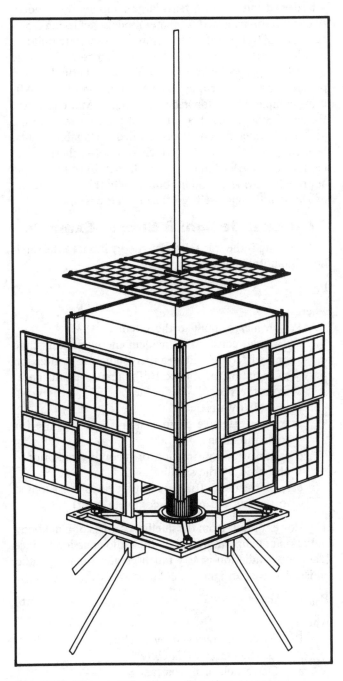

Fig 15-8—Microsat structure. (Drawing courtesy WD4FAB)

while during other seasons it may be eclipsed for several hours each day. Since excessive temperature extremes, either too hot or too cold, may damage the battery or electronic or mechanical subsystems permanently, the thermal design must keep the temperatures of susceptible components within bounds at all times.

Once the satellite is in the vacuum of space, heat is transferred only by radiation and conduction; convection need not be considered. The complete energy balance model of Phase 3A depicted the satellite as comprising 121 sub-units, each connected via conduction and radiation links to several other sub-units. To solve the resulting energy-balance equations mathematically, the designers had to manipulate 121 nonlinear coupled equations, each consisting of about three or four terms. This was not a job for pad and pencil. A fairly large computer was needed. Even with sophisticated computer models, achieving a precision of ±10 K is difficult; commercial satellite builders usually resort to testing the thermal balance of full-scale models in space simulators. The Phase-3 thermal design problem was handled by Richard Jansson, WD4FAB, using computer time donated by the Martin Marietta Corp. Earlier OSCARs used a far simpler and less accurate approach, which nevertheless provided reasonable results. See, for example, Fig 2-2, which shows the thermal behavior of OSCARs 1 and 2. Since the details of the simple approach provide a good introduction to the science (art?) of thermal design, we'll go through an example.

Thermal Design: A Simple Example

The sun is the sole source of energy input to the satellite. Quantitatively we can write:

$$P_{in} = P_o <A> \alpha\beta \qquad \text{(Eq 15.5)}$$

where

P_{in} = energy input to the satellite

P_o = solar constant = incident energy per unit time on a surface of unit area (perpendicular to direction of radiation) at 1.49×10^{11} m (earth-sun distance) from the sun.

P_o = 1380 watts/m^2

$<A>$ = effective capture area of the satellite for solar radiation

α = absorptivity (fraction of incoming radiation absorbed by the satellite)

β = eclipse factor (fraction of time satellite is exposed to the sun during each complete orbit)

Power output from the satellite consists of blackbody radiation at temperature T, and the radio emissions. Since blackbody radiation is very much greater than the radio emissions, we can ignore the latter.

$$P_{out} = A\sigma e T^4 \qquad \text{(Eq 15.6)}$$

where

P_{out} = energy radiated by satellite

A = surface area of satellite

σ = Stefan-Boltzmann constant = $\dfrac{5.67 \times 10^{-8} \text{ joules}}{K^4 m^2 s}$

e = average emissivity factor for satellite surface

T = temperature (K)

For equilibrium, incoming and outgoing radiation must balance,

$$P_{in} = P_{out} \quad \text{or}$$

$$P_o <A> \alpha\beta = A\sigma e T^4$$
$$\text{(energy balance equation)} \qquad \text{(Eq 15.7)}$$

Solving for temperature, we obtain

$$T = \left(\frac{P_o \alpha\beta <A>}{\sigma e A} \right)^{1/4} \qquad \text{(Eq 15.8)}$$

Reasonable average values for the various parameters are $\sigma = 0.8$, $\beta = 0.8$ and e = 0.5. For AMSAT-OSCAR 7, A = 7770 cm^2 and $<A>$ = 1870 cm^2.[13] Inserting these values in Eq 15.8, we obtain T = 294 K. This is equivalent to 21 °C, which is close to the observed equilibrium temperature of OSCAR 7. Over the course of a year, as β varied from 0.8 to 1.0, the temperature of OSCAR 7 varied between 275 K and 290 K. The transponder final amplifier, of course, ran considerably hotter.

Passive methods used to achieve a desired spacecraft operating temperature include adjusting surface absorptivity (α) and emissivity (e) by roughening or painting and taking thermal conductivity of structural components into account when the spacecraft is designed. Active techniques for temperature control include fitting the spacecraft with shutters, or louvers, that are controlled by bimetallic strips or conducting pipes that can be filled with helium gas or evacuated. To a certain extent, active temperature control has been employed on OSCAR spacecraft; in several instances specific subsystems have been activated primarily because of their effect on spacecraft temperature. Heat-pipe technology is an important component in the design of the Phase-4 spacecraft.[14]

Energy-Supply Subsystems

Communications satellites can be classified as active or passive. An example of a passive satellite is a big balloon (Echo I, launched August 12, 1960, was 30 meters in diameter when fully inflated) coated with conductive material that reflects radio signals. Used as a passive reflector, such a satellite does not need any electronic components or power source. While such a satellite is appealingly simple, the radio power it reflects back to earth is less, by a factor of 10 million (70 dB), than the signal transmitted by a transponder aboard an active satellite (assuming equal uplink signal strength and a comparison based on equal satellite masses in the 50-kg range).[15]

An active satellite (one with a transponder) needs power. The energy source supplying the power should be reliable, efficient, low-cost and long-lived. By efficient, we mean that the ratio of available electrical power to weight and the ratio of available electrical power to waste heat should be large. We examine three energy sources that have been studied extensively: chemical, nuclear and solar.[16]

CHEMICAL POWER SOURCES

Chemical power sources include primary cells, secondary cells and fuel cells. Early satellites such as Sputnik I, Explorer I and the first few OSCARs were flown with primary cells. When the batteries ran down, the satellite "died." Spacecraft of this type usually had lifetimes of a few weeks, although Explorer I, with low-power transmitters (about 70 mW total), ran almost four months on mercury (Hg) batteries. These early experimental spacecraft demonstrated the feasibility of using satellites for communications and scientific exploration and thereby provided the impetus for the development of longer-lived power systems.

Today, batteries (secondary cells in this case) are used mainly to store energy aboard satellites; they are no longer used as the primary power source. Table 15-6 provides a capsule history of OSCAR spacecraft power systems and their mission impact. Batteries for most missions were donated by companies or provided by government agencies from spares remaining when programs ended. Because these sources are not likely to be available in the future, AMSAT has initiated an ongoing battery-qualification program to provide cells for future missions.

In the early 1980s, after looking into the procedures used to "space-qualify" batteries, Larry Kayser (VE3PAZ) concluded that carefully screened commercial-grade NiCds were likely to perform as reliably. Kayser and a group of amateurs in Ottawa purchased a large supply of 6-Ah GE aviation NiCds and put them through a qualification procedure that involved X-raying to look for internal flaws and extensive computer-controlled charge-discharge cycling with extremely detailed computer monitoring. All anomalous cells, whether better or worse than the others, were eliminated. The remaining cells were matched, potted and stored in a freezer to prevent deterioration. Cells of this batch have performed flawlessly on OSCAR 11 since March 1984. The Microsats and UoSATs launched in early 1990 were powered by cells from this supply.

AMSAT continues to monitor emerging technologies such as sealed nickel-hydrogen batteries, which have energy densities about five times the value of NiCd cells. However, cost and reliability remain important selection criteria, so there's good reason to stick to proven technology that has served us well in the past.

Another chemical power system, the fuel cell, has been used as a source of energy on manned space missions, where large amounts of power are required over relatively short time spans. Fuel cells do not appear to be appropriate for OSCAR missions currently being considered.

NUCLEAR POWER SOURCES

One nuclear power source to be flight-tested is the radioisotopic-thermoelectric power plant. In devices of this type, heat from decaying radioisotopes is converted directly to electricity by thermoelectric couples. Some early US Transit navigation satellites, the SNAP 3B and SNAP 9A (25 W), have flown generators of this type. The US is not currently using nuclear power in earth orbit, but develop-

---

Table 15-6

OSCAR Satellite Power Systems

Satellite	Primary Power[1]	Secondary Power	Lifetime	Failure Mode
OSCAR I	Mercury battery	—	21 days	reentry
OSCAR II	Mercury battery	—	19 days	reentry
OSCAR III	Si solar cells (2.5 W)	Silver-zinc battery	several months	battery failure ?
OSCAR IV	Si solar cells (10.0 W)	battery	85 days	? (partial launch failure)
OSCAR 5	Manganese alkaline battery	—	52 days	battery depletion
AMSAT-OSCAR 6	Si solar cells (5.5 W)	NiCd battery	4.5 years	battery failure
AMSAT-OSCAR 7	Si solar cells (15 W, 9%)	NiCd battery	6.5 years	battery failure
AMSAT-OSCAR 8	Si solar cells (15 W, 8%)	NiCd battery	5.3 years	battery failure
AMSAT-PHASE 3A	Si solar cells (50 W, 10 + %)	NiCd battery	—	launch failure
UoSAT-OSCAR 9	Si solar cells (17 W, 12.5%)	NiCd battery	8 years	reentry
AMSAT-OSCAR 10	Si solar cells (50 W, 12.5%)	NiCd battery	3 + years	radiation induced computer failure
UoSAT-OSCAR 11	Si solar cells (25 W)	NiCd battery[2]	—	—
Fuji-OSCAR 12	Si solar cells (8.5 W)	NiCd battery	3 + years	[3]
AMSAT-OSCAR 13	Si solar cells (50 W)	NiCd battery	—	—
UoSAT-OSCAR 14	Si & GaAs solar cells	NiCD battery[2]	—	—
UoSAT-OSCAR 15	Si, GaAs & InP solar cells	NiCd battery[2]	<1 day	unknown
Pacsat-OSCAR 16	Si solar cells (15 W, 15%)	NiCd battery[2]	—	—
DOVE-OSCAR 17	Si solar cells (15 W, 15%)	NiCd battery[2]	—	—
Webersat-OSCAR 18	Si solar cells (15 W, 15%)	NiCd battery[2]	—	—
Lusat-OSCAR 19	Si solar cells (15 W, 15%)	NiCd battery[2]	—	—
Fuji-OSCAR 20	GaAs solar cells (13 W)	NiCd battery	—	—

[1]For solar cells, power is specified with satellite in optimal orientation; beginning of life (BOL) efficiency is reported if known.
[2]Off-the-shelf commercial batteries space qualified by AMSAT.
[3]Operation terminated Nov 5, 1989 due to gradual decrease in power budget resulting from solar cell deterioration and reduced battery capacity.

ment work on reactors for SDI is underway. The USSR currently flies nuclear reactors on Radar Ocean Reconnaissance Satellites (RORSATs). In 1989, some 34 deactivated, but still radioactive, reactors were orbiting the Earth.

Nuclear power sources have a high available-power-to-weight ratio, a very long operational lifetime and the ability to function in a high-radiation environment. But they generate large amounts of waste heat and have a high cost per watt because of the fuel. Nuclear power is most useful on missions where one of the following conditions holds: solar intensity is greatly reduced (deep-space), the radiation environment would quickly destroy solar cells (orbits inside the Van Allen belts) or very large amounts of power are required.

There are strong pressures in the US and USSR to put an end to the use of nuclear power in earth orbit. The primary reason is the danger of nuclear contamination in case of launch failure or satellite reentry.[17,18] Satellites with nuclear power are often placed in near-earth orbit with the intention of boosting them to a higher orbit when they reach the end of their useful life. However, many believe the likelihood of a spacecraft failure before the satellite is moved to a higher orbit is unacceptably high. This has been clearly demonstrated, since three nuclear powered spacecraft have re-entered and spewed radioactive material in either the atmosphere or on the ground (SNAP-9A [1964], COSMOS 954 [1978], COSMOS 1402 [1983]). The growing problem of space debris also poses a risk to these spacecraft. A second reason for outlawing nuclear powered spacecraft is that they cause severe gamma-ray pollution and are having a serious impact on gamma-ray astronomy.[19]

Nuclear power is clearly out of the question for OSCAR satellites. If the previous facts don't convince you, then practical concerns like the cost of insurance and the facilities, security precautions and paperwork associated with handling nuclear material certainly will. This conclusion does not, of course, imply that nuclear power is not suitable for scientific deep-space missions. The Voyager mission to the outer planets would have been impossible without the 400-W nuclear generator aboard.

SOLAR POWER SOURCES

The third power source we consider is solar. The first solar cells were built in 1954 using silicon.[20] Solar cells quickly became the dominant supplier of power to spacecraft. However, they are far from ideal for this application. They compete for mounting space on the outer surface of the satellite with antennas and heat-radiating coatings. Their efficiency decreases with time, especially when the satellite orbit passes through the Van Allen radiation belts (roughly at altitudes between 1600 and 8000 km). They work most efficiently below 0 °C (most electronics systems perform best at about 10 °C). They call for a spacecraft orientation that may conflict with mission objectives. Finally, they produce no output when eclipsed from the sun. Despite these shortcomings, power sources that use solar cells to produce electrical energy and secondary cells to store energy are by far the simplest for long-lifetime spacecraft. They're affordable; they generate little waste heat; and they provide acceptable ratios of available-electric power to weight.

In earth orbit, a 1-meter-square solar panel oriented perpendicular to incoming solar radiation intercepts about 1380 W. The amount of this power that can be used on a spacecraft depends on solar-cell efficiency. The efficiency of cells used on OSCAR satellites has nearly doubled in the past 15 years. The silicon solar cells used on the

Fig 15-9—Long-lifetime communications satellite energy subsystem.

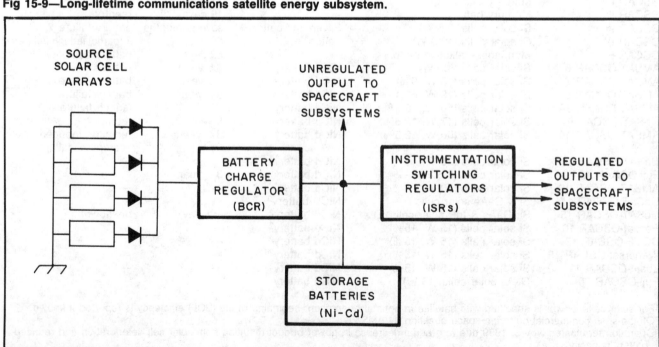

Microsat spacecraft represent another step forward. They employ back-surface reflector technology to produce an efficiency of more than 15%. In the near future, we'll probably use GaAs solar cells having even higher efficiencies.

In 1988, researchers reported the production of two-layer cells using Si and GaAs that had an efficiency of 31%.[21] While these are one-of-a-kind laboratory devices, they do demonstrate that solar-cell technology is continuing to move forward. A few years ago, a satellite the size of a Microsat would not have been capable of producing enough power to supply a useful mission.

Once a satellite is launched, its solar-cell efficiency begins to decrease due to radiation damage. To minimize the rate of decrease, the solar cells are usually covered by glass cover slides. These cover slides tend to reduce initial efficiency and increase spacecraft weight. One of the mission payloads on UoSAT-OSCAR 15 is the Solar-Cell Experiment, which is designed to evaluate the long-term performance of various new solar-cell technologies in the space environment. Cells to be tested include gallium arsenide, indium phosphide and new silicon designs. Radiation degradation over time will be evaluated using several cover slide geometries.

PRACTICAL ENERGY SUBSYSTEMS

The typical AMSAT satellite energy system consists of a source, a storage device and conditioning equipment (shown in Fig 15-9). The source consists of silicon solar cells (future missions may use GaAs cells). A storage unit is needed because of eclipses (satellite in earth's shadow) and the varying load; nickel-cadmium secondary cells are currently being used. The failure-mode column of Table 15-6 points out the critical importance of the energy subsystem to long-term mission success.

Power conditioning equipment typically flown on AMSAT spacecraft includes a battery charge regulator (BCR) and at least one instrument switching regulator (ISR) to provide dc-to-dc conversion with changes of voltage, regulation and protection. Because failures in the energy subsystem could totally disable the spacecraft, special attention is paid to ensuring continuity of operation. BCRs and ISRs usually are built as redundant twin units with switch-over between units controlled automatically, in case of internal failure, or by ground command.

Phase-3 spacecraft carry two separate batteries; a lower-capacity backup, capable of providing full operation through all but the longest eclipses, is kept in cold storage. Solar-cell strings are isolated by diodes, so a failure in one string will lower total output capacity, but will not otherwise affect spacecraft operation. These diodes also prevent the battery from discharging through the cells when the satellite is in darkness. The Microsat approach to reliability does not include the use of redundant power systems.

The Microsats use an interesting method for battery-charge regulation. Spacecraft transmitters are designed to operate efficiently over a wide range of power levels, ranging from a fraction of a watt up to about four watts. The onboard computer selects a power level that places

minimum strain on the battery system. Software is the key element in a feedback loop that operates in an overdamped condition. The software is periodically refined to maximize spacecraft longevity.

When the energy supply subsystem provides sufficient energy to operate a satellite's major systems on a continuous basis, we say the spacecraft has a *positive power budget*. If some subsystems must be turned off periodically for the storage batteries to be recharged, we say the spacecraft has a *negative power budget*. An illustration of how spacecraft geometry can be taken into account when estimating the average power output of a solar-cell array covering a spacecraft can be found in the references.[22] For information on the design and performance of the solar arrays used on a commercial spacecraft, INTELSAT IV-A, see reference 23.

ATTITUDE-CONTROL AND PROPULSION SUBSYSTEMS

Attitude-Control Subsystem

The orientation of a satellite (its attitude) with respect to the earth and sun greatly affects the effective antenna gain, solar-cell power production, thermal equilibrium and scientific-instrument operation. Attitude-control subsystems vary widely in complexity. A simple system might consist of a frame-mounted bar magnet that tends to align itself parallel to the earth's magnetic field; a complex system might use cold-gas jets, solid rockets and inertia wheels, all operating under computer control in conjunction with a sophisticated system of sensors. Attitude-control systems can be used to provide three-axis stabilization, or to point a selected satellite axis in a particular direction—toward the earth, in a fixed direction in inertial space (with respect to the fixed stars), or parallel to the earth's local magnetic field. Fixing a spacecraft's orientation in inertial space is generally accomplished by spinning the spacecraft about its major axis (spin-stabilized).

Attitude-control systems are classified as *active* or *passive*. Passive systems do not require power or sensor signals for their operation. Consequently, they are simpler and more reliable, but also less flexible. Some of the attitude-control systems in general use are described in the following paragraphs.

MASS EXPULSORS

Devices of this type are based on the rocket principle and are classified as active and relatively complex. Examples are cold-gas jets, solid-propellant rockets and ion-thrust engines. Mass expulsors are often used to spin a satellite around its principal axis. The resulting angular momentum of the satellite is then parallel to the spin axis. As a result of conservation of angular momentum, the spin axis will tend to maintain a fixed direction in inertial space.

ANGULAR-MOMENTUM RESERVOIRS

This category includes devices based on the inertia (fly) wheel principle. Assume that a spacecraft contains a flywheel as part of a dc motor that can be powered up on ground command. If the angular momentum of the flywheel is changed, then the angular momentum of the rest of the satellite must change in an equal and opposite direction (conservation of angular momentum). These systems are classified as active.

ENVIRONMENTAL-FORCE COUPLER

Every satellite is coupled to (affected by) its environment in a number of ways. In the two-body central-force model (outlined in Chapter 11), the satellite and earth were first treated as point masses at their respective centers of mass. Further analysis showed that the departure of the earth from spherical symmetry causes readily observable perturbations of the satellite's path. The departure of the satellite's mass distribution from spherical symmetry likewise causes readily observable effects. An analysis of the mass distribution in the spacecraft defines a specific axis that tends to line up pointing towards the geocenter as a result of the earth's gravity gradient.

Gravity-gradient devices exploit this tendency. Anyone who has been on a sailboat, however, knows that gravity can produce two stable states. The gravity gradient effect is greatly accentuated when the spacecraft is in a very low orbit and if one of the satellite dimensions is much longer than the others. Attaching a long boom, with a weight at the far end, to the spacecraft is one way of achieving this configuration.

Another environmental factor that can be tapped for attitude control is the earth's magnetic field. A strong bar magnet carried by the satellite will tend to align itself parallel to the local direction of this field (passive attitude control).

At any point in space, the earth's magnetic field can be characterized by its magnitude and direction. A simple model for the earth's magnetic field employs a dipole offset somewhat from the earth's rotational axis. The magnitude of a dipole field decreases as $1/r^3$, where r is the distance from the center of the earth, so attitude-control systems that depend on this field are most efficient at low altitudes. The direction of the magnetic field is often specified in terms of bearing and inclination (dip) angle. To describe bearing and dip, we imagine a sphere concentric with the earth drawn through the point of interest. Bearing and dip play the same role on this sphere that bearing and elevation play when describing a direction on the surface of the earth. Fig 15-10 provides data on dip angle.

The earth's magnetic field can also be exploited for attitude control via an active system based on electromagnets consisting of coils of wire. By passing current through these coils, one forms a temporary magnet. With proper timing, the coils can produce torques in any desired direction. Devices of this type are often called torquing coils.

Note that even if a satellite designer does not exploit magnetic or gravity-based environmental couplers for attitude control, these forces are always present and their effect on the satellite must be taken into account.

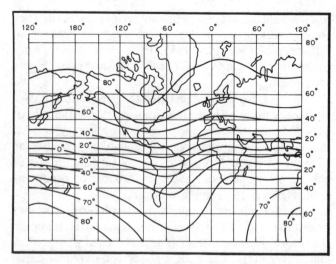

Fig 15-10—Inclination (dip) angle of earth's magnetic field.

ENERGY ABSORBERS

Energy absorbers or dampers convert undesired motional energy into heat. They are needed in conjunction with many of the previously mentioned attitude-control schemes. For example, if dissipative forces did not

Table 15-7

OSCAR Satellite Attitude-Control and Propulsion Systems

Satellite	Attitude Control System Design [propulsion system]
OSCAR I	None
OSCAR II	None
OSCAR III	None
OSCAR IV	Rocket ejection mechanism designed to provide initial spin stabilization
OSCAR 5	Permanent magnets, hysteresis damping rods
OSCAR 6	Permanent magnets, hysteresis damping rods
OSCAR 7	Permanent magnets, hysteresis damping rods, radiometer spin
OSCAR 8	Permanent magnets, hysteresis damping rods
Phase 3A	Spin stabilized, torquing coils, viscous liquid damping [solid-propellant kick motor]
OSCAR 9	Gravity gradient boom and torquing coils
OSCAR 10	Spin stabilized, torquing coils, viscous liquid damping [liquid-fuel kick motor]
OSCAR 11	Gravity gradient boom and torquing coils
OSCAR 12	Permanent magnets, hysteresis damping rods
OSCAR 13	Spin stabilized, torquing coils [liquid-fuel kick motor]
OSCARs 14, 15	Gravity gradient boom and torquing coils
OSCARs 16-19	Permanent magnets, hysteresis damping rods
OSCAR 20	Permanent magnets, hysteresis damping rods
OSCAR 22	Gravity gradient boom and torquing coils
Phase 4A	Earth-oriented, 3-axis stabilized using reaction control system and momentum storage

exist, gravity-gradient forces would cause the satellite's principal axis to swing pendulum-like about the local vertical instead of pointing toward the geocenter. Similarly, a bar magnet carried on a satellite would oscillate about the local magnetic field direction instead of lining up parallel to it. Dampers may consist of passive devices such as springs, viscous fluids or hysteresis rods (eddy-current brakes). At times, torquing coils are used to obtain similar results.

PRACTICAL ATTITUDE CONTROL

The attitude-control systems that have been used on AMSAT spacecraft are summarized in Table 15-7. Passive magnetic stabilization was first tried on OSCAR 5 and has since been used on numerous other Phase-2 spacecraft. When passive magnetic stabilization is used, Permalloy hysteresis damping rods are generally employed to reduce rotation about the spacecraft axis aligned parallel to the earth's local magnetic field and to damp out small oscillations. Note that the principal axis of a spacecraft in a near-polar orbit using this type of stabilization will rotate 720° in inertial space during each revolution of the earth. By using Fig 15-10 and an OSCARLOCATOR to follow the ground track of a satellite using passive magnetic stabilization, you should be able to picture how the spacecraft antennas are oriented with respect to your location.

Because of temperature-regulation concerns, we don't want to allow one side of a spacecraft to face the sun for too long a time. Fig 15-11 shows the residual spin of OSCAR 7 for 15 months following launch. Part of this spin was introduced purposely for temperature regulation. The technique was a novel one—the elements of the canted turnstile antenna were painted with reflective paint on one side and absorbent paint on the other. Solar-radiation pressure then produced a radiometer-like rotation dubbed by users at the time as the "barbecue rotisserie" technique.

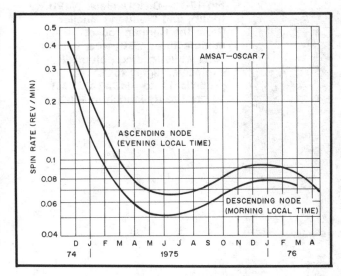

Fig 15-11—The spin rate of AMSAT-OSCAR 7 during the year following launch. (*Data provided by John Fox, WØLER*)

Because the camera on UoSAT must be pointed directly at the earth, a passive magnetic stabilization system wasn't sufficient. To accomplish their mission objectives, UoSAT engineers chose a complex gravity-gradient stabilization system used in conjunction with torquing coils. Equipment modifications over the course of the UoSAT series, coupled with several generations of software optimization, have resulted in major performance improvements of this active control system. The evolution of the system hardware and control software represent an important contribution to spacecraft-stabilization technology.

Elliptical-orbit Phase-3 satellites are spin-stabilized at approximately 20 to 30 RPM. The spin axis is ideally aimed at the geocenter when the spacecraft is at apogee. However, for adequate spacecraft illumination, the orientation will sometimes depart from this ideal state. When the orbital inclination is near 57°, these departures will be relatively small (on the order of 20°). The need for off-aiming occurs periodically. On average, a spacecraft like OSCAR 13 will need to have its attitude adjusted about every 3 months. Attitude information is obtained by sun and earth sensors under the control of the spacecraft computer.

To produce attitude changes, the torquing coils must be pulsed at precisely the correct time. Software loaded by a ground station directs the satellite computer to monitor sun and earth sensors and pulse the torquing coils when the proper conditions are met. Because the magnitude of the earth's magnetic field drops off as $1/r^3$, pulsing takes place near perigee to conserve satellite energy.

Phase-3 satellites employ viscous fluid dampers to discourage nutation (small oscillations of the direction of the spin axis). These dampers consist of a mixture of glycerin and water (about 50/50) contained in thin tubes (about 0.2 cm in diameter and 40 cm long) that run along the far edge of each arm of the spacecraft.

Propulsion Subsystem

The simplest type of space propulsion system consists of a small solid-propellant rocket, which, once ignited, burns until the fuel is exhausted. Rockets of this type are often used to boost a satellite from a near-earth orbit into an elliptical orbit with an apogee close to geostationary altitude (35,800 km), or to shift a satellite from this type of elliptical orbit into a circular orbit near geostationary altitude. Such rockets are known as "apogee kick motors" or simply "kick motors." The first AMSAT satellite to use a kick motor was Phase 3A. The kick motor was intended to shift Phase 3A from the planned transfer orbit (roughly 300 × 35,800 km, 10° inclination) to the target operating orbit (1500 × 35,800 km, 57° inclination).

PHASE 3A: SOLID-FUEL ROCKET

The Phase-3A kick motor was a solid-propellant Thiokol TEM 345-12 containing approximately 35 kg of a mixture of powdered aluminum and organic chemicals in a spherical shell (17-cm radius) with a single exit nozzle. See Fig 15-12. These units were originally designed as retrorockets for the Gemini spacecraft. The TEM 345 was capable of producing a velocity change of 1600 m/s during

Fig 15-12—Thiokol kick motor as used on AMSAT Phase-3A spacecraft.

its single 20-second burn. Because of the launch failure, this kick motor was never fired. Kick motors are dangerous devices. Their use, handling, shipping and storage must conform to rigid safety procedures.

OSCARS 10 AND 13: LIQUID-FUEL ROCKETS

OSCARs 10 and 13 used liquid-fuel rockets that were donated by the German aerospace firm, Messerschmitt-Boelkow-Blohm. These units were significantly more powerful (400-N thrust) than the solid-propellant motor previously used. The added thrust enabled AMSAT to fly a heavier spacecraft (thicker shielding, more electronics modules, and so on) and to compensate for the lower inclination and perigee of the transfer orbit being provided. With a liquid-propellant motor, multiple burns are possible. This greatly increases mission planning flexibility. On the other hand, this type of propulsion system is more complex and potentially more hazardous than the solid motor used on Phase 3A. The fuels used are unsymmetrical dimethyl hydrazine (UDMH) and nitrogen tetroxide (N_2O_4). Both of these chemicals are extremely toxic. Those involved in loading fuel must wear protective suits and breathe filtered air. See Fig 15-13.

While the rocket motor and several of the associated valves were a donation, much of the "plumbing" needed for the fuel system had to be devised and constructed by AMSAT personnel. The OSCAR-10 propulsion system is shown in Fig 15-14. Note the number of components required: filling valves, mixing valves, pressure regulators, check valves to prevent backflow, explosive (pyro) valves to prevent the system from accidentally firing before it is in orbit. Not shown in the figure is the Liquid Ignition Unit, the electronics module that directly controls the various valves and motor firing time. Construction of the two-chamber fuel tank and the high-pressure helium bottle presented significant challenges.

The orbit-transfer strategy planned for OSCAR 10 included two burns. There were important advantages to a two-burn transfer over a single-burn maneuver. With a single-burn transfer, the spacecraft velocity passes through a danger zone where premature termination will cause the spacecraft to re-enter at the next perigee. A second advan-

Fig 15-13—Dick Daniels, W4PUJ (right), and Wolfgang Mueller (from the German rocket manufacturer MBB) wear special suits while loading potentially explosive, highly toxic propellant into AMSAT-OSCAR 10 prior to its launch in 1983.

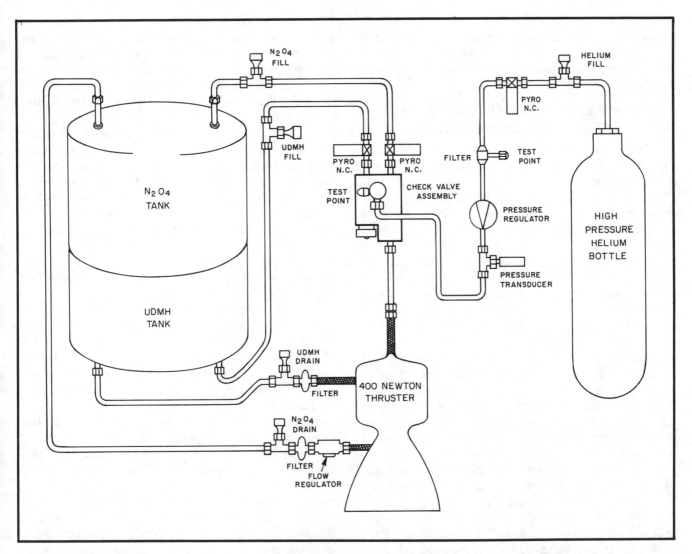

Fig 15-14—OSCAR-10 propulsion system. (From *AMSAT-NA Technical Journal,* Winter 1987-88)

tage of the two-burn approach is that it gives AMSAT an opportunity to compensate for unexpected performance characteristics of the rocket unit; the initial burn serves in part as a calibration run. The first burn went relatively well—there was a small deviation from the expected burn duration, which placed the spacecraft perigee somewhat higher than planned. The second burn could not be accomplished because a slow leak in the high-pressure helium system during the week the spacecraft was being reoriented prevented the opening of valves feeding fuel to the thrust assembly.

OSCAR 13 used a liquid-fuel kick motor similar to the one employed on OSCAR 10. MBB again agreed to donate a rocket, but this time very little support hardware was available. A new propellant flow/storage system was designed to make the system more robust, and to use available hardware. A two-burn strategy was again planned. This time the motor firings went perfectly. Each burn was accomplished exactly as planned. Details of the Phase-3 spacecraft propulsion systems have been described by the "master plumber," Dick Daniels, W4PUJ.[24]

WATER-FUEL ROCKETS

AMSAT has evaluated the potential utility of propulsion systems using water as a fuel. The inherent safety of these systems is a key factor if AMSAT hopes to attain a useful orbit from a Shuttle launch, where it's unlikely that we could obtain permission to fly a solid- or chemical-fuel kick motor. The problem with "water rockets" is that they take a considerable amount of time to affect orbit transfer. For example, it has been estimated that it would require approximately one year to reach a Phase-3-type orbit from a Shuttle launch. The economic consequences of the time involved in orbit change make water rockets impractical for commercial satellites, so this method has never been developed. However, now that AMSAT is constructing spacecraft with long expected lifetimes, this approach may be well matched to our needs.

There are two types of water rockets. Hughes Aircraft Company developed a working model of a rocket based on the electrolysis of water to produce hydrogen and oxygen. A second method involves the production of steam, which is released through a specially designed thruster. The steam method produces less impulse per unit of water than the electrolysis method, but system complexity is extremely low and less external energy is required for operation. The reliability of such a system should therefore be extremely high.[25]

COMPUTER AND GUIDANCE-AND-CONTROL SUBSYSTEMS

On OSCARs 5-8, hard-wired logic was used to interface the various spacecraft modules to both the telemetry system and the command system. As overall spacecraft complexity grows, at some point it becomes simpler and more reliable to use a central computing facility in place of hard-wired logic. Once the decision to incorporate a computer is made, the design of the spacecraft must be totally reevaluated to take advantage of the incredible flexibility provided by this approach.[26] Ground stations need no longer be located in position to send immediate commands; they uplink pretested computer programs. After correct reception is confirmed, these programs take control of the spacecraft and the uploaded directions are executed at designated times or when needed. Phase-3 spacecraft feed data from sun and earth sensors to the computer. Using a simple model of the earth's magnetic field, the computer pulses the torquing coils at the appropriate times to maintain the correct spacecraft attitude. Firing of the apogee kick motor on a Phase-3 spacecraft is also handled by the computer.

Computer programs control telemetry content and format. If we want to change the scale used to monitor a particular telemetry channel or to sample it more frequently, we simply add a couple of bytes to the computer program and it's done. Want to send out a daily Codestore message at 0000 GMT? No problem; uplink the message and control program whenever it's convenient and the message will be broadcast on schedule.

The Phase-3 Series IHU

Each Phase-3 satellite contains a module composed of a central processing unit (CPU) board, a random access memory (RAM) board, and a multiplexer (MUX) and command detector (CMD) board. The entire module is known as the Integrated Housekeeping Unit (IHU). The IHU combines a traditional multi-tasking computer (CPU and RAM) with a telemetry encoder and command decoder, so that a single unit handles all guidance, control and telemetry functions.[27]

The CPU. The 8-bit RCA COSMAC CDP1802 microprocessor was selected for the Phase-3A CPU back in the mid 1970s for a simple reason: It was the only suitable device available when the spacecraft was being designed. The choice turned out to be a good one, since this processor has proven powerful and flexible enough to meet the more complex demands of later missions. A radiation-hardened version (one more resistant to radiation damage) later became available. A novel feature of the spacecraft CPU design is that it does not use any read-only-memory. This was done because radiation damage to ROM was considered a serious possible spacecraft-failure mode.

When the spacecraft recognizes a particular sequence of bits on the command link, a reset is sent to the computer. The next 128 bytes are fed into sequential locations in low memory. When the last of the 128 bytes are received, the processor is automatically toggled into the run mode. A bootstrap loader contained in the 128 bytes then controls the loading of the rest of the operating system.

CPU Language. The CPU runs a high-level language called IPS (Interpreter for Process Structures), a threaded-code language similar to FORTH. IPS was developed by Dr K. Meinzer (DJ4ZC) for multitasking industrial control-type operations. IPS is fast, powerful, flexible and extremely efficient in terms of memory usage. Some say it's also nearly incomprehensible for anyone not brought up using a Hewlett-Packard RPN (colloquially known as Reverse Polish Notation) hand calculator.[28] In any event, users are *not* required to know IPS to recover data from the downlink telemetry beacons.

RAM. Initial plans (1975) were to fly Phase 3A with 2 kbytes of NMOS RAM. By the time Phase 3A was flown, it was possible to include 16k. The unit flown on Phase 3C contains 32k of radiation-hardened CMOS RAM. This may not seem like much when compared to today's personal computers, which typically contain 640k or more RAM. However, the *Voyager* mission to the outer planets ran fine using 32k of memory and memory size hasn't caused any limitations on Phase 3.

Each byte of 8-bit memory is backed up by 4 additional bits in an error-detection-and-correction (EDAC) arrangement. The CPU, as a background task, constantly cycles through RAM, checking each memory cell. If an error (radiation-induced bit-flip) is detected, it is corrected. The memory is thus protected against soft errors (radiation-induced bit-flips), provided that no more than one occurs in a byte in the time it takes for the CPU to check the entire memory, typically less than one minute. The EDAC circuitry is based on a Hamming code.[29]

MUX/CMD Board. The CMD just sits monitoring the uplink. When it identifies a unique bit sequence, it passes data to the CPU. The MUX is an electronic 64-pole switch. When used in conjunction with the analog-to-digital converter on the CPU board, it forms a 64-channel scanning voltmeter with a 0- to 2-volt range. Each parameter to be measured must provide voltage in this range. Temperatures are measured using thermistors. An ingenious technique is used to measure currents without incurring the losses and reliability problems that series resistors would introduce. The system works as follows: A symmetrical low-level ac signal is impressed on a coil wound on a small toroid. A dc measurement across the coil normally reads zero. However, if a wire carrying dc passes through the center of the toroid, a small dc offset voltage will be superimposed on the ac signal. The magnitude of the offset signal is proportional to the current flowing in the wire passing through the center. Small toroids are placed about each wire carrying a current to be measured.

Radiation Shielding. The integrated circuits comprising the electronics modules aboard the spacecraft are susceptible to radiation damage from high-energy particles. This problem is especially acute with Phase-3 spacecraft, because these craft make two passes through the Van Allen radiation belts during each orbit and the high-energy particle density in these belts is a severe threat. Soft errors, ones that simply cause a bit-flip in a memory cell, are not a major problem. The EDAC circuitry will take care of these. The problem is hard errors—permanent destruction of a memory cell. When this occurs in RAM, sections can be placed off limits. However, if too many memory cells are eliminated, or if certain key cells are destroyed, the computer cannot be rescued. There are lots of strategies for minimizing the susceptibility of the spacecraft to such damage, and AMSAT used them all: choosing chips with the best possible radiation properties, placing individual shields on chips, placing shields over groups of chips, and so on. With OSCAR 10, susceptibility to radiation damage was the acknowledged Achilles heel. After slightly more than three years in orbit, the IHU did succumb. When OSCAR 13 was being readied for launch, chips with a much higher resistance to radiation became available, and these were used on the new spacecraft. Radiation damage should not be a limiting factor on OSCAR 13.

UoSAT Series IHU

Spacecraft in the UoSAT series have all been controlled by CDP1802 microprocessors running the IPS language. OSCAR 9 and OSCAR 11 supported the 1802 with 48k of dynamic RAM and carried secondary computers. On OSCAR 9, the backup computer was a Ferranti 16-bit F100L supported by 32k of static CMOS RAM. On OSCAR 11, it was an NSC-800 with 128k of CMOS RAM used in the digital communications experiment. UoSAT-OSCAR 14 carries three computers (1802, 80C31, 80C186) and more than 4 Mbytes of RAM. UoSAT-OSCAR 15 carries the standard 1802 and three Transputer parallel-processing microcomputers. One of the primary objectives of the UoSAT series has been to test and evaluate new hardware and systems approaches. Since Phase 3A never attained orbit, OSCAR 9 gave amateurs their first opportunity to control a spaceborne computer. Many of the techniques used for packet radio satellites were first tested on UoSAT spacecraft.[30,31]

Microsat Series IHU

It has been said, somewhat seriously, that a Microsat is a compact, low-power IBM-computer clone masquerading as a spacecraft. The Microsat CPU uses an NEC CMOS V-40 (similar to the 80C188) and 2k of ROM for a bootstrap loader. EDAC is used for 256k of memory that holds the operating system software. An additional 8 Mbytes of static RAM is used to hold messages. Microsat software is written in assembler and Microsoft C, linked with Microsoft LINK.

The spacecraft computer control system on the Microsats represents a significant change in direction over the 8-bit 1802 architecture running IPS used on Phase 3 and UoSAT spacecraft. The reason for the change is straightforward. The development of the primary Microsat mission subsystem (the mailbox) required a microprocessor more powerful than the 1802. It was decided to use one from the Intel series so that development work could be done on IBM clones, which a great many amateurs have access to. The V-40 had the desired characteristics. It would have been possible to use an 1802 for overall spacecraft control and use the V-40 to manage the mailbox, but this would require that extra circuitry be placed on the satellite and that the spacecraft command and development teams work in IPS in addition to the more familiar languages used for Intel microprocessors. It was simpler to place the V-40 in overall control and treat spacecraft management as one of many tasks that the V-40 was responsible for.

A major innovation of the Microsat series is the introduction of a standard spacecraft bus (interconnection scheme) for linking the onboard computer and the various electronic modules. In the past, an extremely elaborate wiring harness had to be designed and constructed for each spacecraft. This harness provided all the links needed between the various modules. As spacecraft became more complex, the number of interconnections handled by the harness grew and its construction difficulty increased. This has a negative effect on spacecraft reliability. The wiring harness in the Microsats consists mainly of a ribbon cable that each module plugs into. The setup is similar to a very simple local area network, with the onboard computer (OBC) acting as the master and the modules as slaves. Each module contains an AART (Addressable Asynchronous Receiver/Transmitter) chip and associated components for communications with the OBC and telemetry sensor measurements.

The bus-communications orientation provides several additional advantages. With small engineering groups spread around the world, each working on a different spacecraft module, communication has always been a problem. A detailed bus definition greatly reduces this problem and facilitates distributed engineering. The bus orientation also makes it easier for new groups to become involved in spacecraft construction. Finally, it provides a design and control approach that can be efficiently applied to spacecraft of widely varying complexity and with all types of mission objectives.

Guidance and Control Subsystems

A satellite guidance and control subsystem includes components and software involved in the measurement of spacecraft position and orientation, attitude adjustment and in control of all other onboard systems in response to orders issued by telecommand or the spacecraft computer. On OSCAR spacecraft, the command receiver generally uses elements of the transponder (linear or digital) front end. A tap goes to a dedicated IF strip, demodulator and decoder. Data from the decoder is routed to the IHU. An active attitude-control system requires sensors. Since most of the elements of the guidance and control subsystem, except for the sensors, have already been discussed, we'll focus our attention on sensors. The sensors on OSCAR 13 serve as a good example.

OSCAR 13 contains three sets of sensors designed to provide attitude information: a sun sensor, an earth sensor and a top/bottom sensor. In the following discussion, it will help to picture the shape of OSCAR 13 and the fact that it's rotating about its symmetry axis, which is aligned in a fixed direction in space. The attitude-determination strategy consists of first finding the relative orientation of the spacecraft with respect to two celestial bodies, the sun and earth, whose positions are accurately known, and then mathematically reducing this data to absolute orientation. When this is done, there are frequently two solutions, only one of which is correct. The top/bottom sensor is used to eliminate the ambiguity.

The top/bottom sensor consists of a few solar cells mounted on the top and bottom of the spacecraft. When the spacecraft is in sunlight, only one set will be illuminated. This provides a crude estimate of the satellite's orientation that is sufficient to chose between the solutions provided by the sun and earth sensors. The sun and earth sensors are mounted at the end of arm two of the spacecraft, as shown schematically in Fig 15-15.

Because the sun is extremely bright and virtually a point source, construction of the sun sensor is relatively simple. It consists of two slits and two photodiodes. Because of the satellite's spin, the sun sensors will scan a region $\pm 60°$ from the spacecraft's equator once each revolution (about 20 times per minute). If the sun is in this region, we will get a "pip" from each photodiode. The time between these two pips and knowledge of which one pips first provide important information about satellite orientation. There are some real-world complications to this simple model caused by the fact that extraneous pips may be introduced by sunlight reflecting off an antenna. For an excellent discussion of the mathematics involved in reducing this data, see the article by J. Miller.[32] The primary reference used by professional designers of satellite attitude-control systems is by J. Wertz.[33]

The second body we choose to focus on is the earth. The earth is rather dim and its diameter as seen from OSCAR 13 varies from about 18° at apogee to 90° at perigee. Light enters the earth sensor through an anti-glare shield and is focused by a lens on a photodiode. The threshold sensitivity of the diode electronics is set so that a step-like change in output will occur when the diode field of view changes from dark space to sunlight reflected off the earth.

Temporarily assume that the earth is completely bathed in sunlight (of course, this is impossible). During most of the satellite's orbit, the earth sensor will not scan through the earth as the spacecraft rotates. However, there will be two periods, one slightly before perigee and one slightly after perigee, when the sensor will view the earth. As we enter one of these periods, the sensor will scan through the earth very quickly, but the time during which the sensor is focused on the earth will increase, reaching a maximum slightly further along the satellite's orbit, and then decrease back down to zero.

The mathematics needed to convert data from this simple model into information on satellite orientation is

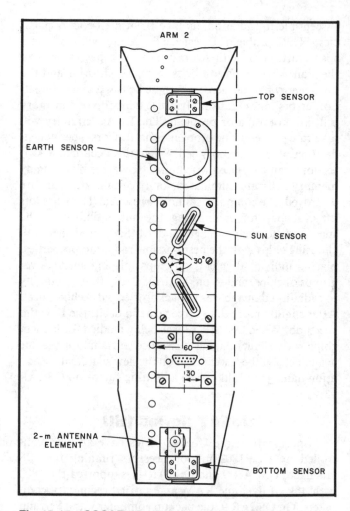

Fig 15-15—OSCAR-10 sun and earth sensors (end of arm 2).

not too horrendous. However, there are complications. The most serious one occurs because only part of the earth may be illuminated. The earth sensors will therefore be reporting illuminated crescent acquisition and loss. Additional uncertainly is introduced by the fact that the transition from light to dark is not very sharp when the earth is partially illuminated. To help with data interpretation, the earth sensor actually includes two photodiodes, each having a beamwidth of about 2°. One points about 4° above the spacecraft's equator and the other points about 4° below.

Using OSCAR-13 telemetry values containing the sensor data, a personal computer can be used to determine the direction of the spacecraft spin axis to about 1°. The absolute spin angle (needed for determining when torquing coils should be pulsed) can be determined to within about 0.1°. This accuracy is completely adequate for orbital transfer maneuvers and for attitude adjustment via the torquing coils.

LAUNCH CONSIDERATIONS

Launch Sites

Launching a satellite takes a lot of energy. The

Table 15-8

Major Launch Sites of the World

Launch Authority	Site Name	Latitude (±0.2°)	Longitude (±0.2°)
Australia	Cape York	12°S	142.5°E
Brazil	Alcantara	2°S	316°E (44°W)
PR China	Shuang Ch'eng Tzu (Jiuquan)	40.4°N	99.7°E
PR China	Xichang (Chengdu)	27.9°N	102.3°E
ESA	Kourou	5.2°N	307.3°E (52.7°W)
India	ISRO Sriharikota	13.7°N	80.2°E
Israel	Palmachim	31.9°N	34.7°E
Italy	San Marco Platform	2.9°S	40.3°E
Japan	Tanegashima	30.4°N	131.0°E
Japan	Kagoshima	31.3°N	131.1°E
USSR	Kapustin Yar	48.3°N	45.9°E
USSR	Plesetsk	62.9°N	40.7°E
USSR	Tyuratam (Baikonur)	45.9°N	63.3°E
US	Cape Canaveral/ Kennedy	28.5°N	279.5°E (80.5°W)
US	Vandenberg	34.6°N	239.4°E (120.6°W)
US	Wallops Island	37.9°N	284.5°E (75.5°W)

Sources

1) S.B. Kramer, *The Satellite Sky*, Graphic Display Chart available from Smithsonian Air and Space Museum, Washington, DC 20560.
2) *Jane's Spaceflight Directory 1986*, Jane's, London 1986, "World Space Centres," pp 345-353.
3) *Aviation Week and Space Technology*, ongoing clipping file.
4) K. Gatland, *The Illustrated Encyclopedia of Space Technology*, 1981, Harmony Books (div of Crown), NY.
5) RAE Table of Earth Satellites, 1987-89.

amount depends on the final orbit, the location of the launch site and the mass of the satellite. Energy constraints related to launch and orbit transfer affect AMSAT's selection of orbits.

To place the largest possible payload in orbit using a specific rocket and launch site, the launch azimuth should be due east, taking full advantage of the relative "boost" given by the earth's rotational velocity. When this is done, the orbital inclination of the satellite will equal the latitude of the launch site. The coordinates of various launch sites are listed in Table 15-8. These locations were chosen for several reasons, including safety (it's best to launch over water or very sparsely populated regions) and energy considerations. Looking at Table 15-8, it's clear that Plesetsk is nearly ideal for placing a payload into a Molniya (63° inclination) orbit, while Kourou is excellent for launch to geostationary orbit.

Changing the inclination of the orbital plane of a spacecraft takes a great deal of energy. If the initial launch azimuth is other than due east (or west), the orbital-plane inclination will be *greater* than the launch-site latitude. Note that it's impossible to place a payload directly into an orbit having an inclination lower than the launch site latitude unless the upper stages of the launch vehicle expend considerable energy to modify the initial trajectory.

Launch Opportunities

When AMSAT secures a ride into space, it must either accept the orbit provided or include a propulsion system on the spacecraft so a new orbit can be attained. The decision to include a propulsion system on a spacecraft is a major one since it requires not only a rocket, but a complex support system of sensors, computer and physical structure, and a much higher level of coordination with the launch agency. With OSCAR 10 and OSCAR 13, AMSAT has clearly demonstrated its ability to maneuver a spacecraft.

The simplest type of propulsion system is a solid-fuel kick motor that fires only once. With such a motor, the perigee height of the final orbit can never be greater than the apogee height of the initial orbit. This is an important constraint when considering launches from the US Shuttle. AMSAT has experience with solid-fuel kick motors, and with liquid-fuel kick motors, which can provide multiple burns. Because of safety concerns, it's unlikely that kick motors of this type will be allowed to fly on the Shuttle. AMSAT has therefore looked into ion propulsion motors using water as a fuel. Conceptual studies were very promising, but development work has been temporarily suspended due to poor prospects for Shuttle launch opportunities.

AMSAT continually explores launch possibilities with existing launch agencies and commercial groups planning to enter the launching business. Free or subsidized launches are becoming exceedingly rare, but they are still possible if a government agency deems such a launch desirable for scientific reasons or to support science education activities. Another opportunity for low-cost launches occurs when an agency is testing a new launch vehicle. Other opportunities may occur when we demonstrate how launch agencies can market new services or space (the Microsats and Phase 4A use space that ESA has not previously thought of as being commercially marketable). Finally, we may have to consider paying the going rate for launches. If commercial launching develops into a viable business, this approach may be feasible. Current estimates are that we could have a MicroSat launched into a useful orbit for about $60,000. The assistance of anyone associated with launch agencies or companies in identifying potential launches is always greatly appreciated by AMSAT.

Experience teaches us to expect the unexpected. If the radio amateur space program remains both flexible and vital, we'll be in an excellent position to take advantage of the launch opportunities that are sure to come.

Notes

[1] W. R. Corliss, *Scientific Satellites* (NASA SP-133), National Aeronautics and Space Administration, Washington, DC (1967), p 78.

[2] P. Klein, J. Goode, P. Hammer and D. Bellair, "Spacecraft Telemetry Systems for the Developing Nations," *1971 IEEE National Telemetering Conference Record*, Apr 1971, pp 118-129.

[3] T. Clark and R. McGwier, "The DSP Project Update," *Proceedings of the AMSAT-NA Sixth Space Symposium*, Atlanta, Nov 1988, ARRL.

[4] K. Meinzer, "Lineare Nachrichtensatellitentransponder Durch Nichtlinear Signalzerlegung" (Linear Communications Satellite Transponder Using Non-linear Signal Splitting), Doctoral Dissertation, Marburg University, Germany, 1974. K. Meinzer, "A Frequency Multiplication Technique for VHF and UHF SSB," *QST*, Oct 1970, pp 32-35.

[5] J. King, "The Third Generation," *Orbit*, Vol 1, no. 4, Nov/Dec 1980, pp 12-18.

[6] G. Hardman, "A Novel Transponder for Mobile Satellite Service," *Telecommunications*, Feb 1986, Vol 20, no. 2, pp 61-62.

[7] K. Meinzer, "The Radio Links to Phase 3-D: An Initial System Concept," *AMSAT-DL Journal*, Vol 14, no. 1, Jan/Feb 1987. For English translation, see *AMSAT-NA Technical Journal*, Vol 1, no. 2, Winter 1987/88, pp 23-26.

[8] "RUDAK—What Is It?," *QST*, Jul 1988, pp 80-81; "The AMSAT RUDAK User Terminals," *QST*, Aug 1988, pp 88-89. These articles, by V. Riportella, were prepared from notes provided by AMSAT-DL RUDAK Program Manager Hanspeter Kuhlen, DK1YQ, and translated by Don Moe, KE6MN/DJØHC. They have been reprinted in *The ARRL Satellite Anthology*, ARRL, 1988.

[9] The main source for information on RUDAK is the *AMSAT-DL Journal* (in German). Several articles have appeared in English translation. See, for example, *AMSAT-NA Technical Journal*, Vol 1, no. 1 (Summer 1987) and Vol 1, no. 2 (Winter, 1987-88).

[10] J. Kraus, *Antennas* (New York: McGraw-Hill, 1950) Chapter 2.

[11] M. Davidoff, *Using Satellites in the Classroom: A Guide for Science Educators*, Catonsville Community College, 1978, pp 6.52-6.56. Out of print. Microfiche copies are available from ERIC (Educational Resources Information Center), Computer Microfilm Corp, 3900 Wheeler Ave, Alexandria, VA 22304. Specify Document # ED 162 635.

[12] See Note 7.

[13] See Note 11, pp 6.24-6.31.

[14] D. Jansson, "Spaceframe Design Considerations for the Phase IV Satellite," *AMSAT-NA Technical Journal*, Vol 1, no. 2, Winter 1987/88.

[15] G. Mueller and E. Spangler, *Communications Satellites*, New York: John Wiley & Sons, 1964, p 12.

[16] See Note 1, Section 9.5.

[17] J. R. Primack, "Let's Ban Nuclear Reactors from Orbit," *Technology Review*, May/June 1989, pp 27-28.

[18] R. Leifer et al, "Detection of Uranium from Cosmos-1402 in the Stratosphere," *Science*, Vol 238, 23 Oct 1987, pp 512-514.

[19] T. Beardsley, *Scientific American*, Feb 1989, pp 15-16.

[20] D. Chapin, C. Fuller, and G. Pearson, "A New Silicon P-N Junction Photocell for Converting Solar Radiation into Electrical Power," *J Applied Physics*, Vol 25, May 1954, p 676.

[21] H. M. Hubbard, "Photovoltaics Today and Tomorrow," *Science*, Vol 244, 21 April 1989, pp 297-304.

[22] See Note 11, pp 6.24-6.31, 6.40-6.41.

[23] J. Lyons III and A. Ozkul, "In-orbit performance of INTELSAT IV-A spacecraft solar arrays," *COMSAT Technical Review*, Vol 17, no. 2, Fall 1987, pp 403-419.

[24] R. Daniels, "The Propulsion Systems of the Phase-III Series Satellites," *AMSAT-NA Technical Journal*, Vol 1, no. 2, Winter, 1987-88, pp 9-15.

[25] J. King, "Using Water as a Primary Method of Propulsion for Spacecraft Modifying Standard STS Orbits," *Orbit*, no. 19, Nov/Dec 1984, pp 5-8.

[26] P. Stakem, "One Step Forward—Three Steps Backup, Computing in the US Space Program," *BYTE*, Vol 6, no. 9, Sep 1981, pp 112, 114, 116, 118, 122, 124, 126, 128, 130, 132-134, 138, 140, 142, 144.

[27] G. Hardman, "The Integrated Housekeeping Unit—A Method of Telemetry, Command and Control for Small Spacecraft," *AMSAT-NA Technical Journal*, Vol 1, no. 2, Winter, 1987-88.

[28] K. Meinzer, "IPS, An Unorthodox High Level Language," *BYTE*, Jan 1979, pp 146, 148-152, 154, 156, 158-159.

[29] R. J. McEliece, "The Reliability of Computer Memories," *Scientific American*, Jan 1985, pp 88-92, 94,95.

[30] T. Jeans and C. Traynar, "The Primary UoSAT Spacecraft Computer," *The Radio and Electronic Engineer*, Vol 52, no. 8/9, Aug/Sep 1982, pp 385-390.

[31] C. Haynes, "A Low-Power 16-bit Computer for Space Application," *The Radio and Electronic Engineer*, Vol 52, no. 8/9, Aug/Sep 1982, pp 391-397.

[32] J. Miller, G3RUH "OSCAR-10 Attitude Determination," *Proceedings of the 4th Annual AMSAT Space Symposium*, Dallas, 1986, ARRL.

[33] J. R. Wertz, *Spacecraft Attitude Determination and Control*, Reidel Publishing Co, 1984. ISBN 90-277-1204-2.

Chapter 16

So You Want to Build a Satellite

Many people view satellite construction as meticulously assembling a huge pile of mechanical and electrical components into an OSCAR. They're about two percent right. The visible part of the satellite program, the flight hardware, is only the tip of a massive iceberg. Without an effective support structure, there wouldn't be any amateur spacecraft: no iceberg, no tip. A partial list of the countless necessary support activities that lead to a finished OSCAR is given in Table 16-1.

CONSTRUCTION SUPPORT ACTIVITIES

The radio amateur satellite program has been, and will probably always be, understaffed. AMSAT attracts people who are both doers and dreamers, doing the nearly impossible while dreaming about what they could accomplish if they only had access to a few more resources. If you share in the dream, you'll probably want to help out in some way. Of the many avenues open to you, the first is to become an active AMSAT volunteer. If this appeals to you, the following steps are in order:

1) Learn all you can about the Amateur Radio space program;

2) Consider seriously how much time and effort you're willing to commit to satellite activities;

3) Pick an area where your personal skills and interests mesh with the needs of the program, identify an unmet

---

Table 16-1

Support Activities Involved in the Production of an OSCAR

1) Design of flight hardware
2) Construction of prototype hardware
3) Construction of flight hardware—Electrical: modules and wiring, PCB layout—Mechanical: machining, sheet metal work, potting, construction of handling fixtures, shipping crates
4) Testing of flight hardware—Includes arranging for test facilities and people to oversee vibration, environmental, burn-in and performance tests
5) Drafting—Mechanical and electrical subsystems
6) Interfacing with launch agency—Providing documentation related to satellite/rocket interface, safety and protection of primary payload (outgassing tests, and so on)—Attending coordination meetings as required
7) Identifying and procuring future launches
8) Construction management—Parts procurement (includes locating special components and ensuring timely arrival of long-lead-time items)—Allocating available resources (financial and human)—Locating volunteers with special expertise
9) Launch information nets
10) Providing user information and membership services
    Production of periodical (at least bi-monthly)
    Production of technical journal (once or twice per year)
    Production of other written materials
    Coordinate weekly AMSAT Nets (provide information)
    Produce information programs (slide shows, videotapes)
    Provide support to professional educators
    Coordinate operating awards (ZRO tests, operating events)
    Area Coordinator program
    Encourage and support production of texts and magazine articles

11) Fundraising
    Sales of QSL cards, software, T-shirts, patches, and so on
    Handling special contribution campaigns
    Artwork preparation for magazine ads, T-shirts, QSL cards, and so on
12) Coordination with international AMSAT affiliates
13) Technical studies focusing on future spacecraft systems—thermal design, orbit selection, orbit determination, attitude control, designs of all subsystems
14) Launch operations—Travel to launch site, shipping satellite and rocket engine, interface satellite to launch vehicle, checkout
15) Command station network—Arranging for construction and operation of worldwide network—Design of special hardware and computer programs
16) Miscellaneous needs
    Language translation: French, German, Italian, Japanese, Russian, Spanish, other
    Legal: procurement contracts, trademark concerns, corporation papers, insurance of various types
17) Financial and business
    Financial record keeping
    Auditing as required
    Maintaining membership files
    Filing corporation reports as required
    Estimating future needs and cash-flow situation
    International cash transfers
18) Maintaining historical records: general, spacecraft telemetry
19) Construction of test equipment and special test facilities

need where you feel you can make a special contribution, and then present your ideas to AMSAT.

Step 1 doesn't need much explanation—you'll be more of an asset than a drain if you know what's going on.

The importance of Step 2 cannot be overemphasized. Space activities have a certain aura of excitement that attracts many of us initially. But the kind of personal involvement AMSAT needs often leads to long hours of tedious work with hardly even a "thank you." For their efforts, most volunteers receive little more than indigestion, a continual drain on their petty cash and an ever-growing sleep deficit! Seriously, you have to be the kind of person who can be satisfied simply with seeing that an important job gets done well and on schedule. If you're after glamour and personal recognition, you've probably chosen the wrong field. Bringing a new volunteer onboard involves a big investment of effort by current workers who are probably already up to their apogees in work. The decision to volunteer should be given very serious consideration.

Step 3 needs further explanation. In truth, many volunteers are attracted initially by the idea of designing and building flight hardware. After learning as much as possible about the OSCAR program, however, they may realize that contributing their special skills in other areas could make a more significant impact on amateur space efforts. While a few immediate needs are often announced on the AMSAT nets, many, many other important tasks go unadvertised. Why? Because long-term efforts to locate the right person to undertake them have been unsuccessful, or because the idea hasn't yet occurred to the AMSAT directors.

Some potential volunteers hesitate to step forward because they fear that their lack of special spacecraft expertise means there aren't any important jobs for them. Nothing could be further from the truth. A glance at Table 16-1 should make it clear that people with any one of a surprisingly large variety of skills or areas of expertise, from graphic arts, writing and editing, language translation and videotape production, to accounting and law, can contribute significantly to the success of the satellite program. In fact, many tasks don't require specialized skills, but are nonetheless important to AMSAT's success. These are often the most difficult jobs to find volunteers for, since a person must be very committed to undertake them.

If, after due consideration, you still want to become part of the team, it takes only an informal note or proposal to AMSAT to get started. Indeed, volunteers are usually amazed at how quickly they can take on major responsibilities.

When I asked several long-term workers if they'd like to pass along any hints to new volunteers, two closely related themes were repeated: Don't be afraid to say no, and never agree to a schedule you feel it isn't possible to meet (and do everything possible to live up to any schedule you've committed to). It's often difficult, especially for a newcomer, to say no when asked to take on some extra assignment. Saying no, however, is best for both the long-term satellite program and everyone involved, if saying yes

would lead to unmet schedules or severe personal sacrifices that destroy the satisfaction that working for the program can provide. Since effective workers are likely to attract additional tasks, one either learns to say no or suffers early burnout. Only the individual knows where the critical overload point is.

The need to meet schedules is absolutely essential. Satellite construction is a team effort to meet deadlines imposed by a launch agency, by a laboratory providing special test facilities on certain dates, or by another volunteer who has scheduled personal vacation time so it can be devoted to a specific AMSAT task. Under these circumstances, one person's late project can be disastrous.

One outstanding characteristic of almost all long-term AMSAT volunteers is the seriousness with which they accept commitments. Once they agree to a task or schedule, they do everything possible to deliver as promised. Over the years, this sense of commitment has led to a very special camaraderie, trust and respect among those involved. It's a spirit that I've never seen anywhere else in the academic, scientific, industrial or sports communities.

You don't have to have a formal title or be willing to invest a big chunk of time to be an AMSAT supporter. Everyone who helps a newcomer get started in satellite communications, provides information on the satellite program to other segments of the Amateur Radio community, or makes a modest financial contribution to AMSAT is filling an important need. A great many people helping in a lot of small ways will keep the satellite program vital, so let's all try to be conscious of the little things.

For example, if you need something from AMSAT headquarters or a volunteer worker, an SASE will save a few minutes, as will a request phrased so that it can be answered yes or no or with an article reprint. Similarly, providing information by telephone requires only a fraction of the time that preparing a written answer does. The point is: The time of key AMSAT volunteers is a very valuable commodity; small efforts by all of us to lighten their workload will have a cumulative pay-off.

By now, you probably have some idea as to what level of support you'd be comfortable with. For those who are determined to become involved in hardware construction (flight, flight-related or ground-command), we now look at the steps involved in the spacecraft-construction aspect of the OSCAR program.

SPACECRAFT HARDWARE

If you want to become directly involved in satellite construction, one of your first steps should be to familiarize yourself with satellite system design. You don't have to become an expert in all areas, but you do need a good practical understanding of the functions of the various subsystems, the tradeoffs involved in their selection, and how they affect each other. Chapter 15 is a good starting point for obtaining this information.

From a project-management perspective, the construction of a satellite can be thought of as consisting of six major stages: (1) preliminary design, (2) system specification, (3) subsystem design and fabrication, (4) integration

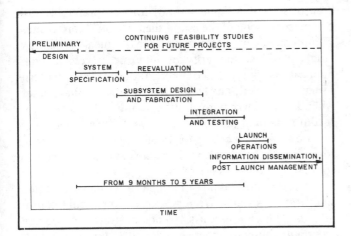

Fig 16-1—Time frame for satellite construction from project management perspective.

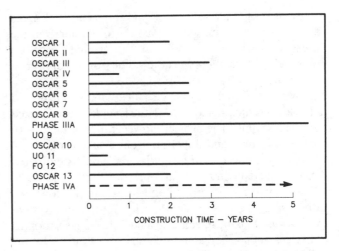

Fig 16-2—Amateur satellite construction time (adapted from D. Jansson, ''Spacecraft Technology Trends in the Amateur Satellite Service,'' *AMSAT-NA Technical Journal*, Vol 1, no. 2, Winter 1987-88, pp 3-8).

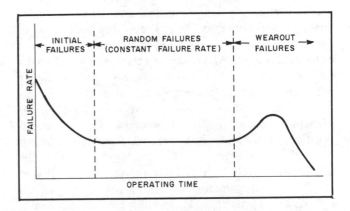

Fig 16-3-Typical component failure curve. The object of AMSAT's testing and stress program is to have the satellite operate in the central flat region of the curve. For information on designing for reliability, see W. C. Williams, ''Reliability: Lessons from NASA,'' *IEEE Spectrum*, Vol 18, no. 10, Oct 1981, pp 79-84.

and testing, (5) launch operations, and (6) information dissemination and post-launch management. The time frame for these activities is outlined roughly in Fig 16-1. Fig 16-2 presents the elapsed time, from commitment to spacecraft construction to launch, for OSCAR spacecraft.

The preliminary-design stage involves feasibility studies of new approaches to satellite design. The availability of recently developed high-performance, low-cost components, launch access to unusual orbits, new sources of financial support and the like are always presenting new design options. In a long-term multi-satellite program, feasibility studies are continually taking place.

At some point, usually in response to a specific launch opportunity, the decision is made to construct a spacecraft. A set of system specifications must then be agreed upon and the subsystem requirements defined. Initial system specifications include selecting a primary-mission subsystem, determining the spacecraft size, shape and attitude-stabilization system, and estimating the available power and allocating this power to various subsystems. The design goes through a couple of iterations until everyone feels the system design is solid.

Subsystems are then designed, built, tested and refined. With AMSAT satellites, subsystem design and construction is usually handled by small core groups scattered around the world. An electronic subsystem generally is built in a number of versions: an *engineering development model*, a *flight prototype* and a *flight unit*, the latter using the most reliable components available. Each subsystem must be tested thoroughly under extreme conditions of temperature, over- and under-voltage, and intense RFI, so that potential weak spots can be identified and corrected.

Next, the subsystems are integrated into a spacecraft so additional stress tests, operational checks and RF-compatibility tests can be performed. The stresses include a *burn-in* period for electronics systems, during which electrical parameters and temperatures are similar to those expected in space, but with the system at atmospheric pressure; *environmental tests* that involve operating the spacecraft in a vacuum chamber under temperatures con-

siderably more severe than those expected in space (for example, $-20°C$ and $+60°C$); and a vibration test to ensure that the satellite will survive the launch. The objectives of the vacuum test include (1) checking for material sublimation that could contaminate spacecraft systems, (2) testing for corona discharge and (3) verifying the predicted thermal behavior in the absence of convective heat flow. Vacuum and vibration tests are usually performed at large government or commercial laboratories that have the special facilities required.

The electrical testing strategy is based on the fact that high temperatures and overvoltages tend to compress the time scale of the failure curves for most electrical components. (A typical curve is shown in Fig 16-3.) Temperature cycling and vibration tests have a similar effect on mechanical components. Consequently, one month of actual testing might be equivalent to two years of testing under normal operating conditions. The aim of the testing program is to discover and correct all weak spots and then bring the spacecraft past the initial hump in the failure curve while it's still on the ground. NASA's experience has clearly proved the validity of this approach to ensuring reliability.

When the satellite has passed all tests, it's transported to the launch site, mated to the launch vehicle and checked out one last time. The project, however, doesn't end with the launch. Command stations must be available when and where they are needed; information must be disseminated to users; and data on spacecraft operation must be collected to assist in current operation and for use in the design of future spacecraft. The entire procedure, from system specification to launch, can take anywhere from six months to five years, depending on the complexity of the spacecraft and the available personnel resources.

Once you know something about satellite systems and the stages of satellite construction, it's time to pick a particular project, subsystem or aspect of construction to focus on. This may involve refining a specific subsystem you're particularly knowledgeable about (for example, analog-digital converter design), applying your skills to various subsystems (for example, optimizing PCB layouts) or looking into spacecraft subsystems that appear to need improvement, even though you have no prior knowledge in the area (when no "expert" is available, as with water-fuel rockets, someone has to start from scratch). Or perhaps you see a technique for accomplishing a spacecraft function that's simpler or more reliable than the approach currently being used. You'll probably be entering the construction cycle (Fig 16-1) at the preliminary design stage, so try to maintain a broad, long-term perspective.

As a first project, pick something modest with clearly defined interfaces. For example, you might elect to work on a high-efficiency linear Mode A/J transponder (146-MHz input, outputs at 29.5 and 435 MHz) designed to fit the TSFR tray in a MicroSat. Initial objectives would probably be limited to producing and testing an engineering development model.

One important area of spacecraft construction that is frequently overlooked is the need for special test equipment and detailed procedures for satellite evaluation and check out. If you're interested in this area, contact AMSAT to find out what's needed.

Once you identify a specific aspect of construction that you'd like to work on, send a brief memo to the person who is coordinating the core group currently handling that aspect. If you're not sure who to write, contact one of the AMSAT directors. It's important to realize that AMSAT workers and directors don't see each other at the office each morning: They're spread around the world and most have full-time jobs. Circulating a letter can take months and the chances of it getting lost are, unfortunately, high. Therefore, the best strategy may be to make several copies and send one to each person you think may be interested.

The object is to establish direct contact with the person or persons responsible for the work you're interested in. Because of AMSAT's geographical scatter, there are always problems in internal communication. It's therefore best to establish a single point of contact, preferably with one person at one of the established AMSAT core groups. The choice can depend on geographical proximity, language skills, similarity of interests, access to communications channels (commercial or amateur), patterns of business travel, and so on. It's anticipated that pacsats will take on a large part of our communications needs, so reading the mail on OSCARs 16 and 19 will give you a picture of who's working on what project, and setting up for pacsat (either direct or via a gateway) may provide the communications links needed.

Before undertaking a major effort that will require significant assistance from AMSAT, you should plan to take on some simpler tasks to demonstrate your competence and ability to adhere to schedules. Initiating a large project generally involves preparing a proposal focusing on the concerns listed in Table 16-2. Note that these aren't formal guidelines; they're merely suggestions. AMSAT is most definitely not a huge, faceless bureaucracy. Every proposal is treated individually. It's the content that counts.

When estimating the amount of time that a project is going to require, don't forget paperwork. Communication is a two-way street. You must be willing to provide

---

Table 16-2

Construction Project Proposals: Topics to be Addressed

1) *General Conceptual Plan*—A general description of the proposed subsystem or project.
2) *Trade-off Discussion*—A more detailed discussion of the proposed subsystem, including a candid evaluation of (1) its advantages and disadvantages with respect to prior subsystems having a similar function, and (2) its impact on other subsystems.
3) *Interface Considerations*—A detailed specification of how the proposed subsystem will interface with other subsystems.
4) *Environmental Design*—A discussion of how the experimenter will attempt to guarantee that the completed unit will perform satisfactorily under anticipated extremes of temperature, vacuum, power variation, radiation and the RF environment, including analysis of waste heat, potential RFI and steps taken to prevent RFI problems.
5) *Component Selection and Construction Techniques* If the experimenter is planning to construct a flight unit, describe steps to be taken to ensure component and construction quality.
6) *Testing Program*—Detailed description of all tests to be performed on the completed unit.
7) *Required Support*—If the experimenter anticipates calling on AMSAT to provide assistance in financing, design or parts procurement, the required type and level of support should be specified.
8) *Experience of Project Personnel*—List the people expected to work on the project, and the expertise and the expected time commitment of each project member.
9) *Schedule and Delivery*—Set up a timetable for the project, indicating dates for milestones and specifying a realistic delivery date that includes allowances for unanticipated delays.

clear documentation to other groups that have an immediate need to know, and eventually to users as well. You must also maintain good records relating to electronic design and testing, and to direct and indirect expenses. Try to anticipate as many of these little loose ends as possible when estimating the total resources needed to accomplish a particular job. Frankly, no matter how thorough you think your estimates are, you'll probably grossly underestimate the real effort needed. Psychologically this might be a good thing: If we knew what we were really committing ourselves to, far fewer might volunteer.

Over the years, several individuals in the academic and educational communities have been able to fuse their vocations with their satellite interests. Students in the undergraduate Electrical Engineering Technology program at Trenton State College (NJ) often chose senior projects, such as a 70-cm to 23-cm linear transponder, related to satellite systems. Key systems for OSCAR 18 (Webersat) and Phase 4A have been (and are being) built by students and faculty at the Center for Aerospace Technology at Weber State University in Utah. Programs are also in place at the University of Surrey (England) and the University of Marburg (Germany). Several students have received Master's and PhD degrees for projects that relate to OSCAR satellite design. Science and engineering educators have, at times, received government grants to work on particular aspects of the OSCAR program.

AMSAT has, on occasion, been able to sponsor an internship program. The object of the program is to train young radio amateurs who have strong backgrounds in science and engineering in all aspects of satellite construction. This is the wisest investment we can make to ensure the continuation of the Amateur Radio space program. Interns may be paid employees, but the wages are terrible and the hours ridiculous. Interns nonetheless receive invaluable experience in all phases of satellite design and construction, and a measure of responsibility usually achieved only by senior engineers. In return, AMSAT receives the services of very bright, very committed scientists and engineers at modest cost. Internships are flexible: Appointments can last from a few months to several years. Don't send for a formal application form—there aren't any. If you'd like to apply, just submit a letter that details your background and explains why you feel you could contribute to the amateur space program. Better yet, volunteer to spend a few weeks as an unpaid intern so that both you and the AMSAT team can better evaluate the desirability of making a longer-term commitment.

Though most of the comments in this section have focused on building flight hardware, the construction of

telecommand equipment and gateways offer many similar satisfactions, as does the design of software for operating spacecraft, and testing spacecraft during the design stages. Many people working directly on today's spacecraft got their start by building telecommand stations. Most notable are Larry Kayser (VE3PAZ) who almost single-handedly kept OSCAR 6 operating during its early days; A. Gschwindt (HA5WH) who set up a command station for OSCAR 6 and later built switching regulators for Phase 3A and OSCARs 10 and 13; and Martin Sweeting (G3YJO) who went from commanding OSCARs 6 and 7 to directing the University of Surrey Satellite Program, where UoSAT-OSCARs 9, 11, 14 and 15 were constructed.

Constructing a local satellite gateway is an excellent way to start a "career" in satellite hardware construction. The gateway might be designed to provide local 2-meter FM voice stations with access to a linear transponder on a high-altitude spacecraft, or to provide local 2-meter packet operators with access to the pacsat system. In either case, you'll be acquiring valuable design experience, generating new support for the OSCAR program, initiating a local support group and demonstrating your ability to oversee a project from conception to completion.

Another avenue for getting involved in the construction of flight hardware that has proven very effective in the past is to organize balloon launches. The payload could consist of a transponder (digital or linear), or an FM or ATV repeater. Many of the design problems associated with balloon launchings are closely related to those involved in spacecraft construction. Of course, you also have to handle your own launch and recovery operations. Several AMSAT groups, in Germany, England, Holland, South Africa and the US, have engaged in this activity. These groups often go on to construct flight hardware. Karl Meinzer (DJ4ZC) a key designer of several AMSAT spacecraft, started this way nearly two decades ago.

If nothing short of building satellite hardware interests you, you're not shut out of the picture. MicroSat provides one possible path. If you construct a prototype experimental module for the MicroSat TSFR tray that's of interest to AMSAT, you're likely to be able to find support for the construction of a flight model.

AMSAT is a flexible, vital organization. Its very informality—the lack of straightforward procedures for submitting proposals or volunteering—can make it difficult for a newcomer to become involved. The same lack of formal structure, however, makes it possible for the competent, committed individual to assume important responsibilities quickly. We'd like to hear from you.

Appendix A

Radio Amateur Satellite History: Dates and Frequencies

Satellite Launch date Operating life	Launch agency, Licensing authority	Transponder input/output freq (bandwidth)	Beacon frequencies	Transmitting power (peak)	Apogee height
OSCAR I Dec 12, 1961 21 days	US, US	——	144.983 MHz	0.1 W	430 km
OSCAR II June 2, 1962 19 days	US, US	——	145.000 MHz	0.1 W	390 km
OSCAR III March 9, 1965 transponder: 18 days beacon: several months	US, US	145.10/145.90 MHz (50 kHz)	145.85 MHz 145.95[1]	1.0 W	940 km
OSCAR IV Dec 21, 1965 85 days	US, US	144.100/431.938 MHz (10 kHz)	431.928 MHz	3.0 W	33,600 km
Australis-OSCAR 5 Jan 23, 1970 52 days	NASA, US	——	144.05 MHz 29.45 MHz	0.2 W	1480 km
AMSAT-OSCAR 6 Oct 15, 1972 4.5 years	NASA, US	145.95/29.50 MHz (100 kHz)	29.450 MHz 435.100 MHz	1.5 W	1460 km
AMSAT-OSCAR 7 Nov 15, 1974 6.5 years	NASA, US	145.90/29.45 MHz (100 kHz) 432.15/145.95 MHz (40 kHz)	29.502 MHz 145.972 MHz 435.100 MHz 2304.1 MHz[2]	8.0 W	1460 km
AMSAT-OSCAR 8 March 5, 1978 5.3 years	NASA, US	145.90/29.45 MHz (100 kHz) 145.95/435.15 MHz (100 kHz)	29.402 MHz 435.095 MHz	1.5 W	910 km
RS-1 RS-2 Oct 26, 1978 several months	USSR, USSR	145.89/29.37 MHz (40 kHz) 145.89/29.37 MHz (40 kHz)	29.401 MHz 29.401 MHz	1.5 W 1.5 W	1700 km 1700 km
AMSAT-Phase 3-A May 23, 1980 launch failure	ESA, FRG	435.22/145.90 MHz (180 kHz) 435.22/145.90 MHz (180 kHz)	145.99 MHz 145.81 MHz	50 W	35,800 km (target)
UoSAT-OSCAR 9 Oct 6, 1981 8+ years[3]	NASA, UK	——	145.825 MHz 435.025 MHz 7.050, 14.002 MHz 21.002, 29.510 MHz 2.401, 10.47 GHz	0.8 W	544 km
RS-3—RS-8 Dec 17, 1981	USSR, USSR				
RS-3 2 years			29.321, 29.401 MHz	1.5 W	1690 km
RS-4 2 years			29.360, 29.403 MHz	1.5 W	1690 km
RS-5 6.5 years		145.93/29.43 MHz (40 kHz) 145.826 MHz[4]	29.331, 29.452 MHz	1.5 W	1690 km
RS-6 3 years		145.93/29.43 MHz (40 kHz)	29.411, 29.453 MHz	1.5 W	1690 km
RS-7 6.5 years		145.98/29.48 MHz (40 kHz) 145.836 MHz[4]	29.341, 29.501 MHz	1.5 W	1690 km
RS-8 4 years		145.98/29.48 MHz (40 kHz)	29.461, 29.502 MHz	1.5 W	1690 km
Iskra 2 May 17, 1982 53 days	USSR, USSR	21.25/29.60 MHz (40 kHz)	29.578 MHz	1.0 W	335 km
Iskra 3 Nov 18, 1982 37 days	USSR, USSR	21.25/29.60 MHz (40 kHz)	29.583 MHz	1.0 W	335 km

Satellite / Launch date / Operating life	Launch agency, Licensing authority	Transponder input/output freq (bandwidth)	Beacon frequencies	Transmitting power (peak)	Apogee height
AMSAT-OSCAR 10 June 16, 1983 OBC failure late 1986 transponder operating sporadically (2/92)	ESA, FRG	435.103/146.901 MHz (152 kHz) 1269.45 MHz/436.55 MHz (800 kHz)	145.810 MHz 145.987 MHz 436.02 MHz 436.04 MHz	50 W	35,500 km
UoSAT-OSCAR 11 March 1, 1984 operating (2/92)	NASA, UK	——	145.826 MHz 435.025 MHz 2401.5 MHz	1.0 W	690 km
Fuji-OSCAR 12 Aug 12, 1986 3+ years[5]	NASDA, Japan	145.95/435.85 MHz (100 kHz) 145.85, .87, .89, .91 MHz[6]	435.797 MHz	2.0 W	1510 km
RS-10/11 June 23, 1987 operating (2/92)	USSR, USSR	21.18/29.38 MHz (40 kHz) 21.18/145.88 MHz (40 kHz) 145.88/29.38 MHz (40 kHz) 21.18/29.38 and 145.88 MHz (40 kHz) 21.18 and 145.88/29.38 MHz (40 kHz) 21.120 and 145.820 MHz[4] 21.23/29.43 MHz (40 kHz) 21.23/145.93 MHz (40 kHz) 145.93/29.43 MHz (40 kHz) 21.23/29.43 and 145.93 MHz (40 kHz) 21.23 and 145.93/29.43 MHz (40 kHz) 21.130 and 145.830 MHz[4]	29.357, 29.403 MHz 145.857, 145.903 MHz 29.407, 29.453 MHz 145.907, 145.953 MHz	5 W	1000 km
AMSAT-OSCAR 13 June 15, 1988 operating (2/92)	ESA, FRG	435.498/145.900 MHz (150 kHz) 1269.496/435.860 MHz (290 kHz) 144.448/435.965 MHz (50 kHz) 435.620/2400.729 MHz (36 kHz) 1269.710/435.677 MHz (RUDAK digital)	145.812 MHz 145.985 MHz 435.652 MHz 2400.664 MHz	50 W	36,265 km
UoSAT-OSCAR 14 Jan 22, 1990 operating (2/92)	ESA, UK	145.975[6] MHz	435.070 MHz	10 W	805 km
UoSAT-OSCAR 15 Jan 22, 1990 < one day	ESA, UK	——	435.120 MHz	10 W	805 km
Pacsat-OSCAR 16 Jan 22, 1990 operating (2/92)	ESA, US	145.900, .920, .940, .960[6] MHz	437.051 MHz 437.026 MHz 2401.143 MHz	4 W 4 W 1 W	805 km
DOVE-OSCAR 17 Jan 22, 1990 operating (2/92)	ESA, Brazil	——	145.824 MHz 145.825 MHz 2401.221 MHz	4 W 4 W 1 W	805 km
Webersat-OSCAR 18 Jan 22, 1990 operating (2/92)	ESA, US	Modes J,L (digital)[6, 7]	437.102 MHz 437.075 MHz	4 W	805 km
Lusat-OSCAR 19 Jan 22, 1990 operating (2/92)	ESA, Argentina	145.840, .860, .880, .900[6] MHz	435.154 MHz 435.126 MHz 437.127 MHz	4 W 4 W 0.8 W	805 km
Fuji-OSCAR 20 Feb 7, 1990 operating (2/92)	NASDA, Japan	145.95/435.85 MHz (100 kHz) 145.85, .87, .89, .91 MHz[6]	435.797 MHz	2.0 W	1745 km
BADR-1 Jul 16, 1990 operating (9/90)	PRC (China) SUPARCO[8]	——	145.028[9] MHz 145.825 MHz		992 km
RS-14/AMSAT-OSCAR 21 Jan 29, 1991	USSR, USSR	435.06/145.89 MHz (80 kHz) 435.08/145.91 MHz (80 kHz) 435.016,.041,.155,.193 MHz[6]	145.822 MHz 145.952 MHz 145.983 MHz	10 W	1000 km
RS-12/13 Feb 5, 1991	USSR, USSR	145.93/29.43 MHz (40 kHz) 21.23/29.43 MHz (40 kHz) 21.23/145.93 MHz (40 kHz) 21.23 and 145.93/29.43 MHz (40 kHz) 21.23/29.43 and 145.93 MHz (40 kHz) 21.129 and 145.830 MHz[4] 145.98/29.48 MHz (40 kHz) 21.28/29.48 MHz (40 kHz) 21.28/145.98 MHz (40 kHz) 21.28 and 145.98/29.48 MHz (40 kHz) 21.28/29.48 and 145.98 MHz (40 kHz) 21.138 and 145.840 MHz[4]	29.408, 29.454 MHz 145.912, 145.958 MHz	8 W	1000 km
UoSAT-OSCAR 22 July 7, 1991	ESA, UK	145.900 MHz[6]	435.120 MHz	5 W	790 km

Notes

[1] Never activated due to technical problems.
[2] Never activated due to regulatory constraints.
[3] Re-entered on Oct 13, 1989.
[4] Autotransponder (ROBOT) uplink, downlink on beacon frequency.
[5] Withdrawn from service Nov 5, 1989 due to deteriorating power budget.
[6] Digital transponder, downlink on beacon frequency.
[7] To be announced.
[8] SUPARCO (Space an Upper Atmosphere Research Commission—Pakistan).
[9] At the fourth Surrey (UK) International Satellite Colloquium in 1989, satellite-construction groups and representatives of the IARU agreed to procedures for coordinating satellite frequencies. BADR-1 frequencies were not coordinated with either group.

Radio Amateur Operations from Space (updated March 1992)

Spacecraft	Dates	Operator(s)	Notes
US Shuttle Columbia (STS-9)	Nov 28- Dec 8, 1983	Owen Garriott (W5LFL)	>250 QSOs on 2-meter FM voice
US Shuttle Challenger (STS-51F)	July 29- Aug 6, 1985	Tony England (WØORE) Gordon Fullerton[2] John Bartoe (W4NYZ)	SAREX.[1] Approximately 6000 students participated. 2-meter FM voice and slow-scan TV (space-to-earth).
US Shuttle Columbia (STS-61A)	Oct 30- Nov 6, 1985	Reinhard Furrer (DD6CF) Ernst Messerschmidt (DG2KM) Wubbo Ockels (PE1LFO)	70-cm FM uplink, 2-meter FM downlink Call sign used: DPØSL SPACELAB-D1
MIR	Nov 6- Dec 26, 1988	Vladimir Titov (U1MIR) Musa Manarov (U2MIR) Valery Polyakoav (U3MIR)	2-meter FM voice
MIR	Feb 9- Apr 27, 1989	Alexander Volkov (U4MIR) Sergei Krikalev (U5MIR)	2-meter FM voice
MIR	1990...[3]		2-m FM voice, ongoing 2-m packet (1991...), ongoing Austrian packet experiment (AREMIR) began Oct 4, 1991 under direction of Franz Viehbock, U/OEØMIR
US Shuttle Columbia (STS-35)	Dec 2-11, 1990	Ron Parise (WA4SIR)	SAREX.[1] 2-meter FM voice, packet
US Shuttle Atlantis (STS-37)	Apr 5-11, 1991	Ken Cameron (KB5AWP) Steven Nagel (N5RAW) Linda Godwin (N5RAX) Jay Apt (N5QWL) Jerry Ross (N5SCW)	SAREX.[1] 2-meter FM voice, packet, fast-scan TV uplink
US Shuttle Atlantis (STS-45)	Note 4	Brian Duffy (N5WQW) Dave Leetsma (N5WQC) Dick Frimout (ON1AFD) Kathy Sullivan (N5YYV)	SAREX.[1] 2-meter FM voice
US Shuttle Columbia (STS-50)	Note 4	Dick Richards	SAREX.[1] 2-meter FM voice, fast-scan TV uplink
US Shuttle Endeavor (STS-47)	Note 4	Jay Apt (N5QWL) Robert "Hoot" Gibson	
US Shuttle Columbia (STS-55)	Note 5	Jerry Ross (N5SCW) Steven Nagel (N5RAW)	
US Shuttle Columbia (STS-62)	Note 6	Ron Parise (WA4SIR)	

Notes
[1] SAREX: Abbreviation for Shuttle Amateur Radio Experiment
[2] No call. Operated under supervision of WØORE as per US regulations.
[3] As MIR operation is frequent and ongoing in the early 1990s, dates are not specified.
[4] Scheduled for 1992 launch.
[5] Scheduled for 1993 launch.
[6] Planned

Spacecraft Profiles

Phase 3 Series
  AMSAT-OSCAR 10
  AMSAT-OSCAR 13
UoSAT Series
  UoSAT-OSCAR 11
  UoSAT-OSCAR 14
  UoSAT-OSCAR 15
JAS Series
  Fuji-OSCAR 20

RS Series
  RS-10/11
  RS-12/13*
  RS-14*
Microsat Series
  Pacsat-OSCAR 16
  DOVE-OSCAR 17
  Webersat-OSCAR 18
  Lusat-OSCAR 19

*Not yet launched

Each section of this appendix contains a succinct description of a radio amateur satellite, or group of satellites, either currently in orbit and operational or soon to be launched. This information has been organized, insofar as possible, using the standard format outlined in Table B-1. Though detailed, the profiles are by no means complete. Additional sources of information have been referenced when available.

Table B-1

Standard Format Used to Describe Satellites Listed in Appendix B

SPACECRAFT PRIMARY NAME
Note
General
  1.1 Identification: international designation, NASA catalog number, pre-launch designation, also known as, radio license call sign
  1.2 Launch: date, vehicle, agency, site
  1.3 Orbital Parameters (as of date specified): general designation, period, apogee and perigee altitude (specified over mean radius of earth [6371 km]), inclination, eccentricity, longitude increment, maximum access distance (see Eq 12.4), maximum access time, expected lifetime in orbit if less than 10 years, transfer orbit characteristics.
  1.4 Ground Track Data
  1.5 Operations: Group(s) responsible for coordination and scheduling
  1.6 Design/Construction Credits: project management, spacecraft subsystems
  1.7 Primary References
Spacecraft Description
  2.1 Physical Structure: shape, mass, thermal design
  2.2 Subsystem Organization: block diagram

Subsystem Description
  3.1 Beacons: frequency, power level, telemetry format, maximum Doppler shift (Table 13-1), data sources
  3.2 Telemetry: formats available, description of each format (including decoding information, sample data, and so on); decoding hints for ground stations
  3.3 Telecommand System
  3.4 Transponder: type, uplink passband, downlink passband, translation equation, output power, suggested uplink EIRP, maximum Doppler shift (Table 13-1)
  3.5 Attitude Stabilization and Control: primary control, secondary control, damping, sensors
  3.6 Antennas: description, gain, polarization
  3.7 Energy Supply and Power Conditioning: solar-cell characteristics and configuration, storage battery, switching regulators, and so on
  3.8 Propulsion System
  3.9 Integrated Housekeeping Unit (IHU) or Onboard Computer (OBC)
  3.10 Experimental Systems

Phase 3 Series

SPACECRAFT NAME:
AMSAT-OSCAR 10

Note: The main computer onboard OSCAR 10 failed in Dec 1986. As a result, ground command stations have very limited control over the orientation or operation of this spacecraft. However, when the orientation is favorable (with respect to the earth and sun) OSCAR 10 continues to provide good Mode-B service. This transponder is open for use when available except when AMSAT announcements request a temporary suspension due to poor sun angles or if FMing occurs on the downlink. If users cooperate, OSCAR 10 may provide many more years of service. Because AMSAT has very limited control over this spacecraft and no telemetry is available, the following profile has been greatly abbreviated.

General

1.1 Identification

International designation: 1983 058 B
NASA Catalog number: 14129
Pre-launch designation AMSAT Phase IIIB, φ3B

1.2 Launch

Date: 16 June 1983
Vehicle: Ariane-2
Agency: ESA
Site: Kourou, French Guiana

1.3 Orbital Parameters (July 1989)

General designation: high-altitude, elliptical, synchronous-transfer, Phase 3, Molniya
Period: 699.4 minutes
Apogee altitude: 35,500 km
Perigee altitude: 3955 km
Eccentricity: 0.605
Inclination: 26.1°
Longitude Increment: 175°W/orbit
Argument of Perigee: changing
12 July 1989 at 0900 UTC: 58.19°
Rate of change: 0.27047°/day, 8.22°/month, 98.8°/year
Maximum access distance: 9034 km

1.4 Ground Track Data

The following information is useful for rough tracking. Given the time, latitude and longitude of a reference apogee, apogee *two* orbits later will 40.9 minutes earlier (next day) at a longitude 9.4° further east at approximately the same latitude.

1.6 Design/Construction Credits

Project Management: AMSAT-NA (Jan King, W3GEY) and AMSAT-DL (Karl Meinzer, DJ4ZC)
Spacecraft subsystems: Contributed by groups in Canada, Hungary, Japan, United States, West Germany

1.7 Primary References

See references for Phase 3A (which follow) and OSCAR 13 (next section): J. A. King, "Phase III: Toward the Ultimate Amateur Satellite," Part 1, *QST*, June 1977, pp 11-14; Part 2, *QST*, July 1977, pp 52-55; Part 3, *QST*, Aug 1977, pp 11-13.

Spacecraft Description

2.1 Physical Structure

Shape: Tri-star [see Fig 1 (AO10)]
Mass: approximately 90 kg + fuel

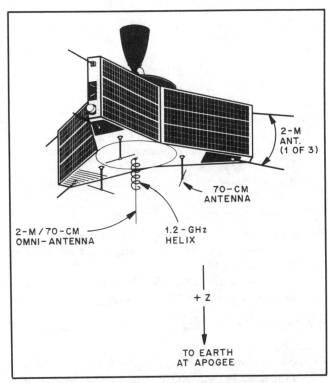

Fig 1 (AO10)—Pictorial view of AMSAT-OSCAR 10.

2.2 Subsystem Organization

(see OSCAR 13)

Subsystem Description

3.4 Transponders

General
Design/Construction credits: K. Meinzer, DJ4ZC, Ulrich Mueller, DK4VW, and Werner Haas, DJ5KQ, University of Marburg, Germany
Transponder I: Mode B (70 cm/2 meters)
Type: linear, inverting
Uplink passband: 435.027-435.179 MHz
Downlink passband: 145.825-145.977 MHz
Translation equation:
Downlink freq [MHz] =
581.004 − uplink freq [MHz] ± Doppler
Maximum Doppler (at perigee): 5.0 kHz

SPACECRAFT NAME: AMSAT-OSCAR 13

General

1.1 Identification

International designation: 1988 051 B
NASA Catalog number: 19216
Pre-launch designation: AMSAT Phase 3C

1.2 Launch

Date: 15 June 1988 (11:19:04.33 UTC)
Vehicle: Ariane-4
Agency: ESA
Site: Kourou, French Guiana

1.3 Orbital Parameters (July 1989)

General designation: high-altitude, elliptical,
   synchronous-transfer, Phase 3, Molniya
Period: 686.7 minutes
Apogee altitude: 36,265 km
Perigee altitude: 2545 km
Eccentricity: 0.656
Inclination: 57.4°
Longitude increment: 187.8°E/orbit
Argument of Perigee: changing
   5 July 1989 at 0400 UTC: 207.04°
   Rate of change: 0.0514°/day, 1.542°/month,
   18.77°/year
Maximum access distance: 9050 km

Notes: OSCAR 13 was initially placed in a transfer orbit of:
   Apogee altitude: 36,077 km; Perigee altitude: 223 km;
   Inclination: 10.0°.
   Resonant perturbations of OSCAR 13's orbit may
lead to reentry in the mid or late 1990s.

1.4 Ground Track Data

See Appendix C
The following information is useful for rough tracking.
Given the time, latitude and longitude of a reference apogee, the apogee two orbits later will occur 66.6 minutes earlier (next day) and 15.6° further east at approximately the same latitude.

1.5 Operations

Coordinating Group: AMSAT-DL
Scheduling:
   Mode JL near apogee (about 2 hours per orbit)
   Mode B remainder of orbit when spacecraft in sunlight
   Mode S near apogee (several hours per week)

1.6 Design/Construction Credits

Project Management: AMSAT-NA (Jan King, W3GEY)
   and AMSAT-DL (Karl Meinzer, DJ4ZC)

1.7 Primary References

1) V. Riportella, "Introducing Phase 3C: A New,
   More Versatile OSCAR," *QST*, Jun 1988,
   pp 22-30; reprinted in *The ARRL Satellite
   Anthology*.

2) J. Miller, "A PSK Telemetry Demodulator for
   OSCAR 10," *Ham Radio*, Apr 1985, pp 50-51,
   53-55, 57-62; also appeared in *Electronics and
   Wireless World*, Part 1, Oct 1984, pp 37-41,
   59-60, Part 2, Nov 1984, pp 37-38. Be sure to
   see Miller's revisions: *OSCAR News*, no. 70,
   Apr 1988, p 13.

3) "Phase 3C System Specifications," Part 1 and
   Part 2, *Amateur Satellite Report*, no. 177, 8 Jun
   1988, pp 3-4.

4) P. Gulzow, DB2OS, "AMSAT OSCAR-13
   Telemetry Block Format," *OSCAR News*, no.
   73, Oct 1988, pp 8-14.

Spacecraft Description

2.1 Shape

Tri-star as shown in Fig 1 (AO13)
Mass: 92 kg + 50 kg fuel

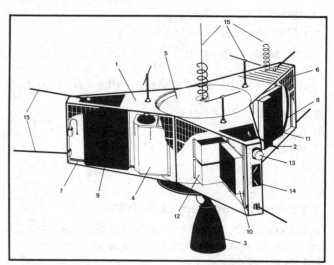

Fig 1 (AO13)   *Graphic by AMSAT-DL*

1—Aluminum space frame
2—S-band transponder
3—Kick motor
4—Helium tank container
5—Fuel and oxidizer tank
6—Solar panel
7—Magnetorquer coil
8—Nutation dampener
9—Integrated housekeeping unit
10—Battery charge regulator
11—Modulator
12—Auxiliary battery
13—Earth sensor
14—Sun sensor
15—Antennas

2.2 Subsystem Organization

See Fig 2 (AO13)

Subsystem Description

3.1 Beacons

	Frequency	Max Doppler (at perigee)
Mode B General	145.812 MHz	2.6 kHz
Mode B Engineering	145.985 MHz	2.6 kHz
Mode JL General	435.652 MHz	7.6 kHz
Mode S General	2400.664 MHz	41.9 kHz

3.2 Telemetry

The general beacon usually operates continuously,

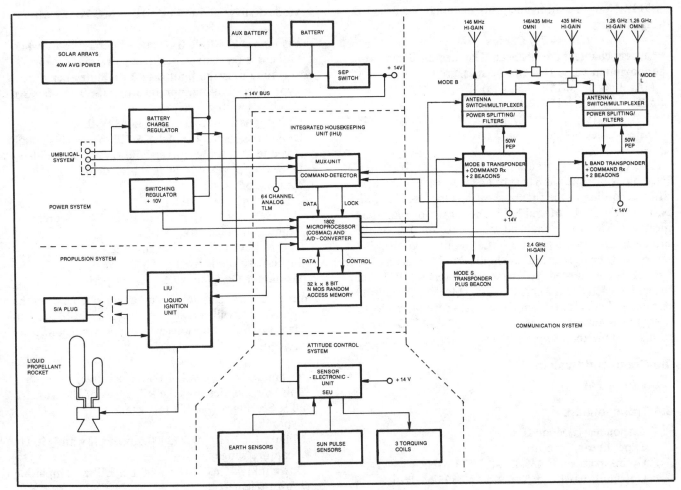

Fig 2 (AO13)—AMSAT-OSCAR 13 functional block diagram.

Table 1 (AO13)
OSCAR 13 RTTY Telemetry

50-baud RTTY data	*Explanation*
Z HI. THIS IS AMSAT OSCAR 13 01.47.27.3827	Time and date of frame. Day 3827 is 11 Jul 88
64 2 0 1 16 226 0	Miscellaneous binary values
	Telemetry channels in hex
245 7 147 132 192 7 138 48 200 126	Telemetry channels 00 thru 09
137 7 96 134 143 47 7 143 138 7	Telemetry channels 0A thru 13
14 99 136 56 7 7 135 7 116 7	Telemetry channels 14 thru 1D
137 7 206 136 134 7 69 131 138 35	Telemetry channels 1E thru 27
107 140 138 7 226 137 129 7 178 134	Telemetry channels 28 thru 31
122 7 111 133 137 137 15 122 129 206	Telemetry channels 32 thru 3C
	(Channel 3B is skipped)

---

providing text messages and selected telemetry data. Formats include CW (5 minutes on the hour and half hour), RTTY (5 minutes at 15 and 45 minutes past the hour) and 400-bps BPSK.

The RTTY is 50-baud Baudot using 170-Hz shift. If you use a multimode digital interface, use the following settings:

1) Mode: RTTY
2) Speed: 60 WPM or 45 bauds (50 bauds if available)
3) HF or VHF: HF (this is 200-Hz shift, but will work at 170 Hz)

4) Unshift-on-Space option: Set off or disabled
5) Receiver sideband: Upper

A frame of OSCAR 13 RTTY telemetry is shown in Table 1 (AO13).

A PSK demodulator by G3RUH is described in Reference 2. IBM PC decoding software for use with the G3RUH demodulator is available from Project OSCAR (Box 1136, Los Altos, CA 94023). Detailed telemetry-decoding information is contained in Reference 4. The telemetry has access to 64 analog channels and 64 digital status points. The analog channels are listed in Table 2

(AO13). Note that all temperature channels are decoded using the formula:

$T = (n - 120)/(1.71)$ Celsius

Current channels are all linear. The formula depends on the maximum value that can be measured.

1 A: $I = (n - 15)(4.854$ mA$)$
2.5 A: $I = (n - 15)(12.14$ mA$)$
5 A $I = (n - 15)(24.27$ mA$)$

Other values are per Table 2 (AO13)

BPSK telemetry is transmitted in blocks of 512 bytes. Each block is treated as 8 lines of 64 bytes. Blocks are identified according to the first byte of the block as K, L, M, N, Q or Y. K, L, M and N blocks consist of plain text messages from one command station to another or for broadcast. All characters use ASCII representation. Bit 7 set to 1 may be used to indicate highlighted characters. No CR/LF is transmitted, so users must insert a CR/LF at the end of each line (64 characters). Q blocks carry all 128 telemetry channels in a compressed data format. Y blocks carry the 64 analog telemetry channels. AMSAT-DL plans to discontinue the use of Y blocks.

3.3 Command System

See Fig 2 (AO13)

3.4 Transponders

Transponder 1: Mode B
  Type: linear, inverting
  Uplink passband: 435.423-435.573 MHz
  Downlink passband: 145.975-145.825 MHz
  Translation equation:
    Downlink freq [MHz] =
        581.398 − uplink freq [MHz] ± Doppler
  Bandwidth (at 3 dB): 150 kHz
  Output power: 50 W PEP, 12.5 W average
  Suggested uplink EIRP: 500 W (27 dBW)
  Maximum Doppler: 5.1 kHz (at perigee)

Transponder 2: Mode JL (Mode J can be selectively disabled). Mode J was included on OSCAR 13 for amateurs in countries where the Mode-L uplink is not available. The IARU Region 1 recommends that the Mode J transponder on OSCAR 13 not be used by amateurs in Region 1. (IARU Region 1 meeting, Spain, April 1990.)
  Type: linear, inverting
  Mode-L uplink passband: 1269.641-1269.351 MHz
  Mode-J uplink passband: 144.423-144.473 MHz
  Mode-L downlink passband: 435.715-436.005 MHz
  Mode-J downlink passband: 435.990-435.940 MHz
  Mode-L translation equation:
    Downlink freq [MHz] =
        1705.356 − uplink freq [MHz] ± Doppler
  Mode-J translation equation:
    Downlink freq [MHz] =
        580.413 − uplink freq [MHz] ± Doppler
  Mode-L bandwidth (at 3 dB): 290 kHz
  Mode-L output power: 50 W PEP, 12.5 W average
  Mode-L suggested uplink EIRP: 4-8 kW (36-39 dBW)
  Maximum Doppler: 14.6 kHz (Mode L at perigee)

Mode-J suggested uplink EIRP: 800 W (29 dBW)

Transponder 3: Mode S (Can be operated only when Mode B is on. Due to negative power budget, operation time must be limited to 30 minutes per orbit.)
  Type: hard-limiting, non-inverting (SSB can be used below limiting level)
  Uplink passband: 435.602-435.638 MHz
  Uplink passband: 435.480-435.516 MHz (This uplink is due to spurious mixing and should be avoided.)
  Downlink passband: 2400.711-2400.747 MHz
  Translation equation:
    Downlink freq [MHz] =
        1965.109 + uplink freq [MHz] ± Doppler
  Bandwidth (at 3 dB): 36 kHz
  Output power: 1.25 W continuous
  Suggested uplink EIRP: 500 W (27 dBW)
  Maximum Doppler: 49.6 kHz (at perigee)

Note: SSB users should employ upper sideband on the uplink. This will enable those listening on Mode S to identify Mode B stations entering the transponder by accident.

Transponder 4: RUDAK-I (Not operative).
  This transponder uses the RF stages of the Mode-L unit. See Fig 3 (AO13) for details.
  Type: digital
  Uplink: 1269.710 MHz (2400-bps BPSK with 7.5-kHz capture range)
  Downlink: 435.677 MHz (400-bps BPSK compatible with telemetry)
  Output power: 6 W
  Suggested uplink EIRP: 800 W (29 dBW)

3.5 Attitude Stabilization

The satellite is spin-stabilized at roughly 20 to 30 r/min about the Z axis. The preferred direction aims the +Z at the SSP when the satellite is at apogee. Spin-up and direction changes are accomplished using three magnetorquer coils, one inside the perimeter of each arm. Pulsing of these coils, which generally takes place near perigee, is under the control of IHU software. The Sensor Electronics Unit employs sun and earth sensors at the end of arm 2 and a simple solar-cell top/bottom sensor. The sun sensor is a cross-slit unit with two detectors. The earth sensor employs a dual-beam unit (See Chapter 15). Tube-shaped nutation dampers containing glycerine and water (about 50/50) are located at the end of each arm. These measure about 40 cm by 0.2 cm.

3.6 Antennas

See Fig 1 (AO13) and Table 3 (AO13)

3.7 Energy Supply and Power Conditioning

Solar Arrays: The silicon solar arrays are designed to initially provide about 50 watts average power. At the end of three years this will decrease to about 35 watts.
Batteries: The primary battery is rated at 10 Ah; the auxiliary battery is rated at 6 Ah.

Table 2 (AO13)†
OSCAR 13 Analog Telemetry Channels

Telemetry Channel (Hex Desig)	Function	Equation	Units
00	Solar panel out and BCR input voltage	$(n-10)*167$	mV
01	70-cm xmtr average power output	$(261-n)^2/724$	W
02	70-cm rcvr temperature	$(n-120)/1.71$	C
03	(Reserved)		
04	BCR output and main battery voltage	$(n-10)*79.5$	mV
05	Special purpose	XXXXXXXXX	
06	2-m xmtr power amplifier temperature	$(n-120)/1.71$	C
07	+14 V rail current to xponder	$(n-15)*24.27$	mA
08	+10 V regulator voltage	$(n-10)*53.2$	mV
09	Helium tank high pressure	$(n-14)*6.56$	Bar
0A	IHU temperature	$(n-120)/1.71$	C
0B	+14 V rail current to magnetorquers and antenna relay	$(n-15)*4.854$	mA
0C	BCR oscillator #1 status	$(0=$ Off; $n>10=$ On	
0D	He tank low side pressure control volt	$(n-15)*0.117$	Bar
0E	BCR temperature	$(n-120)/1.71$	C
0F	+10 V regulator current	$(n-15)*4.854$	mA
10	BCR oscillator #2 status	$0=$ Off; $n>10=$ On	
11	N_2O_4 tank pressure	$(n-106)*0.733$	Bar
12	SEU temperature	$(n-120)/1.71$	C
13	Battery charge current	$(n-15)*12.135$	mA
14	Top (+Z) photocell sun sensor	$(n-10)*8.53$	mV
15	Motor valve status	$102=$ Closed; $118=$ Open	
16	Auxiliary battery #1 temperature	$(n-120)/1.71$	C
17	Active BCR output current	$(n-15)*24.27$	mA
18	Bottom (−Z) photocell sensor	$(n-10)*8.53$	mA
19	XXXXXXXXXXXX		
1A	Auxiliary battery #2 temperature	$(n-120)/1.71$	C
1B	Active BCR input current on 28 V bus	$(n-15)*12.135$	mA
1C	Spin rate		rpm
1D	Mode-L rcvr AGC	$(n-75)^2/-1125$	dB
1E	Main battery temperature	$(n-120)/1.71$	C
1F	Solar panel #6 current	$(n-15)*4.854$	mA
20	2-m xmtr average power output	$(287-n)^2/1796$	W
21	He tank temperature	$(n-120)/1.71$	C
22	Solar panel #1 temperature	$(n-120)/1.71$	C
23	Solar panel #5 current	$(n-15)*4.854$	mA
24	70-cm rcvr AGC	$(n-71)^2/2465$	dB
25	70-cm xmtr PA temperature	$(n-120)/1.71$	C
26	Solar panel #3 temperature	$(n-120)/1.71$	C
27	Solar panel #4 current	$(n-15)*4.854$	mA
28	Special purpose	XXXXXXXXX	
29	24-cm rcvr temperature	$(n-120)/1.71$	C
2A	Solar panel #5 temperature	$(n-120)/1.71$	C
2B	Solar panel #3 current	$(n-15)*4.854$	mA
2C	+14 V regulator voltage	$(n-10)*66.8$	mV
2D	RUDAK temperature	$(n-120)/1.71$	C
2E	Top (+Z) skin temperature of arm #1	$(n-120)/1.71$	C
2F	Solar panel #2 current	$(n-15)*4.854$	mA
30	Mode-B xponder +9 V supply voltage	$(n-10)*54$	mV
31	Wall temperature in arm #2	$(n-120)/1.71$	C
32	Bottom (−Z) skin temperature of arm #1	$(n-120)/1.71$	C
33	Solar panel #1 current	$(n-15)*4.854$	mA
34	Special purpose	XXXXXXXXX	
35	Wall temperature in arm #1	$(n-120)/1.71$	C
36	N_2O_4 tank temperature	$(n-120)/1.71$	C
37	Reserved		
38	Auxiliary battery voltage	$(n-10)*78.5$	mV
39	Mode-S xponder temperature	$(n-120)/1.71$	C
3A	+Z platform temperature (SERI exp)	$(n-120)/1.71$	C
3B	Reserved		
3C	Mode-L xponder +9 V supply voltage	$(n-10)*45.4$	mV
3D	AZ-50 tank temperature	$(n-120)/1.71$	C
3E	Nutation damper temperature	$(n-120)/1.71$	C
3F	Reserved		

†This table courtesy AMSAT-NA.

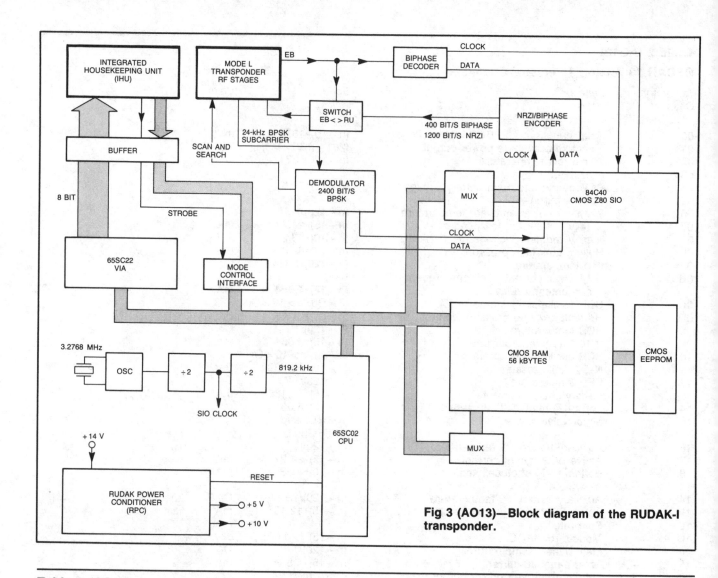

Fig 3 (AO13)—Block diagram of the RUDAK-I transponder.

Table 3 (AO13)

Frequency MHz	Up/ down	Type	Gain[1]	Half-power beamwidth	Polarization
146	down	monopole [2]	2	linear
146	down	[3]	6.0	100°	RHCP
437	up/down	monopole [2]	2	linear
437	up/down	[4]	9.5	67°	RHCP
1269	up	5-t helix	12.2	49°	RHCP
2400	down	8-t helix	14.2	39°	LHCP (?)

[1]dBi or dBic as appropriate
[2]Toroidal pattern concentric with Z axis
[3]Three 2-element beams (ZL special) phased for circular polarization
[4]Three 2-element beams (dipole over reflector) phased for circular polarization

3.8 Propulsion System

The propulsion system uses a liquid-fuel, bi-propellant rocket engine.

Thrust: 400 N

Specific Impulse: 293 seconds

Delta V (velocity change) with 142-kg spacecraft: 1480 m/s

Fuel: Aerozine 50 (50% unsymmetrical dimethyl hydrazine [UDMH], 50% hydrazine)

Oxidizer: Nitrogen tetroxide

Ignition System: none (fuel is self-igniting [hypergolic])

Pressurization: Helium (high side: 400 Bars; low side: 14 Bars)

3.9 Integrated Housekeeping Unit

CPU: 1802 COSMAC

Memory: 32k of error-detecting-and-correcting memory (Hamming code) using Harris HS-6564RH radiation-hardened chips. The Hamming code requires 12 bits for each stored 8-bit byte.

Operating System: Multitasking using IPS language

UoSAT Series

SPACECRAFT NAME: UoSAT-OSCAR 11

General

1.1 Identification

International designation: 84-021B
NASA Catalog number: 14781
Prelaunch designation: UoSAT-B, UoSAT-2

1.2 Launch

Date: 1 March 1984
Vehicle: Delta 3920
Agency: NASA (US)
Site: Western Test Range, Lompoc, CA
(Vandenberg Air Force Base)

1.3 Orbital Parameters (July 1989)

General designation: low-altitude, circular, sun-synchronous, near-polar
Period: 98.3 minutes
Apogee altitude: 688 km
Perigee altitude: 670 km
Eccentricity: 0.0012 (nominally circular)
Inclination: 98.0°
Longitude increment: 24.6°W/orbit
Maximum access distance: 2846 km

1.4 Ground Track Data

See Appendix C

1.5 Operations

Coordinating Group: UoSAT Project, Dr Martin Sweeting (G3YJO), Department of Electronic and Electrical Engineering, University of Surrey, Guildford, Surrey GU2 5XH, England

1.6 Design/Construction Credits

Project manager: Dr Martin Sweeting (G3YJO)
Construction: Built at University of Surrey, England

1.7 Primary References

1) UoSAT-2 Project Summary, UoSAT Datasheet, Sheet 5, UoSAT Spacecraft Engineering Research Unit, Department of Electronic and Electrical Engineering, University of Surrey.
2) M. Sweeting, "The UoSAT-B Experimental Amateur Spacecraft," *Orbit*, Vol 5, no. 1, Jan/Feb 1984, pp 12-16.
3) J. Bloom, "A Profile of the UoSAT-OSCAR 11 Satellite," *The ARRL Satellite Anthology*, ARRL, 1988, pp 52-53.

Spacecraft Description

2.1 Shape

Rectangular solid core approximately 58.5 cm (height) × 35.5 cm × 35.5 cm (see Fig 1 [UO11]). The base (the side with the launch fitting) has a "wing" extending about

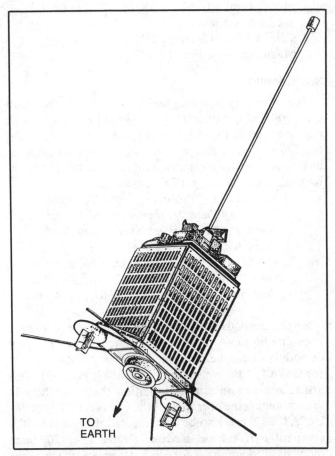

Fig 1 (UO11)—Pictorial view of UoSAT-OSCAR 11.

16 cm that holds two SHF helical antennas (bottom side) and the Navigation Magnetometer and Space Dust experiments (top side).
Mass: just over 60 kg

Subsystem Description

3.1 Beacons

Beacon 1 (nearly identical to UO-9 unit)
Frequency: 145.826 MHz
Power output: 400 mW (nominal)
(TTY channel no. 35)
Modulation: NBFM (AFSK) ± 5 kHz
Total RF/dc efficiency: 45%
Maximum Doppler: 3.6 kHz

Beacon 2 (employs phase-locked synthesizer)
Frequency: 435.025 MHz
Power output: 600 mW (nominal)
(TTY channel no. 45)
Modulation: NBFM (AFSK) ± 5 kHz (PSK option available)
Total RF/dc efficiency: greater than 40%
Maximum Doppler: 10.5 kHz

Beacon 3 (design by Colin Smithers, G4CWH)
Frequency: 2401.5 MHz

Power output: 500 mW (nominal)
  (TTY channel no. 55)
Modulation: NBFM (AFSK) ± 10 kHz (PSK
  option available)
Total RF/dc efficiency: 20%
Maximum Doppler: 57.9 kHz

3.2 Telemetry

The telemetry system, under control of the main spacecraft computer, has been designed to provide a high degree of flexibility with respect to what is measured and the downlink data format. Provisions for monitoring 60 analog and 96 digital status points are included. For information on decoding TTY content, see:

(1) R. Diersing, "Processing UoSAT Whole-Orbit Telemetry Data," *Proceedings of the 4th Annual AMSAT Space Symposium*, Dallas, 1986, ARRL, pp 55-76.

(2) R. Diersing, "Microcomputer Processing of UoSat-OSCAR 9 Telemetry," *The ARRL Satellite Anthology*, 1988, ARRL, pp 46-51.

Several methods for decoding UoSAT 11 telemetry have been employed. A detailed description of an excellent demodulator can be found in J. Miller, "Data Decoder for UoSAT," *Wireless World*, May 1983, pp 28-33. This article contains tutorial information that is of help in understanding the operation of the UoSAT TTY systems. UoSAT TTY can also be demodulated using a Bell-202-compatible 1200-baud modem. (**Note:** The 1200-baud modems in common use for computer access via telephone are *not* Bell 202 types.) The article by Diersing in *The ARRL Satellite Anthology* (see references above) includes directions for a simple demodulator that provides good reception when signal strength is high and QRM is low. The AEA PK-232 (firmware revisions on or after 25 June 87) can also be used. Use the following settings for the PK-232—Mode: ASCII; ABAUD: 1200; WIDE: ON; MFILTER: $80; AFILTER: ON. The PK-232 contains a spare inverter (U15, pins 1 and 2) that can be employed to invert the incoming signal (see next section).

Technical data: UoSAT TTY data is generally transmitted at 1200 bauds using a character format of 1 start bit, 7 data bits, 1 even parity bit and 2 stop bits. Other speeds and modulation schemes (such as CW or RTTY) are available via the spacecraft computer. In 1989, 4800 bauds was regularly scheduled for Wednesday and Sunday mornings (UTC). A frame of data contains a one-line header preceded by a cursor-home character ($1E) and followed by seven rows of 10 channels each. This is similar to OSCAR 9 TTY except that (1) the OSCAR-11 header uses the identifier UOSAT-2 and includes date/time, and (2) status point data has been moved to the end of the frame (channels 60-69). A sample TTY frame is shown in Fig 2 (UO11). The analog telemetry channels are listed in Table 1 (UO11).

One major difference between OSCAR 9 and OSCAR 11 is that the tones used to represent digital "1"'s and "0"'s have been exchanged. For OSCAR 11 (which differs from the normal Bell-202 format):

one cycle of 1200-Hz tone = "0"
two cycles of 2400-Hz tone = "1"

When using a Bell-202 modem or the PK-232, the incoming data stream must be inverted. This can be accomplished by wiring an exclusive-OR gate in series with the data stream. The other input to the XOR gate is wired to a logic +1 voltage level.

3.3 Telecommand System

The telecommand system consists of three redundant uplink receivers and demodulators operating in the 2-meter, 70-cm and 23-cm amateur bands.

3.4 Transponders

The spacecraft does not contain any open-access transponders. It does include an experiment that has been used to develop transponders for Pacsat, however. (See Digital Communications Experiment.)

3.5 Attitude Stabilization and Control

Navigation Magnetometer: Three-axis flux gate device with 14-bit resolution.

Magnetorquers: Three-axis; six coils, one around edge of each face of spacecraft.

Sun Sensors: Six sensors utilizing grey-code masking provide 360° of coverage.

Horizon sensor: Edge detector using two photodetectors mounted at ends of narrow tubes. Capable of responding to earth, moon or sun.

UOSAT-2				8901264102808					
00510	01517	02188	03408	04052	05039	06026	07051	08045	09033
10403	11336	12000	13064	14148	15458	16241	17588	18646	19565
20124	21207	22660	23635	24001	25239	26192	27524	28469	29512
30519	31090	32291	33576	34000	35271	36320	37431	38473	39500
40765	41120	42633	43054	44167	45000	46000	47491	48491	49472
50607	51119	52682	53290	54892	55000	56000	57497	58492	59497
60828	615FD	6201F	63F24	64440	651C0	66601	67700	68000	69000

Fig 2 (UO11)—One type of UoSAT-OSCAR 11 telemetry data frame. Each telemetry channel is sent as five characters (may be followed by a checksum digit). The first two characters are the channel number; the next three contain the encoded telemetry data. (Telemetry provided by J. Miller, G3RUH)

Table 1 (UO11)

UoSAT OSCAR 11 Analog Telemetry Channel Conversion

To calculate the analog value, replace N with the received 3-digit telemetry value.

Channel	Parameter	Equation
00	Solar array current − Y	I = 1.9 (516 − N) ma
01	Nav mag X axis	H = (0.1485N − 68) μT
02	Nav mag Z axis	H = (0.1523N − 69.3) μT
03	Nav mag Y axis	H = (0.1507N − 69) μT
04	Sun sensor 1	N uncalibrated
05	Sun sensor 2	N uncalibrated
06	Sun sensor 3	N uncalibrated
07	Sun sensor 4	N uncalibrated
08	Sun sensor 5	N uncalibrated
09	Sun sensor 6	N uncalibrated
10	Solar array current + Y	I = 1.9 (516 − N) mA
11	Nav mag (Wing) temp	T = (330 − N)/3.45 C
12	Horizon sensor	N uncalibrated
13	435 MHz Beacon VCO control	V = N/200 V
14	DCE RAMUNIT current	I = (N − 70.4)/6.7 mA
15	DCE CPU current	I = (N − 187.1)/2.0 mA
16	DCE GMEM current	I = (N − 121.3)/2.1 mA
17	Facet temp + X	T = (480 − N)/5 C
18	Facet temp + Y	T = (480 − N)/5 C
19	Facet temp + Z	T = (480 − N)/5 C
20	Solar array current − X	I = 1.9 (516 − N) mA
21	+ 10V line current	I = 0.97N mA
22	PCM voltage + 10V	V = 0.015N V
23	P/W logic current (+ 5V)	I = 0.14 mA (N <= 500)
24	P/W Geiger current (+ 14V)	I = 0.21N mA
25	P/W Elec sp. curr (+ 10V)	I = 0.096N mA
26	P/W Elec sp. curr (− 10V)	I = 0.093N mA
27	Facet temp − X	T = (480 − N)/5 C
28	Facet temp − Y	T = (480 − N)/5 C
29	Facet temp − Z	T = (480 − N)/5 C
30	Solar array current + X	I = 1.9 (516 − N) mA
31	− 10V line current	I = 0.48N mA
32	PCM voltage − 10V	V = − 0.036N V
33	1802 comp curr (+ 10V)	I = 0.21N mA
34	Digitalker current (+ 5V)	I = 0.13N mA (N <= 500)
35	145 MHz beacon power O/P	P = (2.5N − 275) mW (N > 200)
36	145 MHz beacon current	I = 0.22N mA
37	145 MHz beacon temp	T = (480 − N)/5 C
38	Command decoder temp (+ Y)	T = (480 − N)/5 C
39	Telemetry temp (+ X)	T = (480 − N)/5 C
40	Solar array voltage (+ 30V)	V = (0.1N − 51.6) V
41	+ 5V line current	I = 0.97N mA
42	PCM voltage + 5V	V = 0.0084N V
43	DSR current (+ 5V)	I = 0.21N mA (N <= 500)
44	Command RX current	I = 0.92N mA
45	435 MHz beacon power O/P	P = (2.5N − 200) mW (N > 175)
46	435 MHz beacon current	I = 0.44N mA
47	435 MHz beacon temp	T = (480 − N)/5 C
48	P/W temp (− X)	T = (480 − N)/5 C
49	BCR temp (− Y)	T = (480 − N)/5 C
50	Battery charge/dischg curr	I = 8.8 (N − 513) mA
51	+ 14V line current	I = 5N mA
52	Battery voltage (+ 14V)	V = 0.021N V
53	Battery cell volts (MUX)	V = N uncalibrated
54	Telemetry current (+ 10V)	I = 0.02N mA
55	2.4 GHz beacon power O/P	P = ((N + 50)^2)/480 mW
56	2.4 GHz beacon current	I = 0.45N mA
57	Battery temp	T = (480 − N)/5 C
58	2.4 GHz beacon temp	T = (480 − N)/5 C
59	CCD imager temp	T = (480 − N)/5 C

3.7 Energy Supply and Power Conditioning

Solar Arrays: The spacecraft contains four arrays attached to the four faces of the main framework. The arrays, manufactured by Solarex, each measure 49.5 cm × 29.5 cm.

Storage Battery: Consists of ten "F"-size commercial NiCd cells in series. These cells are from a group space-qualified by AMSAT (see Chapter 15). Provides 12 V at 6.4 Ah.

Power Conditioning

Two redundant Battery Charge Regulators (BCR) accept 28 V supplied by solar arrays and provide proper voltage and current to charge battery. Power Conditioning Module (PCM) accepts poorly regulated 12-14 V from battery/BCR bus and provides regulated 10, 5 and −10 V to spacecraft systems. Power Distribution Module (PDM) controls and monitors power distributed to various spacecraft systems and experiments. Safety latch thresholds under ground control via telecommand system.

3.9 Integrated Housekeeping Unit

CPU: 1802 COSMAC
RAM: 36k Using Error Detection and Correction (Hamming code)
Digitalker™ speech synthesizer with more than 550 words in ROM

3.10A Digital Communications Experiment (DCE)

The DCE was designed and built by AMSAT and VITA groups in the US and Canada. It consists of an NSC-800 CPU and nearly 128k of CMOS RAM. Communication with the IHU 1802 is via two serial ports. The DCE is being used to investigate various packet-radio protocols for use with future digital store-and-forward radio amateur satellites. In addition, the DCE interfaces with the navigation magnetometer and the telemetry system to provide mass long-term data storage. Since use of the DCE requires knowledge of the spacecraft command system and codes, access is restricted.

3.10B Space Dust Experiment

Detectors employing dielectric diaphragm (for large particles) and piezo-crystal microphone (for small particles) provide information on number of impacts and particle momentum.

3.10C CCD Camera

This camera is an improved version of the unit flown on OSCAR 9. It contains an active area of 384 × 256 pixels, with each pixel having 128 grey levels. Image is stored in 96k of RAM in DSR experiment.

3.10D Particle Detectors and Wave Correlator Experiment

Sensors include three Geiger counter tubes with different electron-energy detection thresholds and a multi-channel electron spectrometer running under the control of an NSC-800 CPU. Electron flux spectrum is measured at eight energy levels. The wave correlator experiment is designed to help identify (1) wave-modes responsible for accelerating electrons into the auroral beam, and (2) wave modes which limit the further growth of the auroral beam.

3.10E Digital Store and Readout Experiment (DSR)

The DSR contains two banks of 96k CMOS RAM that can be used in conjunction with the CCD Camera and other experiments.

3.10F Digitalker

The Digitalker speech synthesizer contains more than 550 words in ROM. Operating under the control of the IHU, it is principally used to announce telemetry for educational demonstrations.

SPACECRAFT NAME: UoSAT-OSCAR 14

1.1 Identification

International designation: 90-005B
NASA Catalog number: 20437
Pre-launch designation: UoSAT-D
Also known as: UO-14, UoSAT-3

1.2 Launch

Date: 22 Feb 1990
Time: 01:35:31 UTC
Vehicle: Ariane (version: 40, family: 4, mission: V35)
Agency: ESA
Site: Kourou, French Guiana

1.3 Orbital Parameters

General Designation: low-altitude, circular, sun-synchronous, near-polar
Period: 100.8 minutes
Mean altitude: 800 km
Eccentricity: 0.0012
Inclination: 98.7°
Longitude increment: 25.3°W/orbit
Maximum access distance: 3038 km

1.4 Ground Track Data

See Appendix C

1.5 Operations

Coordinating Group: UoSAT Project, Dr Martin Sweeting (G3YJO), Department of Electronic and Electrical Engineering, University of Surrey, Guildford, Surrey GU2 5XH, England.

1.6 Design/Construction Credits

Project manager: Dr Martin Sweeting
Construction: Built at University of Surrey

1.7 Primary References

1) M. Sweeting and J. Ward, "UoSAT-D and UoSAT-E Spacecraft to Fly on Ariane," *OSCAR News*, Oct 1988, pp 15-19.
2) "University of Surrey Launches Two New

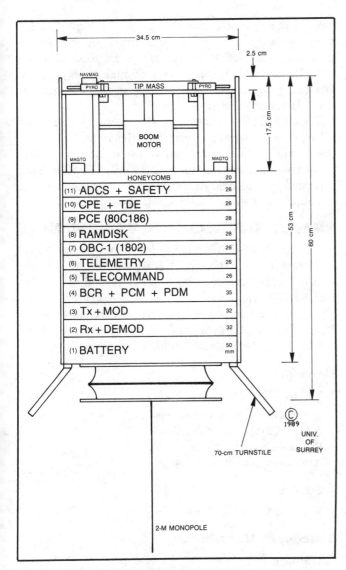

Fig 1 (AO14)—UoSAT-OSCAR 14 structure

Satellites," *OSCAR News*, Feb 1990, pp 16-22.
3) J. Kasser, M. Sweeting and J. Ward, "The UoSAT-OSCAR 14 and 15 Spacecraft," *The AMSAT Journal*, May 1990, pp 9-12.

Spacecraft Description

Note: When the UoSAT-C mission was delayed, mission subsystems were divided between two smaller spacecraft, UoSAT-C and UoSAT-E.

2.1 Physical Structure

See Fig 1 (AO14)

The spacecraft is constructed using a modular design. Subsystems are housed in standard module trays, which are stacked to form the satellite body. Each tray contains a single PCB measuring 320 × 300 mm.

Shape: rectangular solid approximately 60.0 cm (height) × 34.5 cm × 34.5 cm.

Mass: 45 kg

Subsystem Description

3.1 Beacons

Frequency: 435.070 MHz
Modulation: 1200-bit/s AFSK (FM) or 9600-bit/s FSK (switchable)
Maximum Doppler: 10.3 kHz

3.2 Telemetry

The telemetry system monitors 32 analog channels. Telemetry data can be provided as either AX.25 packets (primary system) or through a hardware interface (back-up system).

3.4 Transponder

OSCAR 14 is the first UoSAT spacecraft to carry an open-access transponder for use by radio amateurs. Known as the Packet Communications Experiment (PCE), it consists of one uplink and one downlink using Mode-J frequencies. The links are 9600-baud FSK and are compatible with modems by G3RUH and K9NG. (A comprehensive article describing the G3RUH design can be found in the *7th ARRL Computer Networking Conference* papers, pp 135-140.) RF communications links were designed to provide consistent service to ground stations with omnidirectional antennas. Tests by GØ/K8KA and G3RUH in early April 1990 verified that the PCE is working correctly at 9600 bauds.

UO-14 and the OSCAR 16 and 19 Pacsats use the same frequency bands and software interface. However, different ground-station equpment is required to access the two systems. Ground stations for the UO-14 FSK links generally use modified FM equipment. The modifications involve connecting a G3RUH or K9NG 9600-baud modem directly to the receiver discriminator and transmitter oscillator. G3RUH and K8KA used FT-736R receivers for their initial contacts. The UO-14 2-meter receiver uses AFC, so ground stations do not have to adjust their transmit frequency during a pass to compensate for Doppler.

The flight of the PCE on OSCAR 14 was funded by the University of Surrey and AMSAT-UK. Development of the PCE (hardware and software) was funded by VITA. Additional (non-amateur) frequency allocations will enable experimental use of the transponder by VITA in support of relief communications to remote areas in developing countries.

The PCE is controlled by an 80C186 microprocessor running at 8 MHz and contains 4 Mbytes of RAM. The RAM consists of a variety of CMOS devices from different manufacturers. RAM operation will be carefully monitored to evaluate performance in the space radiation environment. The system has been designed so that segments that fail can be switched off-line. The PCE uses AX.25 packets and also serves as a platform for experimentation with higher-level packet protocols.

Uplink frequency:	(primary)	145.975 MHz
	(secondary)	to be announced
Downlink frequency:	(primary)	435.070 MHz
	(secondary)	to be announced

3.5 Attitude Stabilization and Control

Earth-pointing (+Z axis) using a gravity-gradient boom augmented by computer-controlled magnetorquing. The boom tip mass points away from the earth (−Z axis). Attitude is calculated from measurements of the earth's local magnetic field using a 3-axis flux-gate magnetometer. Direct interface of the magnetometer to the 1802 and 80C186 CPUs provides resolution to 8 nanoTesla over a 100 microTesla range. A slightly reduced resolution is available through the satellite's standard telemetry system.

Since ferrous material distorts the local magnetic field, it is desirable to mount the magnetometer sensor away from the body of the satellite. This is achieved by using a spring-loaded hinged arm. Once the spacecraft is in orbit, the pyro electric device which releases the main gravity-gradient boom also allows the magnetometer arm to extend 30 cm above the top of the spacecraft.

3.7 Energy Supply and Power Conditioning

Solar Cells: gallium-arsenide solar arrays produced by Mitsubishi.

Storage Battery: 14-V, 6-Ah, 10-cell NiCd using commercial cells space-qualified by AMSAT (see Chapter 15).

3.9 Integrated Housekeeping Unit

The primary spacecraft microprocessor is an 80C186 with a full 16-bit bus running at 8 MHz. The spacecraft also contains an 1802 CPU (as in earlier UoSATs) for spacecraft control, and an 80C31 in the Cosmic Particle Experiment. Plans are to have the 1802 control overall spacecraft operation using a multitasking operating system. All CPUs are linked through a multiple-access serial data-sharing bus. This bus eliminates many of the dedicated serial links present in UO-9 and UO-11 while increasing the number of redundant data paths available.

3.10A Packet Communications Experiment (PCE)

See section 3.4

3.10B Cosmic Particle and Total Dose Experiments (CPE/TDE)

This is a follow-on experiment to the UO-9 Radiation Detector Experiment and the UO-11 Particle Detector and Wave Correlator Experiment.

The Total Dose Experiment (TDE) provides, for the first time, a direct measurement of the absorbed radiation doses at various points in a satellite. This allows assessment of shielding provided by the spacecraft structure. The TDE uses seven radiation-detecting transistors (RADFETs). Each RADFET consists of a pair of specially designed MOSFETs with a thick gate oxide which makes them sensitive to radiation. One MOSFET is biased so that positive charges, proportional to ionizing dose, are trapped in the gate oxide, causing a gradual shift in threshold voltage. The shift can be observed by comparing it to that occuring in the unbiased MOSFET on the same die.

The Cosmic Particle Experiment (CPE) responds to cosmic particles as they pass through a large-area silicon PIN diode array by monitoring the charge they deposit using a charge-integrator circuit monitored by an 80C31 microprocessor. Two detector arrays are used, allowing particles to be assigned to one of nine energy bands. Detailed analysis will show how these particles are distributed around the orbit.

The CPE/TDE package, along with Single Event Upset (SEU) monitoring of soft errors on the 4 Mbytes of PCE RAM, will provide a great deal of information to designers of computer systems for satellite use. The CPE/TDE is funded by the Royal Aerospace Establishment and built in conjunction with the Harwell Laboratory (UK).

SPACECRAFT NAME: UoSAT-OSCAR 15

Note: UoSAT-OSCAR 15 appears to have suffered a catastrophic failure shortly after its third orbit. This section has therefore been abbreviated. Several systems on UoSAT-OSCAR 14 and UoSAT-OSCAR 15 are identical. These include mechanical structure, interconnection bus, IHU, telemetry, telecommand, and power generation and conditioning. Only the differences are covered here.

General

1.1 Identification

International designation: 90-005C
NASA Catalog number: 20438
Pre-launch designation: UoSAT-E
Also known as: UoSAT-4

1.2-1.7

See UoSAT-OSCAR 14

Spacecraft Description

2.1 Shape

See Fig 1 (AO15)

Subsystem Description

3.1 Beacons

Frequency: 435.120 MHz

3.4 Transponders

UoSAT-OSCAR 15 does *not* contain a transponder

3.7 Energy Supply and Power Conditioning

Solar Cells: gallium-arsenide solar arrays manufactured by FIAR/CISE (Italy) and EEV/MSS/RAE (UK).

3.10A Transputer Data Processing Experiment (TDPE)

UoSAT-OSCAR 15 is the first spcecraft to test the Transputer in the space environment. The Transputer is a single-chip microprocessor, built by INMOS Britain, that is particularly suitable for parallel processing. The TDPE experiment contains three Transputers. The three units can be used on different parts of a single task simultaneously to improve processing speed, or they can be configured to each run the same task using a fault-tolerant architecture. When the three Transputers run identical programs, ex-

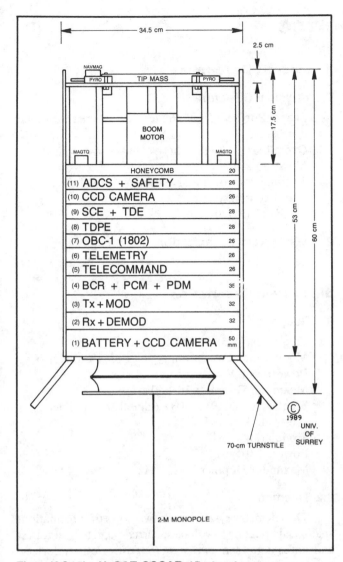

Fig 1 (AO15)—UoSAT-OSCAR 15 structure.

ternal logic monitors Transputer operation, watching for erratic behavior that might result from radiation-induced soft errors. If such operation is observed, the two Transputers in agreement can outvote and reset the third one. By correlating the occurrence of TDPE single-event upsets with observations of the radiation environment provided by the TDE and CPE on UoSAT-OSCAR 14, which is in a nearly identical orbit, Transputer operation can be characterized accurately in the radiation environment of space.

Certain types of processing can be greatly sped up by having each Transputer operate on a different aspect of a problem at the same time. This feature will be demonstrated by image data compression software operating on pictures of earth from the UoSAT-OSCAR 15 CCD camera. The Transputers will also be used to experiment with onboard data processing. For example, CCD images will be corrected for geometric distortion and coastal outlines will be added.

The TDPE was designed and built at ESTEC, ESA's research center. Results from the TDPE will be used to help ESA design high-performance onboard data handling systems for future satellites. Both the increased performance of the parallel processing arrangement and the increased reliability of the "watchdog" arrangement will be studied.

3.10B Advanced Solar Panel Technology Experiment (SCE)

UoSAT-OSCAR 15 contains a set of sample solar cells from several manufacturers, representing the latest advances in solar-cell technology. Most of these cells are being tested in space for the first time. The cells will be constantly monitored for changes in performance caused by radiation, temperature and other environmental effects. Materials include silicon, gallium arsenide, indium phosphide and composite designs. Various cover-cell geometries and materials designed to improve efficiency or reduce degradation due to radiation are also being tested.

3.10C Camera Imaging Experiment (CIE)

UoSAT-OSCAR 15 carries a new CCD camera optimized for meteorological scale imaging. The earth surface resolution of the system is on the order of 2-3 km, with a field of view of approximately 1000 km^2. Images should be comparable in detail to those produced by NOAA spacecraft.

The camera uses a Fairchild charge-coupled-device (CCD) detector similar to those found in home video cameras. Video from the CCD is digitized by a flash analog-to-digital converter and stored in low-power static RAM. The 96-kbyte raw image is sent to the TDPE, where it is compressed before transmission. The data compression will amount to a 50% to 90% reduction in the amount of memory required to store an image, and a similar reduction in the transmission time. The flight CCD sensor was donated by Vision/Fairchild.

JAS Series

SPACECRAFT NAME: Fuji-OSCAR 20

General

1.1 Identification

International designation: 1990-013C
NASA Catalog number: 20480
Prelaunch designation: JAS-1B
Also known as: FO-20, Fuji-2
Call sign: 8J1JBS

1.2 Launch

Date: 7 February, 1990 (01:33 UTC)
Vehicle: H1
Agency: NASDA (Japanese National Space Agency)
Site: Tanegashima Space Center, Japan

1.3 Orbital Parameters (April 1990)

General designation: low-altitude, elliptical, non-sun-synchronous
Period: 112.23 minutes
Apogee: 1745 km
Perigee: 912 km
Eccentricity: 0.0541
Inclination: 99.05°
Longitude increment: 28.1°W/orbit
Maximum access distance: 4257 km

1.4 Ground Track Data

See Appendix C

1.5 Operations

Coordinating Group: JAMSAT (Japanese AMSAT)

1.6 Design/Construction Credits

JARL (Japanese Amateur Radio League)
NASDA (Japanese National space Agency)
NEC (Nippon Electric Company): space frame, power supply, etc
JAMSAT (Japanese AMSAT): transponders, telemetry/command, IHU, ground-support systems.

1.7 Primary References

Note: Except for its orbit, Fuji-OSCAR 20 is very similar to Fuji-OSCAR 12 (aka JAS-1 and Fuji-1). Although much of the detailed information currently available focuses on FO-12, it is valid for FO-20. All known differences are specified in this section and reference 3.

1) *JAS-1 Handbook*, JARL, August 1985. English translation available from Project OSCAR, PO Box 1136, Los Altos, CA 94023.

2) V. Riportella, "Introducing Japanese Amateur Satellite Number One (JAS-1)," *QST*, Jun 1986, pp 71-72; "Operating the Flying Mailbox: FO-12 Mode JD," *QST*, Nov 1986, pp 66-67.

3) "Introduction of JAS-1b," by JARL, *QEX*, Sep 1989, pp 8-11.

Spacecraft Description

2.1 Physical Structure

Shape: Polyhedron of 26 faces, 25 covered with solar cells.
Overall size: 470 (height) × 440 × 440 mm
Mass: 50 kg
Thermal control: Passive using solar cell cover glass and coatings.
See Fig 1 (FO20)

2.2 Subsystem Organization

See Fig 2 (FO20)

Subsystem Description

3.1 Beacons

Frequency: 435.795 MHz
Modulation: CW (primary) or PSK (secondary)
Power: 60 mW
Maximum Doppler: 10.1 kHz
Frequency: 435.910 MHz (digital transponder downlink)
Modulation: PSK
Power: 1.0 W
Maximum Doppler: 10.1 kHz

3.2 Telemetry

The telemetry system sends two types of information: analog data relating to voltages, currents and temperatures; and status data from two-state (on/off or other) devices. The spacecraft has sensors placed at 29 analog test points and 33 status points. There's a considerable amount of

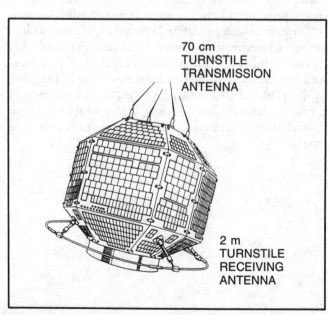

70 cm
TURNSTILE
TRANSMISSION
ANTENNA

2 m
TURNSTILE
RECEIVING
ANTENNA

Fig 1 (FO20)—Drawing of Fuji-OSCAR 20 (from *OSCAR News*, Dec 1989, p 36).

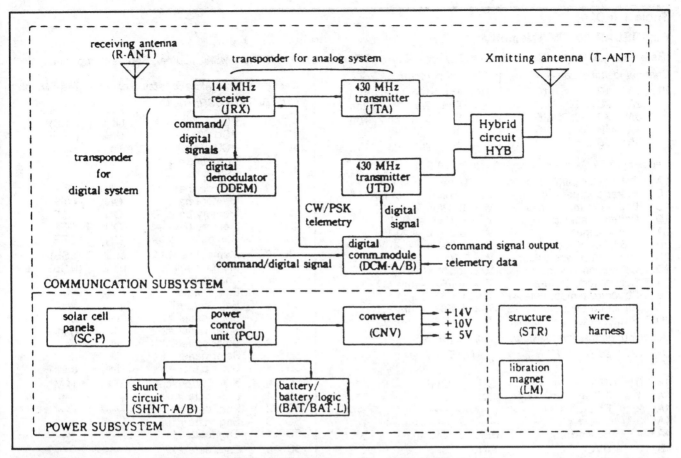

Fig 2 (FO20)—Subsystem block diagram (from *OSCAR News*, Dec 1989, p 38).

flexibility in how this data is sent, but there are two primary formats: PSK packets and CW.

CW telemetry is sent on the 435.795-MHz beacon at 20 WPM. A frame contains 20 channels, which may be thought of as forming a 5-row by 4-column array. The frame marker is the standard HI. Each channel consists of three digits. The first digit is a row marker. The first 12 channels contain analog data, per Table 1 (FO20). The remaining two digits per channel contain the raw data. The last eight channels contain encoded status data. (See Chapter 15 for additional information on CW telemetry decoding.)

PSK telemetry is generally sent on 435.910 MHz in the form of AX.25 packets. Three types of frames are used: real-time TTY (ASCII or binary), stored TTY (ASCII or binary) and messages (numbered 0 to 9). The real-time and stored ASCII frames contain information on all 29 analog data points and 33 status points. Decoding information is contained in Table 2 (FO20).

3.3 Telecommand System

The telecommand system recognizes five operating modes. Numbers in parentheses refer to operating power required.

Mode D. All loads off except command receiver. (2.5 W)

Mode JA on. (5.3 W)

Mode JD on with 1 to 4 256-k memory banks

activated. (JD-1 [6.2 W], JD-2 [6.7 W], JD-3 [7.0 W], JD-4 [7.3 W])

Mode DI. All loads off except command receiver, CPU and memory.

Mode JAD. Both JA and JD on. (10.2 W)

3.4 Transponders

Transponder 1: Mode JA
  Type: linear, inverting
  Uplink passband: 145.900-146.000 MHz
  Downlink passband: 435.800-435.900 MHz
  Translation equation:
    Downlink freq [MHz] =
      581.800 − uplink freq. [MHz] ± Doppler
  Power output: 2 W PEP
  Recommended uplink EIRP: 100 W
  Bandwidth: 100 kHz
  Maximum Doppler: 6.7 kHz

Transponder 2: Mode JD
  Type: digital
  Uplink channels: #1 145.850 MHz
                   #2 145.870 MHz
                   #3 145.890 MHz
                   #4 145.910 MHz
  Downlink channel: 435.910 MHz
  Power output: 1 W RMS
  Recommended uplink EIRP: 100 W

Table 1 (FO20)
Fuji-OSCAR 20 CW Telemetry

(Reference: JR1NVU, "The Telemetry Formats of JAS-1B/Fuji-OSCAR 20," *The AMSAT Journal*, Sep 1990, pp 20-21.

Channel identification

HI	1A	1B	1C	1C
	2A	2B	2C	2D
	3A	3B	3C	3D
	4A	4B	4C	4C
	5A	5B	5C	5D	HI

Channel contents

HI	1nn	1nn	1nn	1nn
	2nn	2nn	2nn	2nn
	3nn	3nn	3nn	3nn
	4jj	4jj	4jj	4jj
	5jj	5jj	5jj	5jj	HI

Decoding equations for analog channels (1A-3D)

1A	Total solar panel current	$I = 19(nn + 0.4)$ mA
1B	Batter charge/discharge	$I = -38(nn - 50)$ mA
1C	Battery terminal voltage	$V = 0.22(nn + 4)$ V
1D	Battery center tap	$V = 0.1(nn + 4)$ V
2A	Bus voltage	$V = 0.20(nn + 4)$ V
2B	+5 V regulator output	$V = 0.062(nn + 4)$ V
2C	Mode JA power output	$P = 2.0(nn + 4)^{1.618}$ mW
2D	Calibration voltage #1	$V = (nn + 4)/50$ V
3A	Battery temperature	$T = 1.4(67 - nn)$ °C
3B	Structure temperature #1	$T = 1.4(67 - nn)$ °C
3C	Structure temperature #2	$T = 1.4(67 - nn)$ °C
3D	Structure temperature #3	$T = 1.4(67 - nn)$ °C

Decoding information for status channels (4A-5D)

CH	BIT	DESCRIPTION	STATE 1	0
4A	0	JTA Power	ON	OFF
4A	1	JTD Power	ON	OFF
4A	2	Eng. data #1	—	—
4A	3	Eng. data #3	—	—
4A	4	Beacon	PSK	CW
4B	0	UVC	ON	OFF
4B	1	UVC level	1	2
4B	2	Battery	tric	full
4B	3	Battery logic	tric	full
4B	4	Main relay	ON	OFF

Decoding information for status channels (4A-5D con't)

CH	BIT	DESCRIPTION	STATE	
4C	0	PCU	bit 1	(LSB)
4C	1	PCU	bit 2	(LSB)
4C	2	PCU	manual	auto
4C	3	Eng. data #3	—	—
4C	4	Eng. data #4	—	—
4D	0	Memory bank #0	ON	OFF
4D	1	Memory bank #1	ON	OFF
4D	2	Memory bank #2	ON	OFF
4D	3	Memory bank #3	ON	OFF
4D	4	Computer power	ON	OFF
5A	0	Memory select	bit 1	(LSB)
5A	1	Memory select	bit 2	(MSB)
5A	2	Eng. data #5	—	—
5A	3	Eng. data #6	—	—
5A	4	Eng. data #7	—	—
5B	0	Solar panel #1	lit	dark
5B	1	Solar panel #2	lit	dark
5B	2	Solar panel #3	lit	dark
5B	3	Solar panel #4	lit	dark
5B	4	Solar panel #5	lit	dark
5C	0	JTA CW beacon	CPU	TLM
5C	1	Eng. data #8	—	—
5C	2	Eng. data #9	—	—
5C	3	Eng. data #10	—	—
5C	4	Eng. data #11	—	—
5D	0	Eng. data #12	—	—
5D	1	Eng. data #13	—	—
5D	2	Eng. data #14	—	—
5D	3	Eng. data #15	—	—
5D	4	Eng. data #16	—	—

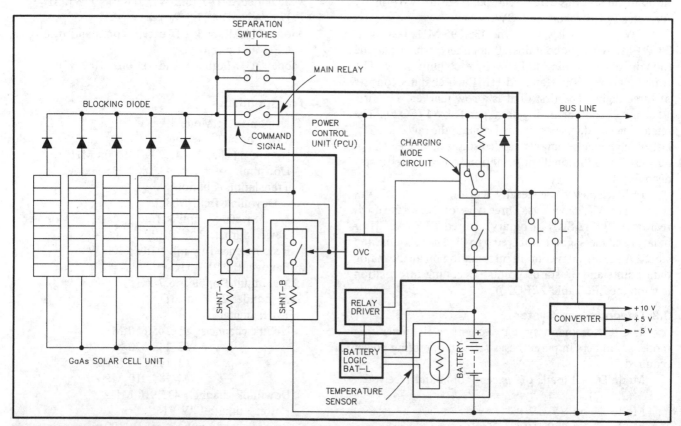

Fig 3 (FO20)—Fuji-OSCAR 20 power subsystem (from *OSCAR News*, June 1989, p 11).

Table 2 (FO20)
Fuji-OSCAR 20 PSK Telemetry

Channel Identification
```
... Header ...
00 01 02 03 04 05 06 07 08 09
10 11 12 13 14 15 16 17 18 19
20 21 22 23 24 25 26 27 28 29
30 31 32 33 34 35 36 37 38 39
```

Sample TTY Frame (as received from FO-20)

8J1JBS>BEACON:
JAS1b RA 90/03/08 11:02:00
```
596 375 692 698 750 837 849 831 001 686
618 001 507 510 532 527 530 532 655 001
662 654 666 677 999 647 879 960 199 000
010 111 000 000 111 100 001 110 111 000
```

Frame contents

```
JAS1b  FF  YY/MM/DD  HH:MM:SS
xxx xxx xxx xxx xxx xxx xxx xxx xxx xxx
xxx xxx xxx xxx xxx xxx xxx xxx xxx xxx
xxx xxx xxx xxx xxx xxx xxx hhh hhh hhh
jjj  jjj  jjj  jjj  jjj  jjj  jjj  jjj  jjj  jjj  jjj
```

FF = Frame identifier
  RA: Real-time TTY (ASCII)
  RB: Real-time TTY (binary)
  SA: Stored TTY (ASCII)
  SB: Stored TTY (binary)
  Mn: Message (n = 0 to 9)

YY/MM/DD   HH:MM:SS = Date & Time (UTC)

Following valid only for RA and SA frames
  xxx = 3-digit decimal number occurring in
         channels 00 to 26.
         Represented by N in calibration equations.
  hhh = series of three individual hex numbers
  jjj = series of three individual binary numbers

Decoding equations for analog channels (00-26)

CH	DESCRIPTION	CALIBRATION
#00	total solar array current	$1.91 \times (N - 4)$ mA
#01	battery charge/discharge	$-3.81 \times (N - 508)$ mA
#02	battery voltage	$N \times 0.022$ V
#03	battery center voltage	$N \times 0.009961$ V
#04	bus voltage	$N \times 0.02021$ V
#05	+5 V regulator voltage	$N \times 0.00620$ V
#06	−5 V regulator voltage	$-N \times 0.00620$ V
#07	+10 V regulator voltage	$N \times 0.0126$ V
#08	JTA output power	$5.1 \times (N - 158)$ mW
#09	JTD output power	$5.4 \times (N - 116)$ mW
#10	calibration voltage #2	N/500 V
#11	offset voltage #1	N/500 V
#12	battery temperature	$0.139 \times (669 - N)$ deg. C
#13	JTD temperature	$0.139 \times (669 - N)$ deg. C
#14	Temperature #1	$0.139 \times (669 - N)$ deg. C
#15	Baseplate Temperature #2	$0.139 \times (669 - N)$ deg. C
#16	Baseplate Temperature #3	$0.139 \times (669 - N)$ deg. C

Decoding equations for analog channels (00-26, con't)

CH	DESCRIPTION	CALIBRATION
#17	Baseplate Temperature #4	$0.139 \times (669 - N)$ deg. C
#18	temperature calibration #1	N/500 V
#19	offset voltage #2	N/500 V
#20	Solar Cell Panel Temp #1	$0.38 \times (N - 685)$ deg. C
#21	Solar Cell Panel Temp #2	$0.38 \times (N - 643)$
#22	Solar Cell Panel Temp #3	$0.38 \times (N - 646)$
#23	Solar Cell Panel Temp #4	$0.38 \times (N - 647)$
#24	———	
#25	temperature calibration #2	N/500 V
#26	temperature calibration #3	N/500 V

Decoding information for hex status bytes (27-29)

CH	DESCRIPTION
#27a	Spare (TBD)
#27b	Spare (TBD)
#27c	Spare (TBD)
#28a	Spare (TBD)
#28b	Spare (TBD)
#28c	error count of memory unit #0
#29a	error count of memory unit #1
#29b	error count of memory unit #2
#29c	error count of memory unit #3

Decoding information for binary status bytes (30-39)

CH	DESCRIPTION	STATE	
#30a	JTA power	on	off
#30b	JTD power	on	off
#30c	JTA beacon	PSK	CW
#31a	UVC status	on	off
#31b	UVC level	1	2
#31c	main relay	on	off
#32a	eng. data #1	——	
#32b	battery status	tric	full
#32c	battery logic	tric	full
#33a	eng. data #2	——	
#33b	PCU status	bit 1	(LSB)
#33c	PCU status	bit 2	(MSB)
#34a	memory unit #0	on	off
#34b	memory unit #1	on	off
#34c	memory unit #2	on	off
#35a	memory unit	on	off
#35b	memory select	bit 1	(LSB)
#35c	memory select	bit 2	(MSB)
#36a	eng. data #3	——	
#36b	eng. data #4	——	
#36c	computer power	on	off
#37a	eng. data #5	——	
#37b	solar panel #1	lit	dark
#37c	solar panel #2	lit	dark
#38a	solar panel #3	lit	dark
#38b	solar panel #4	lit	dark
#38c	solar panel #5	lit	dark
#39a	eng. data #6	——	
#39b	CW beacon source	CPU	TLM
#39c	eng. data #7	——	

Comments: Uplink must be Manchester-coded FM. The protocol is AX.25, Level 2, Version 2. Data rate is 1200 bauds. The downlink is 1200-baud BPSK. For information on modems by TAPR and G3RUH, see Chapters 9 (hardware), 10 (operating notes) and 15 (additional information).

Note: When the power budget permits, the JA and JD transponders may be operated simultaneously. When this occurs, some spurious signals may be observed in the JA passband, especially when stations employing too much power are accessing the mode JA transponder.

3.5 Attitude Stabilization

Two 1 TAm^2 permanent magnets mounted parallel to spacecraft Z axis align along earth's local magnetic field.

3.6 Antennas

2-meter receive: Turnstile mounted on bottom of spacecraft. (This antenna replaces the 2-meter ¼-wavelength monopole mounted on top of the Fuji-1 spacecraft.)

70-cm transmit (Mode JA and JD): Canted turnstile on top of spacecraft. A circulator and phase shifter added to Fuji-2 enables a single antenna to be used for both links.

3.7 Energy Supply and Power Conditioning

Power Subsystem: See Fig 3 (FO20). Battery control logic monitors voltage, current and temperature of storage battery. Software will attempt to utilize spacecraft systems to minimize activation of overvoltage control (OVC) and undervoltage control (UVC) functions.

Solar cells: Gallium arsenide. Spacecraft contains approximately 1530 cells, 900 measuring 1 × 2 cm and 630 measuring 2 × 2 cm. The total cell surface area is about 4320 cm^2. When first launched, these cells generated about 13 W (optimum satellite orientation) and about 11 W average when spacecraft is in sunlight. Compared to Fuji-1, these figures represent an increase of 9% in cell area and 70% in available power.

Bus voltage: +11 to 18 (14 V average)

Battery: 11 NiCd cells providing about 14 V at 6 Ah.

Instrument Switching Regulators are used to provide +10 V, +5 V and −5 V for spacecraft systems. Efficiency >70%.

3.9 IHU

CPU: NSC-800 (Controls spacecraft and digital transponder)

RAM: 48 256-kbit NMOS DRAMs. These are configured using a hardware-based error-detection-and-correction scheme to provide 1 Mbyte of EDAC 8-bit memory. The operating system resides in approximately 32 k; the rest is used for message storage.

RS Series

SPACECRAFT NAME: RS-10/11

Note: COSMOS 1861 carries RS-10 and RS-11 as secondary payloads. Spacecraft on this launch were designed to operate independently, but remain attached to a common space platform.

General

1.1 Identification

International designation: 1987 054A
NASA Catalog number: 18129
Prelaunch designation: BRTK-10 (BTRK is an abbreviation for "Equipment for Radio Amateur Satellite Communication")

1.2 Launch

Date: 23 June 1987
Site: Plesetsk
Launch Vehicle: C-1 (SL-8)

1.3 Orbital Parameters

General designation: low-altitude, circular, near-polar, non-sun-synchronous
Period: 104.9 minutes
Apogee altitude: 997 km
Perigee altitude: 982 km
Eccentricity: 0.0010
Inclination: 82.9°
Longitude increment: 26.23 °W/orbit
Maximum access distance: 3360 km

1.4 Ground Track Data

See Appendix C.

1.5 Operations

The primary control station is RS3A in Moscow. Although both spacecraft can operate concurrently, the usual practice is for only one spacecraft to be on at any given time.

1.6 Design/Construction Credits

Project Managers: Alexandr Papkov (UA3XBU) and Viktor Samkov of the Tsiolkovskiy Museum for the History of Cosmonautics in Kaluga (180 km southwest of Moscow).

Spacecraft Description

2.1 Shape

RS-10 and RS-11 are independent satellites mounted on the COSMOS 1861 spaceframe. A photograph of the RS-10 transponder package and a functional block diagram appeared in the October 1987 issue of the Soviet magazine *Radio*.

2.2 Subsystem Organization

See 2.1.

Subsystem Description

3.1 Beacons

RS-10 beacon frequencies: See Table 1 (RS-10/11)
RS-11 beacon frequencies: See Table 2 (RS-10/11)

3.2 Telemetry

Telemetry is sent in CW at 20 WPM on the beacon frequencies. A frame consists of 16 channels. The frame marker is RS10 (or RS11). The frame can be thought of as two rows by eight columns. Each channel consists of four characters (two letters and two numbers). The first letter is a status bit; the second letter is a column identifier. The two numbers encode analog sensor data. See Tables 3 and 4 (RS-10/11) for decoding information. A high-speed telemetry stream has been observed on the beacons at times when the spacecraft are in view of command

---

Table 1 (RS-10/11)
RS-10 Frequency Chart

	Uplink Band (MHz)	Downlink Band (MHz)	Translation Constant (MHz)
Mode K	21.160-21.200	29.360-29.400	8.200
Mode T	21.160-21.200	145.860-145.900	124.700
Mode A	145.860-145.900	29.360-29.400	− 116.500
Mode KT	21.160-21.200	29.360-29.400 & 145.860-145.900	8.200 124.700
Mode KA	21.160-21.200 & 145.860-145.900	29.360-29.400	8.200 − 116.500

Translation Eq: downlink freq [MHz] = translation constant + uplink freq [MHz] ± Doppler
Beacon frequencies: 29.357, 29.403, 145.857 and 145.903 MHz
Robot uplink frequencies: 21.120 and 145.820 MHz

Table 2 (RS-10/11)

RS-11 Frequency Chart

	Uplink Band (MHz)	Downlink Band (MHz)	Translation Constant (MHz)
Mode K	21.210-21.250	29.410-29.450	8.200
Mode T	21.210-21.250	145.910-145.950	124.700
Mode A	145.910-145.950	29.410-29.450	−116.500
Mode KT	21.210-21.250	29.410-29.450 & 145.910-145.950	8.200 124.700
Mode KA	21.210-21.250 & 145.910-145.950	29.410-29.450	8.200 −116.500

Translation Eq: downlink freq [MHz] = translation constant + uplink freq [MHz] ± Doppler
Beacon frequencies: 29.407, 29.453, 145.907 and 145.830 MHz
Robot uplink frequencies: 21.130 and 145.830 MHz

Table 3 (RS-10/11)

RS-10/11 Binary Status Telemetry

Ch	Col	Binary Parameter	Character/Meaning[2]
1	S	TTY sampling period	I, S, D = 90 minutes N, R, G = 10 minutes)
2	R	RX attenuator #1	I, S, D = 20 dB[1] N, R, G = 0 dB
3	D	RX attenuator #2	I, S, D = 10 dB[1] N, R, G = 0 dB
4	G	15-m RX status	I, S, D = off N, R, G = on
5	U	2-m RX status	I, S, D = off N, R, G = on
6	W	command station ch	I, S, D = off N, R, G = on
7	K	10-m beacon power output setting	I, S, D = 1000 mW N, R, G = 300 mW
8	O	2-m beacon power output setting	I, S, D = 1000 mW N, R, G = 300 mW
9	S	1st memory board	A, U, K = off M, W, O = on
10	R	2nd memory board	A, U, K = off M, W, O = on
11	D	memory loading ch	A, U, K = open M, W, O = closed
12	G	code store memory	A, U, K = open M, W, O = closed
13	U	memory dump via	A, U, K = 10 m M, W, O = 2 m
14	W	attenuator 15-m Robot RX	A, U, K = 10 dB M, W, O = 0 dB
15	K	attenuator 2-m Robot RX	A, U, K = 10 dB M, W, O = 0 dB
16	O	command channel 2-m power output	A, U, K = 1000 mW M, W, O = 300 mW

[1]The attenuation values given by channels 2 and 3 must be added.
[2]The primary binary status characters for the first 8 channels are I and N. The primary binary status characters for the last 8 channels are A and M. When a command station is accessing the satellite via 15 meters, a leading "dit" is added to the Morse code letter representing the primary status character, yielding S, R, U or W. When a command station is accessing the satellite via 2 meters, a leading "dah" is added to the Morse code character, yielding D, G, K or O.

stations. This may be Robot QSO logs being downlinked from the memory boards referred to in channels 9 and 10 of the binary telemetry.

3.4 Transponders

Type: linear, noninverting
Power output: up to 5 W PEP
Frequencies: See Tables 1 (RS-10/11) and 2 (RS-10/11)
Passband: 40 kHz
Note: According to a report in Oct 1987 *Radio*, the passband is divided into 10 4-kHz segments, each with its own AGC circuit, to prevent overloading by high-power users. This has not been confirmed, however.

Table 4 (RS-10/11)

RS-10/11 Analog Telemetry

Ch	Col	Parameter	Decoding Equation
1	S	power supply voltage over sampling period	$V = n/4$ V
2	R	power output of 2-m TX	$P = n/10$ W
3	D	power output of 10-m TX	$P = n/10$ W
4	G	15-m RX AGC voltage	$V = n/5$ V
5	U	2-m RX AGC voltage	$V = n/5$ V
6	W	command channel AGC voltage	$V = n/5$ V
7	K	10-m beacon power output	$P = n/10$ W
8	O	2-m beacon power output	$P = n/10$ W
9	S	10-m TX temperature	$T = n - 10$°C
10	R	2-m TX temperature	$T = n - 10$°C
11	D	20-V power supply temp	$T = n - 10$°C
12	G	9-V power supply temp	$T = n - 10$°C
13	U	backup 9-V power supply	$V = n/5$ V
14	W	IF voltage of 15-m Robot	$V = n/5$ V
15	K	IF voltage of 2-m Robot	$V = n/5$ V
16	O	Robot QSO counter	See note 1.

[1]A value of 00 indicates 32 or fewer QSOs. A value between 80-99 indicates between 33 and 128 QSOs

3.6 Antennas

2 meters: dipole
10 and 15 meters: a single dipole supports both bands

3.7 Energy-Supply and Power Conditioning

Power is provided by the primary payload.

3.10 Robot Transponder

RS-10 and RS-11 carry Robot or autotransponders. Robot reception frequencies are listed in Tables 1 and 2 (RS-10/11). Transmission frequencies may be on any beacon, but the higher frequency 10-meter beacon is usually employed. When the Robot is active, it will call CQ. For example:

CQ CQ DE RS10 QSU 21120 KHZ $\overline{AR}$

indicates that it is monitoring 21.120 MHz for replies. Hints on QSOs with the RS Robots are given in Chapter 10.

SPACECRAFT NAME: RS-12/13

Note: RS-12/13 is similar to RS-10/11. It was integrated into its parent COSMOS navigation satellite in late 1988. Launch date depends on the needs of the maritime navigation system, but is expected in 1990. The orbit will probably be similar to that of RS-10/11. The following data summarizes the information that has been announced.

General

1.3 Orbital Parameters

The orbit is expected to be similar to that of RS-10/11.

1.6 Design/Construction Credits

Built at the Tsiolkovskiy Museum for the History of Cosmonautics in Kaluga (about 180 km southwest of Moscow). The project managers for the spacecraft were Aleksandr Papkov and Viktor Samkov.

Spacecraft Description

2.1 Physical Structure

RS-12/13 is integrated into a COSMOS maritime navigation spacecraft, from which it receives power.

2.2 Subsystem Organization

RS-12/13 contains several linear transponders that can be interconnected. The spacecraft also contains a ROBOT and several beacons.

Subsystem Description

3.1 Beacons

See Tables 1 and 2 (RS-12/13)
Power output: 0.45 W or 1.2 W (low/high option)

Table 1 (RS-12/13)
RS-12 Frequency Chart

	Uplink Band (MHz)	Downlink Band (MHz)	Translation Constant (MHz)
Mode K	21.210-21.250	29.410-29.450	8.200
Mode T	21.210-21.250	145.910-145.950	124.700
Mode A	145.910-145.950	29.410-29.450	−116.500
Mode KT	21.210-21.250	29.410-29.450 & 145.910-145.950	8.200 124.700
Mode KA	21.210-21.250 & 145.910-145.950	29.410-29.450	8.200 −116.500

Translation Eq: Downlink freq [MHz] = translation constant + uplink freq [MHz] ± Doppler
Beacon frequencies: 29.4081, 29.4543, 145.9125 and 145.9587 MHz
Robot uplink frequencies: 21.1291 and 145.8308 MHz

Table 2 (RS-12/13)
RS-13 Frequency Chart

	Uplink Band (MHz)	Downlink Band (MHz)	Translation Constant (MHz)
Mode K	21.260-21.300	29.460-29.500	8.200
Mode T	21.260-21.300	145.960-146.000	124.700
Mode A	145.960-146.000	29.460-29.500	−116.500
Mode KT	21.260-21.300	29.460-29.500 & 145.960-146.000	8.200 124.700
Mode KA	21.260-21.300 & 145.960-146.000	29.460-29.500	8.200 −116.500

Translation Eq: Downlink freq [MHz] = translation constant + uplink freq [MHz] ± Doppler
Beacon frequencies: 29.4582, 29.5043, 145.8622 and 145.9083 MHz
Robot uplink frequencies: 21.1385 and 145.8403 MHz

3.4 Transponders

See Tables 1 and 2 (RS-12/13)
Power output: 8 W PEP

3.7 Energy-Supply and Power Conditioning

The following power requirements have been listed
RS-12 all systems off: 4.6 W
RS-12 all systems on: 35 W
RS-13 all systems off: 3.5 W
RS-13 all systems on: 25 W

SPACECRAFT NAME: RS-14

Note: RS-14 (AMSAT-OSCAR 21) was launched Jan 29, 1991. Like other recent RS spacecraft, it is integrated into a parent satellite that supplies power. The parent satellite is part of the GEOS series.

General

1.1 Identification

Prelaunch designation: RADIO-M1

1.2 Launch

Site: Plesetsk
Launch vehicle: Proton

1.3 Orbital Parameters (tentative)

General designation: low-altitude, circular, near-polar
Apogee Altitude: 1000 km
Inclination: 83°

1.6 Design/Construction Credits

RS-14 was a joint project of AMSAT-DL (Germany), AMSAT-U-Orbita (Molodetschno [near Minsk], USSR) and the Adventure Club (Moscow).

1.7 Primary Reference

AMSAT-DL Journal, Mar/May 1990.

Spacecraft Description

2.1 Physical Structure

RS-14 is integrated into a GEOS spacecraft, from which it receives power.
Mass: 6.2 kg
Shape: Rectangular solid
Dimensions: 230 × 320 × 120 mm

2.2 Subsystem Organization

RS-14 contains two primary subsystems: a linear Mode-B transponder and a digital Mode-B transponder, known

Fig 1 (RS-14)—Block diagram of RS-14 (from *AMSAT-DL Journal*, Mar/May 1990, p 4).

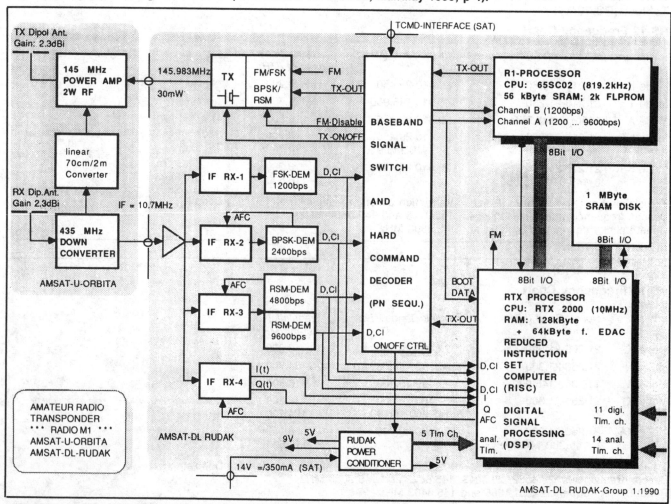

as RUDAK II, which is extremely accommodating in its ability to handle different modulation techniques and transmission speeds on its various links.

Subsystem Description

3.1 Beacon

Linear Transponder 1:
   145.822 MHz (CW)
   145.952 MHz (1100-bps PSK)

Linear Transponder 2:
   145.948 MHz (CW)
   145.838 MHz (1100-bps PSK)

3.4 Transponders

Linear transponder 1: Mode B
   Uplink passband: 435.102-435.022 MHz
   Downlink passband: 145.852-145.932 MHz
   Passband: 80 kHz
   Power: 12 W PEP

Linear transponder 2: Mode B
   Uplink passband: 435.123-435.043 MHz
   Downlink passband: 145.866-145.946 MHz
   Passband: 80 kHz
   Power: 12 W PEP

Transponder 3: Mode-B digital (RUDAK II)
   Type: digital
   Nominal bandwidth: 20 kHz (uplinks 2-4 employ AFC)
   Uplink 1: 435.016 MHz (1200-bit/s Manchester FSK [Pacsat/FO-20 compatible])
   Uplink 2: 435.155 MHz (2400-bit/s BPSK [RUDAK I])
   Uplink 3A: 435.193 MHz (4800-bit/s RSM NRZI)
   Uplink 3B: 435.193 MHz (9600-bit/s RSM NRZI)
   Uplink 4: 435.041 MHz DSP (digital speech processor) experiment
   Downlink: 145.983 MHz
   One of eight modes selectable
   1. Pacsat/FO-20 compatible
   2. AO-13 400-bit/s PSK compatible
   3. AO-13 RUDAK I compatible
   4. 4800 bit/s
   5. 9600 bit/s
   6. CW
   7. FSK (F1 or F2B) RTTY, FAX, SSTV, and so on
   8. FM using D/A converter and DSP (digital speech, and so on)

3.9 Integrated Housekeeping Unit

CPU: two (1802 COSMAC)
RAM: 1 M configured as RAM disk for mailbox

Microsat Series

SPACECRAFT NAME: OSCARs 16-19

Note: The four Microsats listed are very similar in design and were launched together. This section includes data common to all four spacecraft; it is followed by sections containing information specific to each satellite. For additional information on Microsat systems, see Chapter 15.

General

1.2 Launch

Date: 22 Jan 1990
Time: 01:35:31 UTC
Vehicle: Ariane (version: 40, family: 4, mission: V35)
Agency: ESA
Site: Kourou, French Guiana

1.3 Orbital Parameters

General designation: low-altitude, circular, sun-synchronous, near-polar
Period: 100.8 minutes
Mean altitude: 800 km
Eccentricity: 0.0012
Inclination: 98.7°
Longitude increment: 25.3°W/orbit
Maximum access distance: 3038 km

1.4 Ground Track Data

See Appendix C.

1.6 Design/Construction Credits

Project manager: J. King, W3GEY
Accomplishing a task of this magnitude requires major efforts by a large number of people, many of whom are credited in the references. Space doesn't permit repeating the list here, but it would be remiss not to mention the immense contributions of Tom Clark, Dick Jansson, Lyle Johnson and Bob McGwier.

1.7 Primary References

1) T. Clark, C. Duncan, J. King, B. McGwier "The First Flock of MicroSats," *The AMSAT Journal*, May 1989, pp 3-10.
2) D. Loughmiller and B. McGwier "Microsat: The Next Generation of OSCAR Satellites," Part 1, *QST*, May 1989, cover, pp 37-40; Part 2, *QST*, Jun 1989, pp 53-54.
3) T. Clark "AMSAT's MicroSat/Pacsat Program," *Proceedings of the AMSAT-NA Sixth Space Symposium,* Atlanta, Nov 1988, ARRL.

Spacecraft Description

2.1 Physical Structure

Shape: Cube approximately 23 cm on edge. See Fig 1 (MSat)
Mass: Approximately 9 kg

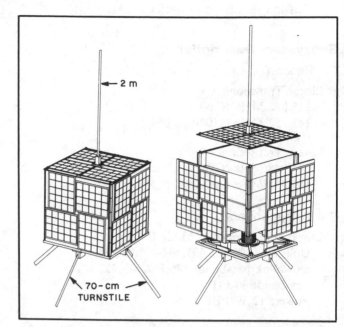

Fig 1 (MSat)—Microsat structure. Exploded view at right. (drawings courtesy WD4FAB)

Thermal design: Passive; spacecraft coatings are designed to minimize heat input from the sun and earth. The objective is to keep the spacecraft temperature between −5°C and +5°C to promote as high an efficiency from the solar cells as possible and to extend the life of the NiCd storage batteries.

Note: A Microsat spacecraft is composed of five aluminum trays formed into a stack held together with stainless-steel tie bolts. The frame stack assembly measures 230 × 230 × 213 mm. Tray slots are numbered from 1 to 5, starting at the −Z face (bottom) of the spacecraft. Honeycombed side panels manufactured from 4.8-mm-thick aluminum stock hold antennas and solar cells. This structure has been flight qualified to standards meeting the requirements of all the world's currently available launchers.

2.2 Subsystem Organization

Table 1 (MSat) lists the standard functions of the various MicroSat trays.

Subsystem organization is based on a bus structure. Each module mates to a wiring harness consisting of a 25-conductor ribbon cable. The ribbon cable carries dc power (40% of conductors), digital data or control signals (40%) and analog voltages for telemetry sensors and direct CPU control of modules. Each module (other than the main computer) contains a Motorola MC14469 AART (Addressable Asynchronous Receiver/Transmitter). The AART provides a standardized computer-to-module interface for commands and data. The AART uses ordinary ASCII communications at 4800 bps over what is essentially a 6-inch-long Local Area Network (LAN).

Table 1 (MSat)
Microsat Module Functions
Listed from top to bottom tray

Module No.	Pacsat & Lusat	DOVE	Webersat
+Z face (top)			
05	FSK Packet RX	Command RX	TSFR
04	TSFR	Flight Computer	Flight Computer
03	Power Module	Power Module	Power Module
02	Flight Computer	TSFR	FSK Packet RX
01	BPSK Packet TX	FM Voice TX	BPSK Packet TX
−Z face (bottom)			

TSFR (This Space For Rent):
   Pacsat: 13-cm transmitter
   Lusat: 70-cm beacon with dedicated microcomputer control
   DOVE: D/A Buffer/Converter, 13-cm transmitter
   Webersat: Camera, multiple experiments

Subsystem Description

3.2 Telemetry

With Microsat spacecraft, the concept of a fixed telemetry frame is obsolete. Telemetry functions are managed by the onboard computer. Telemetry is downlinked over regular beacon or transponder downlink channels in the form of packets. Unconnected packets contain basic telemetry and bulletins. Users can connect to the satellite via the transponder and request additional telemetry information. A transmitted telemetry channel contains a unique identifier and a data value. The identifier references the module of origin and the module's multiplexer address. If spare computing power exists (and if someone is willing to develop software), it should be possible to initiate onboard data processing so that telemetry may be downlinked in the form of plain text packets which identify the parameter being measured and provide a value. The telemetry system allows for up to 200 telemetry points. Preliminary sensor identification and decoding equations are shown in Table 2 (MSat). Updated telemetry decoding information is available on disk for IBM-compatible computers from AMSAT.

3.4 Transponder (All but DOVE)

The standard pacsat transponder consists of a receive unit and a transmit unit operating under the control of the onboard computer. The design is being optimized for file-transfer (mailbox) type operation (as opposed to bulletin-board type operation). During the Microsats' first few months in orbit, while the number of users is modest and mailbox software is being developed, digipeating will be encouraged. After several months of operation, however, it is expected that digipeating will be discouraged, since it does not use resources efficiently. Please be sure to disconnect before the satellite leaves your access circle.

The standard transponder configuration consists of five uplinks (four for users and one for commanding) in the 2-meter band and one downlink at 70 cm. A block diagram of the transponder FM receiver may be found in Chapter 15. It simultaneously monitors four 16-kHz-wide channels, spaced at intervals of 20 kHz, and a command channel. When a properly formatted AX.25 packet is received, it sends a connect acknowledgment. Uplink signals must be Manchester-encoded FM (AFSK) using ±3-kHz deviation. To receive signals from the spacecraft, the ground station must use a special PSK modem and an SSB receiver. A description of the equipment needed to transmit to, and receive signals from, this transponder is included in Chapter 9; operating procedures are discussed in Chapter 10. The downlink is BPSK; uplink and downlink will initially operate at 1200 bps. Transmissions are digital, NRZ-I, BPSK, HDLC and compatible with AX.25 Level Two protocol.

At a later date, selected uplink channels will be switched to 4800 bps. It is expected that the higher-speed channels will primarily be used by gateway ground stations. Because the baseband frequency window is four times wider at this speed, these transmissions will be straight FSK (not Manchester-encoded) so as to match the standard SSB filter passband.

The 437-MHz transmitter section of the transponder contains two units for redundancy; only one operates at a time. Each unit can be set to one of 16 power levels, producing up to 4 W output. One transmitter operates with straight PSK modulation, while the other uses raised-cosine modulation, which is compatible but has lower-level out-of-band modulation products. (Second harmonic subcarriers from the straight PSK transmitter are only down 14 dB from the main lobe; those from the raised-cosine transmitter are down 38 dB.) Experiments will determine if there is a bit-error-rate advantage to using one form of modulation over the other. The 70-cm raised-cosine transmitters use HELAPS techniques to produce up to 9 W PEP. At high power settings, each transmitter achieves an RF/dc efficiency of about 65% (BPSK) and 56% (raised cosine). Efficiency falls off slowly at lower power settings.

Maximum values for Doppler shift on the various Microsat links are as follows:

2 meters: 3.5 kHz

70 cm: 10.3 kHz
23 cm: 30.1 kHz
13 cm: 56.7 kHz

3.5 Attitude Stabilization and Control

Passive magnetic stabilization is used. Each spacecraft contains four small permanent magnets aligned along the Z axis. With this type of stabilization and the near-polar inclination, the satellite Z axis rotates twice per orbit (see Chapter 15). In order to minimize the buildup of thermal gradients, the spacecraft is spun slowly about its Z axis using the 2-meter antenna as a radiometer (one side of each element is painted black and the other side is painted white). This method was used successfully on both AMSAT-OSCAR 7 and AMSAT-OSCAR 8. The spin rate is expected to be between 0.25 and 1.0 RPM. Hysteresis damping rods are used to limit the spin rate and to minimize oscillations of the Z axis about the local magnetic field.

3.6 Antennas

The following antennas are used on Pacsat, Lusat and Webersat. Each spacecraft includes additional antennas described under the section for the specific satellite.

70-cm antenna: The 70-cm transmit antenna is a canted turnstile consisting of four radiating elements mounted on the −Z surface (bottom) of the spacecraft. The signal is circularly polarized along the Z axis only. Antenna elements are made of flexible, springy, semi-cylindrical metal approximately 1.0 cm in width.

2-meter antenna: The 2-meter receive antenna is a stub slightly shorter than ¼ wavelength mounted on the +Z

Table 2 (MSat)
Microsat Telemetry

All decoding equations are of the form

Parameter = $An^2 + Bn + C$
where n is the telemetry count.

Sample coefficients (A, B, C) are given for DOVE. These are early prelaunch values (1/7/90). Updated values and telemetry-decoding software is available on IBM-compatible disks from AMSAT.

Channel hex	dec	Description	C	B	A	Units

All Microsats except as noted below

hex	dec	Description	C	B	A	Units
0	0	RX D Disc				kHz
1	1	RX D S meter				counts
2	2	RX C Disc				kHz
3	3	RX C S meter				counts
4	4	RX B Disc				kHz
5	5	RX B S meter				counts
6	6	RX A Disc	+10.472	−0.09274	0.000	kHz
7	7	RX A S meter	+0.000	+1.000	0.000	counts
8	8	RX E/F Disc	+9.6234	−0.09911	0.000	kHz
9	9	RX E/F S meter	+0.000	+1.000	0.000	counts
A	10	+5-V Bus	+0.000	+0.0305	0.000	Volts
B	11	+5-V Rx current	+0.000	+0.000100	0.000	Amps
C	12	+2.5-V Ref	+0.000	+0.0108	0.000	Volts
D	13	+8.5-V Bus	+0.000	+0.0391	0.000	Volts
E	14	IR detector	+0.000	+1.000	0.000	counts
F	15	LO monitor I	+0.000	+0.000037	0.000	Amps
10	16	+10-V bus	+0.000	+0.05075	0.000	Volts
11	17	GASFET Bias I	+0.000	+0.000026	0.000	Amps
12	18	Ground Ref	+0.000	+0.0100	0.000	Volts
13	19	+Z Array V	+0.000	+0.1023	0.000	Volts
14	20	RX Temp	+101.05	−0.6051	0.000	°C
15	21	+X (RX) Temp	+101.05	−0.6051	0.000	°C
16	22	Bat 1 V	+1.7932	−0.0034084	0.000	Volts
17	23	Bat 2 V	+1.7978	−0.0035316	0.000	Volts
18	24	Bat 3 V	+1.8046	−0.0035723	0.000	Volts
19	25	Bat 4 V	+1.7782	−0.0034590	0.000	Volts
1A	26	Bat 5 V	+1.8410	−0.0038355	0.000	Volts
1B	27	Bat 6 V	+1.8381	−0.0038450	0.000	Volts
1C	28	Bat 7 V	+1.8568	−0.0037757	0.000	Volts
1D	29	Bat 8 V	+1.7868	−0.0034068	0.000	Volts
1E	30	Array V	+7.205	+0.07200	0.000	Volts
1F	31	+5-V Bus	+1.932	+0.0312	0.000	Volts
20	32	+8.5-V Bus	+5.265	+0.0173	0.000	Volts

face of the spacecraft. It is made of the same material used for the 70-cm antenna elements. This antenna produces a linearly polarized signal and a toroidal pattern.

3.7 Energy Supply and Power Conditioning

Solar-Cell Characteristics

Type: Silicon, using back-surface reflective technology

Efficiency: 15% + at beginning of life

Size: 20 × 20 mm

Total area: 1760 cm^2

Peak output: 15.7 W (optimal orientation)

Minimal output: 6 W (poorest orientation, averaged over orbit with 34% eclipse time)

Solar-Cell Configuration (slight differences in Webersat)

Basic unit: a "clip" of 20 cells arranged in a 4 × 5 pattern

Total number of modules: 20 clips + 4 half clips

Location: −Z face contains 4 half clips. The other five surfaces each contain 4 clips.

Storage Battery

Consists of 8 6-Ah commercial aviation NiCd cells manufactured by GE/Gates and space qualified by AMSAT (See Chapter 15).

Battery bus voltage at 100% charge: 11.7 V

Battery bus voltage at 30% charge (maximum safe discharge level): 9.2 V

Power Module

The power module is a slightly oversize tray containing storage battery, battery charging regulator (BCR) and instrument switching regulators (ISRs). It is the only module designed to fit in a specific slot (slot 3). The BCR downconverts the nominal 22-V output

Channel hex	dec	Description	C	B	A	Units
21	33	+10-V Bus	+7.469	+0.021765	0.000	Volts
22	34	BCR Set Point	−8.762	+1.1590	0.000	Counts
23	35	BCR Load Cur	−0.0871	+0.00698	0.000	Amps
24	36	+8.5-V Bus Cur	−0.00920	+0.001899	0.000	Amps
25	37	+5-V Bus Cur	+0.00502	+0.00431	0.000	Amps
26	38	−X Array Cur	−0.01075	+0.00215	0.000	Amps
27	39	+X Array Cur	−0.01349	+0.00270	0.000	Amps
28	40	−Y Array Cur	−0.01196	+0.00239	0.000	Amps
29	41	+Y Array Cur	−0.01141	+0.00228	0.000	Amps
2A	42	−Z Array Cur	−0.01653	+0.00245	0.000	Amps
2B	43	+Z Array Cur	−0.01137	+0.00228	0.000	Amps
2C	44	Ext Power Cur	−0.02000	+0.00250	0.000	Amps
2D	45	BCR Input Cur	+0.06122	+0.00317	0.000	Amps
2E	46	BCR Output Cur	−0.01724	+0.00345	0.000	Amps
2F	47	Bat 1 Temp	+101.05	−0.6051	0.000	°C
30	48	Bat 2 Temp	+101.05	−0.6051	0.000	°C
31	49	Baseplate Temp	+101.05	−0.6051	0.000	°C
32	50	PSK TX RF Out				Watts
33	51	RC TX RF Out				Watts
34	52	PSK TX HPA Temp	+101.05	−0.6051	0.000	°C
35	53	+Y Array Temp	+101.05	−0.6051	0.000	°C
36	54	RC PSK HPA Temp	+101.05	−0.6051	0.000	°C
37	55	RC PSK BP Temp	+101.05	−0.6051	0.000	°C
38	56	+Z Array Temp	+101.05	−0.6051	0.000	°C
39	57	S-Band HPA Temp	+101.05	−0.6051	0.000	°C
3A	58	S-Band TX RF Out	−0.0451	+0.00403	0.000	Watts

DOVE

Channel hex	dec	Description	C	B	A	Units
0	0	E/F Audio (Wide)	+0.000	+0.0246	0.000	V (P-P)
1	1	E/F Audio (Nar)	+0.000	+0.0246	0.000	V (P-P)
2	2	Mixer Bias V	+0.000	+0.0102	0.000	Volts
3	3	Osc Bias V	+0.000	+0.0102	0.000	Volts
4	4	RX A Audio (W)	+0.000	+0.0246	0.000	V (P-P)
5	5	RX A Audio (N)	+0.000	+0.0246	0.000	V (P-P)
32	50	FM TX #1 RF Out	+0.0256	−0.000884	+0.0000836	Watts
33	51	FM TX #2 RF Out	−0.0027	+0.001257	+0.0000730	Watts

Lusat

Channel hex	dec	Description	C	B	A	Units
39	57	Lu Beacon Temp A				°C
3A	58	Lu Beacon Temp D				°C
3B	59	Coax Relay Status				Counts
3C	60	Coax Relay Status				Counts

Webersat

Channel hex	dec	Description	C	B	A	Units
39	57	Not available				
3A	58	Not available				

of the solar array to the battery bus voltage of about 10 V. Two ISRs supply 8.5 V and 5.0 V. Switching regulators using pulse-width modulation provide an efficiency of about 88%. The power module is heavily instrumented with telemetry sensors for measuring voltages, currents and temperatures. Current-sensing technology is based on the highly efficient method developed for Phase 3, which uses toroids biased into saturation by an ac signal (see Chapter 15).

Software continuously monitors power module operation and controls transmitter power level to optimize use of available power and prolong battery life. This is a new method of power management for AMSAT, and it is expected that software will have to be refined over the life of the mission.

3.9 Onboard Computer (OBC)

The OBC is the primary payload on all Microsats. It controls, and forms an essential component of, the packet transponder, telemetry system, power system and other mission-specific modules. The OBC module also contains an 8-bit analog-to-digital (A/D) converter with a measurement range of 0-2.55 V and a resolution of 10 mV. This A/D converter monitors the analog bus lines for telemetry sensor data. A watchdog timer guards against problems which could cause the OBC to "lock up."

CPU: NEC V40 (similar to 80C188)

ROM: 2k for restart

RAM (area #1): 256-k static RAM. Employs error detection and correction (Hamming code with 12-bits per 8-bit byte). Used for operating system and program storage. To reduce power consumption, EDAC circuitry does not run continuously in background.

RAM (area #2): 2M static RAM. Used for message storage. Divided into 4 ½-M bank-switched regions that may be individually powered down.

RAM (area #3): 8 M static RAM. Organized as serial-interface mass storage that operates like a RAM disk.

Operating System: Quadron multitasking (appears similar to MS-DOS to each application). (Donated by Quadron, Inc.) This allows ground stations to develop software on standard IBM-compatible PCs.

Languages (onboard applications): Assembler and Microsoft C linked with Microsoft LINK.

Power requirements: 1.5 W peak, 0.5 W average (expected).

SPACECRAFT NAME: Pacsat-OSCAR 16

Mission Objective

To provide the worldwide community of radio amateurs with a satellite-based digital store-and-forward message system compatible with terrestrial packet communications.

1.1 Identification

International designation: 90-005D

NASA Catalog number: 20439
Prelaunch designation: Pacsat, Microsat-A

1.5 Operations

Coordinating Group: AMSAT-NA

Spacecraft Description

2.2 Subsystem Organization

See Table 1 (MSat)

Subsystem Description

3.2 Telemetry

See Table 2 (MSat)

3.4 Transponder

uplink frequencies:	145.900 MHz
	145.920 MHz
	145.940 MHz
	145.960 MHz
downlink frequencies	437.051 MHz (primary, raised cosine)
	437.026 MHz (secondary, BPSK)
	2401.143 MHz (secondary, BPSK, 1 W)
connect address:	Pacsat-1

3.5 Attitude Stabilization and Control

OSCAR 16 contains the standard Microsat stabilization system. In addition, it has an infrared sensor mounted on the +Z face of the spacecraft, which is connected to the module containing the 2-meter receiver. This sensor can provide information relating to spacecraft attitude. Its field of view is about 8° and its sensitivity is adjustable in 16 steps. Whole-orbit data collected from this sensor will be used to study spacecraft dynamics of the Microsat structure.

3.6 Antennas

The 13-cm transmit antenna is a bifilar helix (volute).

3.10 TSFR Module

The TSFR module on OSCAR 16 contains a 13-cm transmitter based on the 70-cm pacsat design, with a chain of four multipliers added to the local-oscillator strip. The final amplifier is an Avantek AV-8140. Check current periodicals for an operating schedule.

Downlink frequency: 2401.143 MHz BPSK
Power: 1 W
RF/dc efficiency: 32%

SPACECRAFT NAME: DOVE-OSCAR 17

Mission Objective

The primary objective of the DOVE mission is to provide an easily receivable signal suitable for educational demonstrations involving young children. The downlink is digitally synthesized speech containing messages in varied languages interspersed with spoken telemetry. Groups of young schoolchildren around the world are being encouraged to prepare tapes containing messages on themes relating to peace, brotherhood and environmental preservation. Command stations digitize these messages and uplink them for later broadcast.

Note: DOVE = Digital Orbiting Voice Encoder

1.1 Identification

International designation: 90-005E
NASA Catalog number: 20440
Prelaunch designation: Microsat-B
Call sign: PT2PAZ (Brazil)

1.5 Operations

Coordinating Groups: AMSAT-Brazil and AMSAT-NA

1.6 Design/Construction Credits

The idea for DOVE was originated by Dr Junior Torres DeCastro, PY2BJO. As a result of his leadership, AMSAT-Brazil has provided financial and technical support for this project.

Spacecraft Description

2.2 Subsystem Organization

See Table 1 (MSat)

Subsystem Description

3.1 Beacons

Beacon 1 frequency: 145.824 MHz (NBFM)
  Power: up to 4 W
  RF/dc efficiency: 73%
Beacon 2 frequency: 145.825 MHz (NBFM)
  Power: up to 4 W
  RF/dc efficiency: 73%
Beacon 3 frequency: 2401.221 MHz
  Power: (1 W)
  RF/dc efficiency: 32%

The beacons can be fed conventional 2-meter FM audio using the Telemetry Voice Transmitter Module (TVTM, see 3.10). Signals will usually consist of digitally synthesized speech or AFSK AX.25 packets, which can be received using unmodified 2-meter terrestrial packet systems.

3.2 Telemetry

When active, the DOVE telemetry system transmits AFSK AX.25 packets. DOVE is the only Microsat that transmits packets that can be received on a standard 2-meter terrestrial packet station. A sample of DOVE

Table 3 (MSat)

DOVE Telemetry

Raw data packets and decoded information

```
          DOVE  DAYTIME  TELEMETRY
   RAW DATA
DOVE-1>TLM:00:5A 01:5A 02:88 03:32 04:59 05:58 06:6C 07:4A 08:6C 09:68 0A:A2
0B:EC 0C:E8 0D:DC 0E:3F 0F:24 10:D8 11:93 12:00 13:D1 14:9B 15:AE
16:83 17:7C 18:76 19:7E 1A:7C 1B:45 1C:84 1D:7B 1E:C4 1F:6C 20:CF
DOVE-1>TLM:21:BB 22:79 23:26 24:22 25:26 26:01 27:04 28:02 29:3A 2A:02 2B:73
2C:01 2D:7C 2E:58 2F:A2 30:D0 31:A2 32:17 33:6B 34:AC 35:A2 36:A6
37:A8 38:86 39:A2 3A:01
DOVE-1>STATUS: 80 00 00 85 B0 18 77 02 00 B0 00 00 B0 00 00 00 00 00 00 00
DOVE-1>BCRXMT:vary= 21.375 vmax= 21.774 temp=  7.871
DOVE-1>BCRXMT:vbat= 11.539 vlo1= 10.627 vlo2= 10.127 vmax= 11.627 temp=  3.030
DOVE-1>WASH:wash addr:26c0:0000, edac=0x61
DOVE-1>TIME-1:PHT: uptime is 086/01:14:32.  Time is Sat Mar 10 15:43:26 1990

   DECODED TELEMETRY

DOVE    uptime is 086/01:14:32.  Time is Sat Mar 10 15:43:26 199
```

Rx E/F Audio(W)	2.21 V	Rx E/F Audio(N)	2.21 V	Mixer Bias V:	1.39 V
Osc. Bisd V:	0.51 V	Rx A Audio (W):	2.19 V	Rx A Audio (N):	2.16 V
Rx A DISC:	0.41 k	Rx A S meter:	74.00 C	Rx E/F DISC:	-1.08 k
Rx E/F S meter:	104.00 C	+5 Volt Bus:	4.94 V	+5V Rx Current:	0.02 A
+2.5V VREF:	2.51 V	8.5V BUS:	8.60 V	IR Detector:	63.00 C
LO Monitor I:	0.00 A	+10V Bus:	10.96 V	GASFET Bias I:	0.00 A
Ground REF:	0.00 V	+Z Array V:	21.38 V	Rx Temp:	7.26 D
+X (RX) temp:	-4.24 D	Bat 1 V:	1.35 V	Bat 2 V:	1.36 V
Bat 3 V:	1.38 V	Bat 4 V:	1.34 V	Bat 5 V:	1.37 V
Bat 6 V:	1.57 V	Bat 7 V:	1.36 V	Bat 8 V:	1.37 V
Array V:	21.32 V	+5V Bus:	5.30 V	+8.5V Bus:	8.85 V
+10V Bus:	11.54 V	BCR Set Point:	131.48 C	BCR Load Cur:	0.18 A
+8.5V Bus Cur:	0.06 A	+5V Bus Cur:	0.17 A	-X Array Cur:	-0.01 A
+X Array Cur:	-0.00 A	-Y Array Cur:	-0.01 A	+Y Array Cur:	0.12 A
-Z Array Cur:	-0.01 A	+Z Array Cur:	0.25 A	Ext Power Cur:	-0.02 A
BCR Input Cur:	0.45 A	BCR Output Cur:	0.29 A	Bat 1 Temp:	3.02 D
Bat 2 Temp:	-24.81 D	Baseplt Temp:	3.02 D	FM TX#1 RF OUT:	0.05 W
FM TX#2 RF OUT:	0.97 W	PSK TX HPA Temp	-3.03 D	+Y Array Temp:	3.02 D
RC PSK HPA Temp	0.60 D	RC PSK BP Temp:	-0.61 D	+Z Array Temp:	19.97 D
S band HPA Temp	3.02 D	S band TX Out:	-0.04 W		

telemetry has therefore been selected to illustrate Microsat telemetry. See Tables 2 (MSat) and 3 (MSat).

3.3 Telecommand

As with the other Microsats, the telecommand reception system on DOVE operates in the 2-meter band. In order for the command receiver to coexist with the 2-meter beacon, the beacon is turned off for 30 seconds after every 2.5 minutes of operation and the receiver is simultaneously activated. If a software error allows the transmitter to operate continuously, the spacecraft computer can be reset by a ground station with an extremely high EIRP.

3.6 Antennas

2 meters: Canted turnstile antenna mounted on the −Z surface of the spacecraft. It is made of the same material used for the 70-cm canted turnstiles on the other Microsats.

13 cm: bifilar helix (volute).

3.10 Telemetry Voice Transmitter Module (TVTM) (TSFR tray)

The TVTM can transmit either stored voice messages, spoken telemetry or 1200-bps AFSK telemetry (Bell-202 standard). This module contains a voice synthesizer and a digital-to-analog (D/A) buffer/converter. Messages are digitized on the ground and sent up as packets, which are stored on the OBC RAM disk. The digitalker uses a VOTRAX SC-02 chip. A Motorola 68HC11 in the module acts as a smart UART.

SPACECRAFT NAME: Webersat-OSCAR 18

Mission Objective

The primary mission objective is to register, store and downlink earth images in a format compatible with radio amateur satellite packet-radio techniques. Secondary objectives include the operation of several additional scientific/educational experiments.

1.1 Identification

International designation: 90-005F
NASA Catalog number: 20441
Prelaunch designation: Microsat-C

1.5 Operations

Coordinating Group: The Center for Aerospace Technology (CAST) at Weber State University, Ogden, Utah

1.7 Primary References

1) Imaging Experiment
S. Sjol, "Webersat," *AMSAT Journal*, Nov 1989, p 30.
2) Non-imaging Experiments
C. Williams, *A Brief User's Manual for Webersat's Ancillary Experiments* (rev 0.0, pre-launch). Center for Aerospace Technology, Weber State University, Ogden, UT 84408-1805.

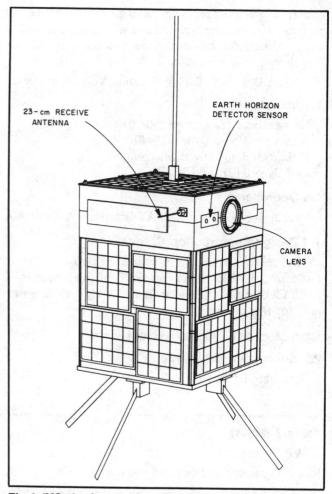

Fig 2 (MSat)—An exterior view of Webersat. (drawing courtesy WD4FAB)

Spacecraft Description

2.1 Physical Structure

Webersat is slightly taller and heavier than the other Microsats due to an oversize tray 5, containing the camera and other experiments, which is mounted on the top of the spacecraft. See Fig 2 (MSat).

Shape: Rectangular solid, 23 cm × 23 cm × 32 cm (height)
Mass: approximately 12 kg

2.2 Subsystem Organization

See Table 1 (MSat).

Subsystem Description

3.4 Transponder

Webersat carries a fully functional pacsat transponder. The primary objective of the mission, however, involves the imaging experiment. Since opening the transponder for general use may interfere with imaging and other educational activities, scheduling of the transponder and the publication of specific uplink frequencies is being left to the discretion of Weber State University.

uplink frequencies: between 144.300 and 144.500 MHz
downlink   437.102 MHz (primary, raised cosine)
frequencies: 437.075 MHz (secondary, BPSK)

3.6 Antennas

In addition to the normal antenna systems, OSCAR 16 carries a 23-cm, ¼-wavelength stub antenna mounted on module 5 perpendicular to the Z axis. This antenna is used for receiving ATV transmissions from ground stations.

3.10 Webersat Experiment Module

The Experiment module (tray 5), which is about 13 cm tall, sits on top of the the normal microsat structure. It contains a color CCD camera, a high-speed flash digitizer, a 23-cm receiver, a particle-impact detector, an earth-horizon detector, a visible-light spectrometer and a magnetometer.

3.10A Camera

The primary spacecraft payload is a standard miniature CCD color TV camera modified to operate in space. The camera produces either a standard color NTSC analog signal with color burst, or separate red, green and blue signals with horizontal and vertical sync. Either product is sent to a high-speed flash analog-to-digital converter operating at a 10-MHz sampling rate. The picture is compressed and stored in the RAM disk area. Pictures are downlinked in compressed form in AX.25 packets. Software will be made available for the display of these pictures (Weberware). The CCD element in the camera has an effective resolution of about 700 pixels × 400 lines. The camera lens covers a ground area of about 300 × 300 km when pointed at the SSP. The lens looks out the side of the spacecraft (in the +X direction, perpendicular to the Z axis). The operational brightness range of the camera extends from starlight to direct sun viewing. Camera operation will generally be controlled by the earth-horizon detector to produce earth images. Since a typical picture will occupy about 200 kbytes of memory, downlinking at 1200 baud will take more than 20 minutes (more than one pass). Plans are to have the spacecraft record and downlink about one picture per day. Since the camera iris is a mechanical device, it is difficult to predict its useful life in space. For this reason, the camera experiment will be given a very high priority during the early days of Webersat's sojourn in space.

3.10B Earth-Horizon Detector

The horizon detector consists of two visible-light-sensing photocells (Siemens BPW21) mounted just behind the face of the spacecraft through which the camera looks out. Both sensors are located to one side of the centrally positioned camera lens and mounted pointing in the XY plane (perpendicular to the spacecraft Z axis). The field of view of each sensor is a cone of about 10 degrees. Each photocell is offset 11 degrees from the camera axis towards the other photocell. As a result, their fields of view cross about 30 cm from the spacecraft and diverge from that point on. The photocell closest to the outside edge of the face and pointing inward towards the camera lens is called sensor 1. The photocell closest to the camera lens and pointing slightly outward is called sensor 2. The mounting geometry was designed so that the only object that can illuminate both photocells equally is the earth. As the spacecraft slowly rotates about the Z axis, there will be times when both cells are illuminated. When this occurs, the camera will be looking directly at the earth.

3.10C Turbo Download and ATV Uplink

The turbo download experiment is based on the flash digitizer and associated digital/analog converter. Data in the flash digitizer memory can be fed through the D/A converter to produce an analog waveform that can then be used to frequency modulate the 70-cm downlink. On the ground, the discriminator of an FM receiver could be connected to an analog-to-digital converter and the recovered data stored in a computer memory. Software could then reconstruct the image. Calculations show that a picture could be downlinked in about 7 seconds using this method, with approximately the same resolution as a 20-minute packet download. Data from various other experiments on the spacecraft could also be downloaded this way. It has even been suggested that it is possible for a ground station to take an audio signal, translate it into a video signal, uplink the video signal, then have the turbo download system produce a synthesized voice FM signal—''With software all things are possible'' [Bob McGwier, 1989].

Webersat carries a 1265-MHz ATV receiver. This unit receives standard analog NTSC pictures uplinked in the 23-cm amateur band. The pictures are digitized by the same circuitry used to digitize images from the onboard CCD camera (the flash digitizer). The pictures can be stored aboard the spacecraft for later transmission as packets or by the turbo downlink system.

3.10D Spectrometer Experiment

The spectrometer experiment is designed to observe the spectrum of sunlight reflected off the Earth's atmosphere and surface. The observations will be used to identify and measure relative concentrations of different atmospheric components.

Light entering a slit in the −Y face of the spacecraft is focused by a lens and reflected from a diffraction grating. The resulting spectrum is recorded on a 2k × 1 byte CCD sensor (linear array camera). The system is responsive throughout the visible light spectrum with some extension at both ends (roughly 300 to 1100 microns). The spectrometer is accompanied by a set of earth sensors like those described in 3.10B.

3.10E Particle Impact Detector

The main sensor for the impact detector is a piezoelectric crystal (about 15 × 3 cm) mounted on the side of the spacecraft. The sensor produces a voltage pulse when a microparticle impact occurs. (Anything larger than a microparticle could have disastrous effects. A particle with a diameter of 0.3 cm moving at 10 km/s has the same kinetic energy as a bowling ball going 60 mi/h. A 0.02-cm particle, roughly the size of the period at the end of this sentence, may cause serious damage.) A counter keeps track of the total number of pulses until reset. A second sensor is mounted inside the spacecraft at right angles to the external sensor. Sensor control electronics are configured so that simultaneous pulses from both sensors are not recorded as an impact, since such events are most likely due to flexing of the structure caused by thermal effects. Since the voltage level of the pulse is not recorded, the detector does not provide direct information on impact magnitude. However, a clever experimenter might be able to infer some information on impact magnitude by exploiting various characteristics of the impact detector system.

3.10F Magnetometer

Webersat contains two orthogonal flux gate magnetometers. Magnetometer 1 is oriented to sense flux lines in the YZ plane of the spacecraft; magnetometer 2 is oriented to sense flux lines in the XY plane. A small permanent magnet mounted internally in the spacecraft is used to locally cancel much of the magnetic field produced by the attitude-control bar magnets. Since no provisions have been made for absolute calibration of the magnetometers, the output readings are mainly useful as relative measures of magnetic field.

SPACECRAFT NAME: Lusat-OSCAR 19
Mission Objective

To provide the worldwide community of radio amateurs with a satellite-based digital store-and-forward message system compatible with terrestrial packet communications.

1.1 Identification

International designation: 90-005G
NASA Catalog number: 20442
Prelaunch designation: Microsat-D

1.5 Operations

Coordinating Group: AMSAT-LU (Argentina) under direction of Arturo Caru, LU1AHC, and Carlos Huertas, LU4ENQ

Spacecraft Description
2.2 Subsystem Organization

See Table 1 (MSat)

Subsystem Description
3.4 Transponder

uplink frequencies: 145.840 MHz
                   145.860 MHz
                   145.880 MHz
                   145.900 MHz
downlink frequencies: 437.153 MHz (primary, BPSK)
                   437.125 MHz (secondary, raised cosine)
connect address: Lusat-1

3.5 Attitude Stabilization and Control

An infrared sensor mounted on the +Z face of the spacecraft is connected to the module containing the 2-meter receiver. This sensor can provide information relating to spacecraft attitude. Its field of view is about 8° and its sensitivity is adjustable in 16 steps.

The four permanent bar magnets mounted parallel to the Z axis have their poles oriented opposite to the direction used on OSCARs 16-18. This orientation should slightly favor southern hemisphere users.

3.10 70-cm CW Beacon

The experimental module on OSCAR 17 contains a 70-cm CW data beacon built by AMSAT-LU. This transmitter is controlled by a dedicated microprocessor and is designed to provide telemetry and bulletins to stations who do not have packet-radio capabilities. Morse telemetry at 12 WPM is sent in an abbreviated format where multiple dashes in numeric data are sent as a single dash (·— = 1, ··— = 2, —· = 9, and so on).
    frequency: 437.127 MHz
    power: 750 mW

Appendix C
Tracking Data and Overlays

Assembling a graphic tracking aid, such as the *OSCAR-LOCATOR* or *P3 Tracker*, requires data for drawing ground tracks and spiderwebs. This appendix contains the necessary material in two formats: (1) tables that can be used in conjunction with any map and (2) tracing masters for use with the map board included here or as part of the ARRL OSCARLOCATOR package. See Chapters 6 and 12 for a complete description of how the information presented in this appendix is used.

Notes:

1) Ground tracks should be drawn directly on the map using data presented. They may then be traced on clear plastic. The plastic overlay is then repositioned for each orbit.

2) Spiderwebs should be drawn directly on the maps using data presented. They may then be traced on clear plastic. The plastic overlay is then repositioned for each orbit.

3) Spiderweb azimuth radials and elevation circles are drawn at 30° intervals.

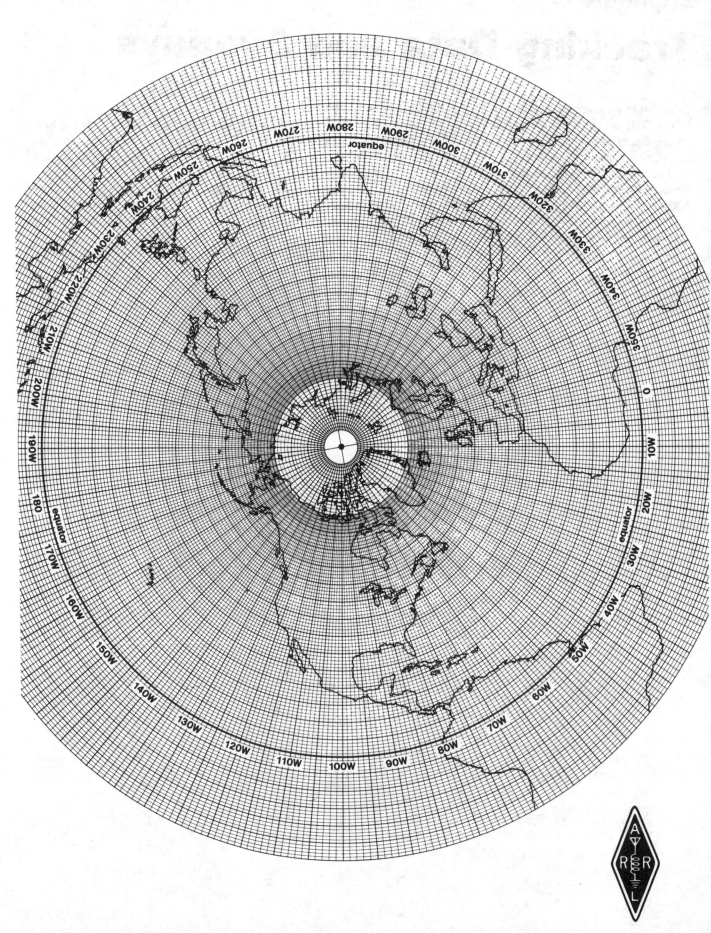

AMSAT-OSCAR 13

A spiderweb produced from the data of this table employs range circles in place of elevation angle circles as explained in the discussion of the P3 TRACKER in Chapter 12. Although presented here for use with AMSAT-OSCAR 13 this spiderweb is not keyed to a specific orbit. It can therefore be used with any satellite. The spiderweb should be drawn directly on the map using data presented. It may then be traced on clear plastic. The plastic overlay is then repositioned at the user's location.

LOCATION OF CENTER — **30.0 DEGREES NORTH LATITUDE** — **90.0 DEGREES WEST LONGITUDE**

AZIMUTH

DISTANCE		NORTH	30	60	EAST	120	150	SOUTH	210	240	WEST	300	330
1000 KM	LATITUDE	39.0	37.7	34.2	29.6	25.2	22.1	21.0	22.1	25.2	29.6	34.2	37.7
	LONGITUDE	90.0	84.3	80.6	79.6	81.4	85.2	90.0	94.8	98.6	100.4	99.4	95.7
2000 KM	LATITUDE	48.0	45.0	37.5	28.4	20.0	14.1	12.0	14.1	20.0	28.4	37.5	45.0
	LONGITUDE	90.0	77.4	70.3	69.4	73.5	80.8	90.0	99.2	106.5	110.6	109.7	102.6
3000 KM	LATITUDE	57.0	51.8	39.9	26.5	14.4	6.0	3.0	6.0	14.4	26.5	39.9	51.8
	LONGITUDE	90.0	68.5	59.2	59.6	66.1	76.8	90.0	103.2	113.9	120.4	120.8	111.5
4000 KM	LATITUDE	66.0	57.7	41.2	23.9	8.6	-2.1	-6.0	-2.1	8.6	23.9	41.2	57.7
	LONGITUDE	90.0	56.7	47.4	50.0	59.0	72.9	90.0	107.1	121.0	130.0	132.6	123.3
4500 KM	LATITUDE	70.5	60.1	41.4	22.4	5.7	-6.1	-10.5	-6.1	5.7	22.4	41.4	60.1
	LONGITUDE	90.0	49.3	41.5	45.4	55.6	71.0	90.0	109.0	124.4	134.6	138.5	130.7
5000 KM	LATITUDE	75.0	62.1	41.3	20.7	2.7	-10.2	-15.0	-10.2	2.7	20.7	41.3	62.1
	LONGITUDE	90.0	41.0	35.5	40.9	52.2	69.0	90.0	111.0	127.8	139.1	144.5	139.0
6000 KM	LATITUDE	84.0	64.2	40.1	17.1	-3.2	-18.2	-24.0	-18.2	3.2	17.1	40.1	64.2
	LONGITUDE	90.0	21.5	23.7	32.2	45.5	64.8	90.0	115.2	134.5	147.8	156.3	158.5
7000 KM	LATITUDE	87.0	63.6	37.8	13.1	-9.1	-26.1	-33.0	-26.1	9.1	13.1	37.8	63.6
	LONGITUDE	270.0	1.0	12.5	23.9	38.6	60.3	90.0	119.7	141.4	156.1	167.5	179.0
8000 KM	LATITUDE	78.1	60.2	34.5	8.9	-14.9	-33.9	-41.9	-33.9	14.9	8.9	34.5	60.2
	LONGITUDE	270.0	-16.8	2.1	15.8	31.6	55.0	90.0	125.0	148.4	164.2	177.9	196.8
9000 KM	LATITUDE	69.1	55.0	30.4	4.5	-20.4	-41.4	-50.9	-41.4	20.4	4.5	30.4	55.0
	LONGITUDE	270.0	-30.5	-7.4	7.9	24.1	48.8	90.0	131.2	155.9	172.1	187.4	210.5

LOCATION OF CENTER — **40.0 DEGREES NORTH LATITUDE** — **90.0 DEGREES WEST LONGITUDE**

AZIMUTH

DISTANCE		NORTH	30	60	EAST	120	150	SOUTH	210	240	WEST	300	330
1000 KM	LATITUDE	49.0	47.6	44.0	39.4	35.1	32.1	31.0	32.1	35.1	39.4	44.0	47.6
	LONGITUDE	90.0	83.3	79.2	78.3	80.5	84.7	90.0	95.3	99.5	101.7	100.8	96.7
2000 KM	LATITUDE	58.0	54.7	46.9	37.7	29.5	24.0	22.0	24.0	29.5	37.7	46.9	54.7
	LONGITUDE	90.0	74.5	67.0	67.0	72.1	80.3	90.0	99.7	107.9	113.0	113.0	105.5
3000 KM	LATITUDE	67.0	60.9	48.3	34.9	23.5	15.8	13.0	15.8	23.5	34.9	48.3	60.9
	LONGITUDE	90.0	62.2	53.8	56.4	64.6	76.4	90.0	103.6	115.4	123.6	126.2	117.8
4000 KM	LATITUDE	76.0	65.5	48.2	31.3	17.2	7.5	4.0	7.5	17.2	31.3	48.2	65.5
	LONGITUDE	90.0	44.9	40.3	46.5	57.8	72.8	90.0	107.2	122.2	133.5	139.7	135.1
4500 KM	LATITUDE	80.5	66.9	47.5	29.3	13.9	3.3	-0.5	3.3	13.9	29.3	47.5	66.9
	LONGITUDE	90.0	34.3	33.7	41.9	54.6	71.0	90.0	109.0	125.4	138.1	146.3	145.7
5000 KM	LATITUDE	85.0	67.5	46.5	27.1	10.6	-0.8	-5.0	-0.8	10.6	27.1	46.5	67.5
	LONGITUDE	90.0	22.8	27.2	37.5	51.5	69.3	90.0	110.7	128.5	142.5	152.8	157.2
6000 KM	LATITUDE	86.0	66.2	43.5	22.2	3.9	-9.1	-14.0	-9.1	3.9	22.2	43.5	66.2
	LONGITUDE	270.0	0.1	15.2	29.1	45.4	65.8	90.0	114.2	134.6	150.9	164.8	179.9
7000 KM	LATITUDE	77.0	62.0	39.3	17.0	-2.8	-17.4	-23.0	-17.4	2.8	17.0	39.3	62.0
	LONGITUDE	270.0	18.3	4.6	21.4	39.4	62.2	90.0	117.8	140.6	158.6	175.4	198.3
8000 KM	LATITUDE	68.1	56.1	34.3	11.5	-9.5	-25.6	-31.9	-25.6	9.5	11.5	34.3	56.1
	LONGITUDE	270.0	-31.5	-4.7	14.0	33.4	58.2	90.0	121.8	146.6	166.0	184.7	211.5
9000 KM	LATITUDE	59.1	49.1	28.7	5.8	-16.1	-33.6	-40.9	-33.6	16.1	5.8	28.7	49.1
	LONGITUDE	270.0	-41.0	-13.0	7.0	27.1	53.6	90.0	126.4	152.9	173.0	193.0	221.0

LOCATION OF CENTER — **50.0 DEGREES NORTH LATITUDE** — **90.0 DEGREES WEST LONGITUDE**

AZIMUTH

DISTANCE		NORTH	30	60	EAST	120	150	SOUTH	210	240	WEST	300	330
1000 KM	LATITUDE	59.0	57.5	53.8	49.2	44.9	42.0	41.0	42.0	44.9	49.2	53.8	57.5
	LONGITUDE	90.0	81.6	76.8	76.2	79.0	84.0	90.0	96.0	101.0	103.8	103.2	98.4
2000 KM	LATITUDE	68.0	64.2	55.9	46.8	39.0	33.8	32.0	33.8	39.0	46.8	55.9	64.2
	LONGITUDE	90.0	69.2	61.5	63.2	69.9	79.3	90.0	100.7	110.1	116.8	118.5	110.8
3000 KM	LATITUDE	77.0	69.3	55.9	43.1	32.5	25.5	23.0	25.5	32.5	43.1	55.9	69.3
	LONGITUDE	90.0	50.2	45.4	51.6	62.2	75.4	90.0	104.6	117.8	128.4	134.6	129.8
4000 KM	LATITUDE	86.0	71.3	54.0	38.3	25.5	17.0	14.0	17.0	25.5	38.3	54.0	71.3
	LONGITUDE	90.0	24.0	30.1	41.5	55.7	72.1	90.0	107.9	124.3	138.5	149.9	156.0
4500 KM	LATITUDE	89.5	70.7	52.3	35.6	22.0	12.8	9.5	12.8	22.0	35.6	52.3	70.7
	LONGITUDE	270.0	10.2	23.2	37.0	52.7	70.6	90.0	109.4	127.3	143.0	156.8	169.8
5000 KM	LATITUDE	85.0	69.3	50.3	32.8	18.4	8.5	5.0	8.5	18.4	32.8	50.3	69.3
	LONGITUDE	270.0	-2.3	16.7	32.8	49.8	69.1	90.0	110.9	130.2	147.2	163.3	182.3
6000 KM	LATITUDE	76.0	64.3	45.3	26.8	11.0	0.0	-4.0	0.0	11.0	26.8	45.3	64.3
	LONGITUDE	270.0	-21.4	5.6	25.1	44.5	66.2	90.0	113.8	135.5	154.9	174.4	201.4
7000 KM	LATITUDE	67.0	57.6	39.4	20.4	3.6	-8.5	-13.0	-8.5	3.6	20.4	39.4	57.6
	LONGITUDE	270.00	-33.8	-3.6	18.2	39.4	63.2	90.0	116.8	140.6	161.8	183.6	213.8
8000 KM	LATITUDE	58.1	50.1	32.9	13.7	-3.9	-17.0	-21.9	-17.0	3.9	13.7	32.9	50.1
	LONGITUDE	270.0	-42.2	-11.3	11.8	34.4	60.2	90.0	119.8	145.6	168.2	191.3	222.2
9000 KM	LATITUDE	49.1	42.1	26.0	6.9	-11.3	-25.4	-30.9	-25.4	11.3	6.9	26.0	42.1
	LONGITUDE	270.0	-48.3	-17.9	5.9	29.3	56.9	90.0	123.1	150.7	174.1	197.9	228.3

AMSAT-OSCAR 13

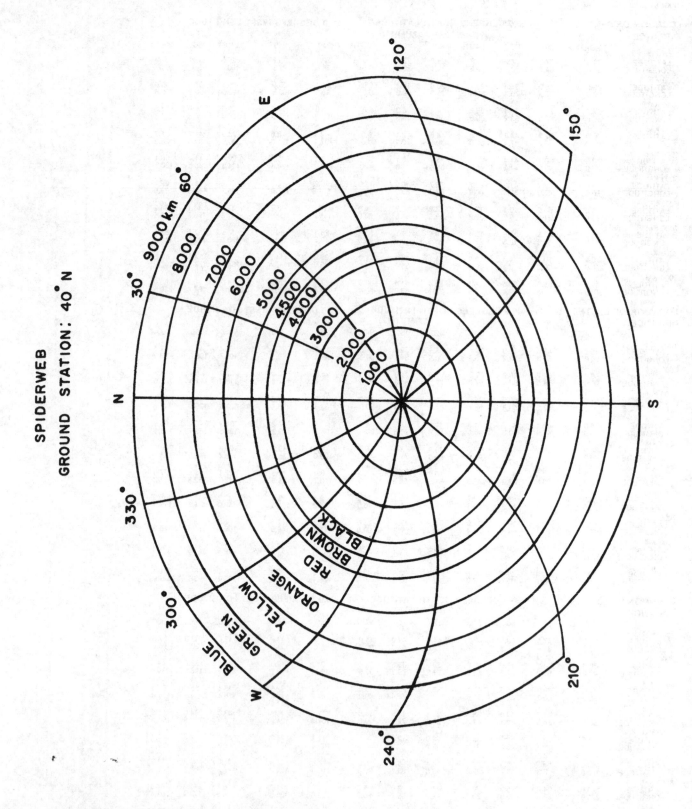

SPIDERWEB
GROUND STATION: 40° N

AMSAT-OSCAR 13

AMSAT-OSCAR 13;
Inclination = 57.05 deg;
eccentricity = 0.68;
arg perigee as shown;
time step = 15 min;
Mean motion = 2.097;

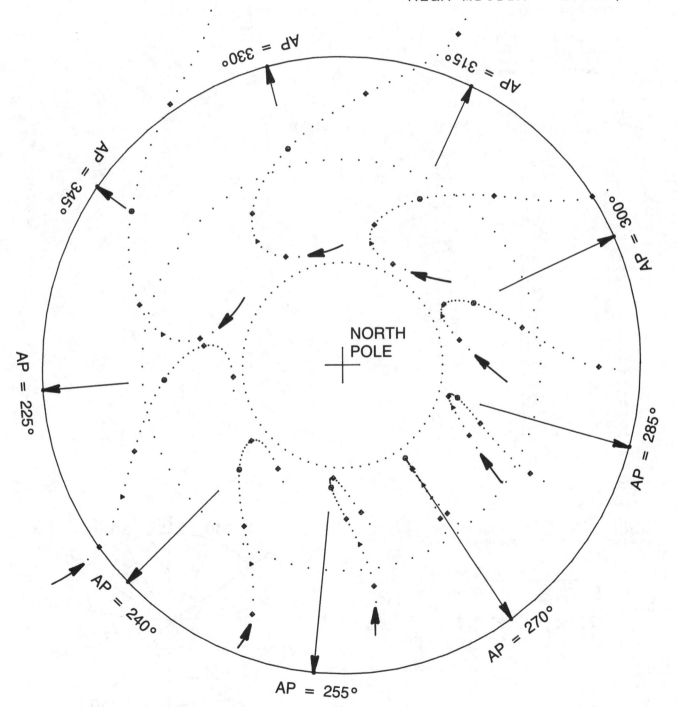

UoSat-OSCAR 11

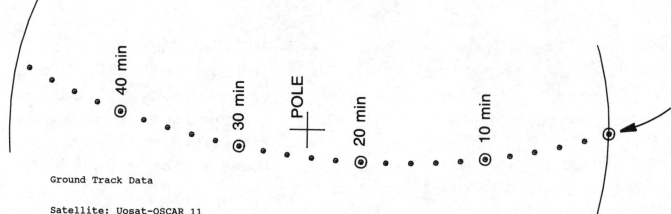

40 min   30 min   POLE   20 min   10 min

Ground Track Data

Satellite: Uosat-OSCAR 11
Mean altitude 680   km
Period: 98.3 Minutes
Inclination: 98 degrees

Time after ascending node (minutes)	Subsatellite point Lat. (N)	Long. (W)
0.0	0.0	0.0
2.0	7.3	1.5
4.0	14.5	3.1
6.0	21.7	4.7
8.0	29.0	6.5
10.0	36.2	8.4
12.0	43.4	10.6
14.0	50.6	13.3
16.0	57.7	16.8
18.0	64.7	21.8
20.0	71.5	29.8
22.0	77.7	45.5
24.0	81.7	81.2
24.6	82.0	96.1
26.0	80.5	129.8
28.0	75.2	155.0
30.0	68.6	166.4
32.0	61.7	172.8
34.0	54.7	177.1
36.0	47.5	180.2
38.0	40.4	182.6
40.0	33.1	184.7
42.0	25.9	186.6
44.0	18.7	188.3
46.0	11.4	189.9
48.0	4.2	191.4
49.2	0.0	192.3

SPIDERWEB DATA

Satellite: UoSAT-OSCAR 11
Satellite mean altitude: 680 km
Ground Station location: 30 North, 90 West

Azimuth	El = 0 deg. 2821. km		El = 30 deg. 946. km		El = 60 deg. 349. km	
	Latitude (N)	Longitude (W)				
0.	55.4	90.0	38.5	90.0	33.1	90.0
30.	50.6	70.3	37.3	84.7	32.7	88.1
60.	39.6	61.2	34.0	81.1	31.5	86.8
90.	26.9	61.3	29.6	80.2	30.0	86.4
120.	15.4	67.4	25.5	81.8	28.4	86.9
150.	7.5	77.5	22.6	85.4	27.3	88.2
180.	4.6	90.0	21.5	90.0	26.9	90.0
210.	7.5	102.5	22.6	94.6	27.3	91.8
240.	15.4	112.6	25.5	98.2	28.4	93.1
270.	26.9	118.7	29.6	99.8	30.0	93.6
300.	39.6	118.8	34.0	98.9	31.5	93.2
330.	50.6	109.7	37.3	95.3	32.7	91.9

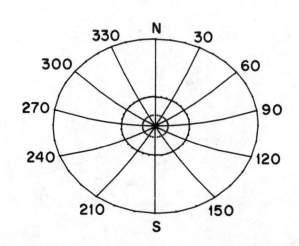

UoSat-OSCAR 11

SPIDERWEB DATA

Satellite: UoSAT-OSCAR 11
Satellite mean altitude: 680 km
Ground Station location: 46 North, 90 West

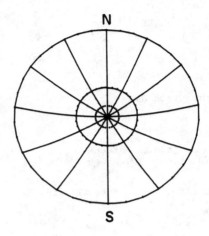

	Latitude (N) / Longitude (W)					
	El = 0 deg.		El = 30 deg.		El = 60 deg.	
	2821. km		946. km		349. km	
Azimuth						
0.	71.4	90.0	54.5	90.0	49.1	90.0
30.	65.2	59.3	53.2	82.9	48.7	87.6
60.	53.0	51.9	49.7	78.6	47.5	86.0
90.	40.5	55.7	45.4	77.8	45.9	85.5
120.	30.1	64.6	41.3	80.2	44.4	86.2
150.	23.1	76.5	38.5	84.6	43.3	87.8
180.	20.6	90.0	37.5	90.0	42.9	90.0
210.	23.1	103.5	38.5	95.4	43.3	92.2
240.	30.1	115.4	41.3	99.8	44.4	93.8
270.	40.5	124.3	45.4	102.2	45.9	94.5
300.	53.0	128.1	49.7	101.4	47.5	94.0
330.	65.2	120.7	53.2	97.1	48.7	92.4

PACSAT

Ground Track Data

Satellite: Pacsat
Mean altitude 800 km
Period: 100.8 Minutes
Inclination: 98.7 degrees

Time after ascending node (minutes)	Subsatellite point Lat. (N)	Long. (W)
0.0	0.0	0.0
2.0	7.1	1.6
4.0	14.1	3.2
6.0	21.2	4.9
8.0	28.2	6.7
10.0	35.2	8.7
12.0	42.2	11.0
14.0	49.2	13.7
16.0	56.1	17.2
18.0	62.9	21.9
20.0	69.6	29.2
22.0	75.7	42.3
24.0	80.3	69.6
25.2	81.3	96.3
26.0	80.8	114.8
28.0	76.8	146.4
30.0	70.8	161.4
32.0	64.3	169.5
34.0	57.5	174.6
36.0	50.6	178.3
38.0	43.6	181.1
40.0	36.6	183.5
42.0	29.6	185.5
44.0	22.6	187.4
46.0	15.5	189.1
48.0	8.5	190.7
50.0	1.4	192.5
50.4	0.0	192.6

SPIDERWEB DATA

Satellite: Pacsat (use for OSCARs 14-19)
Satellite mean altitude: 800 km
Ground Station location: 30 North, 90 West

	Latitude (N) / Longitude (W)					
	El = 0 deg. 3038. km		El = 30 deg. 1078. km		El = 60 deg. 403. km	
Azimuth						
0.	57.3	90.0	39.7	90.0	33.6	90.0
30.	52.0	68.1	38.3	83.8	33.1	87.8
60.	40.0	58.7	34.5	79.8	31.8	86.3
90.	26.4	59.2	29.5	78.8	29.9	85.8
120.	14.2	65.8	24.8	80.7	28.1	86.4
150.	5.7	76.7	21.5	84.8	26.8	88.0
180.	2.7	90.0	20.3	90.0	26.4	90.0
210.	5.7	103.3	21.5	95.2	26.8	92.0
240.	14.2	114.2	24.8	99.3	28.1	93.6
270.	26.4	120.8	29.5	101.2	29.9	94.2
300.	40.0	121.3	34.5	100.2	31.8	93.7
330.	52.0	111.9	38.3	96.2	33.1	92.2

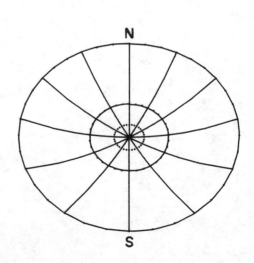

PACSAT

SPIDERWEB DATA

Satellite: Pacsat (use for OSCARs 14-19)
Satellite mean altitude: 800 km
Ground Station location: 46 North, 90 West

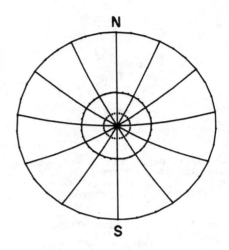

	Latitude (N) / Longitude (W)					
	El = 0 deg. 3038. km		El = 30 deg. 1078. km		El = 60 deg. 403. km	
Azimuth						
0.	73.3	90.0	55.7	90.0	49.6	90.0
30.	66.2	55.3	54.1	81.7	49.1	87.2
60.	53.0	48.7	50.1	76.8	47.7	85.3
90.	39.7	53.4	45.2	76.2	45.9	84.8
120.	28.7	63.1	40.6	78.9	44.1	85.6
150.	21.3	75.7	37.4	83.9	42.8	87.5
180.	18.7	90.0	36.3	90.0	42.4	90.0
210.	21.3	104.3	37.4	96.1	42.8	92.5
240.	28.7	116.9	40.6	101.1	44.1	94.4
270.	39.7	126.6	45.2	103.8	45.9	95.2
300.	53.0	131.3	50.1	103.2	47.7	94.7
330.	66.2	124.7	54.1	98.3	49.1	92.8

Fuji-OSCAR 20

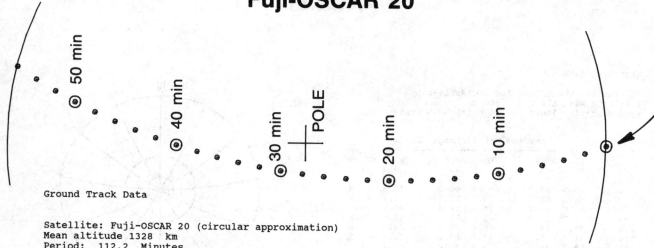

Ground Track Data

Satellite: Fuji-OSCAR 20 (circular approximation)
Mean altitude 1328 km
Period: 112.2 Minutes
Inclination: 99.05 degrees

Time after ascending node (minutes)	Subsatellite point Lat. (N)	Long. (W)
0.0	0.0	0.0
2.0	6.3	1.5
4.0	12.7	3.1
6.0	19.0	4.6
8.0	25.3	6.3
10.0	31.6	8.1
12.0	37.9	10.1
14.0	44.2	12.4
16.0	50.5	15.1
18.0	56.6	18.5
20.0	62.7	23.0
22.0	68.7	29.6
24.0	74.2	40.3
26.0	78.8	60.3
28.0	80.9	96.0
28.1	81.0	97.0
30.0	79.0	132.4
32.0	74.5	153.0
34.0	68.9	164.1
36.0	63.0	170.8
38.0	56.9	175.3
40.0	50.8	178.8
42.0	44.5	181.5
44.0	38.3	183.8
46.0	32.0	185.8
48.0	25.6	187.6
50.0	19.3	189.3
52.0	13.0	190.9
54.0	6.7	192.4
56.0	0.3	194.0
56.1	0.0	194.0

SPIDERWEB DATA

Satellite: Fuji-OSCAR 20 (at apogee)
Satellite altitude: 1745 km
Ground Station location: 30 North, 90 West

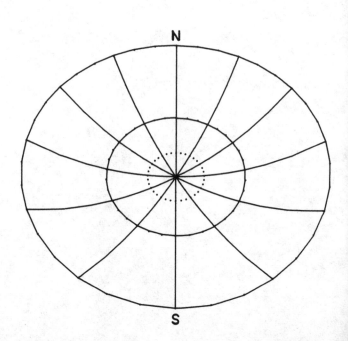

	Latitude (N) / Longitude (W)					
Azimuth	El = 0 deg. 4257. km		El = 30 deg. 1909. km		El = 60 deg. 766. km	
0.	68.3	90.0	47.2	90.0	36.9	90.0
30.	59.0	53.0	44.4	78.1	35.9	85.8
60.	41.4	44.4	37.3	71.3	33.3	82.9
90.	23.1	47.7	28.5	70.4	29.8	82.1
120.	7.1	57.3	20.5	74.2	26.4	83.3
150.	-4.1	71.9	14.9	81.2	24.0	86.2
180.	-8.3	90.0	12.8	90.0	23.1	90.0
210.	-4.1	108.1	14.9	98.8	24.0	93.8
240.	7.1	122.7	20.5	105.8	26.4	96.7
270.	23.1	132.3	28.5	109.6	29.8	97.9
300.	41.4	135.6	37.3	108.7	33.3	97.1
330.	59.0	127.0	44.4	101.9	35.9	94.2

Fuji-OSCAR 20

Satellite: Fuji-OSCAR 20 (at perigee)
Satellite altitude: 912 km
Ground Station location: 30 North, 90 West

	El = 0 deg.		El = 30 deg.		El = 60 deg.	
	3223. km		1195. km		452. km	
Azimuth						
0.	59.0	90.0	40.7	90.0	34.1	90.0
30.	53.2	66.1	39.1	83.1	33.5	87.6
60.	40.3	56.6	34.9	78.6	32.0	85.9
90.	25.9	57.4	29.4	77.6	29.9	85.3
120.	13.2	64.5	24.2	79.8	27.9	86.0
150.	4.2	75.9	20.6	84.3	26.5	87.7
180.	1.0	90.0	19.3	90.0	25.9	90.0
210.	4.2	104.1	20.6	95.7	26.5	92.3
240.	13.2	115.5	24.2	100.2	27.9	94.0
270.	25.9	122.6	29.4	102.4	29.9	94.7
300.	40.3	123.4	34.9	101.4	32.0	94.1
330.	53.2	113.9	39.1	96.9	33.5	92.4

The table above is headed "Latitude (N) / Longitude (W)".

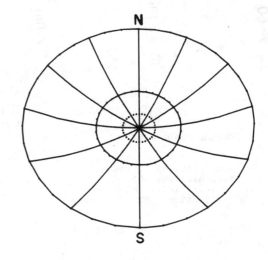

Satellite: Fuji-OSCAR 20 (at apogee)
Satellite altitude: 1745 km
Ground Station location: 46 North, 90 West

Latitude (N) / Longitude (W)

	El = 0 deg.		El = 30 deg.		El = 60 deg.	
	4257. km		1909. km		766. km	
Azimuth						
0.	84.3	90.0	63.2	90.0	52.9	90.0
30.	69.6	27.2	59.9	72.9	51.8	84.4
60.	51.2	31.0	52.2	65.4	49.1	80.9
90.	34.4	41.4	43.4	66.0	45.6	80.1
120.	20.5	55.1	35.8	71.6	42.3	81.9
150.	11.1	71.6	30.6	80.1	39.9	85.5
180.	7.7	90.0	28.8	90.0	39.1	90.0
210.	11.1	108.4	30.6	99.9	39.9	94.5
240.	20.5	124.9	35.8	108.4	42.3	98.1
270.	34.4	138.6	43.4	114.0	45.6	99.9
300.	51.2	149.0	52.2	114.6	49.1	99.1
330.	69.6	152.8	59.9	107.1	51.8	95.6

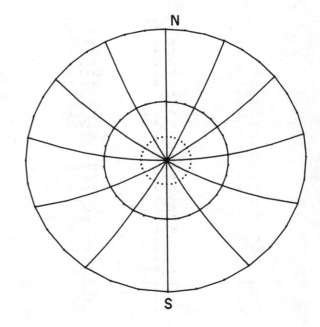

Satellite: Fuji-OSCAR 20 (at perigee)
Satellite altitude: 912 km
Ground Station location: 46 North, 90 West

Latitude (N) / Longitude (W)

	El = 0 deg.		El = 30 deg.		El = 60 deg.	
	3223. km		1195. km		452. km	
Azimuth						
0.	75.0	90.0	56.7	90.0	50.1	90.0
30.	67.0	51.6	55.0	80.6	49.5	86.9
60.	52.9	45.9	50.5	75.3	47.9	84.7
90.	39.0	51.4	45.0	74.7	45.9	84.2
120.	27.4	61.8	39.9	77.8	43.9	85.1
150.	19.7	75.1	36.5	83.3	42.4	87.2
180.	17.0	90.0	35.3	90.0	41.9	90.0
210.	19.7	104.9	36.5	96.7	42.4	92.8
240.	27.4	118.2	39.9	102.2	43.9	94.9
270.	39.0	128.6	45.0	105.3	45.9	95.8
300.	52.9	134.1	50.5	104.7	47.9	95.3
330.	67.0	128.4	55.0	99.4	49.5	93.1

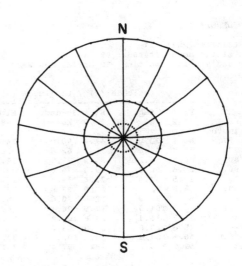

RS-10/11

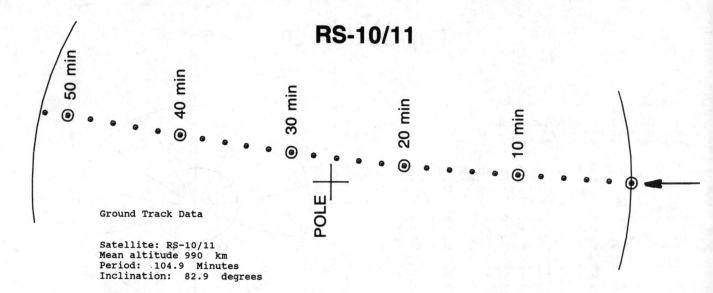

Ground Track Data

Satellite: RS-10/11
Mean altitude 990 km
Period: 104.9 Minutes
Inclination: 82.9 degrees

Time after ascending node (minutes)	Subsatellite point Lat. (N)	Long. (W)
0.0	0.0	0.0
2.0	6.8	359.6
4.0	13.6	359.3
6.0	20.4	358.8
8.0	27.2	358.3
10.0	34.0	357.7
12.0	40.8	356.8
14.0	47.6	355.7
16.0	54.3	354.0
18.0	61.0	351.5
20.0	67.5	347.5
22.0	73.9	340.0
24.0	79.6	323.3
26.0	82.9	282.7
26.2	82.9	276.6
28.0	80.7	236.2
30.0	75.3	215.7
32.0	69.0	206.9
34.0	62.5	202.3
36.0	55.8	199.6
38.0	49.1	197.8
40.0	42.3	196.5
42.0	35.5	195.6
44.0	28.8	194.9
46.0	22.0	194.4
48.0	15.2	193.9
50.0	8.3	193.5
52.0	1.5	193.0
52.5	0.0	193.1

SPIDERWEB DATA

Satellite: RS-10/11
Satellite mean altitude: 990 km
Ground Station location: 30 North, 90 West

	Latitude (N) / Longitude (W)					
	El = 0 deg. 3342. km		El = 30 deg. 1273. km		El = 60 deg. 485. km	
Azimuth						
0.	60.1	90.0	41.4	90.0	34.4	90.0
30.	53.9	64.8	39.7	82.6	33.7	87.4
60.	40.5	55.2	35.2	77.9	32.1	85.5
90.	25.6	56.2	29.3	76.8	29.9	85.0
120.	12.5	63.6	23.8	79.2	27.8	85.7
150.	3.3	75.5	19.9	83.9	26.2	87.6
180.	-0.1	90.0	18.6	90.0	25.6	90.0
210.	3.3	104.5	19.9	96.1	26.2	92.4
240.	12.5	116.4	23.8	100.8	27.8	94.3
270.	25.6	123.8	29.3	103.2	29.9	95.0
300.	40.5	124.8	35.2	102.1	32.1	94.5
330.	53.9	115.2	39.7	97.4	33.7	92.6

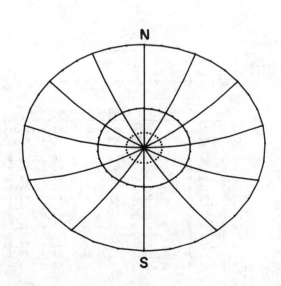

RS-10/11

SPIDERWEB DATA

Satellite: RS-10/11
Satellite mean altitude: 990 km
Ground Station location: 46 North, 90 West

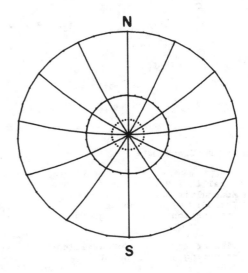

	Latitude (N) / Longitude (W)					
	El = 0 deg.		El = 30 deg.		El = 60 deg.	
	3342. km		1273. km		485. km	
Azimuth						
0.	76.1	90.0	57.4	90.0	50.4	90.0
30.	67.5	49.1	55.5	79.9	49.7	86.6
60.	52.8	44.1	50.7	74.2	48.0	84.4
90.	38.5	50.2	44.8	73.7	45.8	83.7
120.	26.7	61.0	39.5	77.1	43.7	84.8
150.	18.7	74.7	35.8	83.0	42.2	87.1
180.	15.9	90.0	34.6	90.0	41.6	90.0
210.	18.7	105.3	35.8	97.0	42.2	92.9
240.	26.7	119.0	39.5	102.9	43.7	95.2
270.	38.5	129.8	44.8	106.3	45.8	96.3
300.	52.8	135.9	50.7	105.8	48.0	95.6
330.	67.5	130.9	55.5	100.1	49.7	93.4

MIR

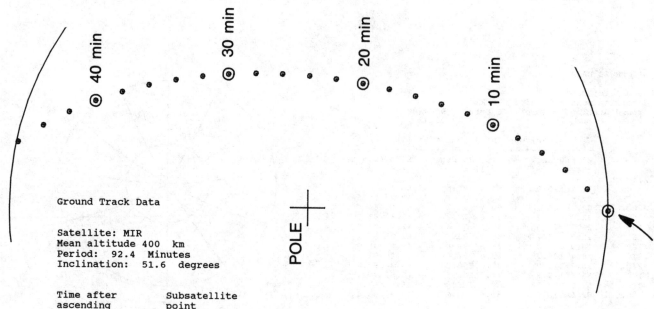

Ground Track Data

Satellite: MIR
Mean altitude 400 km
Period: 92.4 Minutes
Inclination: 51.6 degrees

Time after ascending node (minutes)	Subsatellite point Lat. (N)	Long. (W)
0.0	0.0	0.0
2.0	6.1	355.6
4.0	12.2	351.2
6.0	18.1	346.5
8.0	23.9	341.4
10.0	29.5	335.8
12.0	34.8	329.6
14.0	39.7	322.4
16.0	44.0	314.2
18.0	47.5	304.7
20.0	50.0	294.0
22.0	51.4	282.4
23.1	51.6	275.8
24.0	51.5	270.4
26.0	50.2	258.7
28.0	47.8	247.9
30.0	44.3	238.3
32.0	40.1	229.9
34.0	35.3	222.7
36.0	30.1	216.3
38.0	24.5	210.7
40.0	18.7	205.6
42.0	12.8	200.8
44.0	6.7	196.3
46.0	0.6	192.0
46.2	0.0	191.6

SPIDERWEB DATA

Satellite: MIR
Satellite mean altitude: 400 km
Ground Station location: 30 North, 90 West

Azimuth	El = 0 deg. 2201. km Lat.(N)	Long.(W)	El = 30 deg. 603. km Lat.(N)	Long.(W)	El = 60 deg. 215. km Lat.(N)	Long.(W)
0.	49.8	90.0	35.4	90.0	31.9	90.0
30.	46.4	75.8	34.7	86.7	31.7	88.9
60.	38.1	68.1	32.6	84.4	31.0	88.0
90.	28.1	67.4	29.9	83.7	30.0	87.8
120.	18.9	71.9	27.2	84.7	29.0	88.1
150.	12.5	80.0	25.3	87.0	28.3	88.9
180.	10.2	90.0	24.6	90.0	28.1	90.0
210.	12.5	100.0	25.3	93.0	28.3	91.1
240.	18.9	108.1	27.2	95.3	29.0	91.9
270.	28.1	112.6	29.9	96.3	30.0	92.2
300.	38.1	111.9	32.6	95.6	31.0	92.0
330.	46.4	104.2	34.7	93.3	31.7	91.1

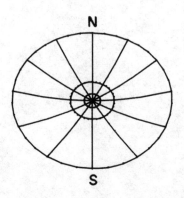

MIR

SPIDERWEB DATA

Satellite: MIR
Satellite mean altitude: 400 km
Ground Station location: 46 North, 90 West

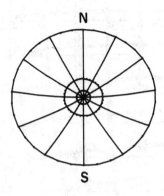

```
                  Latitude (N) / Longitude (W)
         -----------------------------------------------
         El = 0 deg.      El = 30 deg.     El = 60 deg.
         2201. km           603. km          215. km
Azimuth
    0.   65.8  90.0       51.4  90.0       47.9  90.0
   30.   61.7  69.1       50.6  85.7       47.7  88.6
   60.   52.6  61.1       48.5  82.9       46.9  87.5
   90.   42.6  62.6       45.7  82.2       46.0  87.2
  120.   34.0  69.3       43.1  83.6       45.0  87.6
  150.   28.2  78.9       41.2  86.4       44.3  88.6
  180.   26.2  90.0       40.6  90.0       44.1  90.0
  210.   28.2 101.1       41.2  93.6       44.3  91.4
  240.   34.0 110.7       43.1  96.4       45.0  92.4
  270.   42.6 117.4       45.7  97.8       46.0  92.8
  300.   52.6 118.9       48.5  97.1       46.9  92.5
  330.   61.7 110.9       50.6  94.3       47.7  91.4
```

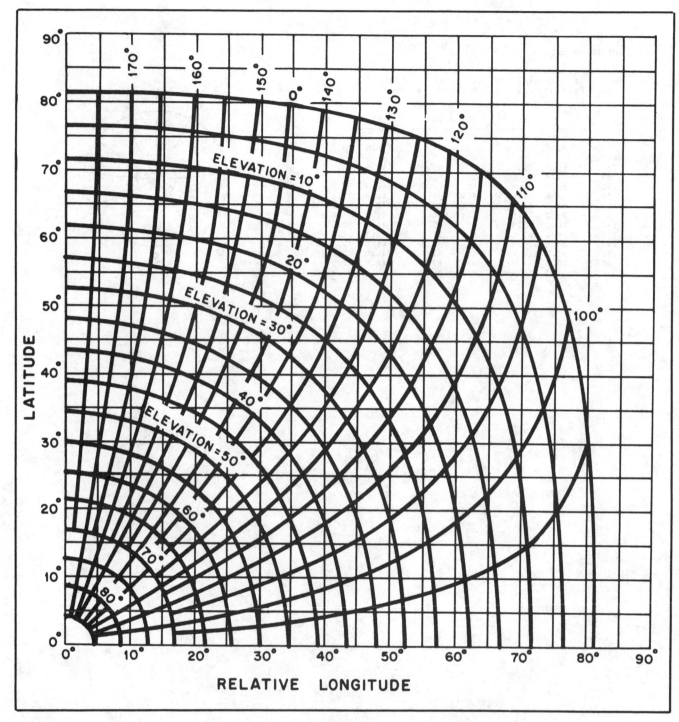

Chart for obtaining azimuth and elevation directions from a ground station to a geostationary satellite. See Chapter 6 for instructions.

Appendix D

Computer Programs

1. Distance and Bearing (DISTBEAR)
2. Spiderweb construction data (SWEB)
3. Ground Track data for circular orbit (GTCIRC)
4. Ground Track data for elliptical orbit (GTEL)

This appendix contains four BASIC programs which illustrate several fundamental problems related to satellite tracking. The programs were designed to be used by those having at least a modest familiarity with the BASIC language. My objective was to construct small modules which present the underlying computations as clearly as possible. If speed or space conflicted with clarity, the latter won. As a result, the algorithms selected may be slow and/or use more memory than is really necessary. Descriptive variable names are used throughout and remark statements are placed at key points. Remarks are also used to cross-reference specific program lines to equations in the text. The algorithms used in the spiderweb program were first described in the Amateur Radio literature by Dimitrios Zachariadis (NJØW), in an article in Feb 1986 *QST*, "Spiderweb—The Range Circle Calculation," pp 36-38.

All programs are written in a limited subset of Microsoft BASIC, a language which runs on IBM and IBM-compatible microcomputers, as well as a number of other popular microcomputers. This language was selected because it's the most widely available version of BASIC, and because nearly everyone familiar with BASIC can trans-late from it to other dialects. It is certainly not the preferred language for this type of work. The coding was restricted to a subset of the available commands—those which appear in most other versions of BASIC. As a result, translating to other BASICs should be very straightforward.

The programs presented here will probably be most useful to (1) those interested in modifying them for special applications, (2) those trying to understand the related mathematics presented in earlier chapters and (3) those looking for little tricks involved in the various calculations. For example, the program DISTBEAR can be easily modified to provide antenna-aiming information (azimuth and elevation) for a ground station at P1 and a subsatellite point at P2. Note that changing any of the programs (satellite orbit characteristics, ground station location, step sizes for the various parameters involved) requires modifying the lines marked by a double asterisk (**). A triple asterisk (***) marks lines which you may also wish to modify. Sample output is provided for each program to help you confirm that it is operating correctly.

All four programs use a spherical earth model and suffer from the inaccuracies this causes. The ground track data for circular and elliptical orbits produced by GTCIRC and GTEL is fine for producing overlays for an *OSCAR-LOCATOR*. Long-term predictions require corrections for sidereal time; GTEL requires adjustments for nodal period and changes in argument of perigee.

```
list
10    '              *** DISTBEAR (April 8, 1990) ***
20    '  This program provides terrestrial distance and bearing
30    '  from point 1 (P1) to point 2 (P2).
40    '  Source: The Satellite Experimenter's Handbook 2/e, Appendix D
50    '
60   P1$ = "Baltimore " : LATITUDE1 = 39.3 : LONGITUDE1 = -76.6 'deg. E **
70   P2$ = "Moscow    " : LATITUDE2 = 56.1 : LONGITUDE2 =  37.5 'deg. E **
80   PRINT : PRINT : PRINT "Program DISTBEAR"
90   PRINT : PRINT "From "; P1$; " to "; P2$
100   '
110  PI = 3.141593 : RTOD = 180/PI : DTOR = PI/180  : KTOM = .6214
120   '                Radians to deg.: Deg. to radians: km to miles
130   '
140  RE = 6371  'Mean Earth radius (km)
150  DEF FNARCCOS(Z) = PI/2 - ATN(Z/SQR(1-Z*Z))
160  LAT1 = LATITUDE1*DTOR : LNG1 = LONGITUDE1*DTOR
170  LAT2 = LATITUDE2*DTOR : LNG2 = LONGITUDE2*DTOR
180  IF ABS(LATITUDE1) > 89.99 THEN GOTO 410   'P1 is North or South Pole
190  COSBETA = SIN(LAT1)*SIN(LAT2)+COS(LAT1)*COS(LAT2)*COS(LNG2-LNG1) 'Eq. 12.2
200  IF COSBETA >  .999999 THEN PRINT "points coincide" : GOTO 480
210  IF COSBETA < -.999999 THEN GOTO 380   'Antipodes
220  BETA = FNARCCOS(COSBETA)
230  DISTSP = BETA*RE                ' Distance Short Path
240  DISTLP = 2*PI*RE - DISTSP  ' Distance Long Path
250  COSAZ = (SIN(LAT2) - SIN(LAT1)*COS(BETA))/(COS(LAT1)*SIN(BETA))  'Eq. 12.3
260  IF COSAZ >  .999999 THEN AZ = 0  : GOTO 290
270  IF COSAZ < -.999999 THEN AZ = 180 : GOTO 290
280  AZ = FNARCCOS(COSAZ) * RTOD
290  IF SIN(LNG2-LNG1) >= 0 THEN AZSP = AZ : AZLP = 180+AZ
300  IF SIN(LNG2-LNG1) <  0 THEN AZSP = 360 - AZ : AZLP = 180-AZ
310  AZLP = INT(AZLP*10 + .5)/10 : AZSP = INT(AZSP*10 + .5)/10 'round off
320  DSP = INT(DISTSP + .5) : DLP = INT(DISTLP + .5)          'round off
330  DSPMI = INT(DISTSP*KTOM+.5) : DLPMI = INT(DISTLP*KTOM+.5) ' km --> miles
340  PRINT "   Short Path:";DSP; "km ("; DSPMI; "mi.)", AZSP; "Deg. E of N"
350  PRINT "   Long Path:" ;DLP; "km ("; DLPMI; "mi.)", AZLP; "Deg. E of N"
360  GOTO 480
370   '
380  PRINT "Antipodes, azimuth not defined, distance ="; INT(RE*PI); "km"
390  GOTO 480
400   '
410  PRINT "If P1 is North or South Pole then azimuth is not defined."
420  PRINT "Point antenna along P2 longitude (short path) or"
430  PRINT "P2 longitude + 180 degrees (long path).
440  IF SGN(LAT1) =  SGN(LAT2) THEN DSP = RE*(PI/2 - ABS(LAT2))
450  IF SGN(LAT1) <> SGN(LAT2) THEN DSP = RE*(PI/2 + ABS(LAT2))
460  PRINT "   Short path:"; INT(DSP); "km"
470  PRINT "   Long path:"; INT(2*PI*RE - DSP); "km"
480  PRINT
490  END  'DISTBEAR
Ok




run


Program DISTBEAR

From Baltimore  to Moscow
    Short Path: 7733 km ( 4805 mi.)        32.9 Deg. E of N
    Long Path: 32297 km ( 20069 mi.)       212.9 Deg. E of N

Ok
```

```
list
10     '                *** SWEB (April 8, 1990) ***
20     ' This program provides spiderweb data around station located
30     ' at P1.  Azimuth and elevation increments are initially set
40     ' to 30 degrees.  (Step sizes are easily modified).
50     ' Source: The Satellite Experimenter's Handbook 2/e, Appendix D
60     '
70     LATITUDE =     46  ' ground station latitude in degrees N     **
80     LONGITUDE =    90  ' ground station longitude in degrees W    **
90     SAT$ = "Fuji-OSCAR 12"  '                                     **
100    H = 1490     'satellite height in km                          **
110    PRINT :   PRINT : PRINT "Spiderweb Data (Program SWEB)" : PRINT
120    PRINT "    Ground Station Location"
130    PRINT "       latitude:"; LATITUDE; "degrees N"
140    PRINT "       longitude:"; LONGITUDE; "degrees W"
150    PRINT : PRINT : PRINT "    Satellite name:"; SAT$
160    PRINT "       Height:"; H; "km"
170    PI = 3.141593 : RTOD = 180/PI  : DTOR = PI/180
180    '               radians to deg. : deg. to radians
190    RE = 6371  'Mean radius of earth
200    DEF FNARCSIN(Z) = ATN(Z/SQR(1-Z*Z))
210    DEF FNARCCOS(Z) = PI/2 - ATN(Z/SQR(1-Z*Z))
220    IF ABS(LATITUDE)> 89.9 THEN LATITUDE = SGN(LATITUDE)*89.9
230    LAT1 = LATITUDE*DTOR : LNG1 = LONGITUDE*DTOR
240    NE = 3  'elevation circle step size (in degrees) = 90/NE      ***
250    NAZ = 12 'azimuth step size (in degrees) = 360/NAZ            ***
260    FOR K = 0 TO NE-1
270       ELEV = K*90/NE
280       E = ELEV*DTOR
290       B(K) = FNARCCOS(COS(E)*RE/(RE+H)) - E            ' See Eq. 12.7
300       PRINT "      The"; ELEV;
310       PRINT "degree elevation circle occurs at a distance of";
320       PRINT INT(RE*B(K)+.5);"km"
330    NEXT K
340    PRINT : PRINT   ' Lines 340-410 set up heading for spiderweb data.
350    PRINT "                ";
360    FOR M=0 TO NE-1:PRINT USING "##"; M*90/NE;:PRINT " degrees      ";:NEXT M
370    PRINT : PRINT "Azimuth      ";
380    FOR M=0 TO NE-1 : PRINT "Lat./Long.      "; : NEXT M
390    PRINT : PRINT "              ";
400    FOR M=0 TO NE-1 : PRINT "----------      "; : NEXT M
410    PRINT
420    FOR I = 0 TO NAZ-1  'Loop to calculate spiderweb data
430       AZ = (I*360/NAZ)*DTOR
440       FOR J = 0 TO NE-1
450          SINLAT2 = SIN(LAT1)*COS(B(J))+COS(LAT1)*SIN(B(J))*COS(AZ)  'Eq. 12.10
460          IF SINLAT2 >  .999999 THEN LAT2 =  PI/2 - .00001 : GOTO 490
470          IF SINLAT2 < -.999999 THEN LAT2 = -PI/2 + .00001 : GOTO 490
480          LAT2 = FNARCSIN(SINLAT2)
490          IF ABS(LAT2)>PI/2-.001 THEN LAT2=SGN(LAT2)*(PI/2 -.001)
500          X=(COS(B(J))-SIN(LAT1)*SIN(LAT2))/(COS(LAT1)*COS(LAT2)) 'see Eq. 12.11
510          IF (AZ=0  AND B(J) > (PI/2-LAT1)) THEN LNG2 = LNG1+PI : GOTO 550
520          IF (AZ=PI AND B(J) > (PI/2+LAT1)) THEN LNG2 = LNG1+PI : GOTO 550
530          IF ABS(X)>.99999 THEN LNG2= LNG1 :                      GOTO 550
540          LNG2 = LNG1 - SGN(PI-AZ)*FNARCCOS(X)
550          IF LNG2<0 THEN LNG2=LNG2+2*PI
560          IF LNG2>2*PI THEN LNG2 = LNG2-2*PI
570          IF J = 0 THEN PRINT USING "####.#"; AZ*RTOD;
580          PRINT " "; : PRINT USING "####.#"; LAT2*RTOD, LNG2*RTOD;
590       NEXT J
600    PRINT
610    NEXT I
620    PRINT
630    END 'sweb
Ok
```

Spiderweb Data (Program SWEB)

Ground Station Location
    latitude: 46 degrees N
    longitude: 90 degrees W

Satellite name:Fuji-OSCAR 12
    Height: 1490 km
    The 0 degree elevation circle occurs at a distance of 3987 km
    The 30 degree elevation circle occurs at a distance of 1715 km
    The 60 degree elevation circle occurs at a distance of 678 km

Azimuth	0 degrees Lat./Long.		30 degrees Lat./Long.		60 degrees Lat./Long.	
0.0	81.9	90.0	61.4	90.0	52.1	90.0
30.0	69.3	34.1	58.6	75.2	51.2	85.1
60.0	51.9	34.8	51.8	68.1	48.8	82.0
90.0	35.7	43.9	43.9	68.3	45.7	81.3
120.0	22.3	56.7	36.9	73.3	42.7	82.8
150.0	13.3	72.5	32.2	81.0	40.6	86.0
180.0	10.1	90.0	30.6	90.0	39.9	90.0
210.0	13.3	107.5	32.2	99.0	40.6	94.0
240.0	22.3	123.3	36.9	106.7	42.7	97.2
270.0	35.7	136.1	43.9	111.7	45.7	98.7
300.0	51.9	145.2	51.8	111.9	48.8	98.0
330.0	69.3	145.9	58.6	104.8	51.2	94.9

```
list
10     '               *** GTCIRC (April 8, 1990) ***
20     ' This program provides ground track data for one revolution of
30     ' a satellite in a circular orbit.  Starting at ascending node,
40     ' latitude and longitude of the subsatellite point are given at
50     ' the specified timestep.
60     ' Source: The Satellite Experimenter's Handbook 2/e, Appendix D
70     '
80     SAT$ = "Pacsat"        '                           **
90     MM = 14.28604          'Mean Motion (rev./day)     **
100    INCLINATION = 98.7 'degrees                        **
110    TIMESTEP = 4           'minutes                    **
120    '
130    PERIOD = 1440/MM    'minutes (anomalistic period)
135    ' Replace anomalistic period with nodal period if available
140    HEIGHT = 331.25 * PERIOD^(2/3) - 6371      '(km)
150    PI = 3.14159 : RTOD = 180/PI : DTOR = PI/180
160    INCL = INCLINATION * DTOR
170    DEF FNASIN(Z) = ATN(Z/SQR(1-Z*Z))
180    DEF FNACOS(Z) = PI/2 - ATN(Z/SQR(1-Z*Z))
190    PRINT : PRINT : PRINT
200    PRINT "Ground Track Data (Program GTCIRC)" : PRINT : PRINT
210    PRINT "Satellite: "; SAT$ : PRINT
220    PRINT "    Mean Motion: ";MM ; " rev./day"
230    PRINT "    Mean altitude:"; INT(HEIGHT*10+.5)/10; " km"   'round off
240    PRINT "    Period: "; PERIOD; " Minutes"
250    PRINT "    Inclination: "; INCLINATION; " degrees"
260    PRINT : PRINT
270    PRINT "Time after    Subsatellite point"
280    PRINT "ascending     ------------------"
290    PRINT "node          Lat.        Long."
300    PRINT "(minutes)     (N)         (W)"
310    PRINT
320    S1 = 1 : IF INCLINATION > 90 THEN S1 = -1
330    FOR T = 0 TO PERIOD STEP TIMESTEP
340      X = SIN(INCL)*SIN(2*PI*T/PERIOD)
350      IF X*X > .9999 THEN LAT = SGN(X)*PI/2 : GOTO 370
360      LAT = FNASIN(X)
370      S2 = 1 : IF T/PERIOD > .5 THEN S2 = -1
380      Y = COS(2*PI*T/PERIOD)/COS(LAT)
390      IF Y >  .99999 THEN LNG = 0  : GOTO 420
400      IF Y <- .99999 THEN LNG = PI : GOTO 420
410      LNG = S1*S2*FNACOS(Y)
420      LNG = LNG*RTOD  : LAT = LAT*RTOD
430      LNG = LNG - T/4  ' Take rotation of earth into account
440      IF LNG <  0 THEN LNG = 360 + LNG : GOTO 440
450      IF LNG <> 0 THEN LNG = 360 - LNG ' Switch to degrees W
460      PRINT USING "###.#"; T,
470      PRINT USING "#########.#" ;LAT; LNG
480    NEXT T
490    PRINT : PRINT : END 'GTCIRC
Ok
```

Ground Track Data (Program GTCIRC)

Satellite: Pacsat

    Mean Motion: 14.28604 rev./day
    Mean altitude: 803.5 km
    Period: 100.7977 Minutes
    Inclination: 98.7 degrees

Time after ascending node (minutes)	Subsatellite point	
	Lat. (N)	Long. (W)
0.0	0.0	0.0
4.0	14.1	3.2
8.0	28.2	6.7
12.0	42.2	11.0
16.0	56.1	17.2
20.0	69.6	29.2
24.0	80.3	69.7
28.0	76.8	146.4
32.0	64.3	169.5
36.0	50.6	178.3
40.0	36.6	183.5
44.0	22.6	187.4
48.0	8.5	190.7
52.0	-5.7	193.9
56.0	-19.8	197.2
60.0	-33.8	200.9
64.0	-47.8	205.7
68.0	-61.6	213.4
72.0	-74.5	231.5
76.0	-81.2	298.4
80.0	-72.1	351.7
84.0	-58.9	6.3
88.0	-45.0	13.2
92.0	-31.0	17.7
96.0	-16.9	21.3
100.0	-2.8	24.6

```
list
10       '              *** GTEL (April 8, 1990) ***
20     ' This program provides ground track data for one revolution of
30     ' a satellite in an elliptical orbit.  Latitude and longitude of
40     ' the subsatellite point are provided at the specified timestep.
50     ' Source: Satellite Experimenter's Handbook 2/e, Appendix D
60     ' Items marked *** are optional inputs.
70     '
80     MM = 2.10075 'Orbits per day  (Mean Motion)    ** see note line 100
90     SMA = 0  'km  (Semi-major axis)                ** see note line 100
100    '  Note: Either Mean Motion or Semi-major axis should be set to zero
110    '          to indicate that it is being treated as an unknown.
120    ETY = .65685  'Eccentricity                    **
130    INCLINATION =  57.05    'degrees               **
140    ARGUMENTPERIGEE = 225  'degrees                **
150    SAT$ = "AMSAT-OSCAR 13"  '                     ***
160    TIMESTEP = 20  'minutes                        ***
170    SETLNG = 85.7  'degrees East (Horizontal shift) ***
180    '   (typical values are zero, or longitude at perigee, etc.)
190    '
200    PI = 3.141593  :  RTOD = 180/PI  :  DTOR = PI/180
210    INCLIN = INCLINATION*DTOR  :  ARGPER = ARGUMENTPERIGEE*DTOR
220    RE = 6371  'km  Mean radius of earth
230    IF MM = 0 THEN MM = 1440/(1.6587E-04 * SMA^(3/2))
240    IF SMA = 0 THEN SMA = 331.25*(1440/MM)^(2/3)
250    PERIOD = 1440/MM    'anomalistic period in minutes
260    DEF FNARCSIN(Z) = ATN(Z/SQR(1-Z*Z))
270    DEF FNARCCOS(Z) = PI/2 - ATN(Z/SQR(1-Z*Z))
280    '
290    PRINT : PRINT
```

```
300  PRINT "Ground Track Data (Program GTEL)" : PRINT : PRINT
310  PRINT "Satellite: "; SAT$ : PRINT
320  PRINT "    Mean Motion: "; MM ; "rev./day"
330  PRINT "    Semimajor axis: "; SMA; "km"
340  PRINT "    Inclination: "; INCLINATION ; "degrees"
350  PRINT "    Eccentricity: "; ETY
360  PRINT "    Argument of Perigee: "; ARGUMENTPERIGEE; "degrees"
370  PRINT : PRINT
380  PRINT "Time after        Subsatellite point"
390  PRINT "apogee          --------------------"
400  PRINT "(minutes)       Lat. (N)    Long. (E)"
410  PRINT
420  '
430  ' lines 450-490:  Solve spherical triangle involving perigee and
440  '            ascending node on non-rotating earth as per SEH
450  SINLATPER = SIN(ARGPER) * SIN(INCLIN)
460  LATPER = FNARCSIN(SINLATPER)
470  COSD1 = COS(ARGPER)/COS(LATPER)
480  D1 = FNARCCOS(COSD1)
490  IF  ARGPER <= PI THEN D1 = - D1
500  '
510  TIME = 0 : GOSUB 600   'Go to subroutine to calculate subsatellie point
520  STARTTIME = PERIOD/2
530  IF STARTTIME >= TIMESTEP THEN STARTTIME = STARTTIME-TIMESTEP : GOTO 530
540  FOR TIME = STARTTIME TO PERIOD STEP TIMESTEP
550    GOSUB 600
560  NEXT TIME
570  TIME =  PERIOD : GOSUB 600
580  GOTO 850 'Go to end of program
590  '
600  'Subroutine to plot subsatellite point (SSP) for given value of time
610    EAINITIAL = 2*PI*TIME/PERIOD     'Eccentric anomaly (initial guess)
620    EA = EAINITIAL
630    FOR I = 1 TO 20 'Loop to find eccentric anomaly
640      DIFFERENCE = (EA - EAINITIAL - ETY*SIN(EA))/(1 - ETY*COS(EA))
650      EA = EA - DIFFERENCE
660      IF ABS(DIFFERENCE) < .0001 THEN GOTO 680
670    NEXT I
680    THETA = 2*ATN(SQR((1+ETY)/(1-ETY))*TAN(EA/2))   'THETA = true anomaly
690    IF THETA < 0 THEN THETA = THETA + 2*PI
700    SINSSPLAT = SIN(THETA+ARGPER) * SIN(INCLIN)
710    SSPLAT = FNARCSIN(SINSSPLAT)
720    COSD2 = COS(THETA+ARGPER)/COS(SSPLAT)
730    D2 = FNARCCOS(COSD2)
740    IF SSPLAT < 0 THEN D2 = - D2
750    D3 = TIME/4
760    IF INCLINATION > 90 THEN S1 = -1 ELSE S1 = 1
770    SSPLAT = SSPLAT*RTOD
780    SSPLNG = S1*(D1+D2)*RTOD - D3 + SETLNG
790    IF SSPLNG >= 360 THEN SSPLNG = SSPLNG - 360  : GOTO 790
800    IF SSPLNG < 0    THEN SSPLNG = SSPLNG + 360  : GOTO 800
810    PRINT USING "####.#"; (TIME - PERIOD/2); : PRINT "      ";
820    PRINT USING "########.#"; INT(SSPLAT*10+.5)/10; INT(SSPLNG*10+.5)/10
830  RETURN
840    '
850  PRINT : END    'GTEL
Ok
```

Ground Track Data (Program GTEL)

Satellite: AMSAT-OSCAR 13

   Mean Motion: 2.10075 rev./day
   Semimajor axis: 25752.23 km
   Inclination: 57.05 degrees
   Eccentricity: .65685
   Argument of Perigee: 225 degrees

Time after apogee (minutes)	Subsatellite point	
	Lat. (N)	Long. (E)
-342.7	-36.4	85.7
-340.0	-42.9	93.5
-320.0	-52.5	173.8
-300.0	-33.3	201.3
-280.0	-19.3	208.3
-260.0	-9.6	210.2
-240.0	-2.3	210.0
-220.0	3.4	208.7
-200.0	8.2	206.8
-180.0	12.3	204.6
-160.0	15.9	202.1
-140.0	19.1	199.5
-120.0	22.1	196.7
-100.0	24.8	193.9
-80.0	27.3	191.0
-60.0	29.7	188.2
-40.0	32.0	185.4
-20.0	34.2	182.7
0.0	36.4	180.0
20.0	38.5	177.5
40.0	40.6	175.2
60.0	42.6	173.1
80.0	44.6	171.3
100.0	46.6	169.8
120.0	48.6	168.9
140.0	50.6	168.5
160.0	52.5	169.1
180.0	54.3	170.7
200.0	55.8	173.8
220.0	56.8	178.9
240.0	56.9	186.5
260.0	55.3	196.9
280.0	50.5	209.7
300.0	39.4	224.3
320.0	15.8	240.9
340.0	-29.4	267.9
342.7	-36.4	274.3

Appendix E

Conversion Factors, Constants and Derived Quantities

Conversion Factors

The following values have been established by international agreement and are exact as shown. There is no round-off or truncation error.

1 foot = 0.3048 meter
1 statute mile = 1609.344 meters
1 nautical mile = 1852 meters

Some additional conversion factors:

Length: 1.000° of arc at surface of earth
= 60.00 nautical miles
= 111.2 km
= 69.10 statute miles

Mass: 1.000 kg = 6.852×10^{-2} slugs

Force: 1.000 N = 0.2248 pounds
1.000 kg (force) = 2.205 pounds (at surface of earth)

Selected Conversion Procedures

(to four significant digits unless indicated otherwise)

1) To convert from statute miles to kilometers, multiply by 1.609
2) To convert from kilometers to statute miles, multiply by 0.6214
3) To convert from inches to meters, multiply by 0.0254 (exact)
4) To convert from meters to inches, multiply by 39.37

Constants

Flattening factor for the Earth:
f = 1/298.257

Geocentric gravitational constant:

$$GM = 3.986005 \times 10^{14} \frac{m^3}{s^2}$$

Mass of earth:
M = 5.976×10^{24} kg
= 4.095×10^{23} slugs

Mean earth-sun distance:
1 AU = 1.49600×10^{11} m

Mean equatorial radius of earth:
R_{eq} = 6378.140 km = 3963.376 statute miles

Mean radius of earth:
R = 6371 km = 3959 statute miles

Mean solar year = 365.24219870 mean solar days

π = 3.1415926535898

Solar day = 1440 minutes (exact)

Sidereal day = 1436.07 minutes
= 23:56:45 (HH:MM:SS)
= 23.9344 hours

Solar constant:
P_o = 1380 W/m^2

Speed of light in vacuum:
c = 299,792.458 km/s

Stefan-Boltzmann constant

$$\sigma = 5.67 \times 10^{-8} \frac{joules}{K^4 m^2 s}$$

Universal gravitational constant:

$$G = 6.672 \times 10^{-11} \frac{m^3}{kg\text{-}s^2}$$

$$= 3.439 \times 10^{-8} \frac{ft^3}{slug\text{-}s^2}$$

Abbreviations

K = kelvin
kg = kilogram
m = meter
N = Newton $\left(kg\frac{m}{s^2}\right)$
s = second
W = watt

Sources

[1] *The Astronomical Almanac For The Year 1989*, issued by the Nautical Almanac Office, United States Naval Observatory, US Government Printing Office.
[2] *The CRC Handbook of Physics and Chemistry*, 68th Ed., Chemical Rubber Co, 1987/88, Cleveland, Ohio.

Appendix F

Rules and Regulatons Governing the Amateur-Satellite Service

The amateur service, of which the amateur-satellite service is a part, is governed by a complex hierarchy of rules. At the top of the pyramid lies the International Telecommunication Union (ITU), with headquarters in Geneva. Member nations of the ITU (most countries) are obligated to see that radio services under their jurisdiction operate in compliance with ITU regulations. ITU member nations meet aperiodically at World Administrative Radio Conferences (WARCs) to consider changes to the existing regulations.

The US agency responsible for administering the radio spectrum is the Federal Communications Commission. The rules governing the amateur service are known as Part 97. In Part 97 the FCC delegates certain frequency management and coordination tasks to the amateur service. In the US (and possibly in other countries) the band plans adopted by the amateur service therefore constitute an informally recognized adjunct to the rules governing the amateur service. This latter authority is based on an FCC letter (April 27, 1983) which stated that "...we [FCC] conclude that any amateur who selects a station transmitting frequency not in harmony with those plans [IARU Region 2 band plans] is not operating in accord with good amateur practice." Sufficient cause would therefore exist for issuance of an Official Notice of Violation. (*The FCC Rule Book*, ARRL, 1989, Chapter 10.)

This appendix contains selected rules and regulations specifically directed at the amateur-satellite service. Keep in mind that this list is in no way complete and that, in addition to the regulations quoted here, amateur-satellite service operations must comply with all regulations governing the amateur service unless specific mention is made otherwise. The rules and regulations presented in the remainder of this appendix are divided into three categories: those attributed to the ITU, which are international in nature; those attributed to the FCC, which affect amateurs operating under its jurisdiction; and those attributed to International Amateur Radio Union (IARU) Region 2 band plans.

ITU International Radio Regulations

Article 1—Terms and Definitions

Section III. Radio Services

§3.34 Amateur Service:

A radiocommunication service for the purpose of self-training, intercommunication and technical investigations carried out by amateurs, this is, by duly authorized persons interested in radio technique solely with a personal aim and without pecuniary interest.

§3.35 Amateur-Satellite Service:

A radiocommunications service using space stations on earth satellites for the same purposes as those of the amateur service.

Article 32—Amateur Service and Amateur-Satellite Service

Section II. Amateur-Satellite Service

Sec. 6. The provisions of Section I of this Article [Amateur Service] shall apply equally, as appropriate, to the Amateur-Satellite Service.

Sec. 7. Space stations in the Amateur-Satellite Service operating in bands shared with other services shall be fitted with appropriate devices for controlling emissions in the event that harmful interference is reported in accordance with the procedure laid down in Article 22. Administrations authorizing such space stations shall inform the IFRB [International Frequency Registration Board] and shall ensure that sufficient earth command stations are established before launch to guarantee that any harmful interference which might be reported can be terminated by the authorizing administration.

Resolution No. 642

Relating to the Bringing into Use of Earth Stations in the amateur-satellite service.

Recognizing

that the procedures of Articles 11 and 13 are applicable to the amateur-satellite service.

Recognizing Further

a) that the characteristics of each station in the amateur-satellite service vary widely.

b) that space stations in the amateur-satellite service are intended for multiple access by amateur earth stations in all countries.

c) that coordination among stations in the amateur and amateur satellite services is accomplished without the need for formal procedures.

d) that the burden of terminating any harmful interference is place upon the administration authorizing a space station in the amateur-satellite service pursuant to the provisions of No. 2741 of the Radio Regulations.

Notes

that certain information specified in Appendices 3 and

4 cannot reasonably be provided for earth stations in the amateur-satellite service.

Resolves

1. that when an administration (or one acting on behalf of a group of named administrations) intends to establish a satellite system in the amateur-satellite service and wishes to publish information with respect to earth stations in the system it may:

1.1 communicate to the IFRB all or part of the information listed in Appendix 3; the IFRB shall publish such information in a special section of its weekly circular requesting comments to be communicated within a period of four months after the date of publication.

1.2 notify under Nos. 1488 to 1491 all or part of the information listed in Appendix 3; the IFRB shall record it in a special list.

2. that this information shall include at least the characteristics of a typical amateur earth station in the amateur-satellite service having the facility to transmit signals of the space station to initiate, modify, or terminate the functions of the space station.

FCC Rules and Regulations

Part 97—Amateur Radio Service (Adopted May 31, 1989)

Subpart A—General Provisions

§97.3 Definitions.

(a) The definitions of terms used in part 97 are:

(3) *Amateur-satellite service.* A radiocommunication service using stations on Earth satellites for the same purpose as those of the amateur service.

(14) *Earth station.* An amateur station located on, or within 50 km of, the Earth's surface intended for communications with space stations or with other Earth stations by means of one or more other objects in space.

(32) *Radio Regulations.* The latest ITU *Radio Regulations* to which the United States is a party.

(36) *Space station.* An amateur station located more than 50 km above the Earth's surface.

(37) *Space telemetry.* A one-way transmission from a space station of measurements made from the measuring instruments in a spacecraft, including those relating to the functioning of the spacecraft.

(39) *Telecommand.* A one-way transmission to initiate, modify, or terminate functions of a device at a distance.

(40) *Telecommand station.* An amateur station that transmits communications to initiate, modify or terminate functions of a space station.

(41) *Telemetry.* A one-way transmission of measurements at a distance from the measuring instrument.

§97.5 Station license required.

(c) When a station is transmitting on any amateur-satellite service frequency from a location more than 50 km above the Earth's surface aboard any craft that is documented or registered in the United States, the person having physical control of the apparatus must hold an FCC-issued written authorization for an amateur station.

Subpart B—Station Operation Standards

§97.113 Prohibited transmissions.

(e) No station shall retransmit programs or signals emanating from any type of radio station other than an amateur station, except communications originating on United States Government frequencies between a space shuttle and its associated Earth stations. Prior approval for such retransmissions must be obtained from the National Aeronautics and Space Administration. Such retransmissions must be for the exclusive use of amateur operators.

Subpart C—Special Operations

§97.207 Space station.

(a) Any amateur station licensed to a holder of an Amateur Extra Class operator license may be a space station. A holder of any class operator license may be the control operator of a space station, subject to the privileges of the class of operator license held by the control operator.

(b) A space station must be capable of effecting a cessation of transmissions by telecommand whenever such cessation is ordered by the FCC.

(c) The following frequency bands and segments are authorized to space stations:

(1) The 17 m, 15 m, 12 m and 10 m bands, 6 mm, 2 mm and 1 mm bands; and

(2) The 7.01-7.1 MHz, 14.00-14.25 MHz, 144-146 MHz, 435-438 MHz, 1260-1270 MHz and 2400-2450 MHz, 3.40-3.41 GHz, 5.83-5.85 GHz, 10.45-10.50 GHz and 24.00-24.05 GHz segments.

(d) A space station may automatically retransmit the radio signals of Earth stations and other space stations.

(e) A space station may transmit one-way communications.

(f) Space telemetry transmissions may consist of specially coded messages intended to facilitate communications or related to the function of the spacecraft.

(g) The licensee of each space station must give two written, pre-space station notifications to the Private Radio Bureau, FCC, Washington, DC 20554. Each notification must be in accord with the provisions of Articles 11 and 13 of the Radio Regulations.

(1) The first notification is required no less than 27 months prior to initiating space station transmissions and must specify the information required by Appendix 4 and Resolution No. 642 of the Radio Regulations.

(2) The second notification is required no less than 5 months prior to initiating space station transmissions and must specify the information required by Appendix 3 and Resolution No. 642 of the Radio Regulations.

(h) The licensee of each space station must give a written, in-space station notification to the Private Radio Bureau, FCC, Washington, DC 20554, no later than 7 days following initiation of space station transmissions. The notification must update the information contained in the pre-space notification.

(i) The licensee of each space station must give a written, post-space station notification to the Private Radio Bureau, FCC, Washington, DC 20554, no later than 3 months after termination of the space station transmissions. When the termination is ordered by the FCC, notification is required no later than 24 hours after termination. The pre-space notifications mentioned in §97.207 (g) should include the following information:

Space operation date. A statement of the expected date space operations will be initiated, and a prediction of the duration of the operation.

Identity of satellite. The name by which the satellite will be known.

Service area. A description of the geographic area on the Earth's surface which is capable of being served by the station in space operation. Specify for both the transmitting and receiving antennas of this station.

Orbital Parameters. A description of the anticipated orbital parameters as follows:

Nongeostationary satellite. (1) Angle of inclination, (2) Period, (3) Apogee (kilometers), (4) Perigee (kilometers), (5) Number of satellites having the same orbital characteristics.

Geostationary satellite. (1) Normal geographical longitude, (2) Longitudinal tolerance, (3) Inclination tolerance, (4) Geographical longitudes marking the extremities of the orbital arc over which the satellite is visible at minimum angle of elevation at 10° at points within the associated service area, (5) Geographical longitudes marking the extremities of the orbital arc within which the satellite must be located to provide communications to the specified service area, (6) Reason when the orbital arc of (5) is less than that of (4).

Technical parameters. A description of the proposed technical parameters for:

(1) The station in space operation; and

(2) A station in earth operation suitable for use with the station in space operation; and

(3) A station in telecommand operation suitable for use with the station in space operation.

The description shall include:

(1) Carrier frequencies if known; otherwise give frequency range where carrier frequencies will be located.

(2) Necessary bandwidth.

(3) Class of emission.

(4) Total peak power.

(5) Maximum power density (watts/Hz).

(6) Antenna radiation pattern. (For both transmitting and receiving antennas.)

(7) Antenna gain (main beam). (For both transmitting and receiving antennas.)

(8) Antenna pointing accuracy (geostationary satellites only). (For both transmitting and receiving antennas).

(9) Receiving system noise temperature. (For station in space operation.)

(10) Lowest equivalent satellite link noise temperature. The total noise temperature at the input of a typical

Amateur Radio station receiver shall include the antenna noise (generated by external sources [ground, sky, etc] peripheral to the receiving antenna and noise re-radiated by the satellite), plus noise generated internally to the receiver. The additional receiver noise is above thermal noise kT_oB. Referred to the antenna input terminals, the total system noise temperature is given by

$$T_s = T_a + (L - 1) T_o + LT_r$$

where

T_a = antenna noise temperature

L = line losses between antenna output terminals and receiver input terminals

T_o = ambient temperature, usually given as 290K

T_r = receiver noise temperature. This is also given as $(NF - 1) T_o$, where NF is receiver noise figure.

§97.209 Earth station.

(a) Any amateur station may be an Earth station. A holder of any class operator license may be the control operator of an Earth station, subject to the privileges of the class of operator license held by the control operator.

(b) The following frequency bands and segments are authorized to Earth stations:

(1) The 17 m, 15 m, 12 m and 10 m bands, 6 mm, 4 mm, 2 mm and 1 mm bands; and

(2) The 7.0-7.1 MHz, 14.00-14.25 MHz, 144-146 MHz, 1260-1270 MHz and 2400-2450 MHz, 3.40-3.41 GHz, 5.65-5.67 GHz, 10.45-10.50 GHz and 24.00-24.05 GHz segments.

§97.211 Space Telecommand station.

(a) Any amateur station designated by the licensee of a space station is eligible to transmit as a telecommand station for that space station, subject to the privileges of the class of operator license held by the control operator.

(b) A telecommand station may transmit special codes intended to obscure the meaning of telecommand messages to the station in space operation.

(c) The following frequency bands and segments are authorized to telecommand stations:

(1) The 17 m, 15 m, 12 m and 10 m bands, 6 mm, 4 mm, 2 mm, and 1 mm bands; and

(2) The 7.0-7.1 MHz, 14.00-14.25 MHz, 144-146 MHz, 1260-1270 MHz and 2400-2450 MHz, 3.40-3.41 GHz, 5.65-5.67 GHz, 10.45-10.50 GHz and 24.00-24.05 GHz segments.

(d) A telecommand station may transmit one-way communications.

§97.303 Frequency sharing requirements.

[Numerous entries in this section pertain to frequencies allocated to the amateur-satellite service. Due to space limitations only two prominent ones will be listed.]

(f) in the 70 cm band:

(3) The 430-440 MHz segment is allocated to the amateur service on a secondary basis in ITU Regions 2 and 3. No amateur station transmitting in this band in ITU Regions 2 and 3 shall cause harmful interfer-

ence to, nor is protected from interference due to the operation of, stations authorized by other nations in the radiolocation service. In ITU Region 1, the basic principle that applies is the equality of right to operate. Amateur stations authorized by the United States and Radiolocation stations authorized by other nations in ITU Region 1 shall operate so as not to cause harmful interference to each other.

(h) No amateur station transmitting in the 23 cm band, the 3 cm band, the 24.05-25.25 GHz segment, the 76-81 GHz segment, the 144-149 GHz segment and the 241-248 GHz segment shall cause harmful interference to, nor is protected from interference due to the operation of, stations authorized by other nations in the radiolocation service.

§97.313 Transmitter power standards.

(a) An amateur station must use the minimum transmitter power necessary to carry out the desired communications.

ARRL Coordinated Band Plans
Amateur Satellite Service Frequencies

Voluntary band plans adopted by Region 2 radio amateurs are summarized in Chapter 5 of *The FCC Rule Book*, ARRL, 1989. Sub-bands allocated to amateur-satellite activities include:

144.30-144.50 MHz
145.80-146.00 MHz
435.00-438.00 MHz
  1260-1270 MHz

Glossary

Note: All terms are defined as they apply to space satellites.

access range: See acquisition distance.

acquisition circle: "Circle" drawn about a ground station and keyed to a specific satellite. When the subsatellite point is inside the circle, the satellite is in range.

acquisition distance: Maximum distance between subsatellite point and ground station at which access to spacecraft is possible.

alligator: Ground station with high-power transmitter and poor receiver.

altitude: The distance between a satellite and the point on the earth directly below it. Same as height.

AMSAT: Registered trademark of Radio Amateur Satellite Corporation. PO Box 27, Washington, DC 20044. Also known as AMSAT-NA (North America).

anomalistic period: The elapsed time between two successive perigees of a satellite.

AOS (Acquisition Of Signal): The time at which a particular ground station begins to receive radio signals from a satellite. For calculations, AOS is generally assumed to occur at an elevation angle of 0°.

apogee: Point on orbit where satellite-geocenter distance is maximum.

argument of perigee: The polar angle locating the perigee point of a satellite in the orbital plane; drawn between the ascending node, geocenter and perigee; and measured from ascending node in direction of satellite motion. When the argument of perigee is between 0 and 180 degrees, perigee is over the northern hemisphere. When the argument of perigee is between 180 and 360 degrees, the perigee is over the southern hemisphere.

ARRL (American Radio Relay League): Membership organization of radio amateurs in the US. 225 Main St, Newington, CT 06111.

ascending node (EQX): Point on satellite orbit (or ground track) where subsatellite point crosses equator with satellite headed north.

ascending pass: With respect to a particular ground station, a satellite pass during which spacecraft is headed in a northerly direction while in range.

AU (Astronomical Unit): Mean sun-earth distance = 1.49×10^{11} m.

autotransponder: A computer-like device aboard a spacecraft designed to receive and respond to uplink signals directed to it. Several RS spacecraft have carried autotransponders. Also known as a ROBOT.

azimuth: Angle in the horizontal plane measured clockwise with respect to north (north = 0°).

Bahn latitude and longitude (ALAT and ALON): angles which describe the orientation of a Phase 3 satellite in its orbital plane. When Bahn latitude is 0° and Bahn longitude is 180°, the directional antennas on OSCAR 13 will be pointing directly at the SSP when the spacecraft is at apogee.

BCR (battery charge regulator): Electronic control unit on satellite placed between solar cells and battery.

bird: Slang for satellite.

BRAMSAT: AMSAT Brazil.

BOL (beginning of life): Usually used in reference to a satellite parameter that changes over time, such as solar-cell efficiency.

boresight: The direction of maximum gain of a spacecraft antenna. Also refers to point on earth where spacecraft antenna produces maximum signal level.

classical orbital elements: A set of *orbital elements*, usually including an epoch time specified at perigee, right ascension of ascending node (RAAN), inclination, eccentricity, argument of perigee and period. Because these parameters are earth-referenced and based on geometrical properties, they're especially useful for intuitively picturing an orbit.

Codestore: A digital memory system aboard early AMSAT spacecraft that could be loaded with data by ground stations for later rebroadcast in Morse or other codes.

coverage circle: (With respect to a particular ground station) the region of earth that is eventually accessible for communication via a specific satellite; (With respect to a particular satellite) the region around the instantaneous SSP that is in view of the satellite.

DBS (Direct Broadcast Satellite): Commercial satellite designed to transmit TV programming directly to the home; (Direct Broadcast Service): Environmental Satellite service designed to be received directly by end user.

decay rate: Short name for rate of change of mean motion. A parameter specifying how atmospheric drag affects a satellite's motion.

delay time: (transponder delay time)—The elapsed time between the instant a signal enters a transponder and the instant it leaves; (path delay time)—the elapsed time between the transmission of an uplink signal and the instant it is received.

descending node: Point on satellite orbit (or ground track) where subsatellite point crosses equator with satellite headed south.

descending pass: With respect to a particular ground station, a satellite pass during which satellite is headed in a southerly direction while in range.

digital transponder: A device which receives a digitally encoded signal, demodulates it and then retransmits it in a digital format. The retransmitted signal may be on the same frequency or a different frequency; may occur after a short time interval or may be on demand; may use the same or different modulation and encoding scheme.

Doppler shift: The observed frequency difference between the transmitted signal and the received signal on a link where the transmitter and receiver are in relative motion.

downlink: A radio link originating at a spacecraft and terminating at one or more ground stations.

eccentricity: A parameter used to describe the shape of the ellipse constituting a satellite orbit.

EIRP: effective isotropic radiated power.

elevation: Angle above the horizontal plane.

elevation circle: On a map or globe, the set of all points about a ground station where the elevation angle to a specified satellite is a fixed value.

EME (earth-moon-earth): Communication mode that involves bouncing signals off the moon.

epoch (epoch time): A reference time at which orbital elements are specified.

equatorial plane: An imaginary plane, extending throughout space, which contains the equator of the primary body (often the earth).

EQX: Ascending node.

ESA (European Space Agency): A consortium of European governmental groups pooling resources for space exploration and development.

footprint: A set of signal-level contours, drawn on map or globe, showing the performance of a high-gain satellite antenna. Usually applied to geostationary satellites.

geocenter: center of the oblate spheroid used to represent the earth.

geostationary satellite: A satellite that appears to hang motionless over a fixed point on the equator.

ground station: A radio station, on or near the surface of the earth, designed to receive signals from, or transmit signals to, a spacecraft.

ground track (subsatellite path): Path on surface of earth traced out by SSP as satellite moves through space.

HELAPS: A technique for producing high-efficiency linear transponders. Developed by Dr Karl Meinzer, DJ4ZC.

IARU: International Amateur Radio Union.

IHU (integrated housekeeping unit): The onboard computer and associated electronics used to control a spacecraft.

inclination: The angle between the orbital plane of a satellite and the equatorial plane of the earth.

Increment: See longitudinal increment.

IPS (Interpreter for Process Structures): A high-level, FORTH-like language employed as an operating system on the computers aboard several OSCAR satellites. Developed by Dr Karl Meinzer, DJ4ZC.

ITU (International Telecommunication Union): International organization responsible for coordinating use of radio spectrum.

JAMSAT: AMSAT Japan.

JARL: Japanese Amateur Radio League.

Keplerian orbital elements: A set of *orbital elements* specified at an arbitrary epoch time, which include mean anomaly, right ascension of ascending node (RAAN), inclination, eccentricity, argument of perigee and mean motion (or period or semi-major axis). Closely related to classical orbital elements.

LEO: Low earth orbit.

line of nodes: The line of intersection of a satellite's orbital plane and the earth's equatorial plane.

linear transponder: A device that receives radio signals in one segment of the spectrum, amplifies them linearly, translates (shifts) their frequency to another segment of the spectrum and retransmits them.

LNA (low-noise amplifier): Term sometimes used for the device that radio amateurs generally refer to as a low-noise preamp.

longitudinal increment: Change in longitude of ascending node between two successive passes of a specified satellite. Measured in degrees west per orbit (°W/orbit).

LOS (loss of signal): The time at which a particular ground station loses radio signals from a satellite. For calculations, LOS is generally assumed to occur at an elevation angle of 0°.

maximum access distance: The maximum distance, measured along the surface of the earth, between a ground station and the subsatellite point at which the satellite enters one's range circle. (Corresponds to a 0° elevation angle.)

mean anomaly (MA): A number that increases uniformly with time, which is used to locate satellite position on orbital ellipse. For OSCAR satellites, MA varies from 0 to 256. When MA is 0 or 256, satellite is at perigee. When MA is 128, satellite is at apogee. When MA is between 0 and 128, satellite is headed up towards apogee. When MA is between 128 and 256, satellite is headed down towards perigee. Astronomers usually work with an MA that varies from 0 to 360.

mean motion: Number of revolutions (perigee to perigee) completed by satellite in a solar day (1440 minutes).

MIR: Soviet Space Station.

Molniya: Series of communications satellites produced by USSR. Orbits selected for AMSAT Phase 3 satellites were patterned after the Molniya series.

NASA: US National Aeronautics and Space Administration.

NASDA: Japanese National Space Development Agency.

nodal period: The elapsed time between two successive ascending nodes of a satellite.

node: Point where satellite ground track crosses the equator.

OBC (onboard computer): The central computer which controls spacecraft functions.

orbital elements: Set of six numbers, specified at a particular time (epoch), which completely describe size, shape and orientation of satellite orbit.

orbital plane: An imaginary plane, extending throughout space, that contains a satellite's orbital track.

OSCAR: Orbiting Satellite Carrying Amateur Radio.

OSCARLOCATOR: A tracking device designed for satellites in circular orbits.

pass: See satellite pass.

PCA (point of closest approach): Point on segment of satellite orbit, or ground track, at which satellite is closest to specific ground station.

perigee: Point on orbit where satellite-geocenter distance is a minimum.

period: The amount of time it takes a satellite to complete one revolution about the earth. See anomalistic period and nodal period.

Phase: See mean anomaly.

Phase3 Tracker: A tracking device related to the OSCARLOCATOR that is designed to be used with a satellite in an elliptical orbit.

point of closest approach: See PCA.

RAAN (right ascension of ascending node): An angle that specifies the orientation of a satellite's orbital plane with respect to the fixed stars. The angular distance, measured eastward along the celestial equator, between the vernal equinox and the hour circle of the ascending node of the spacecraft.

range circle: On a map or globe, ''circle'' of specific radius centered about ground station.

reference orbit: First orbit of UTC day for satellite specified.

Satellabe: A tracking device for circular orbits. Similar to OSCARLOCATOR, but with added features.

satellite pass: Segment of orbit during which satellite passes in range of particular ground station.

s/c: Informal abbreviation for spacecraft.

semi-major axis (SMA): Half the long axis of an ellipse. Can be used to describe the size of an elliptical orbit in place of the orbital element mean motion.

sidereal day: The amount of time it takes the earth to rotate exactly 360° about its axis with respect to the ''fixed'' stars. The sidereal day contains approximately 1436.07 minutes (see solar day).

slant range: Distance between satellite and a particular ground station at time specified.

SMA: See semi-major axis.

solar constant: Incident energy 1 AU from the sun falling on a plane surface of unit area oriented perpendicular to the direction of radiation. Value is approximately 1.38 kW/m².

solar day: The solar day, by definition, contains exactly 24 hours (1440 minutes). During the solar day, the earth rotates slightly more than 360° with respect to ''fixed'' stars (see sidereal day).

spiderweb: On a map or globe, set of azimuth curves radiating outward from, and concentric elevation ''circles'' about, a particular terrestrial location.

SSP: Subsatellite point.

stationary satellite: See geostationary satellite.

subsatellite path: See ground track.

subsatellite point (SSP): Point on surface of earth directly below satellite.

telemetry: Radio signals, originating at a satellite, that convey information on the performance or status of onboard subsystems. Also refers to the information itself.

TCA (time of closest approach): Time at which satellite passes closest to a specific ground station during orbit of interest.

TLM: Short for telemetry.

transponder: See linear transponder and digital transponder.

true anomaly: The polar angle that locates a satellite in the orbital plane. Drawn between the perigee, geocenter and current satellite position, and measured from perigee in direction of satellite motion.

TVRO (TV receive only): A TVRO terminal is a ground station set up to receive downlink signals from 4-GHz or 12-GHz commercial satellites carrying TV programming.

uplink: A radio link originating at a ground station and directed to a spacecraft.

WARC: World Administrative Radio Conference.

window: For a specific satellite, the overlap region between acquisition "circles" of two ground stations. Communication between the two stations via the specified satellite is possible when SSP passes through window.

ZRO test: Contest which involves ground stations attempting to receive very weak reference signals being retransmitted through a satellite transponder. Named in memory of K2ZRO for his important contributions to the OSCAR program.

Index

Notes

Notes

Notes

Notes

SATELLITE
EXPERIMENTER'S
HANDBOOK

PROOF OF
PURCHASE

FEEDBACK

Please use this form to give us your comments on this book and what you'd like to see in future editions.

License class:
- ☐ Novice ☐ Technician ☐ Technician with HF privileges
- ☐ General ☐ Advanced ☐ Extra

Name
_____ Call sign _____

Address _____

City, State/Province, ZIP/Postal Code _____

Daytime Phone (   ) _____ Age _____

If licensed, how long?_____ ARRL member?   ☐ Yes   ☐ No

Other hobbies _____

_____

Occupation _____

From _____

_____

_____

EDITOR, SATELLITE EXPERIMENTER'S HANDBOOK
AMERICAN RADIO RELAY LEAGUE
225 MAIN ST
NEWINGTON CT 06111

please fold and tape